26th Winter Workshop on Nuclear Dynamics 2010

Journal of Physics: Conference Series Volume 230

Ocho Rios, Jamaica
2-9 January 2010

Editors:

Rene Bellweid
Scott Pratt
Roy Lacey

ISBN: 978-1-61738-737-1
ISSN: 1742-6588

Printed from e-media with permission by:

Curran Associates, Inc.
57 Morehouse Lane
Red Hook, NY 12571

proceedings.com

Some format issues inherent in the e-media version may also appear in this print version.

Printed by Curran Associates, Inc. (2010)

For permission requests, please contact the Institute of Physics
at the address below.

Institute of Physics
Dirac House, Temple Back
Bristol BS1 6BE UK

Phone: 44 1 17 929 7481
Fax: 44 1 17 920 0979

techtracking@iop.org

Additional copies of this publication are available from:

Curran Associates, Inc.
57 Morehouse Lane
Red Hook, NY 12571 USA
Phone: 845-758-0400
Fax: 845-758-2634
Email: curran@proceedings.com
Web: www.proceedings.com

TABLE OF CONTENTS

These are the proceedings of the 26th edition of the Winter Workshop on Nuclear Dynamics. As in previous years the unique character of this conference series has allowed us to bring nuclear scientists from very different fields together to discuss their scientific achievements.

At the high-energy frontier results from the first elementary collisions at the LHC were shown. At the lower energies the RHIC beam energy scan is underway to span the gap between the SPS and RHIC and search for critical phenomena in the nuclear matter phase diagram. RHIC is at its peak of scientific output and many of the measurements and theories discussed here will serve as a roadmap for new measurements at the LHC and at the lower RHIC energies. The nuclear structure and astrophysics communities have started to construct dedicated facilities at Michigan State University (F-RIB) and the GSI in Germany (FAIR).

These proceedings of the 26th Winter Workshop on Nuclear Dynamics again provide a snapshot of the status of the field. The articles, many of which are written by some of the most promising young scientists in the field, are conveying the feeling of excitement that permeates all subfields of nuclear science.

Rene Bellwied (Wayne State University)
Roy Lacey (SUNY Stony Brook)
Scott Pratt (Michigan State University)

All papers published in this volume of *Journal of Physics: Conference Series* have been peer reviewed through processes administered by the proceedings Editors. Reviews were conducted by expert referees to the professional and scientific standards expected of a proceedings journal published by IOP Publishing.

Covariant transport approach for strongly interacting partonic systems

W. Cassing[1] **and E. L. Bratkovskaya**[2]

[1]Institut für Theoretische Physik, Universität Giessen, Germany
[2]Institut für Theoretische Physik, Universität Frankfurt am Main, Germany

E-mail: Wolfgang.Cassing@theo.physik.uni-giessen.de

Abstract. The dynamics of partons, hadrons and strings in relativistic nucleus-nucleus collisions is analyzed within the novel Parton-Hadron-String Dynamics (PHSD) transport approach, which is based on a dynamical quasiparticle model for partons (DQPM) matched to reproduce recent lattice-QCD results - including the partonic equation of state - in thermodynamic equilibrium. Scalar- and vector-interaction densities are extracted from the DQPM as well as effective scalar- and vector-mean fields for the partons. The transition from partonic to hadronic degrees of freedom is described by covariant transition rates for the fusion of quark-antiquark pairs or three quarks (antiquarks), respectively, obeying flavor current-conservation, color neutrality as well as energy-momentum conservation. Since the dynamical quarks and antiquarks become very massive close to the phase transition, the formed resonant 'pre-hadronic' color-dipole states ($q\bar{q}$ or qqq) are of high invariant mass, too, and sequentially decay to the groundstate meson and baryon octets increasing the total entropy. When applying the PHSD approach to Pb+Pb colllisions at 158 A·GeV we find a significant effect of the partonic phase on the production of multi-strange antibaryons due to a slightly enhanced $s\bar{s}$ pair production from massive time-like gluon decay and a larger formation of antibaryons in the hadronization process.

1. Introduction

The 'Big Bang' scenario implies that in the first micro-seconds of the universe the entire state has emerged from a partonic system of quarks, antiquarks and gluons – a quark-gluon plasma (QGP) – to color neutral hadronic matter consisting of interacting hadronic states (and resonances) in which the partonic degrees of freedom are confined. The nature of confinement and the dynamics of this phase transition has motivated a large community for several decades and is still an outstanding question of todays physics. Early concepts of the QGP were guided by the idea of a weakly interacting system of partons which might be described by perturbative QCD (pQCD). However, experimental observations at the Relativistic Heavy Ion Collider (RHIC) indicated that the new medium created in ultrarelativistic Au+Au collisions is interacting more strongly than hadronic matter and consequently this concept had to be severely questioned. Moreover, in line with theoretical studies in Refs. [1, 2, 3] the medium showed phenomena of an almost perfect liquid of partons [4, 5] as extracted from the strong radial expansion and the scaling of elliptic flow $v_2(p_T)$ of mesons and baryons with the number of constituent quarks and antiquarks [4].

The question about the properties of this (nonperturbative) QGP liquid is discussed controversially in the literature and dynamical concepts describing the formation of color neutral hadrons from colored partons are scarce. A fundamental issue for hadronization models is the conservation of 4-momentum as well as the entropy problem because by fusion/coalescence of massless (or low constituent mass) partons to color neutral bound states of low invariant mass (e.g. pions) the number of degrees of freedom and thus the total entropy is reduced in the hadronization process [6, 7, 8]. This problem - a violation of the second law of thermodynamics as well as the conservation of four-momentum and flavor currents - has been addressed in Ref. [9] on the basis of the DQPM employing covariant transition rates for the fusion of 'massive' quarks and antiquarks to color neutral hadronic resonances or strings. In fact, the dynamical studies for an expanding partonic fireball in Ref. [9] suggest that the latter problems have come to a practical solution.

A consistent dynamical approach - valid also for strongly interacting systems - can be formulated on the basis of Kadanoff-Baym (KB) equations [10] or off-shell transport equations in phase-space representation, respectively [10, 11, 12]. In the KB theory the field quanta are described in terms of dressed propagators with complex selfenergies. Whereas the real part of the selfenergies can be related to mean-field potentials (of Lorentz scalar, vector or tensor type), the imaginary parts provide information about the lifetime and/or reaction rates of time-like 'particles' [13]. Once the proper (complex) selfenergies of the degrees of freedom are known the time evolution of the system is fully governed by off-shell transport equations (as described in Refs. [10, 13]). The determination/extraction of complex selfenergies for the partonic degrees of freedom has been performed before in Refs. [14, 15] by fitting lattice QCD (lQCD) 'data' within the Dynamical QuasiParticle Model (DQPM). In fact, the DQPM allows for a simple and transparent interpretation of lattice QCD results for thermodynamic quantities as well as correlators and leads to effective strongly interacting partonic quasiparticles with broad spectral functions. For a review on off-shell transport theory and results from the DQPM in comparison to lQCD we refer the reader to Ref. [13].

In the present contribution we will report on an update of the DQPM parameters [16] since more precise lQCD calculations for 2+1 flavors with almost physical quark masses have become available in 2008 [17]. Furthermore, we extend the study in Refs. [14, 15] by fixing independently scalar- and vector-interaction densities for the fermion degrees of freedom. First applications of the PHSD approach to Pb +Pb collisions at SPS energies - in comparison to experimental data from the NA49 Collaboration - are presented, too.

2. The PHSD approach

The Parton-Hadron-String-Dynamics (PHSD) approach is a microscopic covariant transport model that incorporates effective partonic as well as hadronic degrees of freedom and involves a dynamical description of the hadronization process from partonic to hadronic matter. Whereas the hadronic part is essentially equivalent to the conventional Hadron-Strings-Dynamics (HSD) approach [18] the partonic dynamics is based on the Dynamical QuasiParticle Model (DQPM) [14, 15] which describes QCD properties in terms of single-particle Green's functions (in the sense of a two-particle irreducible (2PI) approach).

2.1. Reminder of the DQPM

We briefly recall the basic assumptions of the DQPM: Following Ref. [19] the dynamical quasiparticle mass (for gluons and quarks) is assumed to be given by the thermal mass in the asymptotic high-momentum regime, which is proportional to the coupling $g(T/T_c)$ and the temperature T, i.e. for gluons

$$M_g^2(T) = \frac{g^2(T/T_c)}{6}\left((N_c + \frac{1}{2}N_f)\,T^2 + \frac{N_c}{2}\sum_q \frac{\mu_q^2}{\pi^2}\right), \qquad (1)$$

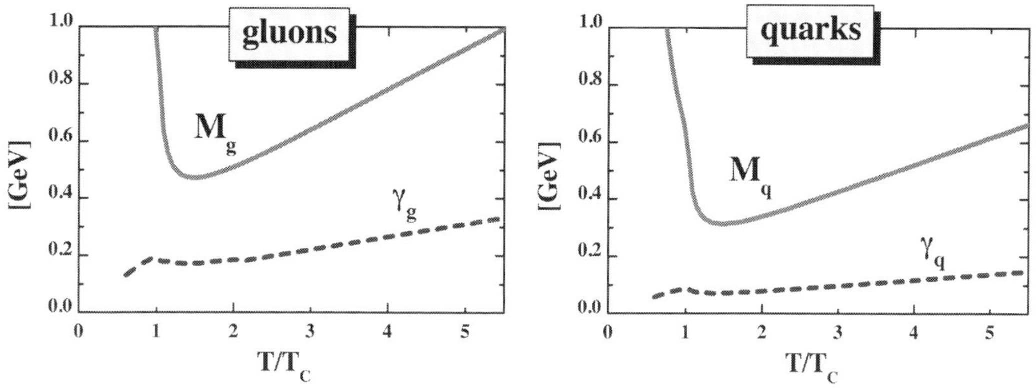

Figure 1. The effective gluon mass M_g and witdh γ_g as function of temperature T/T_c (l.h.s.). The r.h.s. shows the corresponding quantities for quarks.

and for quarks (assuming vanishing constituent masses here) as,

$$M_q^2(T) = \frac{N_c^2-1}{8N_c}\, g^2(T/T_c)\left(T^2+\frac{\mu_q^2}{\pi^2}\right), \tag{2}$$

with a running coupling (squared) (for $T > T_s$),

$$g^2(T/T_c) = \frac{48\pi^2}{(11N_c-2N_f)\ln(\lambda^2(T/T_c-T_s/T_c)^2}, \tag{3}$$

which permits for an enhancement near the critical temperature T_c. Here $N_c = 3$ stands for the number of colors while N_f denotes the number of flavors. The parameters controlling the infrared enhancement of the coupling λ and T_s have been refitted compared to Ref. [19] to the recent lQCD results from Ref. [17] (see below).

The width for gluons and quarks (for $\mu_q = 0$) is adopted in the form [20]

$$\gamma_g(T) = N_c\frac{g^2T}{8\pi}\ln\frac{2c}{g^2}, \qquad \gamma_q(T) = \frac{N_c^2-1}{2N_c}\frac{g^2T}{8\pi}\ln\frac{2c}{g^2}. \tag{4}$$

where the parameter c is related to a magnetic cut-off. We stress that a non-vanishing width γ is the central difference of the DQPM to conventional quasiparticle models [21, 22]. Its influence is essentially seen in correlation functions as e.g. in the stationary limit of the correlation function in the off-diagonal elements of the energy-momentum tensor T^{kl} which defines the shear viscosity η of the medium [3, 23]. Here a sizable width is mandatory to obtain a small ratio in the shear viscosity to entropy density η/s [3].

The actual gluon mass M_g and width γ_g – employed in the further calculations – as well as the quark mass M_q and width γ_q are depicted in Fig. 1 on the lhs and rhs, respectively, as a function of T/T_c. These values for the masses and widths are smaller than those presented in Ref. [19] since the recent lQCD calculations [17] have been performed with much smaller bare fermion masses. This, in fact, leads to an increase in the entropy density $s(T/T_c)$ relative to earlier lQCD calculations, which implies that the quasiparticle mass and width - fitting the lQCD results - become smaller relative to Refs. [14, 15, 19].

In line with Ref. [19] the parton spectral functions are no longer $\delta-$ functions in the invariant mass squared but taken as

$$\rho_j(\omega) = \frac{\gamma_j}{E_j}\left(\frac{1}{(\omega-E_j)^2+\gamma_j^2} - \frac{1}{(\omega+E_j)^2+\gamma_j^2}\right) \tag{5}$$

Journal of Physics: Conference Series 230 (2010) 012001 doi:10.1088/1742-6596/230/1/012001

separately for quarks and gluons ($j = q, \bar{q}, g$). With the convention $E^2(\boldsymbol{p}) = \mathbf{p}^2 + M_j^2 - \gamma_j^2$, the parameters M_j^2 and γ_j are directly related to the real and imaginary parts of the retarded self-energy, e.g.$\Pi\ _j = M_j^2 - 2i\gamma_j\omega$. The spectral function (5) is antisymmetric in ω and normalized as

$$\int_{-\infty}^{\infty} \frac{d\omega}{2\pi}\ \omega\ \rho_j(\omega, \mathbf{p}) = \int_0^{\infty} \frac{d\omega}{2\pi}\ 2\omega\ \rho_j(\omega, \mathbf{p}) = 1\ . \tag{6}$$

With the spectral functions fixed by Eqs. (1)-(5) the total energy density in the DQPM $T_+^{00} + T_-^{00}$ can be evaluated as [15]

$$\begin{aligned} T_{\pm}^{00} &= d_g \int_0^{\infty} \frac{d\omega}{2\pi} \int \frac{d^3p}{(2\pi)^3}\ 2\omega^2 \rho_g(\omega, \mathbf{p})\ n_B(\omega/T)\ \Theta(\pm p^2) \\ &+ d_q \int_0^{\infty} \frac{d\omega}{2\pi} \int \frac{d^3p}{(2\pi)^3}\ 2\omega^2 \rho_q(\omega, \mathbf{p})\ n_F((\omega - \mu_q)/T)\ \Theta(\pm p^2) \\ &+ d_{\bar{q}} \int_0^{\infty} \frac{d\omega}{2\pi} \int \frac{d^3p}{(2\pi)^3}\ 2\omega^2 \rho_{\bar{q}}(\omega, \mathbf{p})\ n_F((\omega + \mu_q)/T)\ \Theta(\pm p^2)\ , \end{aligned} \tag{7}$$

where n_B and n_F denote the Bose and Fermi functions, respectively, while μ_q stands for the quark chemical potential. The number of transverse gluonic degrees of freedom is $d_g = 16$ while the fermion degrees of freedom amount to $d_q = d_{\bar{q}} = 2N_cN_f = 18$ in case of three flavors (N_f=3). The indices $\pm$ stand for the time-like (+) and space-like part (−) of the energy density (7) as defined via $\Theta(\pm p^2)$ with $p^2 = \omega^2 - \mathbf{p}^2$.

The pressure $P(T)$ then can be obtained by integrating the differential thermodynamic relation (for $\mu_q = 0$)

$$P - T\frac{\partial P}{\partial T} = -T^{00} = P - Ts \tag{8}$$

with the entropy density $s(T)$ given by

$$s = \frac{\partial P}{\partial T} = \frac{T^{00} + P}{T}\ . \tag{9}$$

This approach is thermodynamically consistent and represents a two-particle irreducible (2PI) approximation to hot QCD.

For the further presentation of the DQPM it is useful to introduce the shorthand notations (following Ref. [15])

$$\begin{aligned} \tilde{\mathrm{Tr}}_g^{\pm} \cdots &= d_g \int \frac{d\omega}{2\pi} \frac{d^3p}{(2\pi)^3}\ 2\omega\ \rho_g(\omega)\ \Theta(\omega)\ n_B(\omega/T)\ \Theta(\pm p^2) \cdots \\ \tilde{\mathrm{Tr}}_q^{\pm} \cdots &= d_q \int \frac{d\omega}{2\pi} \frac{d^3p}{(2\pi)^3}\ 2\omega\ \rho_q(\omega)\ \Theta(\omega)\ n_F((\omega - \mu_q)/T)\ \Theta(\pm p^2) \cdots \\ \tilde{\mathrm{Tr}}_{\bar{q}}^{\pm} \cdots &= d_{\bar{q}} \int \frac{d\omega}{2\pi} \frac{d^3p}{(2\pi)^3}\ 2\omega\ \rho_{\bar{q}}(\omega)\ \Theta(\omega)\ n_F((\omega + \mu_q)/T)\ \Theta(\pm p^2) \cdots \end{aligned} \tag{10}$$

with $p^2 = \omega^2 - \mathbf{p}^2$ denoting the invariant mass squared. Here the $\Theta(\pm p^2)$-functions project on 'time-like' (+) and 'space-like' (−) sectors of the four-momentum. The traces in (10) then give directly time-like and space-like 'densities' for gluons $N_g^{\pm}$ as well as for quarks $N_q^{\pm}$ (and antiquarks). Note, that only the time-like parts have the physical interpretation of particles per volume. We mention that scalar gluon and quark (+antiquark) densities are evaluated as

$$\rho_g^s(\frac{T}{T_c}) = Tr_g^+\left(\frac{\sqrt{p^2}}{\omega}\right), \qquad \rho_q^s(\frac{T}{T_c}) = Tr_q^+\left(\frac{\sqrt{p^2}}{\omega}\right) + Tr_{\bar{q}}^+\left(\frac{\sqrt{p^2}}{\omega}\right)\ . \tag{11}$$

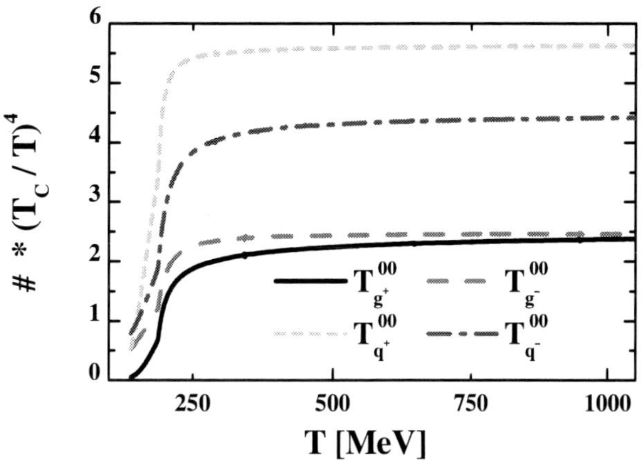

Figure 2. The different time- and space-like contributions to the total energy density as a function of T/T_c in units of GeV/fm^3. All quantities here have been multiplied by $(T_c/T)^4$ to map out the scaling properties with temperature T.

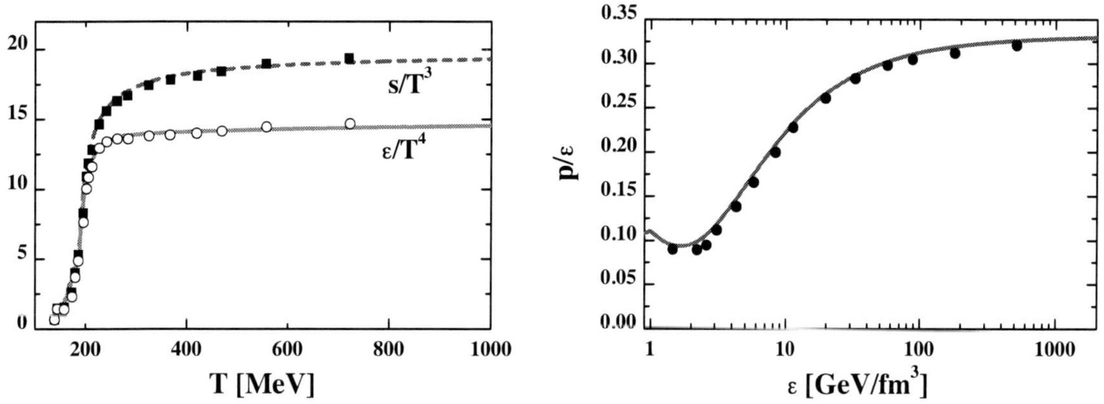

Figure 3. Lhs: The entropy density $s(T)$ (dashed line) and energy density $\epsilon(T)$ (solid line) from the DQPM in comparison to the lQCD results from Ref. [17] (full squares and open dots). Rhs: The equation-of-state (solid line) from the DQPM in comparison to the lQCD results from Ref. [17] (full dots).

As seen from Fig. 2 in Ref. [16] the 'densities' roughly scale with $(T/T_c)^3$ except for the region close to T_c.

The time- and space-like energy densities $T^{00}_{\pm}$ for gluons and quarks (+ antiquarks) are displayed in Fig. 2 as a function of T/T_c for the present parameter set in units of GeV/fm^3. The space-like energy density for gluons is slightly larger than the time-like part whereas the fermion contribution (quarks + antiquarks) is clearly dominated by the time-like sector. All quantities roughly scale as $(T/T_c)^4$ except for the region close to T_c.

A direct comparison of the entropy density $s(T)$ and energy density $\epsilon(T)$ from the DQPM with results from lQCD from Ref. [17] is presented in Fig. 3 (lhs). Both results have been divided by T^3 and T^4, respectively, to demonstrate the scaling with temperature. We briefly note that the agreement is sufficiently good. This also holds for the 'equation of state', i.e. P/ϵ versus ϵ as demonstrated in Fig. 3 (rhs).

As discussed in detail in Refs. [14, 15] and explicitly addressed in Eq. (7), the energy-density functional can be separated in space-like and time-like sectors when the spectral functions acquire a finite width. The space-like (-) part of (7)

$$V_p(T,\mu_q) = T^{00}_{g-} + T^{00}_{q-} + T^{00}_{\bar{q}-} \tag{12}$$

may be interpreted as a potential-energy density V_p since the field quanta involved are virtual and correspond to partons exchanged in interaction diagrams. The time-like part (+) of (7) corresponds to effective field quanta which can be propagated within the light-cone. Without repeating the details we mention that mean-field potentials for partons can be defined by the derivative of the potential-energy density V_p with respect to the time-like parton densities and effective interactions by second derivatives of V_p (cf. Section 3 in Ref. [15]).

In analogy to relativistic effective approaches for nucleonic degrees of freedom [24, 25] we assume the potential energy density (12) to be a sum of scalar and vector parts, i.e.

$$V_p(T,\mu_q) = V_s(T,\mu_q) + V_v(T,\mu_q)\ . \tag{13}$$

In the dynamical quasiparticle picture the pressure P then is a sum of kinetic as well as interaction parts, i.e.

$$P(T,\mu_q) = \langle P_{xx}\rangle_{T,\mu_q} - V_s(T,\mu_q) + V_v(T,\mu_q) \tag{14}$$

with

$$\langle P_{xx}\rangle_{T,\mu_q} = \frac{1}{3}\left(Tr^+_g\left(\frac{\mathbf{p}^2}{\omega}\right) + Tr^+_q\left(\frac{\mathbf{p}^2}{\omega}\right) + Tr^+_{\bar{q}}\left(\frac{\mathbf{p}^2}{\omega}\right)\right)\ . \tag{15}$$

In a similar way the total energy density ϵ can be expressed as

$$\epsilon(T,\mu_q) = \langle\omega\rangle_{T,\mu_q} + V_s(T,\mu_q) + V_v(T,\mu_q) \tag{16}$$

with the time-like quasiparticle energy density given by

$$\langle\omega\rangle_{T,\mu_q} = Tr^+_g(\omega) + Tr^+_q(\omega) + Tr^+_{\bar{q}}(\omega)\ . \tag{17}$$

Note the different signs for the scalar interaction part in (14) and (16) which stem from the metric tensor in the energy-momentum tensor $T^{\mu\nu}$. Since the total energy density as well as pressure are known in the DQPM and the kinetic parts (15) and (16) can be evaluated as well, the vector and scalar interaction densities can be uniquely extracted from the equations above. The corresponding results are displayed in Fig. 4 (lhs) for the total interaction per scalar particle V_p/ρ_s (upper dot-dashed line) and its decomposition in scalar (solid line) and vector (dashed line) parts as a function of the scalar parton density ρ_s,

$$\rho_s = \rho^s_g + \rho^s_q + \rho^s_{\bar{q}}, \tag{18}$$

which is a convenient (Lorentz invariant) quantity to characterize the system instead of the Lagrange parameters T^{-1} and μ_q. In fact, as found in Ref. [13] the thermodynamic quantities for different T and μ_q are very close when representing them as a function of ρ_s or the parton density ρ_p, respectively.

Scalar and vector mean-field potentials are defined by derivatives of V_s and V_v with respect to the densities ρ_s and ρ_v, respectively,

$$U_s(\rho_s) = \frac{\partial V_s}{\partial \rho_s}; \qquad U^0_v = \frac{\partial V_v}{\partial \rho_p}\ . \tag{19}$$

Journal of Physics: Conference Series **230** (2010) 012001 doi:10.1088/1742-6596/230/1/012001

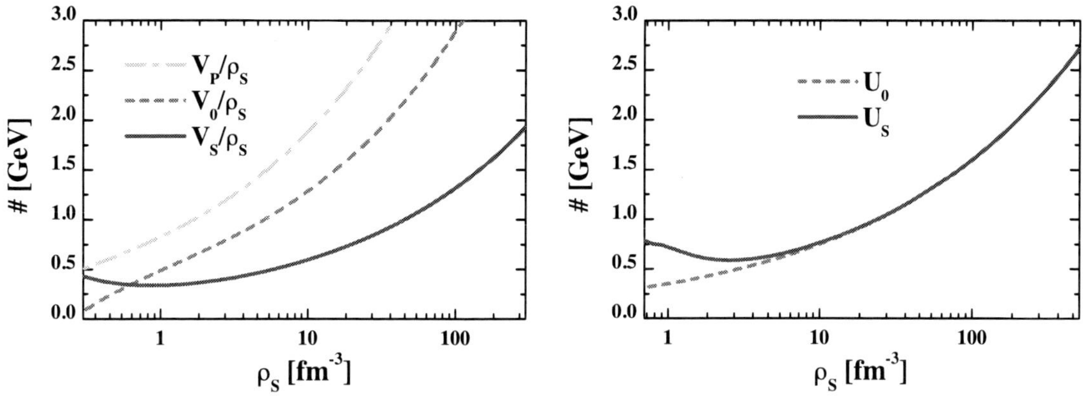

Figure 4. Lhs: The total potential energy per scalar parton (green dashed-dotted line) separated into scalar (solid blue line) and vector (red dashed line) parts (see text) as a function of the scalar parton density ρ_s. Rhs: The scalar (blue solid line) and vector (red dashed line) mean-fields for quarks (19) as a function of the scalar parton density ρ_s.

Both quantities (after differentiation) may be considered as a function of the parton-scalar density ρ_s. The resulting functions are displayed in Fig. 4 (rhs) as a function of ρ_s and show approximately equal results at high scalar density (or energy density, respectively). The low-density properties, however, are very different. Whereas the vector-mean field drops to zero with decreasing ρ_s the scalar-mean field increases substantially below $\rho_s \approx 1$ fm^{-3}; it very smoothly increases with density for $\rho_s > 2$ fm^{-3} [1]. Accordingly scalar forces on the partons $\sim -\partial U_s/\partial\rho_s \nabla\rho_s$ are rather small in the high-density partonic phase and only become stronger in the low-density phase close to hadronization. The DQPM as well as lQCD do not allow to extrapolate U_s down to $\rho_s = 0$ reliably but one might imagine that $U_s(\rho_s \to 0) \to \infty$ thus encoding scalar confinement on the mean-field level. We note that in actual PHSD calculations such low partonic densities are not probed dynamically because the partons hadronize in the region $\rho_s \approx 1$ to 2 fm^{-3} (see below).

With the scalar- and vector-mean fields fixed for the partons (assuming the gluon mean fields to be twice the quark mean fields according to the analysis within the DQPM (cf. Ref. [15]) and with the effective masses and dynamical widths fixed by the DQPM (as a function of ρ_s instead of (T/T_c)) the mean-field propagation of partons in PHSD is fully determined by the off-shell transport equations (cf. the review [13]). Note that these mean-fields are presently momentum independent whereas nuclear physics studies suggest a moderate dependence on the particle momentum [26, 27].

The transport equations still require to specify the elastic and inelastic cross sections of quarks and gluons, which enter the collision terms. For the latter quantities we refer the reader to Ref. [16].

2.2. Hadronization in PHSD

The hadronisation, i.e. the transition from partonic to hadronic degrees of freedom, is described in PHSD by local covariant transition rates as introduced in Ref. [9] e.g. for $q + \bar{q}$ fusion to a meson m of four-momentum $p = (\omega, \mathbf{p})$ at space-time point $x = (t, \mathbf{x})$:

$$\frac{dN_m(x,p)}{d^4x d^4p} = Tr_q Tr_{\bar{q}}\, \delta^4(p - p_q - p_{\bar{q}})\, \delta^4\left(\frac{x_q + x_{\bar{q}}}{2} - x\right)$$

[1] Note the logarithmic scale in ρ_s.

$$\times \omega_q \; \rho_q(p_q) \; \omega_{\bar{q}} \; \rho_{\bar{q}}(p_{\bar{q}}) \; |v_{q\bar{q}}|^2 \; W_m(x_q - x_{\bar{q}}, (p_q - p_{\bar{q}})/2)$$
$$\times N_q(x_q, p_q) \; N_{\bar{q}}(x_{\bar{q}}, p_{\bar{q}}) \; \delta(\text{flavor, color}). \tag{20}$$

In Eq. (20) we have introduced the shorthand notation,

$$Tr_j = \sum_j \int d^4x_j \int \frac{d^4p_j}{(2\pi)^4} , \tag{21}$$

where $\sum_j$ denotes a summation over discrete quantum numbers (spin, flavor, color); $N_j(x, p)$ is the phase-space density of parton j at space-time position x and four-momentum p. In Eq. (20) δ(flavor, color) stands symbolically for the conservation of flavor quantum numbers as well as color neutrality of the formed hadron m which can be viewed as a color-dipole or 'pre-hadron'. Furthermore, $v_{q\bar{q}}(\rho_p)$ is the effective quark-antiquark interaction from the DQPM (displayed in Fig. 10 of Ref. [15]) as a function of the local parton $(q + \bar{q} + g)$ density ρ_p (or energy density). Furthermore, $W_m(x, p)$ is the dimensionless phase-space distribution of the formed 'pre-hadron', i.e.

$$W_m(\xi, p_\xi) = \exp\left(\frac{\xi^2}{2b^2}\right) \; \exp\left(2b^2(p_\xi^2 - (M_q - M_{\bar{q}})^2/4)\right) \tag{22}$$

with $\xi = x_1 - x_2 = x_q - x_{\bar{q}}$ and $p_\xi = (p_1 - p_2)/2 = (p_q - p_{\bar{q}})/2$ (which has been previously introduced in Eq. (2.14) of Ref. [28]). The width parameter b is fixed by $\sqrt{\langle r^2 \rangle} = b = 0.66$ fm (in the rest frame) which corresponds to an average rms radius of mesons. We note that the expression (22) corresponds to the limit of independent harmonic oscillator states and that the final hadron-formation rates are approximately independent of the parameter b within reasonable variations. By construction the quantity (22) is Lorentz invariant; in the limit of instantaneous 'hadron formation', i.e. $\xi^0 = 0$, it provides a Gaussian dropping in the relative distance squared $(\mathbf{r}_1 - \mathbf{r}_2)^2$. The four-momentum dependence reads explicitly (except for a factor 1/2)

$$(E_1 - E_2)^2 - (\mathbf{p}_1 - \mathbf{p}_2)^2 - (M_1 - M_2)^2 \leq 0 \tag{23}$$

and leads to a negative argument of the second exponential in (22) favoring the fusion of partons with low relative momenta $p_q - p_{\bar{q}} = p_1 - p_2$.

Related transition rates (to Eq. (20)) are defined for the fusion of three off-shell quarks $(q_1 + q_2 + q_3 \leftrightarrow B)$ to color neutral baryonic (B or $\bar{B}$) resonances of finite width (or strings) fulfilling energy and momentum conservation as well as flavor current conservation using Jacobi coordinates (cf. Ref. [16]).

On the hadronic side the PHSD transport approach includes explicitly the baryon octet and decouplet, the 0^-- and 1^--meson nonets as well as selected higher resonances as in HSD [18]. Hadrons of higher masses (> 1.5 GeV in case of baryons and > 1.3 GeV in case of mesons) are treated as 'strings' (color-dipoles) that decay to the known (low-mass) hadrons according to the JETSET algorithm [29]. We discard an explicit recapitulation of the string decay and refer the reader to the original work [29] or Ref. [30].

3. Application to nucleus-nucleus collisions

In this Section we employ the PHSD approach - described in Section 2 - to nucleus-nucleus collisions at moderate relativistic energies, i.e. at SPS energies where our approximations are expected to work. Note that at RHIC or LHC energies other initial conditions (e.g. a color-glass condensate [31]) might be necessary. Since this is a slightly different subject we here restrict to bombarding energies below 160 A·GeV where such problems/questions are expected to be not relevant.

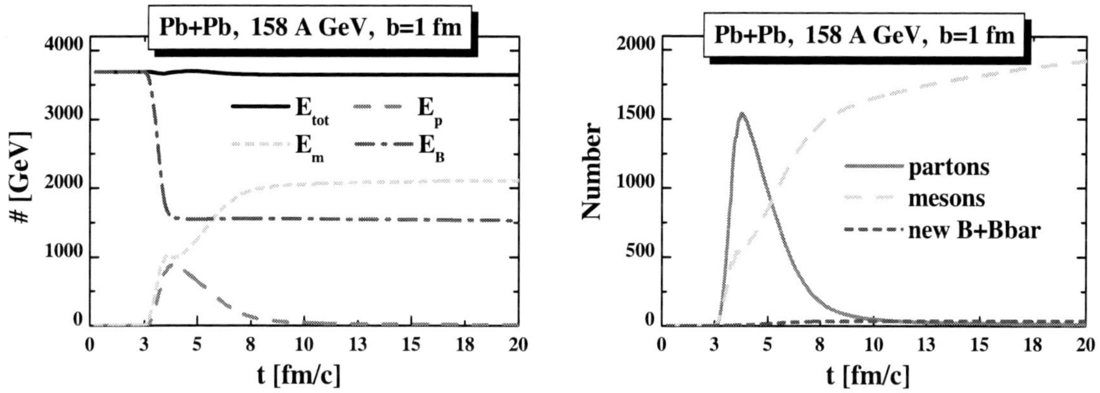

Figure 5. Lhs: The total energy E_{tot} (upper solid line) for central (b=1 fm) collisions of Pb+Pb at 158 A·GeV. The long-dashed (red) line shows the energy contributions E_p from partons while the short- dashed (green) line displays the energy contribution from mesons E_m (including 'unformed mesons' in strings). The dot-dashed (blue) line is the contribution of baryons (and antibaryons) E_B; the difference between the initial baryonic energy $E_B(t = 0)$ and final baryonic energy $E_B(t \to \infty)$ gives the energy that is converted during the heavy-ion collision to final mesonic states. Rhs: The number of produced partons (solid red line), mesons (long dashed green line) and newly produced baryon + antibaryons (blue dashed line) as a function of time for Pb+Pb at 158 A·GeV (for b=1 fm). Note that the number of mesons still increases for $t > 20$ fm/c due to the decay of vector mesons.

3.1. Partonic energy fractions

We start with a consideration of energy partitions in order to map out the fraction of partonic energy in time for relativistic nucleus-nucleus collisions. In Fig. 5 (lhs) we show the energy balance for a central (impact parameter b=1 fm) reaction of Pb+Pb at 158 A · GeV, i.e. at the top SPS energy. The total energy E_{tot} (upper line) - which at $t = 0$ is given by the energy of the colliding nuclei in the cms - is conserved throughout the reaction, i.e. in the partonic and hadronization phase as well as in the hadronic phase. Whereas in the first $\sim$ 3 fm/c the total energy is entirely contained in the impinging nucleons (including about 6 MeV per nucleon of binding energy) a rapid transition to partonic degrees of freedom is seen at $t \approx 3$ fm/c, i.e. when the nuclei have started to overlap and react. We recall that the transition time of the two Pb-nuclei is about $2R_{Pb}/\gamma_{cm} \approx 1.5$ fm/c at the top SPS energy. During this time period about 60% of initial kinetic energy of nucleons is converted to partons (red dashed line) and to mesons (short dashed green line) in the surface region ('corona') of the colliding system. Note that in the 'mesonic' energy E_m also 'unformed mesons' - as fragments of the strings - are accounted for. The energy of residual baryons (including antibaryons) is shown in terms of the dot-dashed (blue) line and is almost constant for $t > 5$ fm/c implying that the various final-state interactions do not show up significantly in the energy fractions. The partonic phase - in a limited space-time region - approximately ends for $t > 9$ fm/c which means that the further time evolution of the system is essentially described by hadronic interactions (HSD). Note that a sizeable fraction of energy is asymptotically still contained in the baryons. Since the baryon-rest masses amount to an energy of about 449 GeV (including newly produced $B\bar{B}$ pairs) this implies that full stopping is not achieved in central Pb+Pb collisions at 158 A· GeV.

Let's have a closer look at the 'particle' composition in time for this reaction. We concentrate on those species that carry the energy transferred during the collision to new degrees of freedom. In this respect we display in Fig. 5 (rhs) the number of produced partons (solid red line), mesons

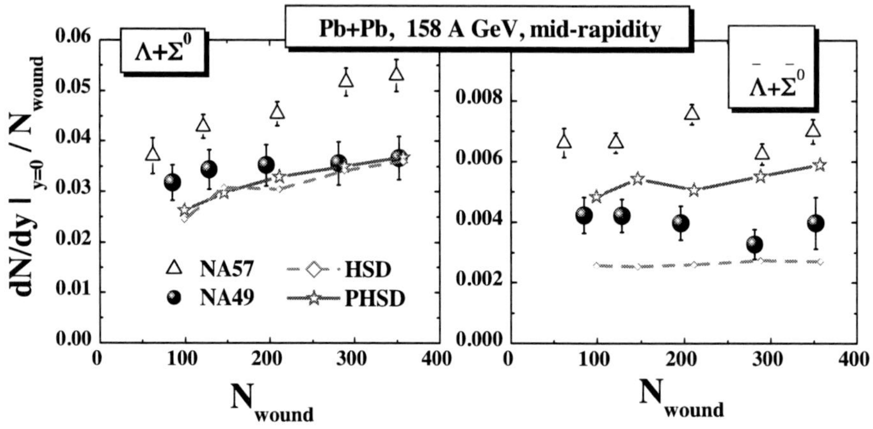

Figure 6. The multiplicities of $(\Lambda + \Sigma^0)/N_{wound}$ (l.h.s.) and $(\bar{\Lambda} + \bar{\Sigma}^0)/N_{wound}$ (r.h.s.) as a function of the number of wounded nucleons N_{wound} for Pb+Pb collisions at 158 A·GeV at mid-rapidity from PHSD (blue solid lines) and HSD (red dashed-dotted lines) in comparison to the experimental data from the NA57 Collaboration [32] (open triangles) and the NA49 Collaboration [33] (solid dots). The HSD and PHSD calculations have an error of about 5−10% due to limited statistics.

(long dashed green line) and newly produced baryons + antibaryons (blue dashed line) as a function of time for the same reaction as before. We recall that the initial number of nucleons is 416 in this case. Slightly more than 1500 partons are produced during the passage time of the nuclei which disappear practically after 9 fm/c and essentially form mesons. The number of newly produced $B + \bar{B}$ pairs is small at this energy but its flavor decomposition is quite interesting. An essential point here is that the number of final hadronic states is larger than the number of partons, i.e. there is a production of entropy in the hadronization process as pointed out before in Ref. [9]. This implies that in PHSD the second law of thermodynamics is not violated in the hadronization process!

3.2. Strangeness in PHSD

As found in Ref. [16] the impact of the partonic degrees of freedom in PHSD on the longitudinal rapidity distribution of protons, pions and kaons is only small in central Pb+Pb collisions at SPS energies from 40 to 158 A·GeV.

Additional experimental information is provided by the centrality dependence of the strange (and antistrange) baryon yield. In this respect we compare in Fig. 6 the multiplicities of $(\Lambda + \Sigma^0)/N_{wound}$ (l.h.s.) and $(\bar{\Lambda} + \bar{\Sigma}^0)/N_{wound}$ (r.h.s.) as a function of the number of wounded nucleons N_{wound} for Pb+Pb collisions at 158 A·GeV at mid-rapidity from PHSD (blue solid lines) and HSD (red dashed-dotted lines) to the experimental data from the NA57 Collaboration [32] (open triangles) and the NA49 Collaboration [33] (solid dots). The (green) full squares correspond to the 10% central data points at midrapidity from Ref. [34]. We mention that we employ the same definition of wounded nucleons N_{wound} as the NA49 Collaboration. Whereas the HSD and PHSD calculations both give a reasonable description of the $\Lambda + \Sigma^0$ yield of the NA49 Collaboration, both models underestimate the NA57 data (open triangles) by about 30%. An even larger discrepancy in the data from the NA49 and NA57 Collaborations is seen for $(\bar{\Lambda} + \bar{\Sigma}^0)/N_{wound}$ (r.h.s.); here the PHSD calculations give results which are in between the NA49 data (solid dots) and the NA57 data (open triangles). We also see that HSD underestimates the $(\bar{\Lambda} + \bar{\Sigma}^0)$ midrapidity yield at all centralities.

The latter results suggest that the partonic phase does not show up explicitly in an enhanced production of strangeness (or in particular strange mesons and baryons) but leads to a different redistribution of antistrange quarks between mesons and antibaryons in the hadronization process.

4. Summary

In this contribution we have presented a brief overview of the DQPM model and its results in comparison to lattice QCD. Furthermore, we have addressed relativistic collisions of Pb+Pb at SPS energies in the PHSD approach which includes explicit partonic degrees of freedom as well as dynamical local transition rates from partons to hadrons (20). The hadronization process conserves four-momentum and all flavor currents and slightly increases the total entropy since the 'fusion' of rather massive partons dominantly leads to the formation of color neutral strings or resonances that decay microcanonically to lower mass hadrons. Since this dynamical hadronization process increases the total entropy the second law of thermodynamics is not violated (as is the case for simple coalescence models incorporating massless partons).

The PHSD approach has been also applied to nucleus-nucleus collisions from 40 to 160 A·GeV in order to explore the space-time regions of 'partonic matter' [16]. We have found that even central collisions at the top SPS energy of $\sim$158 A· GeV show a large fraction of non-partonic, i.e. hadronic or string-like matter, which can be viewed as a 'hadronic corona'. This finding implies that neither purely hadronic nor purely partonic 'models' can be employed to extract physical conclusions in comparing model results with data.

On the other hand - studying in detail Pb+Pb reactions at 40, 80 and 158 GeV·GeV in comparison to the data from the NA49 Collaboration [16]- it is found that the partonic phase has only a very low impact on rapidity distributions of hadrons but a sizeable influence on the transverse-mass distribution of final kaons due to the parton interactions. The most pronounced effect is seen on the production of multi-strange antibaryons due to a slightly enhanced $s\bar{s}$ pair production in the partonic phase from massive time-like gluon decay and a more abundant formation of strange antibaryons in the hadronization process. This enhanced formation of strange antibaryons in central Pb+Pb collisions at SPS energies by hadronization supports the early suggestion by Braun-Munzinger and Stachel [35, 36] in the statistical hadronization model - which describes well particle ratios from AGS to RHIC energies.

Some note of caution has to be stated here with respect to applications of PHSD at FAIR energies (5-40 A·GeV) since the partonic equation-of-state employed so far - as fixed to the lQCD results from Ref. [17] - describes a crossover transition between the hadronic and partonic phase while at lower SPS and FAIR energies a first-order phase transition and the appearance of a critical point in the QCD phase diagram are expected [37]. Such phenomena may not be described by the present realization of PHSD but need subtle extensions. Additionally, we have to mention that the high transition temperature from Ref. [17] might be questioned since related lattice calculations of the Wuppertal group in 2009 lead to lower critical temperatures [38]. Once the question of the 'proper T_c or critical energy density ϵ_c is settled by the lattice community we might have to readjust the partonic equation-of-state in PHSD accordingly. On the other hand an application to RHIC energies is rather straight forward - except for a possible color-glass initial state - and detailed results from PHSD will be presented in a forthcoming study.

Acknowledgement

The authors are grateful to C. Blume, C. Höhne, O. Linnyk and S. Matticllo for valuable discussions.

References

[1] Shuryak E 2004 *Prog. Part. Nucl. Phys.* **53** 273
[2] Thoma M H 2005 *J. Phys.* G **31** L7
[3] Peshier A and Cassing W 2005 *Phys. Rev. Lett.* **94** 172301
[4] Arsene I *et al.* 2005 *Nucl. Phys.* A **757** 1;
Back B B *et al.* 2005 *Nucl. Phys.* A **757** 28;
Adams J *et al.* 2005 *Nucl. Phys.* A **757** 102;
Adcox K *et al.* 2005 *Nucl. Phys.* A **757** 184
[5] Hirano T and Gyulassy M 2006 *Nucl. Phys.* A **769** 71
[6] Hwa R C and Yang C B 2003 *Phys. Rev.* C **67** 034902;
Greco V, Ko C M and Levai P 2003 *Phys. Rev. Lett.* **90** 202302
[7] Fries R J , Müller B , Nonaka C and Bass S A 2003 *Phys. Rev. Lett.* **90** 202303
[8] Lin Z-W *et al.* 2005 *Phys. Rev.* C **72** 064901
[9] Cassing W and Bratkovskaya E L 2008 *Phys. Rev.* C **78** 034919
[10] Juchem S, Cassing W and Greiner C 2004 *Nucl. Phys.* A **743** 92
[11] Cassing W and Juchem S 2000 *Nucl. Phys.* A **665** 377
[12] Ivanov Y B, Knoll J and Voskresensky D N 2000 *Nucl. Phys.* A **672** 313
[13] Cassing W 2009 *E. Phys. J. ST* **168** 3
[14] Cassing W 2007 *Nucl. Phys.* A **791** 365
[15] Cassing W 2007 *Nucl. Phys.* A **795** 70
[16] Cassing W and Bratkovskaya E L 2009 *Nucl. Phys.* A **831** 215
[17] Cheng M *et al.* 2008 *Phys. Rev.* D **77** 014511
[18] Cassing W and Bratkovskaya E L 1999 *Phys. Rep.* **308** 65
[19] Peshier A 2004 *Phys. Rev.* D **70** 034016
[20] Pisarski R D 1989 *Phys. Rev. Lett.* **63** 1129
[21] Peshier A, Kämpfer B , Pavlenko O P and Soff G 1996 *Phys. Rev.* D **54** 2399
[22] Bluhm M, Kämpfer B, Schulze R, Seipt D and Heinz U 2007 *Phys. Rev.* C **76** 034901
[23] Aarts G and Martinez Resco J M 2002 *J. High Energy Phys.* **04** 054
[24] Serot B D and Walecka J D 1988 *Adv. Nucl. Phys.* **16** 1
[25] Maruyama T, Cassing W, Mosel U, Teis S and Weber K 1994 *Nucl. Phys.* A **573** 653
[26] Larionov A B , Cassing W, Greiner C and Mosel U 2000 *Phys. Rev.* C **62** 064611
[27] Bratkovskaya E L, Cassing W, Rapp R and Wambach J 1998 *Nucl. Phys.* A **634** 168
[28] Dover C B , Heinz U, Schnedermann E and Zimanyi J 1991 *Phys. Rev.* C **44** 1636
[29] Bengtsson H U and Sjöstrand T 1987 *Comp. Phys. Commun.* **46** 43
[30] Falter T, Cassing W, Gallmeister K and Mosel U 2004 *Phys. Rev.* C **70** 054609
[31] McLerran L 2007 *Nucl. Phys.* A **787** 1
[32] Antinori F *et al.* 2004 *Phys. Lett.* B **595** 68
[33] Anticic T *et al.* NA49 Collaboration 2009 *Phys. Rev.* C **80** 034906
[34] Alt C *et al.* NA49 Collaboration 2008 *Phys. Rev.* C **78** 034918
[35] Braun-Munzinger P *et al.* 2001 *Phys. Lett.* B **365** 1
[36] Andronic A, Braun-Munzinger P and Stachel J 2006 *Nucl. Phys.* A **772** 167
[37] Senger P *et al.* 2010 CBM Physics Book, *Springer*, in press.
[38] Fodor Z and Katz S D 2009 *arXiv: 0908.334 [hep-ph]*

Journal of Physics: Conference Series **230** (2010) 012002 doi:10.1088/1742-6596/230/1/012002

Recent results from FOPI From the dynamics of heavy ion collisions to the (Anti)Kaon-nucleon potential

Yvonne Leifels for the FOPI Collaboration
GSI Helmholtz Center for Heavy Ion Research, D-64291 Darmstadt, Germany
E-mail: Y.Leifels@gsi.de

Abstract. Strange particles allow to probe the properties of hot and dense baryonic matter. At incident energies close to the production threshold medium effects on the strange hadrons are magnified. New data obtained with the recently completed FOPI upgrade are presented for the system Ni+Ni at 1.91 AGeV and discussed in comparison to earlier findings.

1. Introduction

Modifications of hadron properties in dense baryonic matter are a current subject of intensive research in hadron physics. Various theoretical approaches [1, 2, 3] agree qualitatively in predicting, for example, modifications of masses and coupling constants for kaons and anti-kaons. Due to the density dependence of the KN($\overline{\rm K}$N) potential, the K^- effective mass is expected to drop, whereas the mass of K^+ mesons is predicted to rise with increasing density of nuclear matter. Already two decades ago, Kaplan and Nelson [4] pointed out that due to additional attractive interactions with the surrounding nucleons a condensation of anti-kaons (K^-) may take place in a dense baryonic environment as encountered in the interior of neutron stars.

From general arguments one can conclude how the medium-modifications of the KN($\overline{\rm K}$N) potentials influence particle production yields and phase space distributions. However, access to the depth of those potentials is only possible via microscopic transport codes modeling the dynamical evolution of heavy ion collisions.

At AGS energies clear evidence was found that the directed K^0 flow is opposite in sign to the flow of protons [5]. This experimental observation was explained by microscopic model calculation incorporating a repulsive KN potential [6]. Experimental data for charged kaons where less convincing and still lacking a consistent description [7]. At lower energies the sensitivity to the KN($\overline{\rm K}$N) potential should be enhanced and several investigations by the KAOS, FOPI and lately HADES collaborations have been started at the SIS-18 accelerator at GSI.

The strength of the K^- nucleon potential might give rise to a new class of bound states [8, 9] with quite exotic properties. These bound states are called kaonic clusters and are being searched for in experiments using stopped K^- beams [10, 11] and pp reactions [12]. Due to the predicted large central densities and the large compression reached in the collision of heavy ions the production of kaonic clusters might be favored in such an environment [13]. Therefore a complementary program was started to look for bound multi baryonic strange states in heavy ion reactions [14].

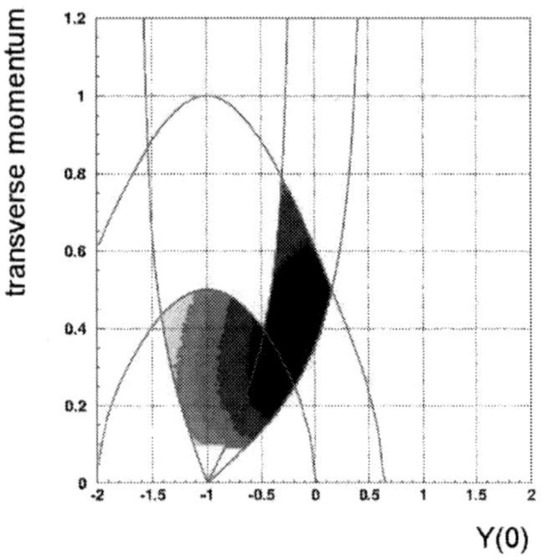

Figure 1. Acceptance for charged kaons making use of CDC - RPC and CDC - Plastic Barrel matches.

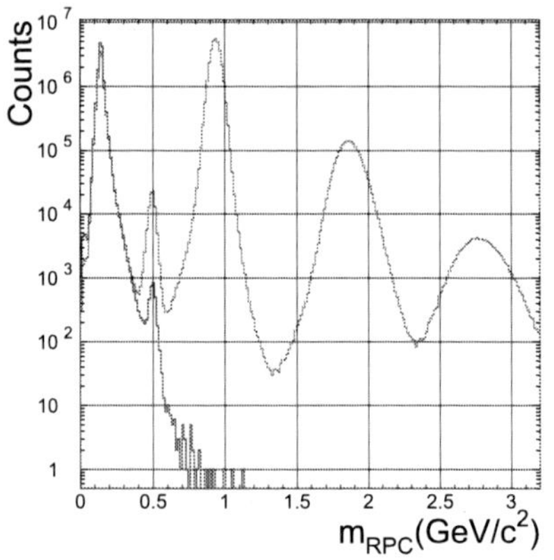

Figure 2. Reconstructed masses of positive (negative) charged particles using CDC–MRPC matches with a momentum cut of 1.0(0.8) GeV/c.

In the following new data obtained with the FOPI detector in 2008 at the SIS18 accelerator of GSI are shown for the reaction Ni+Ni at an incident energy of 1.91 AGeV together with older data from Al+Al collisions at the same energy elucidating some key aspects of strangeness physics in the threshold energy range.

2. Experiment: FOPI - Phase III

FOPI is a large acceptance detector system designed to measure charged particles emitted form heavy-ion collisions. It is installed at the SIS-18 accelerator at the GSI, Darmstadt (Germany). The modular structure allows to cover nearly the full solid angle in the laboratory frame. The core part of FOPI is a supra conducting solenoid with two drift chambers. The Central Drift Chamber CDC covering $30° < \Theta_{lab} < 140°$ and the forward drift chamber Helitron $7° < \Theta_{lab} < 30°$. Both drift chambers are augmented by time–of–flight–detectors for charge identification.

Apart from charged particles FOPI can measure neutral strange particles by their charged decay product ($K^0 \rightarrow \pi^+ + \pi^-$, branching ration BR = 69 %, $\Lambda \rightarrow \pi^- + p$, BR =64 %, $\Phi \rightarrow K^+ + K^-$, BR = 49.1 %) as well as short lived strange resonances $\Sigma^{\pm *}$ and K^{0*}.

In order to improve the detection of charged kaons the FOPI detector was upgraded with a new time-of-flight barrel covering the polar angle range $35° < \Theta < 60°$ based on Multigap–Resistive–Plate–Chambers (MRPC) [15] in 2006. In the polar angle range $60° < \Theta < 140°$ a plastic scintillator barrel is used as a time–of–flight detector. The phase region in which charged kaons are thus identified in the FOPI Phase III setup is depicted in Fig. 1. Since then this detector was used in major data taking runs for which first result become available now. The performance in identifying the emitted particles is depicted for the reaction Ni+Ni at an incident energy of 1.91 AGeV in Fig. 2 where a mass spectrum is shown that is determined from the momentum reconstructed in the Central Drift Chamber (CDC) and the velocity measured in the MRPC barrel. The overall time resolution is 90 ps, in which the MRPC resolution is $\sigma_{tof,MRPC} = 65\,\text{ps}$. This time resolution and the granularity of the detector are sufficient to separate positively (negatively) charged kaons with a signal-to-background ratio in excess of 10 when the maximum momentum is limited to 1.0 (0.8) GeV/c, respectively.

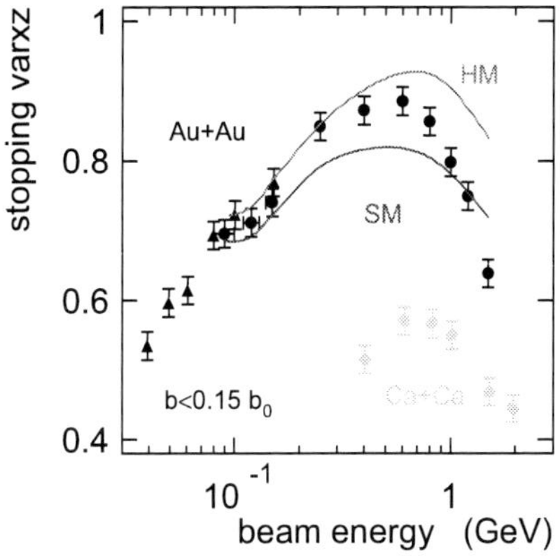

Figure 3. Stopping variable *varxz* as function of beam energy for central Au+Au and Ca+Ca collisions. The data for Au+Au collisions is a compilation of FOPI (filled circles) and INDRA (filled triangles) [34] results. The red and blue lines are prediction from the IQMD model [21] using a hard (HM) or a soft (SM) equation of state with momentum dependent interactions, respectively.

3. Dynamics of Heavy Ion collisions

After production the kaons are interacting with the surrounding matter. Some of these interactions are due to the respective KN or ($\overline{\rm K}$N potentials. Negatively charged Kaons are additionally subject to strong absorption and positive kaons, despite their relatively long mean free path in a nuclear medium, are frequently rescattered. Hence, it is indispensable to measure phase space distributions of nucleons and light fragments to control the properties of the bulk matter and to constrain the input to microscopic transport models.

A relevant observable measuring the stopping in central collisions and the amount of energy redistributed from the primary longitudinal direction into the transverse direction is the ratio of the variances of the rapidity distributions in x–direction (*varx* transverse to the beam direction) and the rapidity distribution in z–direction (*varz* longitudinal). This observable *varxz* (or *vartl*) was introduced in [33]. The integrations for calculating the variances are limited to the range $|y_{z0}|, |y_{x0}| < 1$, where $|y_{z0}|, |y_{x0}|$ are scaled rapidities (i.e. $y_0 = y_{lab}/y_{cm,proj} - 1$). Other observables were introduced to characterise stopping (e.g. [31, 32]). Their common feature is that they assess stopping with respect to the initial conditions, whereas *varxz* refers to a completely stopped scenario (i.e. $varxz = 1$).

For the data points in Fig. 3 all measured charged particles have been added up to try to characterise the global system stopping in central collisions ($\approx 2\%$ of σ_{reac}). The excitation curve of *varxz* for Au+Au [33] was combined with data points from lower energies [34]. The observable *varxz* rises with energy and reaches a plateau between 0.4 and 0.6 AGeV, but obviously the colliding Au+Au system is not completely stopped, i.e. the rapidity distribution is more elongated in tranverse than in longitudinal direction. At energies beyond 0.8 AGeV *varxz* is rapidly dropping. The data points for the smaller Ca+Ca system are shown for comparison. Generally, they depict a similar behaviour with the slight shift of the maximum to higher energies but the elongation of the rapidity distributions in the beam direction is more pronounced for all investigated beam energies.

The Au+Au data are confronted with IQMD [21] predictions incorporating a hard (HM)

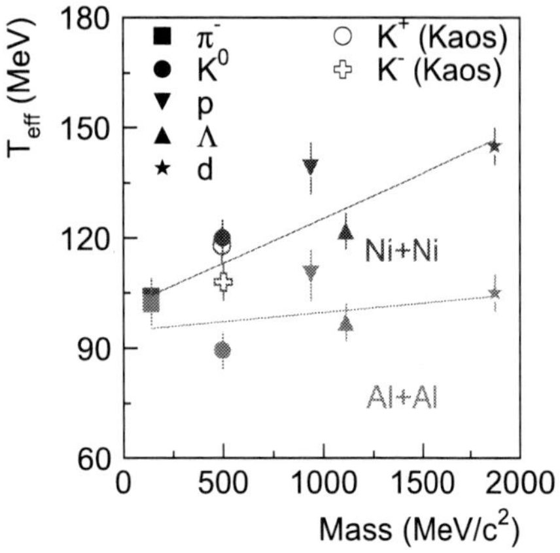

Figure 4. Inverse slope parameter T_{eff} as function of particle mass for Ni+Ni and Al + Al central collisions. Data points measured by the KAOS collaboration [36] are included where measured.

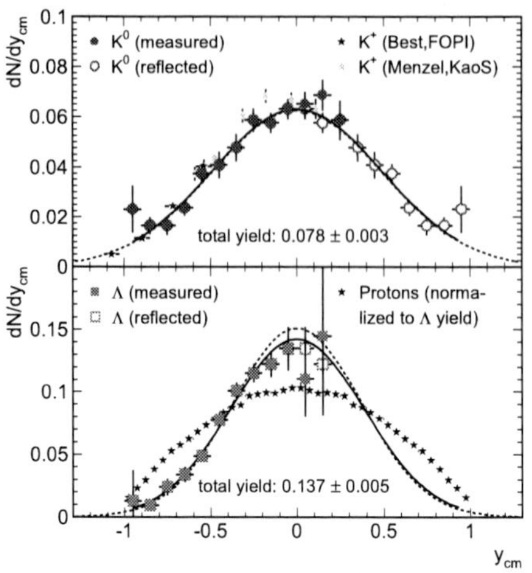

Figure 5. Rapidity density distributions for K^0 (full circles, upper panel) and Λ particles (filled squares, lower panel) in central Ni+Ni collisions at 1.93 AGeV. The open data points are reflected with respect to mid-rapidity. The (green and black) star symbols correspond to previously measured K^+ [36] (upper panel) and to protons (lower panel). The full line corresponds to a Gaussian fit, the dashed line is a prediction of a thermal model with radial flow [37].

and a soft (SM) equation of state with momentum dependent interaction. As can be seen in Fig. 3 $varxz$ is sensitive to the mean-field EOS and opens up a possibility to observe EOS dependence also in central reactions. However, it is also clear that other interactions (i.e. collisions) influence the outcome of the calculations. The present parameterizations of IQMD are not able to reproduce the data at energies beyond 0.8 AGeV but a fair description with a hard equation of state is reached at lower energies.

The observed incomplete stopping in central collisions at SIS beam energies can be identified as partial transparency: This scenario is favoured by the observed system size dependence, and it was demonstrated explicitly by the so-called isospin tracing method combining the rapidity distributions of four collisions systems using ^{96}Zr and ^{96}Ru [35]. These results support the interpretation of central collisions at SIS energies in terms of two remnant counterflowing (not completely stopped) fluids and suggest that an equilibrium is not reached.

4. Strangeness production

The FOPI collaboration investigated the strangeness production for Al+Al (small size) and Ni+Ni (medium size) systems at E_{beam} = 1.93 AMeV. Two neutral strange particles, K^0 and Λ were reconstructed for about 20 % of the most central collisions of the total reaction cross

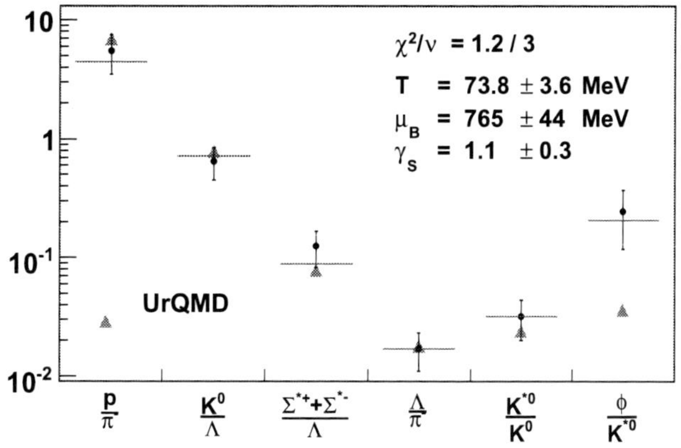

Figure 6. Independent particle ratios for central Al+Al [26] collisions with the results of thermal model fits and UrQMD predictions.

section [24]. The analysis of the phase space occupied by K^0s and Λs at freeze-out revealed that a thermal Boltzmann distribution with a single inverse slope parameter, T, is sufficient to describe the profiles of transverse mass distributions in subsequent rapidity bins. The inverse slope parameters T_{eff} at mid-rapidity for Ni+Ni and Al+Al collisions are shown in Fig. 4 together with the results for other particle species. For the heavier system T_{eff} is rising with particle mass which is commonly interpreted as flow. In a simplified picture the kinetic energy of a particle is the sum of a chaotic 'thermal' motion and flow therm which is linearly rising with particle mass. This is approximately the case for Ni+Ni collision, contrasting the smaller Al+Al system which does not show any significant flow. The less pronounced flow can be linked to the smaller system size and, therefore, smaller pressure build-up during the collision. The individual inverse slope parameters of the different particle species are not following strictly a linear trend: The appearent temperature of the Λ - Hyperons are significantly smaller than the one of protons with a similar mass, but clearly more systematic studies are needed to confirm this observation.

The rapidity density distribution along the beam axis of Λ's is significantly less elongated than the proton rapidity distribution (Fig. 5. Its shape is nearly gaussian and can be reproduced by a simple thermal model incorporating radial expansion [37]. The K^0 rapidity density distributions follows likewise the thermal model predictions. In contrast to the protons still keeping a trace of the initial state, the produced strange particles have lost the reminiscence of the entrance channel for the most part.

Φ mesons are reconstructed via their dominant decay channel $\Phi \rightarrow K^+ + K^-$ and yield have been extraced for central collisions of Al+Al and Ni+Ni at 1.9 AGeV. The strange resonances $\Sigma^{\pm *}(1385)$ and $K^{0*}(892)$ (Σ^* and K^* further on) also decay into charged particles which are accesibble to FOPI: $\Sigma^* \rightarrow \Lambda + \pi^{\pm}$ and $K^* \rightarrow K^+ + \pi^-$. The short lifetimes of those resonances ($c\tau = 5$ fm/c and 4 fm/c) leaves their decay vertices to close to the event vertex to be separated. Hence, their decay products cannot be distinguished from the other charged particles by a vertex cut. Despite the resulting huge background about 3100 ± 500 and 6100 ± 850 resonances were found with a significance of 9 and 10 [26], the invariant masses are in agreement with the values reported in literature. The efficiency corrections necessary to extract the particle yields are described e.g. in [26] or [43].

In total, 6 independent ratios of particle yields were constructed for central collisions of Al+Al at 1.91 AGeV beam energy. These ratios were compared to the statistical model predictions

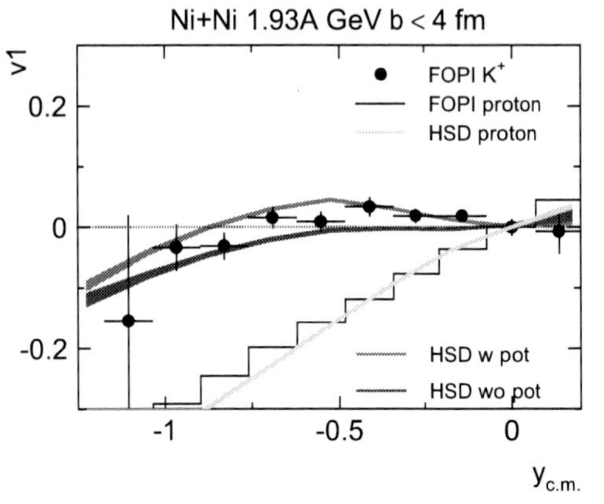

Figure 7. Results for directed flow v_1 of kaons in comparison to the flow of protons as a function of the center-of-mass rapidity $y_{c.m.}$. Solid lines are predictions using the HSD transport code.The v_1 - values are obtained within the acceptance of the CDC, for details see text.

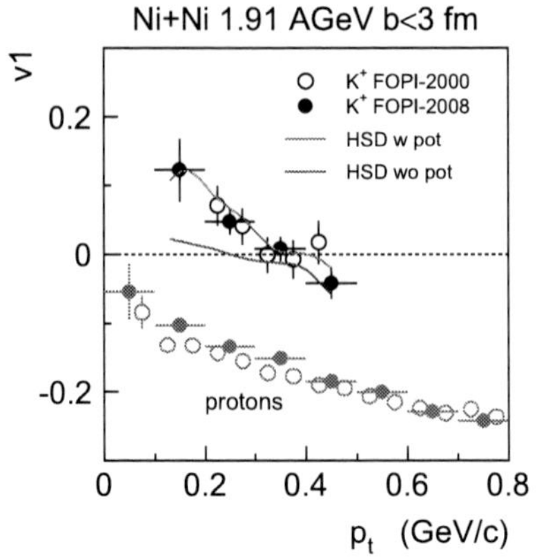

Figure 8. Directed flow v_1 of kaons as function of transverse momentum p_t in comparison to the flow of protons. Data from the new MRPC barrel (full circles) are compared to older results (filled circles) [16]. Solid lines are predictions of the HSD transport code with or without KN potential.

performed with the THERMUS code [38], as shown in Fig. 6. The calculations were done in the frame of a Grand Canonical ensemble for non-strange particles. Due to a necessity to constrain the strangeness number S at low beam energies, a Canonical ensemble was used for particles characterized by S $\neq$ 0 [39]. In general, a good fit–quality of the statistical model parameters Temperature T and baryo–chemical potential μ_B was obtained, expressed by $\chi^2/\nu = 0.3$ for Al+Al. The best fits were found for T of about 74 MeV and μ_B of about 780 MeV respectively. These values match the systematics of T and μ_B extracted at other beam energies and systems [40]. In case of Al+Al it was possible to fit in addition the strangeness undersaturation factor γ_S, being non-equal to unity in case of non-equilibration of the yield of particles containing strange quarks [41]. However, it was found to be consistent with unity within the experimental errors. These results demonstrate that the statistical model with two (three) free fit parameters of T, μ_B (and γ_S) is capable of reproducing the experimental yield ratios at 1.9 AGeV, despite the observation that the underlying assumption of a complete equilibration is not supported by the experimental rapidity profiles of nuclear bulk matter.

Similar observations have been made in Ni+Ni collisions at 1.9 AGeV: All particle ratios are described within a thermal model yielding similar parameters [44]. There is only one prominent exception the η/π ratio, a fact which was already reported in the literature (e.g. [42]).

It is interesting to note that the measured particle ratios in central Al+Al collisions are reasonably well described by the microscopic transport code UrQMD [45]. UrQMD results were found to be in agreement not only with experimental data, but also with the predictions of the statistical model except for the Φ/K^* yield ratio. It is known, however, that the production yield of Φ mesons in UrQMD too low in this energy range [46]. Apart from the Φ meson the qualitity of description is equally good for both models. To draw further conclusions one needs more systematic studies on heavier collision systems and different energies.

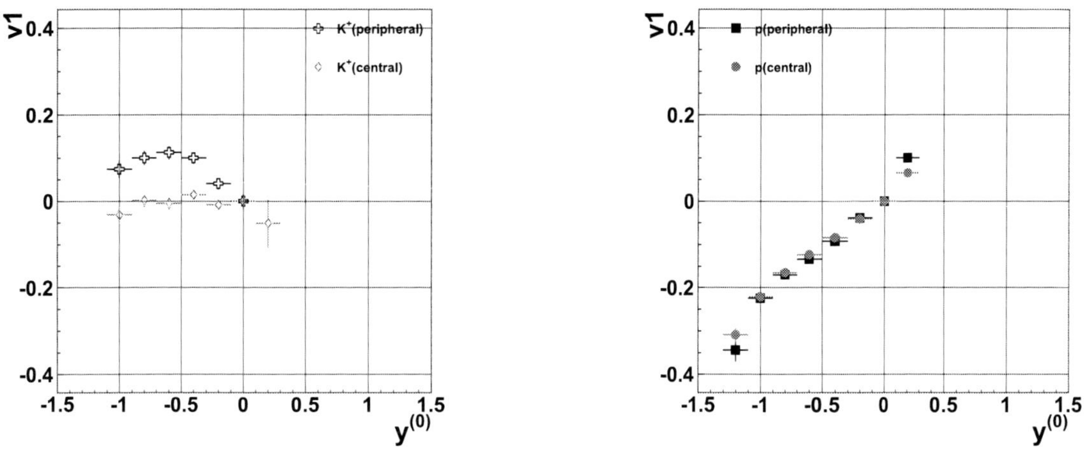

Figure 9. Sideflow of kaons and protons in the reaction Ni+Ni at an incident energy of 1.91 AGeV as function of normalized rapidity $y(0) = y/y_{cm,proj}1$. The first fourier coefficients v_1 are plotted for two different centrality bins for kaons (left panel) and protons (right panel). Black data points are obtained for peripheral collisions corresponding to a mean impact parameter of 6 ± 1 fm, the red ones for central collisions with a geometrical impact parameter b < 3 fm. The v_1–values are obtained within the acceptance of the CDC, for details see text.

5. Kaon flow

The excellent kaon identification capabilities of the MRPC barrel allows to perform differential flow analysis of these rare particles and thus to continue the effort to determine the in-medium potential for kaons and anti-kaons [16].

In Fig. 7 the first fourier coefficients v_1 [18] measured in central Ni+Ni collisions at 1.91 AGeV are plotted as function of the normalized rapidity for positive kaons and protons together with predictions of the HSD model [47] with and without a KN nucleon potential. The direction of the reaction plane was determined by the transverse momentum method [19] making use of particles detected in the forward wall of FOPI only in order to avoid autocorrelations. The data are corrected for the finite reaction plane resolution employing the Ollitrault formalism [20]. For better comparability with different reaction plane reconstruction methods the v_1 data were shifted such that the mean $v_1$value at midrapidity is antisymmetric. In order to compare to the data shown in Fig. 7 model calculations have to take into account the acceptance of the detector, that is limited for this detailed analysis to the acceptance of the CDC covering a polar angle range of $26° < \Theta_{lab} < 120°$. For kaons additional constraints have to be put on the momenta necessary to obtain sufficient particle identification in the data. In the polar angular range from $30.5° < \Theta_{lab} < 52°$ (MRPC acceptance) kaons are included up to a momentum of 1.0 GeV/c, in the range $53° < \Theta_{lab} < 120°$ (plastic barrel acceptance) the laboratory momentum is limited to 0.5 GeV/c. The angular limits correspond to the so called shifted target geometry, where the target was placed 40 cm upstream of its nominal position in the FOPI apparatus.

The microscopic model results agree reasonably well with the experimental data for kaons and protons, however, the sensitivity of this obervable to the in-medium KN potentials is rather small. The repulsive KN potential pushes the kaons away from the bulk matter, but the effect is moderate. The picture changes if one includes the transverse direction and switches to v_1 as a function of rapidity *and* transverse momentum. In Fig. 8 experimental data measured in 2000 [16] is shown together with the new results confirming the old findings. One observes a strong p_t dependence for kaons. The fourier coefficient v_1 is changing sign. For low p_t the kaon flow signal is opposite to the one of the protons which may signal the influence of the KN potential being more relevant for small transverse momenta. At larger transverse momenta the directed flow of kaons tends to be negative like the one of the protons. The HSD transport code is only able to reproduce this observation when a repulsive KN potential is considered. A potential of $U_{pot} = 20MeV$ at normal nuclear matter density is describing the data. This is consistent with results from pion induced reactions measured by the FOPI collaboration [17].

Another effect sensitive to the KN potential is observed in the centrality dependence of v_1 (Fig. 9). The centrality selection was chosen such that for both centrality bins the proton flow is still approximately the same within the detector acceptance. The event samples differ by the amount of stopping, the maximum density reached in the collision zone and the size of the cold spectator matter. The latter manifests itself at rather low transverse momenta which are not accessible in the chosen acceptance of this analysis. While the kaon data in the central event sample confirm quantitatively our earlier findings as shown before, the large sideflow in the most peripheral data sample is suprising: At a normalized rapidity of y(0)$\approx$ 0.5 the antiflow of the kaons has the same magnitude as the flow of the protons. Tentatively, this effect can be attributed to the strong momentum dependence of the kaon potential. This hypothesis awaits its confirmation/rejection by transport model calculations.

The available data sample also allows the reconstruction of the K^- flow. However, this analysis is much more difficult due to the very low yield of K^- that is at the 3 % level of K^+ production. The analysis is in progress.

6. Summary

The observable $varxz$ characterizes stopping (i.e. the redistribution of momentum from the longitudinal into the transverse direction) in heavy ion collisions with respect to a fully stopped system by correlating the variance of the rapiditity distribution in the transverse direction to the one in the longitudinal direction. The $varxz$ reaches a maximum between 0.6 and 0.8 AGeV and drops rapidly at higher energies. This observation was identified as increasing 'partial transparency' of two counterflowing fluids never reaching global equilibration. Nevertheless, yield ratios of nucleons and produced particles can be described by thermal model fits.

New data have been presented making use of the FOPI phase III setup at GSI elucidating the properties of strange particle in dense baryonic matter. For the first time a strong antiflow component for K^+ is seen for peripheral Ni+Ni reactions at 1.91 AGeV. pointing to a strong momentum dependence of the Kaon-Nucleus potential. More data are available from the 2009 measurement campaign of FOPI where more than 10^8 events were recorded for the reactions Ni+Pb at an incident energies of 1.91 AGeV and Ru+Ru at 1.65 AGeV opening new possibilities to study hadronic states in the hadronic environment of a heavy-ion collision.

Acknowledgments
This work was supported by the German BMBF under Contract No. 06HD190I, by the Korea Science and Engineering Foundation (KOSEF) under Grant No. F01-2006- 000-10035-0, by the mutual agreement between GSI and IN2P3/CEA, by the Hungarian OTKA under Grant No. 47168, within the Framework of the WTZ program (Project RUS 02/021), by DAAD (PPP D/03/44611), by DFG (Projekt 446-KOR-113/76/04) and by the EU, 6th Framework Program, Integrated Infrastructure: Strongly Interacting Matter (Hadron Physics), Contract No. RII3-CT-2004-506078.

References

[1] Brown G E et al. 1994 *Nucl. Phys.* A **567** 937
[2] Weise W *Nucl. Phys.* A **610** 35c
[3] Li G Q et al. 1997 *Nucl. Phys.* A **625** 372

[4] Kaplan D B and Nelson A E 1986 *Phys. Lett.* B **175** 57
[5] Chung P et al. [EOS collaboration] 2000 *Phys. Rev. Lett* **85** 340-943
[6] Pal S, Ko C M, Li Z, Bin Z 2000 *Phys. Rev.* C **62** 061903
[7] Pinkenburg C et al. [EOS collaboration] 2002 *Nucl. Phys.* A **698** 495-498
[8] Akaishi Y and Yamazaki T 2002 *Phys. Rev.* C **65** 044005
[9] Yamazaki T, Doté A, Akaishi Y 2004 *Phys. Lett.* B **587** 167
[10] Agnello M et al. 2005 *Phys. Rev. Lett.* **94** 212303
[11] Suzuki T et al. [E549 collaboration] 2008 *Mod.Phys.Lett.* A **23** 2520
[12] Yamazaki T et al. (*Prepint*) arXiv:0810.5182
[13] Yamazaki T et al. 2004 *Nucl. Phys.* A **738** 168
[14] Herrmann N [FOPI collaboration] 2005 *Proceedings of the EXA05 conference, Osterreichische Akademie der Wissenschaften* 73
[15] Schüttauf A et al. 2009 *Nucl. Instr. Meth.* A **602** 679
[16] Crochet P et al. [FOPI collaboration] 2000 *Phys. Lett.* B **486** 6
[17] Benabderrahmane M L et al. [FOPI collaboration] 2009 *Phys. Rev. Lett.* **102** 182501
[18] Voloshin S A and Zhang Y Z 1996 *Z. Phys.* C **70** 665
[19] Danielewicz P and Odyniec G 1985 *Phys. Lett.* B **157** 146
[20] Ollitraut J Y 1997 (*Preprint* nucl-ex/9711003)
[21] Hartnack C et al. 1996 *Eur. Phys. J.* A **1** 151
[22] Ritman J et al. [FOPI collaboration] 1995 *Z. Phys.* A **352** 355
[23] Herrmann N, Wessels J P, Wienold T 1999 *Ann. Rev. Nucl. Part. Sci.* **49** 581
[24] Merschmeyer M et al. [FOPI collaboration] 2007 *Phys. Rev.* C **76** 02490
[25] Lopez X et al. [FOPI Collaboration] 2007 *Phys. Rev.* C **76** 052203
[26] Lopez X et al. [FOPI Collaboration] 2008 *J. Phys.* G **35** 044020
[27] Herrmann N [FOPI Collaboration] 2009 *Prog. Part. Nucl. Phys.* **62** 445
[28] Braun O et al. 1977 *Nucl. Phys.* B **124** 45
[29] Pigeot C et al. 1985 *Nucl. Phys.* B **249** 172
[30] Siebert R et al. 1994 *Nucl. Phys.* A **567** 819
[31] Videbaek F and Hansen O 1995 *Phys. Rev.* C **52** 2684
[32] Ströbele H 2009 *Preprint* nucl-ex:0908.2777
[33] Reisdorf W et al. [FOPI Collaboration] 2004 *Phys. Rev. Lett.* **92** 232301
[34] Andronic A, Lukasik J, Reisdorf W, Trautmann W 2006 *Eur. Phys. J.* A**30** 31
[35] Rami F et al. [FOPI Collaboration] 2000 *Phys. Rev. Lett.* **84** 1120; Hong B, et al. [FOPI Collaboration] 2002 *Phys. Rev.* C **66** 034901
[36] Menzel M et al. [KAOS Collaboratoin] 2000 *Phys. Lett.* B **495** 26
[37] Siemens P J and Rasmussen J 0 1979 *Phys. Rev. Lett.* **42** 880
[38] Wheaton S, Cleymans J, Hauer M 2004 *Preprint* hep-ph/0407174
[39] Braun-Munzinger P, Redlich K, Stachel J 2004 in Quark-Gluon Plasma 3, World Scientific Publishing, Singapore
[40] Braun-Munzinger P, Wambach J 2009 *Rev. Mod. Phys.* **81** 1031
[41] Becattini R et al. 2001 *Phys. Rev.* C **64** 024901
[42] Averbeck R, Holzmann R, Metag V, Simon R S *Phys. Rev* C **67** 024903
[43] Mangiarotti A 2003 *Nucl. Phys.* **A 714** 89
[44] Piasecki K 2009 Proceeding to the Bormio winter workshop on nuclar physics
[45] Bass S et al. 1998 *Prog. Part. Nucl. Phys.* **41** 225-370
[46] Bleicher M 2009 *private communication*
[47] Cassing W, Tolos L, Bratkovskaya E, Ramos A 2003 *Nucl. Phys.* **A 727** 59

Journal of Physics: Conference Series **230** (2010) 012003 doi:10.1088/1742-6596/230/1/012003

Particle Production at the SPS and the QCD Phase Diagram

Christoph Blume

Institut für Kernphysik, J.W. Goethe-Universität,
Max-von-Laue-Str. 1, D-60438 Frankfurt am Main, GERMANY

E-mail: `blume@ikf.uni-frankfurt.de`

Abstract. Recent results of particle production in the energy regime of the CERN-SPS are reviewed. In order to collect information on the properties of the QCD phase diagram systematic studies of the system size and the energy dependence of particle production in heavy ion collisions have been performed. Net-baryon distributions and results on strangeness production are discussed. The system size dependence of many observables can be understood in the core-corona approach, which has implications on the possibility to use system size as a control parameter to study different areas of the phase diagram. Recent attempts to search for a critical point, such as multiplicity fluctuations and the transverse mass dependence of anti-baryon/baryon ratios are reviewed.

1. Introduction

The QCD phase diagram, as it is illustrated in Fig. 1, contains a variety of theoretically predicted features. The most important one is the phase boundary separating a hadron gas and a quark gluon plasma (red dashed line in Fig. 1). Its position at vanishing baryonic chemical potential ($\mu_{\rm B} = 0$) is defined by the critical temperature $T_{\rm C}$. Even though recent attempts to calculate the exact value of $T_{\rm C}$ with lattice QCD still result in different numerical values [1, 2, 3], there is a general consensus that for $\mu_{\rm B} = 0$ this transition is of the type of crossover. However, this crossover line might turn for a certain $\mu_{\rm B}$ into a first order phase transition, which would result in a critical point at that position [4]. However, predictions on the location of the critical point are still very uncertain [5]. It might very well be that it does not exist at all [6]. Recently, it has been conjectured that the phase diagram might contain yet another phase, the so-called quarkyonic phase [7], which would be separated from normal hadronic matter by a different phase boundary line (blue dashed line in Fig. 1).

Experimentally, there are several control parameters available that might allow to study different regions of the phase diagram with heavy ion collisions. One is the variation of the center-of-mass energy which will force the reaction systems to follow different trajectories in the T-$\mu_{\rm B}$ plane, reflected by a change of the chemical freeze-out parameters. The second control parameter is the variation of the system size, which can either be achieved by performing central collisions of nuclei of different size, or by studying centrality selected minimum bias collisions. However, whether this can be considered as a good control parameter for probing different areas of the phase diagram shall be discussed further below.

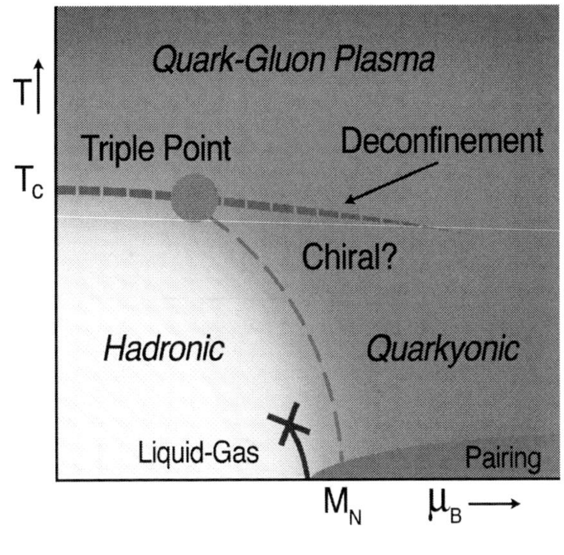

Figure 1. The phase diagram of strongly interacting matter. The picture is taken from [8].

In the following we review some results from systematic studies of particle production in heavy ion collisions performed at the CERN-SPS, and will discuss what can be learned from existing data on the experimental accessibility of different areas in the phase diagram. Finally, we also present some results from existing attempts to search for a critical point.

2. Variation of Energy

Recent results on rapidity spectra of (anti-)protons in central Pb+Pb reactions, measured at energies between 20*A* - 158*A* GeV at the SPS [9, 10], together with central Au+Au data from the AGS [11] and RHIC [12] allow to study the energy evolution of stopping. Based on the measured rapidity spectra for p, $\bar{\mathrm{p}}$, Λ, $\bar{\Lambda}$,Ξ^-, and $\bar{\Xi}^+$, all corrected for feed down from weak decays, the net-baryon distributions $\bar{\mathrm{B}} - \mathrm{B}$ have been constructed [13]. The contribution of unmeasured baryons (n,$\Sigma^{\pm}$, Ξ^0) was estimated using the results of a statistical hadron gas model [14]. In the SPS energy region a clear evolution of the shape can be observed (see left panel of Fig. 2), changing from a single maximum at midrapidity towards a structure with two maxima and a dip at $y = 0$. Concurring with this evolution, also a strong change of the anti-baryon/baryon ratios takes place in the energy region of the SPS [10, 15] and thus of the baryonic chemical potential at freeze-out μ_{B}.

The right panel of Fig. 2 shows a comparison of the energy dependence of mid-rapidityΛ, $\bar{\Lambda}$, Ξ^-, and $\bar{\Xi}^+$ production to several models. While the transport models UrQMD1.3 [16] and HSD [17] provide a reasonable description of the Λ/π and $\bar{\Lambda}/\pi$ ratios, they are clearly below the data points in case of theΞ^- and $\bar{\Xi}^+$. This might indicate that an additional partonic contribution is necessary to reach the production rates observed for multi-strange particles. Statistical models on the other hand, generally provide a better match to the data. These models are based on the assumption that the particle yields correspond to their chemical equilibrium value and can thus be described by the parameters temperature T, baryonic chemical potential μ_{B}, volume V, and, in some implementations, by an additional strangeness under-saturation factor γ_{s}. The curves shown in the right panel of Fig. 2, labeled SHM(B), are taken from [18] and are based on parametrizations of the $\sqrt{s_{\mathrm{NN}}}$ dependence of T and μ_{B}, assuming $\gamma_{\mathrm{s}} = 1$. The results of various

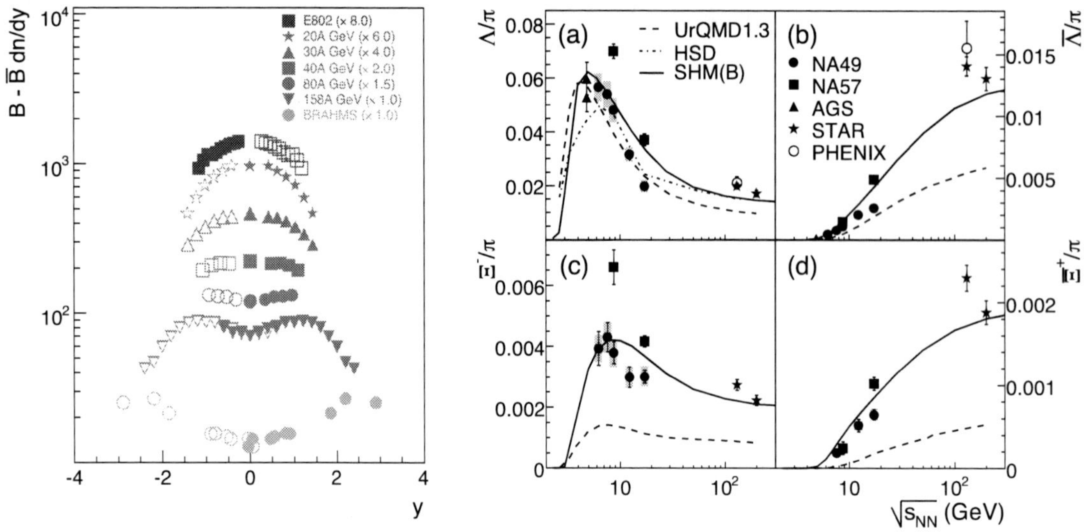

Figure 2. Left: The rapidity distributions of net-baryons at SPS energies [13, 9, 10], together with results from the AGS [11] and from RHIC [12] for central Pb+Pb(Au+Au) collisions. Right: The energy dependence of the rapidity densities dn/dy around mid-rapidity ofΛ, $\bar{\Lambda}$,Ξ^-, and $\bar{\Xi}^+$, divided by the total pion rapidity densities ($\pi = 1.5\,(\pi^- + \pi^+)$) for central Pb+Pb and Au+Au collisions [19, 20, 21, 22, 15, 23, 24, 25, 26, 27, 28, 29, 30, 31]. Also shown are results for the transport models UrQMD1.3 (dashed line) [16] and HSD (dotted line) [17], as well as a statistical hadron gas model (solid line) [18].

analysis of the measured particle abundances with statistical model approaches show that T and μ_B parameter for the SPS energy range vary over a wide region, following an universal freeze-out curve [32], and might be in the vicinity of the possible critical point location [18, 33, 34, 35]. Thus, the variation of $\sqrt{s_{NN}}$ provides a well defined way of selecting different regions in the T-μ_B plane.

3. Variation of System Size

3.1. Centrality Dependence of Net Protons

Recent data on proton and antiproton production in minimum bias Pb+Pb reactions allow to study the system size dependence of stopping. Figure 3 shows the net-proton rapidity distributions for five different centrality classes, selected from minimum bias Pb+Pb interactions at 40A (left panel) and 158A GeV (right panel) [36]. A remarkable feature of this data is that there is no drastic change with centrality of the shapes of the distributions inside the measured region at 158A GeV. However, at 40A GeV the shapes of the distributions change with centrality, going from a two-maxima structure with a narrow dip at midrapidity for central collisions to a shallow valley like distribution for peripheral collisions, similar as observed at the AGS [37].

The $\bar{p}/p$ ratio, being directly sensitive to the baryonic chemical potential μ_B, depends much less on system size than it does depend on energy. While for the mid-rapidity ratios at 158A GeV only a change in the order of a factor of two is observed when comparing central Pb+Pb reactions ($\bar{p}/p = 0.56$, 0 – 5% most central) to very peripheral ones ($\bar{p}/p = 0.11$ for 43 – 100%), this ratio

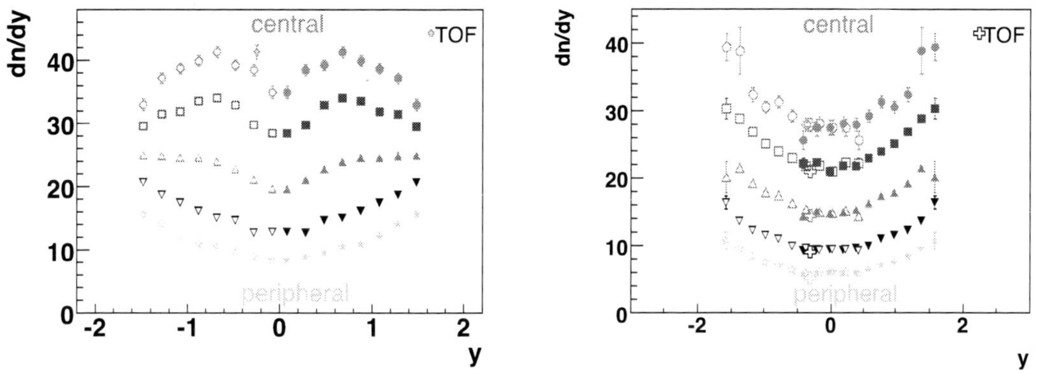

Figure 3. Net-proton rapidity distributions for Pb+Pb collisions at 40A (left) and 158A GeV (right) for different centralities as measured by the NA49 experiment [36]. The full symbols represent the measurements. The open symbols are reflected at midrapidity.

varies by a factor of 43 between central Pb+Pb collisions at 20A GeV ($\bar{\mathrm{p}}/\mathrm{p} = 0.0013$, 0 – 7%) and 158$A$ GeV [10]. Thus, only a small variation of μ_{B} can be achieved when studying reactions of different system sizes.

3.2. Core Corona Model

It has been found that the core-corona approach is able to describe the system size dependence of particle production reasonable well [38, 39, 40, 41, 42]. In this model a nucleus-nucleus collision is decomposed into a central core, which corresponds to the large fireball produced in central A+A collisions, and a peripheral corona, that is equivalent to independent nucleon-nucleon reactions. To quantify the relative contribution of the two components, the fraction of nucleons that scatter more than once $f(N_{\mathrm{part}})$ can be used. $f(N_{\mathrm{part}})$ can simply be calculated with a Glauber model [43]. This quantity allows for a natural interpolation between the yields Y measured in elementary p+p ($= Y_{\mathrm{corona}}$) and in central nucleus-nucleus collisions ($= Y_{\mathrm{core}}$):

$$Y(N_{\mathrm{part}}) = N_{\mathrm{part}}\left[f(N_{\mathrm{part}})\,Y_{\mathrm{core}} + (1 - f(N_{\mathrm{part}}))\,Y_{\mathrm{corona}}\right] \tag{1}$$

The left panel of Fig. 4 shows for example the result of this approach compared to the system size dependence of the strangeness enhancement factors E at 158A GeV [44, 45]. Here, E is defined relative to p+p reactions as baseline measurement:

$$E = \left(\frac{1}{\langle N_{\mathrm{part}}\rangle}\,\frac{\mathrm{d}N(\mathrm{Pb{+}Pb})}{\mathrm{d}y}\bigg|_{y=0}\right) \Bigg/ \left(\frac{1}{2}\,\frac{\mathrm{d}N(\mathrm{p{+}p})}{\mathrm{d}y}\bigg|_{y=0}\right) \tag{2}$$

The observed rapid rise for very small systems and the subsequent saturation can naturally be described within this model.

It is interesting to observe that this approach not only works for yields, but also for dynamical quantities such as the average transverse momenta $\langle m_{\mathrm{t}}\rangle - m_0$ (see right panel of Fig. 4). This suggests that the core-corona approach in general provides a reasonable way for understanding the evolution from elementary p+p to central Pb+Pb collisions.

A more differential look on the core-corona model is shown in the left panel of Fig. 5 for symmetric reaction systems. Due to their different surface to volume ratio, the N_{part} dependence

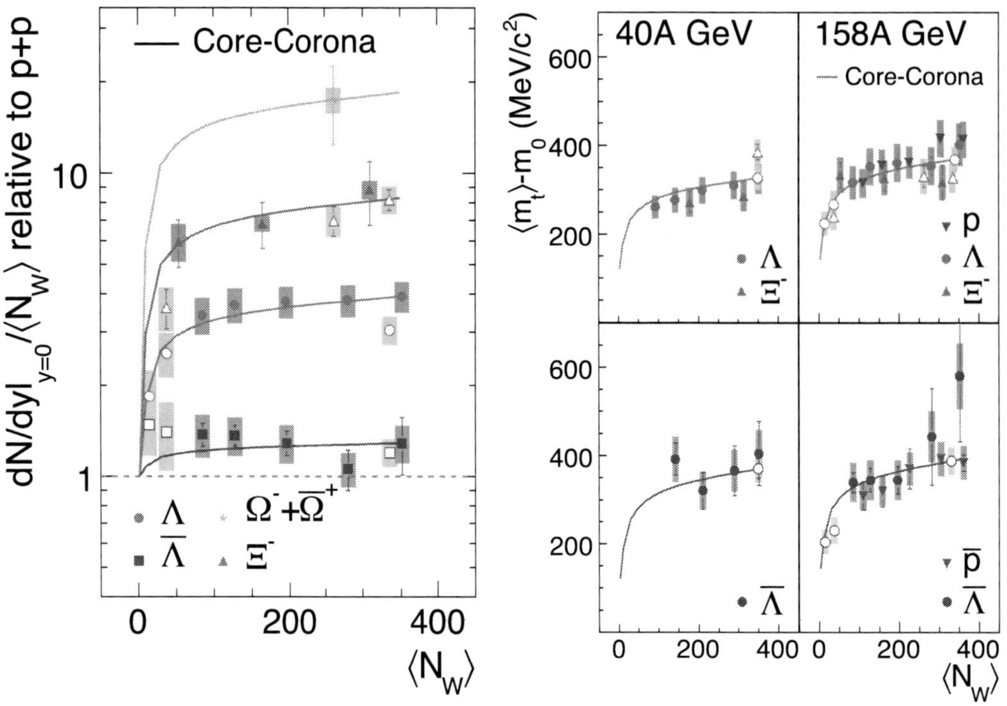

Figure 4. Left: The midrapidity yields per wounded nucleon relative to p+p yields for central C+C, Si+Si and minimum bias Pb+Pb reactions at 158A GeV [44, 45]. Right: The $\langle m_t \rangle - m_0$ values at mid-rapidity for Pb+Pb collisions at 40A and 158A GeV, as well as for near-central C+C and Si+Si reactions at 158A GeV [44]. The (anti-)proton data are taken from [46]. Also shown are the results from a fit for Λ and $\bar{\Lambda}$ with the core-corona approach (solid lines).

of $f(N_{\rm part})$ is much steeper for smaller reaction systems than for larger ones[1] Also, the maximum value of $f(N_{\rm part})$ depends on the size of the nuclei. E.g., while for very central Pb+Pb collisions $f_{\rm max}(N_{\rm part}) \approx 0.9$ can be reached, the maximum value for very central C+C reactions is $f_{\rm max}(N_{\rm part}) \approx 0.65$. This A dependence of $f_{\rm max}$ is illustrated by the dashed red line in the left panel of Fig. 5. This has the consequence that the different relative core corona contributions have to be taken into account when comparing even very central nucleus-nucleus collisions of different size. Especially, this might explain any apparent system size dependence of the chemical freeze-out parameters T and $\mu_{\rm B}$, as determined by statistical model fits to central A+A collisions of different size (e.g. [14]). Therefore, we conclude that, assuming the core-corona model is valid, the system size might not provide a good control parameter to probe different regions of the QCD phase diagram with produced fireballs of different temperature. It is more likely that one only observes a change in the relative admixture of central fireball (core) and peripheral p+p like corona, whose freeze-out parameters are different, but independent of the size of the involved nuclei.

Another interesting aspect of the core-corona model is shown in the right panel of Fig. 5. Here, the $N_{\rm part}$ dependence of $f(N_{\rm part})$ is shown for asymmetric collisions. In this case quite distinct centrality dependences can be observed. While for symmetric collisions this dependence

[1] Please note that the curves shown in Fig. 4 are based on a function $f(N_{\rm part})$ that was calculated for Pb+Pb interactions [41]. Therefore their direct comparison to the smaller systems C+C and Si+Si is not possible.

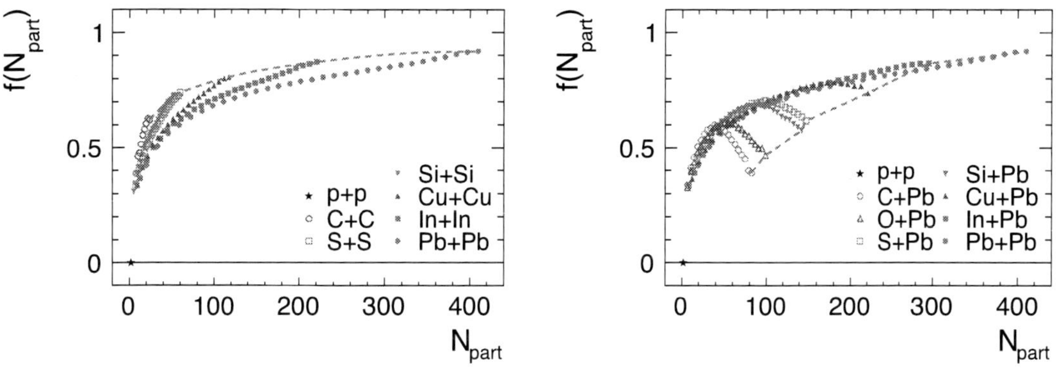

Figure 5. The fraction $f(N_{\text{part}})$ of participating nucleons that scatter more than once as a function of the number of participants N_{part}. $f(N_{\text{part}})$ was calculated with a Glauber model [43, 47]. The left panel shows results for symmetric systems, the right panel for asymmetric ones. The dashed lines connect the values for the most central collisions.

is following a continous rise (left panel of Fig. 5), for asymmetric collision systems with a small projectile nucleus (e.g. O+Pb or Si+Pb) a rapid rise followed by a maximum and a subsequent decrease of $f(N_{\text{part}})$ is seen. Therefore, a similar centrality dependence of particle yields can be expected for these type of collisions and its measurement would constitute a test for the validity of the core-corona model.

4. Search for the Critical Point

4.1. Fluctuations

One of the observables proposed to be sensitive to the presence of a critical point in the QCD phase diagram are fluctuations of either mean transverse momentum or multiplicity [48]. If the chemical freeze-out is happening in the vicinity of the critical point, an enhancement of fluctuations might be visible. The NA49 experiment has therefore performed a systematic study of multiplicity fluctuations in very central (1% most central) Pb+Pb collisions as a function of beam energy [49]. Figure 6 shows the resulting scaled variance $\omega = \text{Var}(n)/\langle n \rangle$ as a function of baryonic chemical potential μ_{B}, that has been derived from statistical model fits to the particle abundances measured at the different beam energies [14].

While for all charged particles taken together ω is close to unity, it is slightly below one for negatively and positively charged particles analyzed separately, i.e. the distributions are in these cases a bit narrower than the corresponding Poissonian. However, in none of the studied charge combinations a significant energy dependence has been so far observed.

Figure 6 also shows some theoretical expectations for ω, which result from the combination of several assumptions [50]. The position of the critical point in terms of T and μ_{B} is taken from a lattice QCD calculation [35]. The magnitude of the fluctuations at the critical point is based on [48, 51], assuming two different correlation lengths $\xi = 3$ fm (solid line) and $\xi = 6$ fm (dashed line), while the widths of the enhancements around the critical point has been chosen according to [52] as $\sigma(\mu_{\text{B}}) \approx 30$ MeV. Both predictions are at variance with the data, so that up to now there is no evidence for a critical point in fluctuation measurements.

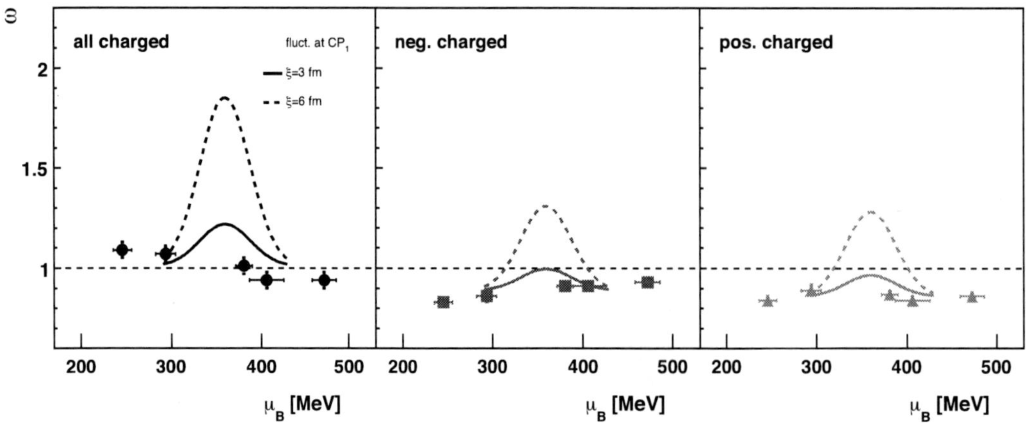

Figure 6. Energy dependence of multiplicity fluctuations, given by the scaled variance ω, for the 1% most central Pb+Pb collisions in the forward rapidity region ($1.1 < y_\pi < y_{\mathrm{beam}}$) as measured by the NA49 experiment [49, 50]. The different beam energies are represented by the corresponding μ_B values, determined by a statistical model fit [14]. The lines correspond to predictions for for critical point (see text).

4.2. Transverse Mass Spectra of Baryons and Anti-baryons

Another possible way of detecting a critical point has been suggested in [53]. Here it was shown that the proximity of a critical point, or rather a critical area, might deform the isentropic trajectories of an expanding fireball in the T-μ_B plane, due to a focussing effect towards the critical point. As a consequence one should see a decrease of the anti-baryon/baryon ratios with increasing transverse mass, while in the absence of a critical point this ratio should rather be flat as a function of m_t.

The left panel of Fig. 7 shows the $\bar{\mathrm{p}}/\mathrm{p}$ ratio versus $m_\mathrm{t} - m_0$ for five different beam energies [10, 50]. In order to quantify any change in transverse mass dependence, a linear fit has been applied to the data. The $\bar{\Lambda}/\Lambda$ and $\bar{\Xi}^+/\Xi^-$ ratios have been analyzed in the same fashion. The resulting energy dependence of the slope parameters a is summarized in the right panel of Fig. 7. No significant energy dependence can be seen and thus this observable currently does not provide any evidence for a critical point.

5. Conclusions

In order to probe different areas in the QCD phase diagram, the variation of the center-of-mass energy can be considered as the best defined control parameter. Especially in the energy regime of the SPS a wide region in the T-μ_B plane can be explored experimentally, as is evident from the strong dependence of the chemical freeze-out parameters on the energy here. The situation is much less well defined when using the system size as control parameter. Since the system size dependences of many observables (e.g. strangeness production and $\langle m_\mathrm{t}\rangle - m_0$) seem to be well described within the core-corona approach, it is unlikely that the freeze-out temperatures of the fireballs vary with their size. Rather one observes a change of the relative admixture of the core and corona contributions. However, the trajectories of the fireballs in the core might be always the same, irrespective of their size. Studying centrality dependences in asymmetric reaction

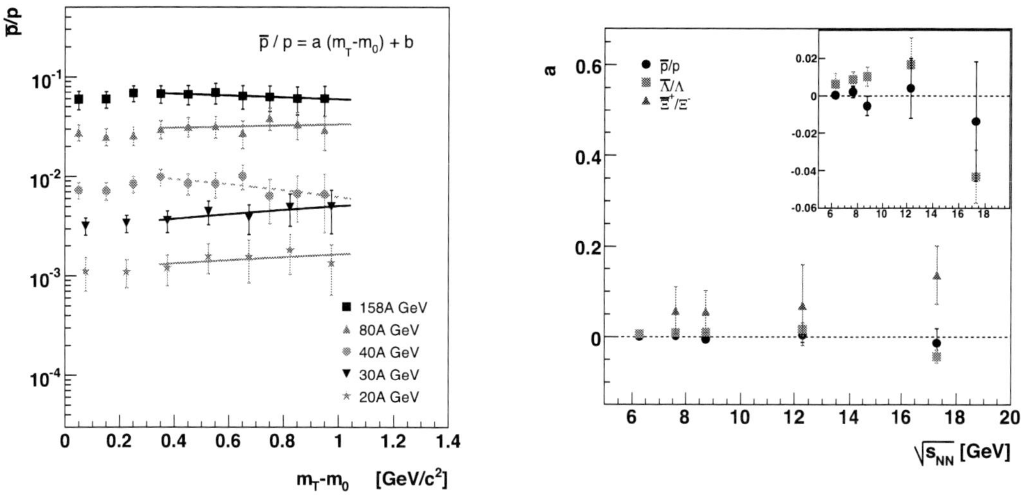

Figure 7. Left: The p̄/p ratio versus the transverse mass $m_t - m_0$ [10, 50]. The solid lines show a linear fit to the ratios. Right: The energy dependence of the slope parameters a from the fits for different $\bar{\mathrm{B}}/\mathrm{B}$ ratios [10, 15, 50]. The insert shows the same data with an expanded vertical scale.

systems could provide an additional experimental test of the core-corona model. In these cases this model predicts for smaller projectile nuclei a maximum in the relative core contribution for semi-central reactions, in contrast to symmetric reaction systems that reach the maximum for very central collisions.

Several attempts have been made to search within existing data for evidences of a critical point, such e.g. as fluctuations and the transverse mass dependence of anti-baryon/baryon ratios. So far non of these searches yielded a positive result. The success of future searches will on one side mainly depend on the development of more refined and differential (i.e. scale and p_t dependent) observables, that allow to distinguish effects due to a critical point from more trivial sources. On the other side, since the magnitude of these effects might be small, experimental progress on reducing systematic and statistical errors will be necessary.

Acknowledgments

The author would like to thank K. Reygers for the help with the Glauber model calculations and H. Ströbele for providing helpful comments and suggestions.

References

[1] Aoki Y et al., 2009 *JHEP* **0906** 088.
[2] Cheng M et al., 2006 *Phys. Rev.* D **74** 054507.
[3] Fodor Z, *Proceedings of this WWND10 workshop.*
[4] Stephanov M, Rajagopal K, and Shuryak E, 1998 *Phys. Rev. Lett.* **81** 4816.
[5] Stephanov M, 2006 *PoS LAT2006* **024**.
[6] de Forcrand P and Philipsen O, 2007 *JHEP* **0701** 077.
[7] McLerran L and Pisarski R D, 2007 *Nucl. Phys.* A **796** 83.
[8] Andronic A et al., 2009 *arXiv*:0911.4806.
[9] Appelshäuser H et al. (NA49 Collaboration), 1999 *Phys. Rev. Lett.* **82** 2471.
[10] Alt C et al. (NA49 Collaboration), 2006 *Phys. Rev.* C **73** 044910.

[11] Ahle L et al. (E802 Collaboration), 1999 *Phys. Rev.* C **60** 064901.
[12] Bearden I G et al., (BRAHMS Collaboration), 2004 *Phys. Rev. Lett.* **93** 102301.
[13] Blume C (for the NA49 Collaboration), 2007 *J. Phys.* G **34** S951.
[14] Becattini F, Manninen J, and Gaździcki M, 2006 *Phys. Rev.* C **73** 044905.
[15] Alt C et al. (NA49 Collaboration), 2008 *Phys. Rev.* C **78** 034918.
[16] Bleicher M et al., 1999 *J. Phys.* G **25** 1859.
[17] Weber H, Bratkovskaya E L, Cassing W, and Stöcker H, 2003 *Phys Rev.* C **67** 014904.
[18] Andronic A, Braun-Munzinger P, and Stachel J, 2006 *Nucl. Phys.* A **772** 167.
[19] Albergo S et al. (E896 Collaboration), 2002 *Phys. Rev. Lett.* **88** 062301.
[20] Back B B et al. (E917 Collaboration), 2001 *Phys. Rev. Lett.* **87** 242301.
[21] Ahmad S et al. (E891 Collaboration), 1996 *Phys. Lett.* B **382** 35.
[22] Ahle L et al. (E802 Collaboration), 1998 *Phys. Rev.* C **57** R466.
[23] Antinori F et al. (NA57 Collaboration), 2006 *J. Phys.* G **32** 427.
[24] Antinori F et al. (NA57 Collaboration), 2004 *Phys. Lett.* B **595** 68.
[25] Adler C et al. (STAR Collaboration), 2002 *Phys. Rev. Lett.* **89** 092301.
[26] Adams J et al. (STAR Collaboration), 2004 *Phys. Rev. Lett.* **92** 182301.
[27] Adcox K et al. (PHENIX Collaboration), 2002 *Phys. Rev. Lett.* **89** 092302.
[28] Adams J et al. (STAR Collaboration), 2007 *Phys. Rev. Lett.* **98** 062301.
[29] Adler C et al. (STAR Collaboration), 2004 *Phys. Lett.* B **595** 143.
[30] Adams J et al. (STAR Collaboration), 2004 *Phys. Rev. Lett.* **92** 112301.
[31] Adcox K et al. (PHENIX Collaboration), 2002 *Phys. Rev. Lett.* **88** 242301.
[32] Cleymans J and Redlich K, 1999 *Phys. Rev.* C **60** 054908.
[33] Becattini F, Gaździcki M, Keränen A, Manninen J, and Stock R, 2004 *Phys. Rev.* C **69** 024905.
[34] Cleymans J, Oeschler H, Redlich K, and Wheaton S, 2006 *Eur. Phys. J.* A **29** 119.
[35] Fodor Z and Katz S D, 2004 *JHEP* **0404** 050.
[36] Ströbele H (for the NA49 Collaboration), 2009 *arXiv*:0908.2777.
[37] Back B B et al. (E917 Collaboration), 2001 *Phys. Rev. Lett.* **86** 1970.
[38] Bozek P, 2005 *Acta Phys. Polon.* B **36** 3071.
[39] Becattini F and Manninen J, 2008 *J. Phys.* G **35** 104013.
[40] Becattini F and Manninen J, 2009 *Phys. Lett.* B **673** 19.
[41] Aichelin J and Werner K, 2009 *Phys. Rev.* C **79** 064907.
[42] Werner K, *Proceedings of this WWND10 workshop.*
[43] Glauber R J, 1955 *Phys. Rev.* **100** 242.
[44] Anticic T et al. (NA49 Collaboration), 2009 *Phys. Rev.* C **80** 034906.
[45] Blume C (for the NA49 Collaboration), 2008 *J. Phys.* G **35** 044004.
[46] Anticic T et al. (NA49 Collaboration), 2004 *Phys. Rev.* C **69** 024902.
[47] Miller M L, Reygers K, Sanders S J, and Steinberg P, 2007 *Ann. Rev. Nucl. Part. Sci.* **57** 205.
[48] Stephanov M, Rajagopal K, and Shuryak E V, 1999 *Phys. Rev.* D **60** 114028.
[49] Alt C et al. (NA49 Collaboration), 2008 *Phys. Rev.* C **78** 034914.
[50] Grebieszkow K (for the NA49 Collaboration), 2009 *Nucl. Phys.* A **830** 547C.
[51] Stephanov M, *private communication.*
[52] Hatta Y and Ikeda T, 2003 *Phys. Rev.* D **67** 014028.
[53] Askawa A, Bass S A, Müller B, and Nonaka C, 2009 *Phys. Rev. Lett.* **101** 122302.

Journal of Physics: Conference Series **230** (2010) 012004 doi:10.1088/1742-6596/230/1/012004

Production and elliptic flow of heavy quarks at RHIC and LHC within a partonic transport model

Jan Uphoff,[1] Oliver Fochler,[1] Zhe Xu,[2,1] and Carsten Greiner[1]

[1] Institut für Theoretische Physik, Johann Wolfgang Goethe-Universität Frankfurt, Max-von-Laue-Str. 1, D-60438 Frankfurt am Main, Germany

[2] Frankfurt Institute for Advanced Studies, Ruth-Moufang-Str. 1, D-60438 Frankfurt am Main, Germany

E-mail: uphoff@th.physik.uni-frankfurt.de

Abstract. Production and elliptic flow of heavy quarks are investigated in nucleus-nucleus collisions at RHIC and LHC within the *Boltzmann Approach of MultiParton Scatterings* (BAMPS). The initial heavy quark yield is estimated with the leading order mini-jet model and compared to PYTHIA for several parton distribution functions. Secondary production of heavy quarks in the quark-gluon plasma is examined within the full 3+1 dimensional BAMPS simulation of heavy ion collisions. At RHIC this yield is negligible, but for LHC charm production in the QGP plays a significant role. In addition, we study the elliptic flow of charm quarks at RHIC with the result that leading order processes are not sufficient to describe the experimental data.

1. Introduction

Charm and bottom quarks are an interesting probe for the hot and dense medium, which is produced in heavy ion collisions at RHIC [1, 2, 3, 4]. Due to their large mass heavy quarks are produced at a very early stage of the collision, revealing unique insight into the initial properties of the quark-gluon plasma (QGP). This large mass also determines the relevant scale of their production processes, which should, therefore, be describable in the framework of perturbative QCD (pQCD).

The production of heavy quarks can be divided in three stages: They are created (i) in hard scatterings during initial nucleon-nucleon collisions, (ii) in the QGP and (iii) during the hadronic phase. We will neglect the latter one since the energy density in the hadronic gas is low and heavy flavor production becomes very unlikely. At RHIC and LHC primary heavy quark production in hard scatterings plays the most important role. At RHIC secondary production in the QGP is nearly negligible, but at LHC it contributes significantly to the total heavy flavor yield.

The experimentally observed elliptic flow [5] and energy loss [6, 5] of heavy quarks indicate that they interact strongly with the medium and thermalize relatively fast, although the "dead cone effect" [7, 8] predicts a delay in their thermalization compared to light quarks. The investigation of this puzzle is still in progress [9, 10, 11, 12, 13, 14].

The present article is organized as follows: In the next section we describe our model, the parton cascade BAMPS. Thereafter, we discuss primary heavy quark production in hard parton scatterings during initial nucleon-nucleon collisions within the leading order mini-jet model and PYTHIA. In Sec. 4 secondary heavy quark production in the QGP at RHIC and LHC is investigated with BAMPS for different initial conditions. Finally, we address elliptic flow of charm quarks at RHIC in Sec. 5, followed by brief conclusions.

2. Parton cascade BAMPS

The full $3+1$ space-time evolution of the QGP is studied within the partonic transport model *Boltzmann Approach of MultiParton Scatterings* (BAMPS) [15, 16], which solves the Boltzmann equation,

$$\left(\frac{\partial}{\partial t} + \frac{\mathbf{p}_i}{E_i}\frac{\partial}{\partial \mathbf{r}}\right) f_i(\mathbf{r}, \mathbf{p}_i, t) = \mathcal{C}_i^{2\to 2} + \mathcal{C}_i^{2\leftrightarrow 3} + \ldots \ , \tag{1}$$

dynamically for on-shell partons with a stochastic transport algorithm and pQCD interactions. $\mathcal{C}_i$ denotes the collision integrals of the $2 \to 2$ and also $2 \leftrightarrow 3$ interactions and $f_i(\mathbf{r}, \mathbf{p}_i, t)$ the one particle distribution function of species $i = g$, Q, $\bar{Q}$ $(Q = c, b)$, since light quarks are not included yet. The following processes are implemented in BAMPS:

$$\begin{aligned}
g + g &\to g + g \\
g + g &\to g + g + g \\
g + g + g &\to g + g \\
g + g &\to Q + \bar{Q} \\
Q + \bar{Q} &\to g + g \\
g + Q &\to g + Q \\
g + \bar{Q} &\to g + \bar{Q}
\end{aligned} \tag{2}$$

For further details on the purely gluonic processes we refer to [15, 16] and to [17] for the processes, in which heavy quarks are involved.

3. Primary heavy quark production in initial nucleon-nucleon scatterings

For the estimation of the primary heavy quark yield during initial nucleon-nucleon scatterings we use the following two approaches.

3.1. Leading order heavy quark production within the mini-jet model

In leading order (LO) heavy quarks are produced in the two processes

$$\begin{aligned}
g + g &\to Q + \bar{Q} \\
q + \bar{q} &\to Q + \bar{Q} \ .
\end{aligned} \tag{3}$$

The differential cross section for heavy quark production in the collision of hadron A and B is essentially given by the product of the parton distribution functions (PDF) f_i and the differential partonic cross sections for both processes, [18]

$$\frac{\mathrm{d}\sigma^{AB}_{Q\bar{Q}}}{\mathrm{d}p_T^2 \mathrm{d}y_Q \mathrm{d}y_{\bar{Q}}} = x_1 x_2 \left[f_g^A(x_1)\, f_g^B(x_2)\, \frac{\mathrm{d}\hat{\sigma}_{gg\to Q\bar{Q}}}{\mathrm{d}\hat{t}} \right. \\
\left. + \sum_q \left[f_q^A(x_1)\, f_{\bar{q}}^B(x_2) + f_{\bar{q}}^A(x_1)\, f_q^B(x_2) \right] \frac{\mathrm{d}\hat{\sigma}_{q\bar{q}\to Q\bar{Q}}}{\mathrm{d}\hat{t}} \right] , \tag{4}$$

y being the rapidity and p_T the transversal momentum of the heavy quark and anti-quark in the center of mass frame. The PDFs are dependent on the factorization scale and the partonic

PDF	Reference	Charm pairs
CTEQ5l (LO)	[23]	8.9
CTEQ6l (LO)	[24]	9.2
CTEQ6m ($\overline{MS}$)	[24]	13.6
MRST2007LOmod	[25]	9.2
HERAPDF01	[26]	12.3
GRV98 (LO)	[27]	3.0

Table 1. Number of charm pairs produced in primary hard scatterings in central Au+Au collisions at RHIC for some parton distribution functions by sampling nucleon-nucleon collisions with PYTHIA and scaling to Au+Au collisions.

		Heavy quark pairs	
PDF	Reference	Charm	Bottom
CTEQ6l (LO)	[24]	62	7.2
CTEQ6m ($\overline{MS}$)	[24]	66	6.9
MRST2007LOmod	[25]	67	8.9

Table 2. As Tab. 1 but for central Pb+Pb collisions at LHC.

cross sections on the renormalization scale. Therefore, heavy quark production in general is sensitive to both scales. Other uncertainties come from the heavy quark mass and the choice of PDFs. Depending on these parameters the total LO heavy quark production cross section in nucleon-nucleono collisions changes by a factor of about 2 at RHIC and even of about 4 at LHC energy [17].

3.2. Heavy quark production with PYTHIA

In addition, we use PYTHIA [19], the LO event generator for nucleon-nucleon collisions, to estimate the number of initially produced heavy quarks in heavy ion collisions. For that we scale heavy quarks from nucleon-nucleon collisions with roughly the number of binary collisions[1]. Details on how this is done can be found in [17]. Tables 1 and 2 list the number of initially produced charm quarks in central Au+Au collisions at RHIC and charm and bottom quarks in central Pb+Pb collisions at LHC, respectively, according to PYTHIA for various parton distribution functions. As a note, there are also large errors due to uncertainties in mass, shadowing, factorization and renormalization scale like in the LO pQCD calculation, although these are not reflected in the tables.

Fig. 1 compares the rapidity distribution of the charm production cross section in one nucleon-nucleon collision simulated with PYTHIA together with the LO pQCD calculations and the experimental data points [28, 29]. The LO calculations are far below both experimental values, which indicates that next-to-leading order processes [31] have to be considered. In contrast, PYTHIA lies much closer at least to the PHENIX data point. Therefore, we decided to use PYTHIA for sampling our initial heavy quark distributions. For the PDFs we chose CTEQ6l, because it is designed for LO event generators such as PYTHIA [24].

[1] Actually, this value is lowered to 1000 at RHIC [20] and 1500 at LHC [21, 22] in order to take shadowing into account.

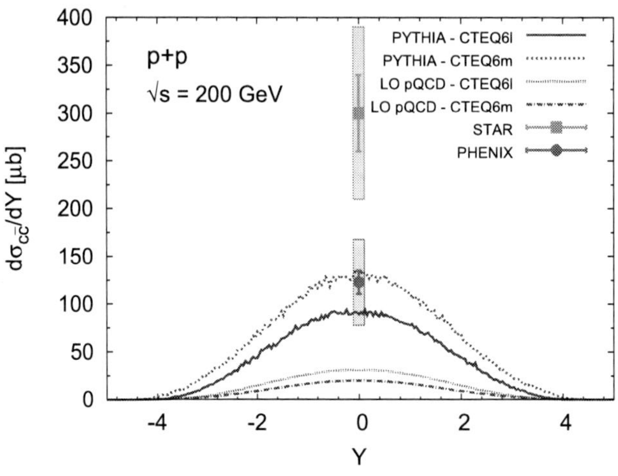

Figure 1. Charm production cross section $d\sigma_{c\bar{c}}^{NN}/dy$ as a function of rapidity y in a nucleon-nucleon collision at RHIC energy simulated with PYTHIA and pQCD, respectively, for the PDFs CTEQ6l and CTEQ6m together with experimental data [28, 29]. The pQCD calculation is done in LO with $\mu_F = \mu_R = \sqrt{p_T^2 + M_c^2}$, $N_f = 3$, $M_c = 1.5\,\text{GeV}$, $\lambda_{\text{QCD}} = 346\,\text{MeV}$ [30] and $K = 1$.

4. Secondary heavy quark production in the QGP

4.1. Charm production at RHIC

We study charm production during the QGP phase in central Au+Au collisions at RHIC within the framework of BAMPS. The initial charm distribution is obtained from PYTHIA. Initial gluons are sampled according to three different models: a color glass condensate (CGC) inspired model [32, 33], the mini-jet model [34, 35], and PYTHIA. The detailed prescription how we scale partons from PYTHIA to heavy ion collisions can be found in [17]. For the partonic cross sections in BAMPS of the processes from (2) we employ a constant coupling of $\alpha_s = 0.3$.

Fig. 2 shows the charm yield in the QGP as a function of time. For all three initial distribution models the charm production during the QGP is very small compared to the initial yield. With PYTHIA initial conditions only 3 % of the total final charm quarks are produced in the QGP. For CGC or mini-jet initial distributions for the gluons this fraction is about 6 %. The higher yield in the latter models is due to the higher initial energy density [17]. The main increase in the charm number takes place at an early stage of the evolution, where the QGP is not fully equilibrated yet.

Changing the charm mass to a smaller value of $M_c = 1.3\,\text{GeV}$ or employing a K factor of 2 to the charm production cross section increase these results by a factor of 2, respectively. As a result, using mini-jet initial conditions, $M_c = 1.3\,\text{GeV}$ and also $K = 2$ gives the maximum value in our model of 3.4 charm pairs produced in the QGP (cf. Fig. 3), which is still just 27 % of the total charm yield.

Our results indicate that the charm production in the QGP at RHIC is nearly negligible. That is also underlined by experiment, where a charm scaling with the number of binary collisions was found [36, 28, 37]. Charm quarks are, therefore, produced in initial hard scatterings during nucleon-nucleon collisions.

At first sight this small charm production is a bit surprising since the charm quark fugacity is always smaller than 1 in the QGP [17]. However, the reason for that lies in the huge timescale of chemical equilibration of charm quarks in the medium produced at RHIC, which is about a factor of 100 larger than the lifetime of the QGP [17].

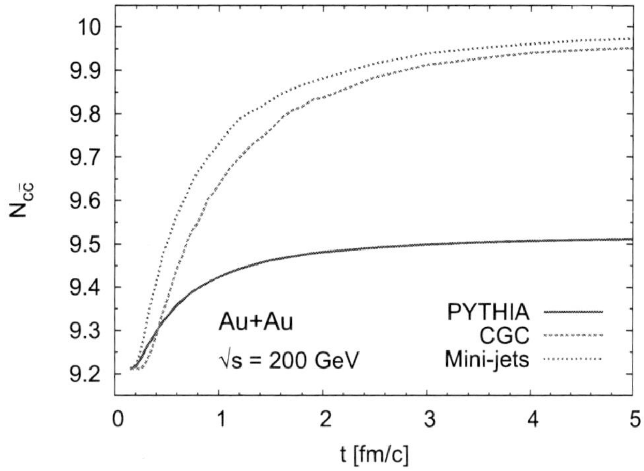

Figure 2. Number of charm quark pairs produced in a central Au+Au collision at RHIC according to BAMPS. The initial parton distribution for gluons is obtained with PYTHIA, CGC, and the mini-jet model. For better comparison the initial charm distribution from PYTHIA is used for all models. Please note the small range of the y axis.

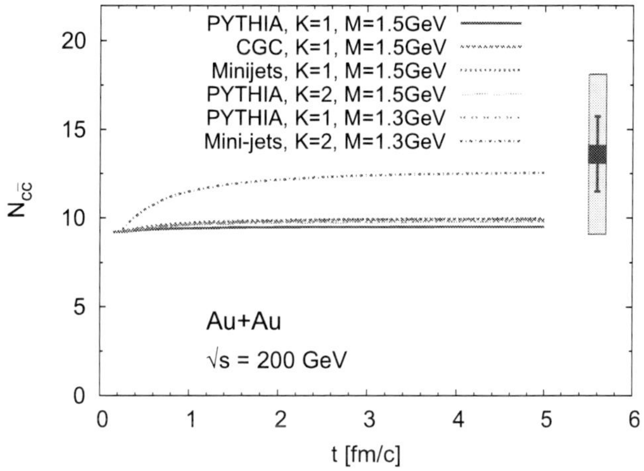

Figure 3. As Fig. 2, varying initial gluon conditions, charm mass M and K factor for $gg \rightarrow c\bar{c}$. In addition, the experimental value for the number of final charm pairs is plotted [36].

4.2. Charm and Bottom production at LHC

At LHC the picture changes compared to RHIC: Due to the larger energy density, charm production in the QGP cannot be neglected anymore. Fig. 4 depicts our predictions for the charm number evolution at LHC. Again, we use PYTHIA for the initial charm yield and PYTHIA, CGC, and mini-jet initial conditions for the gluons. With PYTHIA initial conditions 15 % of the total charm quarks are produced in the QGP. With gluons from CGC this fraction even rises and for mini-jet initial gluon conditions the charm yield during the QGP phase is nearly equal to the initial charm number. Lowering the charm mass or introducing a K factor increase these values again.

Just like at RHIC, the charm quark fugacity at LHC is far below 1 [17]. However, because of

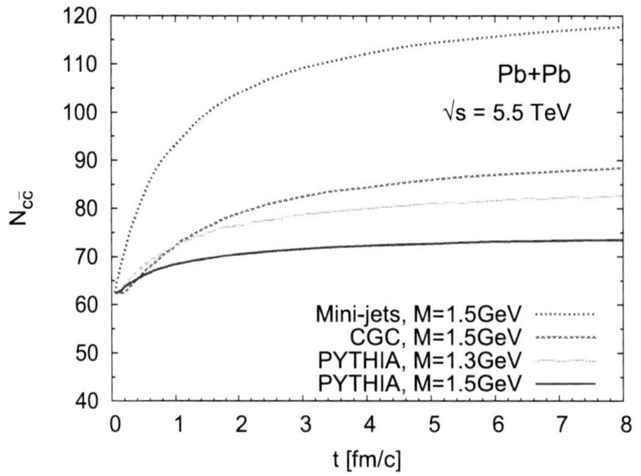

Figure 4. Number of charm pairs in a central Pb+Pb collision at LHC simulated with BAMPS for PYTHIA, CGC, and mini-jet initial conditions for gluons. Charm quarks are sampled with PYTHIA for all three models.

the much smaller equilibration time of charm quarks in such a hot medium as at LHC, which is of the same order of magnitude as the QGP lifetime [17], the charm production is much stronger.

As shown in [17] the bottom production during the QGP phase at LHC and, of course, also at RHIC is completely negligible due to the larger mass. Therefore, bottom quarks are just produced in initial hard scatterings, which renders them as a promising probe of the early stage of the collision due to the precisely determined production time.

5. Elliptic flow of charm quarks at RHIC

The elliptic flow of charm quarks is defined by the average

$$v_2 = \left\langle \frac{p_x^2 - p_y^2}{p_T^2} \right\rangle \tag{5}$$

over all charm quarks at mid-rapidity and is an indicator how strong charm quarks are coupled to the bulk medium. In the present article we study only charm scattering with gluons in leading order, $g + c \rightarrow g + c$. Without considering higher order corrections or employing a K factor for the cross section the interaction between charm quarks and the gluonic medium is rather small. Therefore, although a strong gluonic v_2 is observed in BAMPS [38, 39, 40], which is in agreement with experimental data, the v_2 of charm quarks with $K = 1$ vanishes nearly, as shown in Fig. 5. For a substantial charm flow one must multiply the cross section of $gc \rightarrow gc$ by at least a factor of 10. In order to come in the region of the experimentally observed flow of heavy flavor [5, 41] even an unreasonable $K = 50$ is necessary, which shows that the charm scattering cannot be described just by the LO process. Of course, one has to be careful with comparing the flow from charm quarks to the experimentally observed flow of heavy flavor electrons, which stem from the decay of D or B mesons. However, in [42] it is found that the effect of hadronization and decay on the flow is negligible. In contrast, in the coalescence model the flow of D mesons is considerably larger than that of charm quarks [43]. We also made some calculations with bottom quarks and saw that our here presented results are only slightly modified and that only for high p_T, if one considers both charm and bottom.

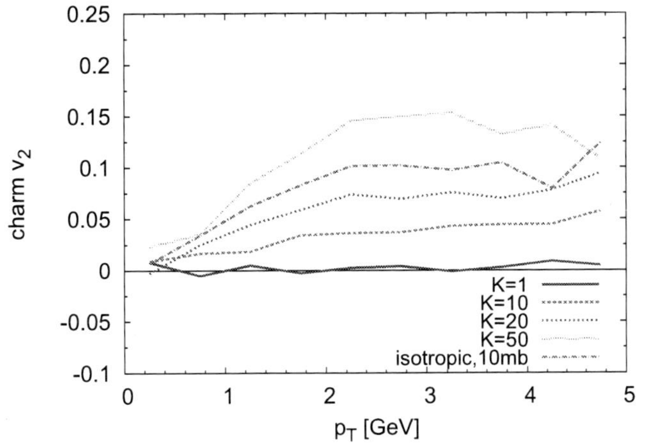

Figure 5. Elliptic flow of charm quarks with pseudo-rapidity $|\eta| < 0.35$ at time $t = 5\,\mathrm{fm/c}$ for RHIC collisions with impact parameter of $b = 8.2\,\mathrm{fm}$. The cross section of $gc \to gc$ is multiplied with different K factors.

In order to improve our calculations we will investigate the effect of a running coupling and modify the evaluation of the screening mass in the $gc \to gc$ cross section [44, 13, 14, 45]. In addition, we want to study the impact of taking higher order corrections like $g + c \to g + c + g$ into account, for which BAMPS is an ideal framework since $2 \leftrightarrow 3$ interactions are already included for gluons [15]. With these improvements we will also extend recent studies on the nuclear modification factor [46, 47] to the heavy flavor sector.

6. Conclusions

We have studied heavy quark production in heavy ion collisions at RHIC and LHC. The initial yield was estimated with LO pQCD processes, which did not agree with the data, and PYTHIA. The latter was then used to sample the initial heavy quark distribution for the QGP phase, which was simulated with BAMPS. In that framework we investigated the dependence of heavy quark production on different initial gluon distributions, namely from PYTHIA, CGC and the mini-jet model.

At RHIC the production of charm and, of course, bottom quarks in the QGP is highly suppressed due to the huge chemical equilibration time scale in the medium created at RHIC. For LHC we predict a significant charm production of 10 to 60 charm pairs in the QGP phase, depending on the initial conditions of the medium. This is due to the high initial energy density and a relatively small chemical equilibration time scale, which is of the same order as the QGP lifetime. In contrast, bottom production in the QGP at LHC is negligible due to the high bottom mass.

The experimentally observed elliptic flow of heavy quarks cannot be described with LO pQCD cross sections. Therefore, we want to take higher orders into account and study the effect of a running coupling and improved Debye screening.

Acknowledgements

The BAMPS simulations were performed at the Center for Scientific Computing of the Goethe University. This work was supported by the Helmholtz International Center for FAIR within the framework of the LOEWE program (Landes-Offensive zur Entwicklung Wissenschaftlichökonomischer Exzellenz) launched by the State of Hesse. We would like to thank the organizers of the 26th Winter Workshop on Nuclear Dynamics for this pleasant and stimulating conference.

References

[1] Adams J *et al.* (STAR) 2005 *Nucl. Phys.* **A757** 102–183 (*Preprint* nucl-ex/0501009)
[2] Adcox K *et al.* (PHENIX) 2005 *Nucl. Phys.* **A757** 184–283 (*Preprint* nucl-ex/0410003)
[3] Arsene I *et al.* (BRAHMS) 2005 *Nucl. Phys.* **A757** 1–27 (*Preprint* nucl-ex/0410020)
[4] Back B B *et al.* (PHOBOS) 2005 *Nucl. Phys.* **A757** 28–101 (*Preprint* nucl-ex/0410022)
[5] Adare A *et al.* (PHENIX) 2007 *Phys. Rev. Lett.* **98** 172301 (*Preprint* nucl-ex/0611018)
[6] Abelev B I *et al.* (STAR) 2007 *Phys. Rev. Lett.* **98** 192301 (*Preprint* nucl-ex/0607012)
[7] Dokshitzer Y L and Kharzeev D E 2001 *Phys. Lett.* **B519** 199–206 (*Preprint* hep-ph/0106202)
[8] Zhang B W, Wang E and Wang X N 2004 *Phys. Rev. Lett.* **93** 072301 (*Preprint* nucl-th/0309040)
[9] Armesto N, Cacciari M, Dainese A, Salgado C A and Wiedemann U A 2006 *Phys. Lett.* **B637** 362–366 (*Preprint* hep-ph/0511257)
[10] van Hees H, Greco V and Rapp R 2006 *Phys. Rev.* **C73** 034913 (*Preprint* nucl-th/0508055)
[11] Moore G D and Teaney D 2005 *Phys. Rev.* **C71** 064904 (*Preprint* hep-ph/0412346)
[12] Wicks S, Horowitz W, Djordjevic M and Gyulassy M 2007 *Nucl. Phys.* **A784** 426–442 (*Preprint* nucl-th/0512076)
[13] Peigne S and Peshier A 2008 *Phys. Rev.* **D77** 114017 (*Preprint* 0802.4364)
[14] Gossiaux P B and Aichelin J 2008 *Phys. Rev.* **C78** 014904 (*Preprint* 0802.2525)
[15] Xu Z and Greiner C 2005 *Phys. Rev.* **C71** 064901 (*Preprint* hep-ph/0406278)
[16] Xu Z and Greiner C 2007 *Phys. Rev.* **C76** 024911 (*Preprint* hep-ph/0703233)
[17] Uphoff J, Fochler O, Xu Z and Greiner C 2010 (*Preprint* 1003.4200)
[18] Gavai R V *et al.* 1995 *Int. J. Mod. Phys.* **A10** 2999–3042 (*Preprint* hep-ph/9411438)
[19] Sjostrand T, Mrenna S and Skands P 2006 *JHEP* **05** 026 (*Preprint* hep-ph/0603175)
[20] Adams J *et al.* (STAR) 2004 *Phys. Rev.* **C70** 054907 (*Preprint* nucl-ex/0407003)
[21] Emel'yanov V, Khodinov A, Klein S R and Vogt R 2000 *Phys. Rev.* **C61** 044904 (*Preprint* hep-ph/9909427)
[22] Accardi A *et al.* 2004 (*Preprint* hep-ph/0308248)
[23] Lai H L *et al.* (CTEQ) 2000 *Eur. Phys. J.* **C12** 375–392 (*Preprint* hep-ph/9903282)
[24] Pumplin J *et al.* 2002 *JHEP* **07** 012 (*Preprint* hep-ph/0201195)
[25] Sherstnev A and Thorne R S 2008 *Eur. Phys. J.* **C55** 553–575 (*Preprint* 0711.2473)
[26] Nagano K (H1) 2008 (*Preprint* 0808.3797)
[27] Glück M, Reya E and Vogt A 1998 *Eur. Phys. J.* **C5** 461–470 (*Preprint* hep-ph/9806404)
[28] Adams J *et al.* (STAR) 2005 *Phys. Rev. Lett.* **94** 062301 (*Preprint* nucl-ex/0407006)
[29] Adare A *et al.* (PHENIX) 2006 *Phys. Rev. Lett.* **97** 252002 (*Preprint* hep-ex/0609010)
[30] Bethke S 2007 *Prog. Part. Nucl. Phys.* **58** 351–386 (*Preprint* hep-ex/0606035)
[31] Cacciari M, Nason P and Vogt R 2005 *Phys. Rev. Lett.* **95** 122001 (*Preprint* hep-ph/0502203)
[32] Adil A, Drescher H J, Dumitru A, Hayashigaki A and Nara Y 2006 *Phys. Rev.* **C74** 044905 (*Preprint* nucl-th/0605012)
[33] Drescher H J and Nara Y 2007 *Phys. Rev.* **C75** 034905 (*Preprint* nucl-th/0611017)
[34] Kajantie K, Landshoff P V and Lindfors J 1987 *Phys. Rev. Lett.* **59** 2527
[35] Eskola K J, Kajantie K and Lindfors J 1989 *Nucl. Phys.* **B323** 37
[36] Adler S S *et al.* (PHENIX) 2005 *Phys. Rev. Lett.* **94** 082301 (*Preprint* nucl-ex/0409028)
[37] Abelev B I *et al.* (STAR) 2008 (*Preprint* 0805.0364)
[38] Xu Z, Greiner C and Stöcker H 2008 *Phys. Rev. Lett.* **101** 082302 (*Preprint* 0711.0961)
[39] Xu Z and Greiner C 2009 *Phys. Rev.* **C79** 014904 (*Preprint* 0811.2940)
[40] Xu Z and Greiner C 2010 (*Preprint* 1001.2912)
[41] Dion A (for the PHENIX collaboration) 2009 Heavy flavor via leptons at PHENIX, Talk at Quark Matter 2009 http://www.phenix.bnl.gov/phenix/WWW/publish/adion/Dion_QM09.pdf
[42] Zhang B, Chen L W and Ko C M 2005 *Phys. Rev.* **C72** 024906 (*Preprint* nucl-th/0502056)
[43] Molnar D 2005 *J. Phys.* **G31** S421–S428 (*Preprint* nucl-th/0410041)
[44] Peshier A 2008 (*Preprint* 0801.0595)
[45] Gossiaux P B, Bierkandt R and Aichelin J 2009 *Phys. Rev.* **C79** 044906 (*Preprint* 0901.0946)
[46] Fochler O, Xu Z and Greiner C 2009 *Phys. Rev. Lett.* **102** 202301 (*Preprint* 0806.1169)
[47] Fochler O, Xu Z and Greiner C 2010 (*Preprint* 1003.4380)

26th Winter Workshop on Nuclear Dynamics IOP Publishing
Journal of Physics: Conference Series **230** (2010) 012005 doi:10.1088/1742-6596/230/1/012005

Heavy Flavor measurements using high-p_T electrons in the ALICE EMCal

Mark T Heinz (for the ALICE collaboration)
Yale University, WNSL, 272 Whitney Ave, New Haven, CT 06520, USA
E-mail: mark.heinz@yale.edu

Abstract. Heavy flavor hadrons, i.e. those containing charm and bottom quarks, will be abundantly produced at the LHC and are important probes of the Quark-Gluon Plasma (QGP). Of particular interest is the investigation of parton energy loss in the medium. Using heavy flavor jets we will have a pure sample of quark jets with which to study the color-charge effects on energy loss. In addition, studies of bottom production in p+p collision at LHC energies will be utilized to further constrain the current parameters used by NLO and FONLL calculations. The talk will focus on the very high-pt electron particle identification using the EMCal detector. We present the electron reconstruction and measurements which can be achieved with 1 nominal year of Pb-Pb running at $\sqrt{s}$=5.5 TeV. We then estimate the rate of non-photonic electrons and present systematic and statistical error bars. Finally, we show preliminary results on B-jet tagging techniques in p+p which utilize jet-finding algorithms (FASTJET) in conjunction with displaced secondary vertices containing high-pt electrons.

1. Introduction

From a heavy ion physics point of view the interest is focussed on the parton energy loss for quarks and gluons in the strongly interacting medium produced by the heavy ion collisions. Heavy flavor quarks provide excellent probes since they are produced at early times in the collisions and need to traverse the Quark-Gluon Plasma (QGP). At intermediate p_T, the prediction that the energy loss of massive quarks in a colored medium is reduced due to the suppression of forward radiation ("dead cone effect" [1]) has not been validated by measurements at RHIC [2, 3], which find heavy flavor production at high p_T to be suppressed at the same level as light flavor quarks and gluons. Beyond the energy range where the dead cone effect is expected to be significant (i.e. $E_T \gg m$), heavy quark jet production may provide a tool to probe the color-charge dependence of energy loss [4]. This is illustrated in Fig. 1 where one can see that the modification, quantified by the nuclear modification factor R_Q, between charm and bottom quarks becomes comparable for $p_T > 20$ GeV/c. However the ratio between light quarks and gluons remains constant at $\sim$9/4 due to the color Casimir factor. In a more speculative vein, modelling of heavy "quark" propagation in a strongly coupled fluid using the techniques of AdS/CFT has received significant recent attention [5], including a suggestion that the relative suppression of charm vs. bottom quarks is different in pQCD and AdS/CFT, with a magnitude that could be resolvable by experiment [6].

In addition, with the first p-p data obtained at 7 TeV basic measurements such as jet cross-sections, jet-profiles and jet fragmentation-functions of heavy flavor jets will allow us to make important tests of NLO and MLLA calculations.

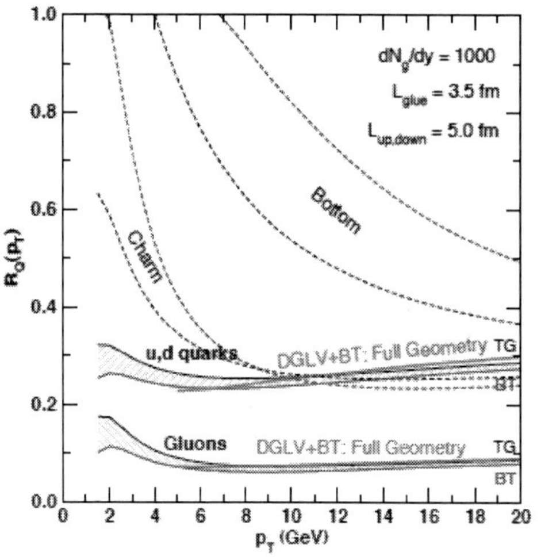

Figure 1. Theory calculations of quark and gluon nuclear modification factors at RHIC from the DGLV model [7]

The ALICE experiment is well suited to perform the measurements of heavy flavor production at high p_T . While exclusive reconstruction of charm mesons will be carried out via the use of the TPC and ITS detectors, measurements of high p_T heavy flavor production require a fast trigger, which is possible only for the semi-leptonic decay mode, with a branching ratio of $\sim 10\%$. The TRD has an efficient electron trigger and good hadron rejection for $p_T < 10$ GeV/c, but at higher p_T additional tools are required for heavy flavor measurements. The EMCal has excellent capabilities for fast triggering and hadron rejection, and provides unique coverage in ALICE for heavy flavor measurements at very high p_T (above 10 GeV/c), and the following studies are therefore focused on that region.

In addition the EMCal together with the central tracking system will also allow for full jet reconstruction. Jets are the primary experimental observable for partons. Combining the jet-finding with heavy flavor tagging techniques will allow us to obtain a reasonably clean sample of heavy flavor jets. These tagging techniques were developed by the Fermilab experiments (see for example [8]). We will present the results from the implementation of the displaced secondary vertex method which uses heavy flavor electrons as "seed" particles.

2. High p_T Electron Rates

The goal of this section is to establish the expected p_T reach of heavy flavor hadrons in a Pb-Pb run and determine the dominant background sources. The primary physics sources of electrons at high p_T are the semi-leptonic decays of charm (C) and bottom (B) hadrons (mostly mesons), and the decay of W-bosons. Electrons from C-hadron decay come both from prompt charm production and secondary decay of C-hadrons produced by the hadronic decay of B-hadrons.

In addition, there are significant backgrounds to these sources, due both to other physical processes and detector effects. The physics backgrounds consist primarily of electrons from Dalitz decays of high p_T π^0 and η hadrons in jet fragmentation and to a lesser extent quarkonia decays (not studied here), while the dominant detector source are photon conversions in the ITS and beampipe which are tracked in the TPC.

To obtain the expected annual yields of single electrons in the EMCal, a sample of PYTHIA

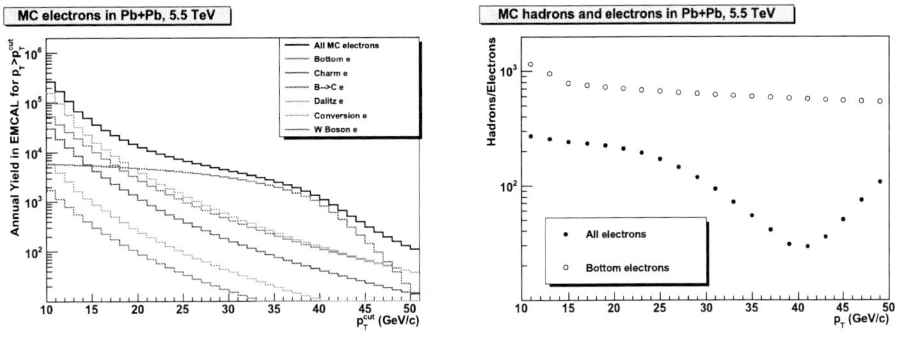

Figure 2. (left) Rates of MC-Electrons from the various physics/detector sources shown as the integrated annual yield above a p_T-cut for a nominal ALICE Pb–Pb year at $\sqrt{s_{\mathrm{NN}}} = 5.5\ TeV$. (right) Ratio of charged hadrons to electrons at the particle level expected in Pb-Pb events. The dip near 40 GeV/c is due to the contribution of W-decay electrons.

p-p collisions at 5.5 TeV, triggered on specific subprocesses, was simulated. A large sample of inclusive jet events, with one jet constrained to point towards the EMCal acceptance, was analyzed to evaluate the dominant physics and detector backgrounds. A smaller sample of W-boson events was generated to estimate the contribution from W-decay. The signal events were created using the ALICE standard PYTHIA heavy-flavor settings and requiring a B-jet in the EMCal acceptance and the B-hadron to decay semi-leptonically.

The left panel of Fig. 2 shows the contributions of the different physics sources to the distribution of electrons in the EMCal acceptance. Yields correspond to the expected Pb–Pb luminosity of 0.5 $\mathrm{mb}^{-1}\mathrm{s}^{-1}$ and one month (10^6 s) of Pb–Pb running to obtain an annual yield.

From this figure it is clear that there is a significant rate of bottom electrons to $p_T \sim 50$ GeV/c in the EMCal. The dominant backgrounds to heavy flavor electrons come from the photon conversions and from W-boson decays for $p_T > 20$ GeV/c. The other significant background to measuring B-electrons will be from charged hadrons misidentified as electrons. In order to evaluate the minimum hadron rejection required by the electron identification algorithm, the ratio of charged hadrons to electrons from the same simulation is shown in the right panel of Fig. 2. The ratio for transverse momentum in the range of interest from 10-50 GeV/c is a few hundred, which sets the scale for the hadron rejection requirement.

3. Electron Identification

The method of identifying electrons in the EMCal relies on the fact that electrons deposit all of their energy in the EMCal while hadrons typically leave only a small fraction (i.e. MIP) of their energy in the EMCal. In order to calculate the ratio of EMCal energy to reconstructed track momentum, tracks are matched to EMCal clusters. The procedure takes a reconstructed track and extrapolates it to the EMCal. If the distance between the extrapolated track-position and any cluster-position is less than a given value (in this study about the size of 1 tower) then the track is considered to be matched. In a second loop the matches are compared and only the closest matches are kept, thereby eliminating competing pairs.

We use the matched cluster-track pairs, the cluster energy, and the reconstructed track momentum to calculate the E/p ratio. This distribution is shown in Fig. 3 for single electron and pion tracks for two different momenta. The normalization of the two distributions is arbitrary and does not reflect the respective ratio of pions to electrons. The effect of particle interactions with the detector material is shown for each species. This was done by comparing the MC-input

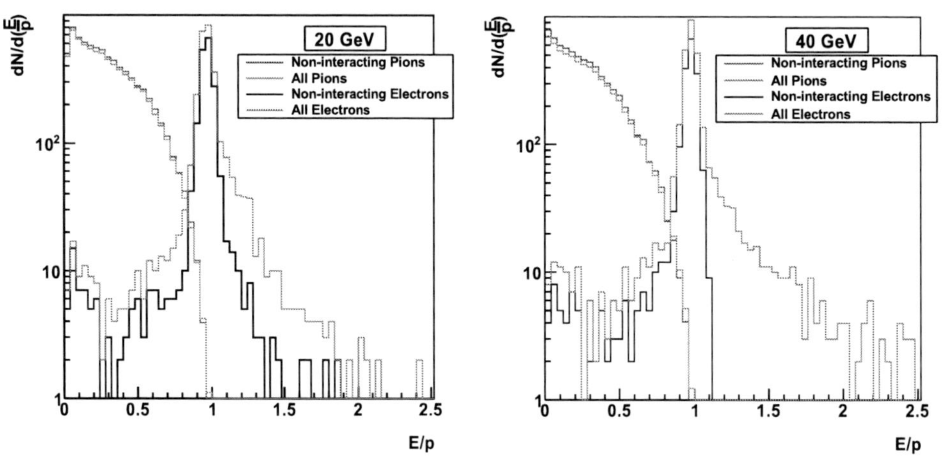

Figure 3. E/p distributions for single electrons and pions at momenta of 20 GeV/c (left) and 40 GeV/c (right). The normalization of the particle species relative to each other is arbitrary. The lines marked "non-interacting" refer to particles that lose less than 10% of their momentum before arriving at the EMCal surface.

to the reconstructed momentum. If the particle had lost more than 10% of its input momentum it was considered to have interacted with the detector material and labeled "interacting". If, on the other hand, the reconstructed momentum was more than 90% of the input momentum then it was considered "non-interacting". Electrons are known to suffer bremsstrahlung in the detector material as can be seen in the slight "tails" of the E/p distribution towards the larger values.

We establish an electron identification criterion by setting a lower limit on the E/p ratio. We define the efficiency, ϵ, as the number of electrons that pass the cut divided by the total number of electrons in our sample. One can then estimate the amount of hadron contamination by integrating the counts above the cut value. This determines the rejection power of the method, defined as ϵ^{-1} for pions. The actual electron purity in a p-p (Pb-Pb) collision will depend on the relative ratio of pions to electrons in the data.

On the left hand side of Fig. 4 the hadron rejection power is plotted as a function of cluster energy and compared to the recently published 2007 CERN-SPS test-beam data [9]. The results between simulation and real data are consistent if one considers non-interacting particles in the simulation.

The right hand side of Fig. 4 shows the results for hadron rejection power using a full GEANT simulation of the ALICE detector material. In simulations with large hadronic and photonic backgrounds (PYTHIA, HIJING) we determined that an additional cut on EMCal cluster shape and size was necessary to ensure that the electron clusters are correctly identified. This comes at the expense of electron efficiency, which drops to $\sim 65\%$. With these cuts the rejection power still reaches 1000 at 40 GeV/c, yielding a resulting electron/hadron ratio of at least 3:1.

4. Reconstructed Electron Spectra

Utilizing the PID capabilities of the EMCal we can now reconstruct inclusive and non-photonic electron (NPE) spectra. In order to remove the photonic conversion electrons from the inclusive sample, all candidate electrons are checked against the list of reconstructed secondary vertices

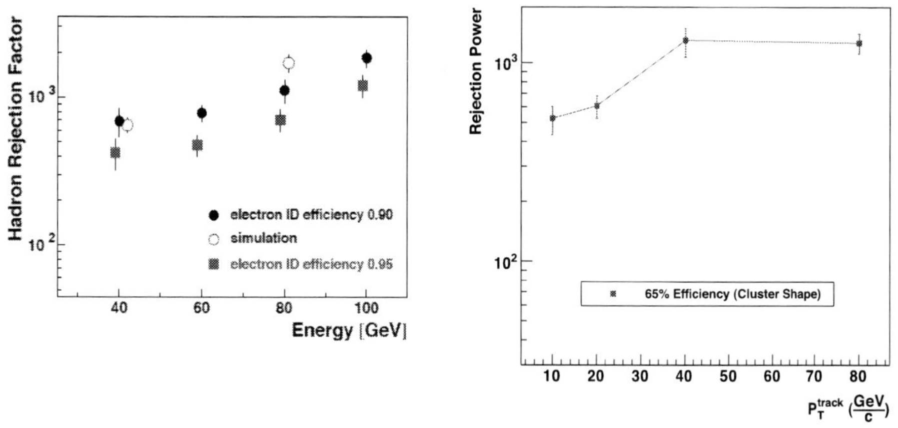

Figure 4. Hadron rejection factor as a function of electron efficiency and energy/momentum. (left) Simulation results compared to 2007 CERN test-beam data [9]. (right) Results from simulation with full detector material and optimized cuts for PYTHIA and HIJING multiplicities using cluster-shape and size criteria. The tracks used are required to have at least 50 TPC hits and 3 ITS hits.

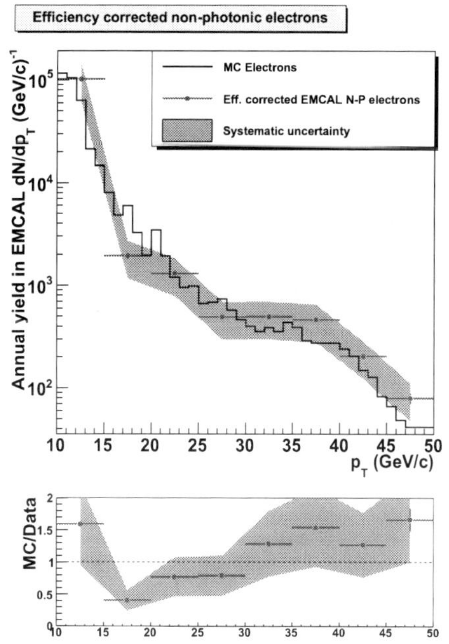

Figure 5. Efficiency corrected signal of non-photonic electrons for EMCal PID compared to MC truth for bottom and W-decay electrons. Systematic errors due to varying the EMCal electron identification criteria are shown.

(V0) created from charged particles. Those matching a V0 with invariant mass near that of a photon, ρ^0, ω or ϕ hadron are considered "photonic" and are subtracted from the inclusive spectrum to obtain the NPE candidate yield.

To determine the total yield of non-photonic electrons, the estimated contamination from misidentified hadrons is subtracted from the inclusive NPE candidate spectrum and the resulting signal distribution is corrected by the efficiency. The result of these operations is presented in Fig. 5. The efficiency corrected spectrum is in agreement with the MC-input distribution within the systematical uncertainties. The systematic uncertainty bands were determined by varying the track-matching and electron identification cuts in several combinations. The decreased efficiency resulting from tighter cuts is compensated by the increased purity and vice versa. With these simulations we obtain point-by-point uncertainties that vary by less than a few percent about an average value of ~40%. Additional systematic effects, in particular in conjunction with the reconstruction of conversion electrons, still need to be evaluated.

5. B-Jet Tagging: Displaced Vertex Method ("DVM")

This tagging method relies on the reconstruction of displaced secondary vertices from semi-leptonic B-decays, which are typically displaced by a few hundred μm from the primary vertex. It has been used by CDF to identify bottom contributions in semi-leptonic muon decays [10]. In addition, bottom decays typically produce a large number of charged particles by decaying via charm mesons to lighter hadrons. The highest p_T hadrons are correlated in phase-space and all point back to a common, displaced vertex. By identifying the semi-leptonic displaced vertex consistent with the B-meson lifetime and at least one additional hadron from the decay we have a powerful tool to discriminate non-bottom electrons from bottom electrons.

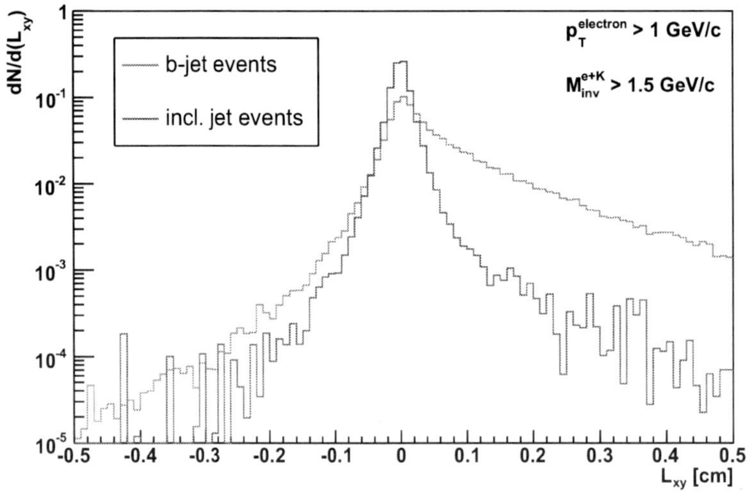

Figure 6. Signed DCA (L_{xy}) distributions from a sample of b-jet events ("Signal") and inclusive jet events ("Background+Signal") at 5.5 TeV. The distributions are normalized to the same integral. The inclusive jet events also contain a small contribution of b-jets that was not subtracted.

The method uses a high p_T electron within a jet as a seed, then searches for intermediate p_T hadrons (above 1.0 GeV/c) from a common secondary, displaced vertex within a cone of $R = \sqrt{\Delta\eta^2 + \Delta\phi^2} < 1.0$ around the trigger. This R value was successfully used by CDF but

Journal of Physics: Conference Series **230** (2010) 012005 doi:10.1088/1742-6596/230/1/012005

still needs to be optimized for the heavy ion environment [10]. Electrons below p_T 10 GeV/c are identified using the combined PID of TPC, TRD and EMCAL, whereas above this cut EMCAL PID only is used to guarantee sufficient purity of the sample. A minimum of 4 (out of 6) ITS hits are required on both tracks to ensure sufficient spatial resolution of the secondary vertex. Once a pair is found and its displaced vertex determined, the quantity L_{xy} (in the bending plane) is calculated:

$$L_{xy} = \frac{r \cdot p_e}{|p_e|} = |r| \cdot cos(\theta) \tag{1}$$

Where **r** is the vector from the primary vertex to the secondary vertex and **p** is the electron momentum. The distribution of this quantity is symmetric around zero for the background, but strongly biased towards positive values for real decays. Fig. 6 shows the distributions of L_{xy} for electrons from inclusive jet ("Background+Signal") and b-jet ("Signal") events. More details can be found in reference [11].

Based on the distribution of L_{xy} we define cuts to select electrons from b-decays. In this study we have adopted preliminary cuts that require at least n (n=1,2,3...) tracks from secondary vertices with $L_{xy} > 0.1$cm and $M_{inv}^{e+K} > 1.5$ GeV. Cutting on the invariant mass is a powerful discriminator to reject semi-leptonic vertices from charm.

The next step consists of identifying such tagged electrons within jets. For this we ran a jetfinder algorithm (FASTJET) on our data and then compared the charged jet constituents to our collection of tagged electrons. If a jet contained a tagged electron it was tagged as a b-jet. All jets first had to pass EMCal acceptance criteria in η, ϕ and neutral energy fraction to ensure complete reconstruction of the neutral and charged components.

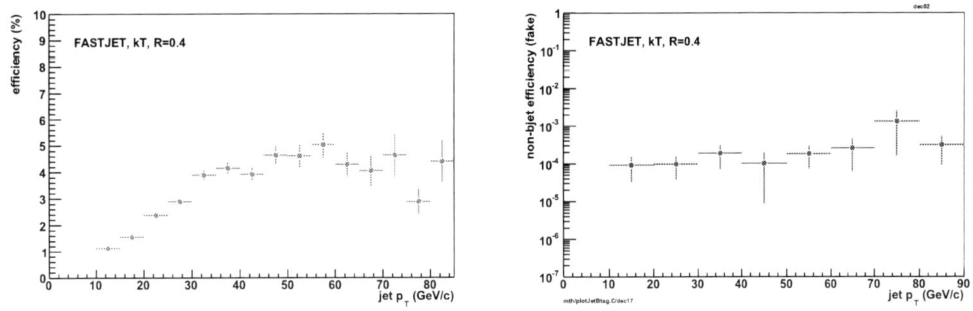

Figure 7. Tagging efficiency (left) and fake rate (right) for DVM B-tagging as a function of reconstructed jet p_T. These results were obtained requiring at least 2 displaced vertices. Jets were reconstructed using the k_T-algorithm of the FASTJET package with R=0.4.

In Fig. 7 the current results of the DVM B-tagging algorithm are shown as a function of reconstructed jet p_T. With the algorithm configured for highest purity we currently achieve an efficiency of about 5% for $p_T > 40\ GeV/c$ as compared to MC-input B-jets. This efficiency is expected since we require one high p_T electron and two hadrons from the secondary displaced vertex. The fake tag rate, defined as the efficiency of tagging a non-B jet, is of the order of 10^{-4} which means that the purity of the tagged sample, depending on the input model, is well above 90%. In figure 8 we show the estimated amount of tagged B-jets from one nominal year of Pb-Pb running which confirms that our statistical reach is up to p_T of $\sim$ 60 GeV/c. These studies are still preliminary and work is ongoing to optimize the efficiency versus fake rate of the algorithm.

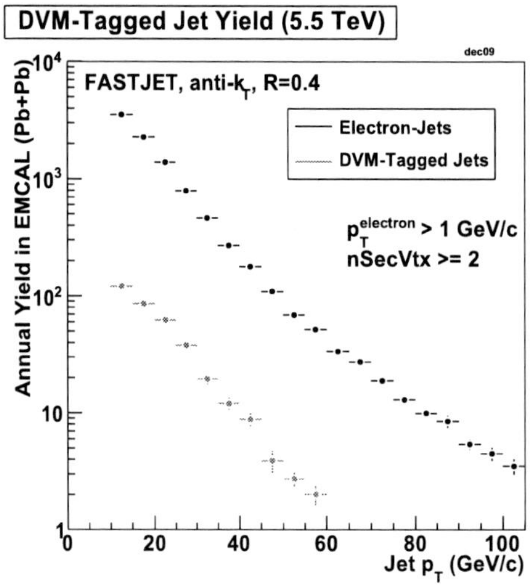

Figure 8. Final yield per Pb-Pb year of B-tagged jets using the FASTJET anti-k_T-algorithm and the DVM tagging method.

6. Conclusions

We have shown that the EMCAL possesses excellent electron PID capabilities and extends ALICE's electron PID in the p_T range up to 80 GeV/c, with hadron rejection factors of the order of several hundreds. This capability will allow to measure non-photonic electrons in Pb–Pb collisions for p_T ranges well above that previously reported at RHIC, and thus gain insight into color-charge and/or quark mass effects of partonic energy loss. Further studies for effectively reducing conversion and W-boson backgrounds are underway. The studies of B-tagging are at a proof-of-principle stage. We have shown that the displaced secondary vertex algorithm yields good results in p-p collisions and allows to measure B-jets up to a jet-p_T of $\sim$ 60 GeV/c. We expect to further optimize the algorithm to find an optimal balance between efficiency and purity.

References

[1] Y.L. Dokshitzer, and D.E. Kharzeev, *Phys. Lett.* **B519** (2001) 199
[2] A. Adare et al. (PHENIX coll), *Phys. Rev. Lett.* **98** (2007) 172301
[3] B.I. Abelev et al. (STAR coll.), *Phys. Rev. Lett.* **98** (2007)192301
[4] N. Armesto et al., *Phys. Rev.* **D71** (2005) 054027
[5] S.S. Gubser, arXiv:0907.4808
[6] W.A. Horowitz and M. Gyulassy, *Phys. Lett.* **B666** (2008) 320
[7] S. Wicks et al., *Nucl. Phys.* **A784** (2007) 426
[8] D. Acosta et al. (CDF coll.), *Phys. Rev.* **D69** (2004) 072004
[9] J. Allen et al. *Nuclear Inst. and Methods in Physics Research* **A 615** (2010) 6-13.
[10] D. Acosta et al. (CDF Coll.), *Phys. Rev.* **D66** (2002) 014003
[11] M. Heinz for the ALICE collaboration, arXiv:0712.2422

Journal of Physics: Conference Series **230** (2010) 012006 doi:10.1088/1742-6596/230/1/012006

Overview of Open Heavy Flavor Measurements in STAR

Sarah LaPointe for the STAR Collaboration

Department of Physics and Astronomy
Wayne State University, Detroit, MI 48201, USA

E-mail: lapointe@physics.wayne.edu

Abstract. We present an overview of measurements related to open heavy flavor production. Results from indirect reconstruction of heavy flavor via semi-leptonic decays in $p + p$ and minimum bias Au+Au collisions at $\sqrt{s_{NN}} = 200$ GeV, and results from D meson measurements via their hadronic decay channel in minimum bias Cu+Cu and Au+Au collisions are discussed. For the semi-leptonic decays the focus is on partonic energy loss, R_{AA}, and the use of e-h and e-D^0 azimuthal correlation measurements to extract the B meson contribution to the non-photonic electron spectrum. For the hadronic decay the focus is on open charm yields in Cu+Cu collisions and first measurements of D meson in heavy ion collisions using a secondary vertexing technique.

1. Introduction

Measurements using data taken at the Relativistic Heavy Ion Collider (RHIC) point to the production of a near perfect Quark Gluon Liquid [1, 2], that is opaque to the passage of light partons [3, 4, 5]. In this hot and dense medium, it is believed that the partons lose energy via gluon radiation before fragmenting into hadrons. A method used to quantify this energy loss is through the nuclear modification factor, R_{AA}. Heavy quarks, because of their large masses, are expected to be produced predominately from the initial gluon fusion in parton-parton hard scatterings in nuclear collisions at RHIC energies [6]. This makes heavy flavor a powerful probe of the QCD properties of the matter created in heavy ion collisions. It was expected that heavy quarks would lose energy in the same manner as their lighter counterparts, via gluon radiation. However, a suppression of radiative energy loss should occur because of the dead cone effect, which states that gluonic radiation of a massive parton should be suppressed at angles $\theta < M_q/E_q$ [7]. This would indicate that the R_{AA} for heavy quarks in central nucleus-nucleus collisions is larger than that of lighter quarks.

RHIC experiments have reported the measurement of heavy quarks using semi-leptonic decays (non-photonic electrons) in $p + p$ and Au+Au collisions [8]. These heavy flavor decay electrons do not reveal the dramatic differences in energy loss for charm and beauty that was originally predicted, and at $p_T > 4$ GeV/c show a similar magnitude of suppression as light flavor in Au+Au collisions. Models using only radiative energy loss predict significantly less supression [9, 10], and those that include other components for energy loss, i.e. collisional [13, 12], also underpredict the observed suppression. The only radiative plus collisional theoretical model that describes the data well is one where only charm is contributing to the non-photonic electron spectrum [11].

Alternatively, it has been shown that the observed energy loss can be described quite well using a model where open heavy mesons can form early and collisionally disassociate in the medium, multiple times. This may be a significant contribution to the non-photonic electron energy loss observed [14].

These theoretical predictions are sensitive to the beauty contribution to the electron spectrum. Recent Fixed Order Next to Leading Log (FONLL) calculations for heavy flavor production in p+p collisions show the beauty contribution to the non-photonic electron spectrum becoming comparable to charm near intermediate p_T (4-5 GeV/c) [15]. The relative contribution is experimentally unknown at p_T above 3-4 GeV/c. In order to understand heavy flavor energy loss observed from the non-photonic electrons the contribution of D and B mesons needs to be determined experimentally.

Another probe of the Quark Gluon Liquid is radial flow, which is described in the so-called Blast-Wave model [16]. STAR has reported the radial flow of the lighter flavor hadrons in Cu+Cu collisions at $\sqrt{s_{NN}} = 200$ GeV [17], and the observed flow suggests a strongly interacting liquid. If heavy flavor quarks also interact strongly with the light flavor medium they will exhibit a similar flow.

In this article, the focus will be on extracting beauty's contribution to the non-photonic electron spectrum using e-h azimuthal correlations and by measuring D mesons directly through their hadronic decay channel. In addition, the D meson measurements can be compared to the Blast-Wave model and the Statistical Hadronization model predictions [18], which state that a large strangeness production and "free" charm in the medium will enhance the D_s yield compared to e^+e^- collisions.

2. Data Analysis

2.1. Single Electrons

Charm and beauty quarks can be identified through the measurement of single electrons that come from semi-leptonic decays of D and B mesons. The electrons will carry a fraction of the total D momentum. The STAR Collaboration has studied the spectrum of single electrons, $e^\pm$. The main background sources in the single $e^\pm$ spectrum are primarily e^+e^- pairs from photon conversions in detector material and π^0 and η Dalitz decays. For this reason the single $e^\pm$ signal is called non-photonic and the background $e^\pm$ is called photonic.

The $e^\pm$ are identified using the Time Projection Chamber (TPC), the Electromagnetic Calorimeter (EMC), and the Shower Maximum Detector (SMD). For $p > 1.5$ GeV/c, $e^\pm$ are identified through measurements of ionization energy loss (dE/dx) in the TPC, the energy (E) deposited into the EMC, and the shape of the shower in the SMD. An energy loss cut of 3.5 keV/cm $< dE/dx <$ 5.1 keV/cm is used to isolate $e^\pm$. Compared to hadrons, $e^\pm$ produce a different shower and deposit more energy into the EMC. Thus, additional cuts on shower size in the SMD and a $p/E < 2$ cut aids in hadron rejection. Finally, the photonic background is rejected using an invariant mass technique. The e^+e^- invariant mass distribution due to photonic conversions and π^0 and η Dalitz decays has a maximum near zero, and a tail at nonzero [20]. A cut of $M_{inv}(e^+e^-) < 150$ MeV/c^2 can be applied to subtract off most of photonic contribution. The non-photonic electrons are identified with an efficiency near 70%.

2.2. Direct Hadronic Reconstruction

D^0 and $\bar{D^0}$ have been reconstructed through their hadronic decay $D^0(\bar{D^0}) \rightarrow K^\mp + \pi^\pm$ (B.R. = 3.91% [20]). The pions and kaons are identified by their energy loss in the TPC. All primary kaons and oppositely charged pions are paired in the same event and each pair is then considered a D^0 candidate. The invariant mass of the D^0 candidate can then be determined from the momenta of the tracks and assuming the daughter masses. The invariant mass spectra contains the contribution from real D^0 and combinatorial background from random positive/negative pairs. This background can be estimated by the rotational or event mixing technique. The impact parameter resolution of the track is much greater than $c\tau(D^0) = 123\ \mu$m [20] so when using the TPC alone, STAR can not obtain a full reconstructed D^0 decay topology.

D^0 has also been measured in 200 GeV Au+Au collisions using a secondary vertexing and tracking technique that utilizes the Silicon Vertex Tracker (SVT) and the Silicon Strip Detector (SSD) alongside the TPC. The inclusion of 2 or more hits from the SVT and SSD improves the impact parameter resolution of the global tracks to the primary vertex (PV). For example, at $p_T = 1$ GeV/c the resolution improves by a factor of 10, compared to the TPC alone [21]. The resolution is comparable to $c\tau(D^0)$, so a full reconstruction of the decay topology can be achieved. The topology contains variables such as, D^0 distance of closest approach (DCA) to the PV, D^0 decay length, DCA of the decay daughters to the PV, and the DCA of the daughters to one another. These geometrical variables are used as cut parameters to distinguish between D^0 signal and combinatorial background.

A secondary vertexing technique reconstructs the D^0 by pairing all oppositely charged tracks and projecting them toward the primary collision vertex. If the two track trajectories cross at some point away from the primary vertex they are considered candidates for the D^0 decay. Again, the D^0 is identified through an invariant mass analysis. The difference here is that a set of decay topology cuts can be applied. These cuts discriminate between the signal and the background.

3. Results and Discussion

3.1. Electron-Hadron Azimuthal Correlations

Since charm and beauty have large masses, pQCD can be used to predict their production. FONLL predicts heavy flavor decay electrons from charm to dominate at low p_T , with beauty's contribution becoming comparable to charm near 5 GeV/c.

Electron-hadron (e-h) azimuthal correlations have been studied with the objective of disentangling the D and B meson contributions to the non-photonic $e^{\pm}$ spectrum. Momentum conservation implies that the heavy quark anti-quark pairs produced in the initial hard scatterings of the collision are correlated azimuth ($\Delta\phi$), here a back-to-back orientation arises (Figure 1 (right)). D^0 mesons will predominately decay hadronically ($D^0 \rightarrow K$+anything, B.R. $\sim$ 55%) and beauty, via B meson, will decays into D^0 mesons. In addition, both the D and B mesons will decay into $e^{\pm}$. With the D^0 also decaying to $K^-\pi^+$, azimuthal correlations can be made between electrons and charged hadrons. The expected charm and beauty contributions are simulated using PYTHIA. Figure 1 (left) shows the predicted distributions using PYTHIA and the correlation for $e^{\pm}$ and hadrons in $p+p$ collisions at $\sqrt{s} = 200$ GeV for a single p_T bin. The width of the near side ($\Delta\phi = 0$) peak for electrons from B decays is much wider compared to D . The energy released in the B meson semi-leptonic decay leads to this broader angular correlation between the decay electron and hadron daughters. The measured distribution is then fit with a linear combination of the PYTHIA curves with B/(B+D) as a parameter in the fit function. Figure 2 (right) shows the relative beauty contribution (blue symbols) to the total non-photonic $e^{\pm}$ yield as a function of p_T using this method.

Additionally, STAR had studied e-D^0 azimuthal correlations in $p+p$ collisions at $\sqrt{s} = 200$ GeV. Here the D^0 is fully reconstructed from the $K\pi$ decay. Azimuthal correlations are made between like sign e-K pairs. Figure 1 (right) shows the fragmentation of charm and beauty, respectively. The figure demonstrates that a like sign e-K correlation with $\Delta\phi$ near π would arise mostly from the $c\bar{c}$ diagram and a $\Delta\phi$ near 0 would come from the $b\bar{b}$ diagram. Figure 2

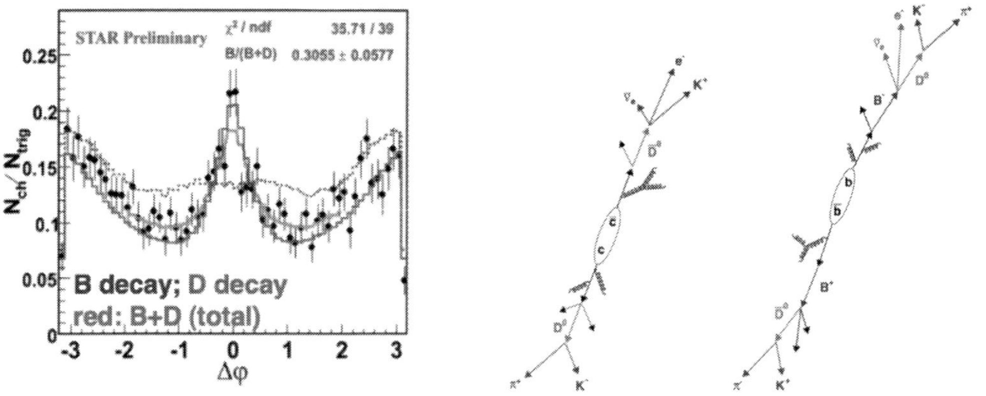

Figure 1. Left: $\Delta\phi$ for non-photonic $e^{\pm}$ and hadron pairs in $p+p$ collisions at $\sqrt{s_{NN}} = 200$ GeV, where $2.5 < p_T$ (trig) < 3.5 GeV/c and p_T (assoc) > 0.3 GeV/c. The vertical axis is the number of charged tracks per $e^{\pm}$ trigger. Right: Diagram of the fragmentation of $c\bar{c}$ and $b\bar{b}$ pair.

(left) shows PYTHIA and MC@NLO simulations [19] as a comparison to data. The away side ($\Delta\phi = \pi$) correlation peak is a feature of prompt charm pair production $\sim 75\%$ D and $\sim 25\%$ B decays. The near side peak largely represents contributions for B decays. From this, the relative beauty contribution to the non-photonic $e^{\pm}$ can be extracted, and the result (red symbols) is shown in Figure 2 (right) along with the results from e-h azimuthal correlations [23, 24].

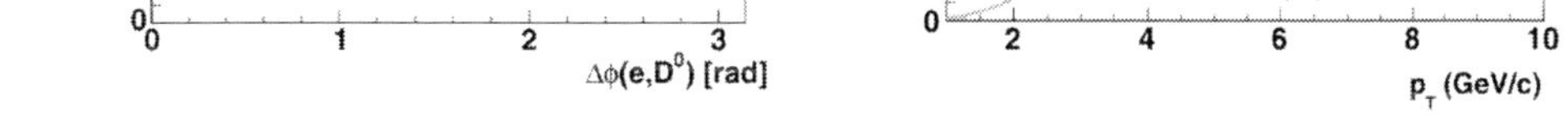

Figure 2. Left: $\Delta\phi$ of non-photonic $e^{\pm}$ and D^0 mesons, for like sign e-K pairs in $p+p$ collisions at $\sqrt{s} = 200$ GeV. Right: Relative beauty contribution to the total non-photonic $e^{\pm}$ yield for e-D^0 (red symbols) and e-h (blue symbols) correlations. The uncertainty band from a FONLL calculation (green curves)

All experimental results are consistent with FONLL estimates. The measurements suggest that the B meson contribution to the non-photonic spectra increases with p_T and is comparable to the D meson contribution at and above 5 GeV/c. This indicates that theoretical models that try to describe the observed non-photonic $e^{\pm}$ suppresion must take beauty's suppression in to account as well.

Due to incomplete kinematics of reconstructing heavy flavor meson from their semi-leptonic decays, there is an uncertainty in the relative fraction of charm and beauty. A direct measurement of charm through the D meson hadronic decay could resolve this. In the next sections measurements of this type will be discussed.

3.2. D^0 in Cu+Cu collisions

D^0 and $\bar{D}^0$ yields have been measured in Cu+Cu collisions at $\sqrt{s_{NN}} = 200$ GeV [25]. Figure 3 (left) shows the invariant mass distribution of K+π pairs after the rotational background subtraction has been applied. The signal shown has been rebinned into 3 p_T bins. The result is the p_T spectrum, Figure 3 (right). A blast wave fit to a p_T spectrum allows the temperature at thermal freeze out, T_{fo}, and the average radial velocity, $< \beta >$, to be extracted. However, the blast wave function, has three free parameters and with only three p_T bins it is difficult to extract all three from the spectrum. Instead, a comparison with the lighter particle species in the Cu+Cu system can be made.

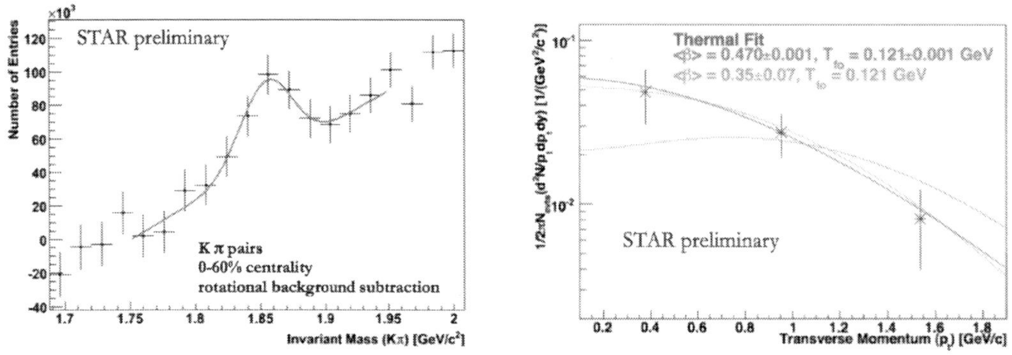

Figure 3. Left: Kπ invariant mass distribution after background subtraction using a rotational method. Right: The D^0+$\bar{D}^0$ spectum in 200 GeV Cu+Cu collisions at 0-60% fit with a thermal fit (red curve), a blast wave fit derived from T_{fo} and $< \beta >$ of pions, kaons, and protons in 0-60% central Cu+Cu collisions (brown), and a blast wave fit from fixing T_{fo} to the light species and allowing $< \beta >$ to be a free parameter (green).

If the D^0 mesons are fully coupled to the lighter species hadrons in the late stages of the quark-gluon medium, they will also take on the parameters, T_{fo} and $< \beta >$, that were found for pions, kaons, and protons. Figure 3 (right) shows that the blast wave curve using the lighter flavor fit parameters is inconsistent with the data. If instead, the freeze out temperature is kept constant, a value of the average radial velocity can be extracted. The $< \beta >$ from the fit is 0.35 $\pm$ 0.07. A comparison to the lighter flavor species value of $< \beta > = 0.470 \pm$ 0.001 suggests that the D mesons do not have as strong a radial flow as the light quark hadrons. One interpretation is that the charm decouples from the medium in a different manner than that of the lighter species.

3.3. D mesons in Au+Au Collisions

D^0 and $\bar{D}^0$ in Au+Au collisions at $\sqrt{s_{NN}} = 200$ GeV have also been measured [26]. The main difference from the previous analysis is that a secondary vertexing technique is used. Figure 4 (left) shows the invariant mass obtained after identifying kaons and pions using dE/dx, requiring both daughter tracks to have 2 or more hits in the SVT, and applying cuts on the decay topology. At this stage the D^0 peak is not easily visible, so the background is subtracted by fitting a polynomial to the side bands of the distribution (see Figure 4 (left)) and subtracting the background under the peak based on the χ^2 minimized polynomial fit. The background obtained by the fit can be confirmed through independent studies using a rotational technique. The preliminary result is Figure 4 (right). The signal has a statistical significance S/$\sqrt{S+B} \simeq$ 4.5.

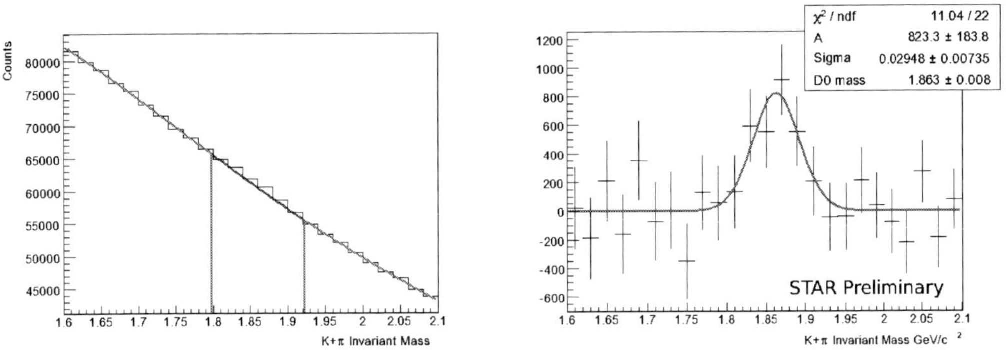

Figure 4. Left: Kπ invariant mass after optimized cuts in Au+Au collisions at $\sqrt{s_{NN}} = 200$ GeV. Right: Kπ invariant mass after background subtraction using a polynomial fit.

In addition, the secondary vertexing technique, that utilizes the SVT+SSD, has also been used to perform a $D_s^{\pm}$ measurement. $D_s^{\pm}$ decays to $\phi + \pi^{\pm}$, with $\phi \to K^+K^-$ (B.R. = 2.18%) [25]. Once again, kaons and pions are identified using dE/dx information from the TPC, and the daughter tracks must have at least 2 hits in the SVT. In addition to applying decay topology cuts, we can require that the reconstructed K^+K^- pair invariant mass is near the ϕ mass peak. The $D_s^{\pm}$ signal, shown in Figure 5 (left), has a statistical significance of 3.1σ. The yield, $dN/d\eta$, can be compared to the Statistical Hadronization Model (SHM) predictions. A comparison is made between the inclusive charm-strange yield, $D_s^{\pm}$, and the inclusive charm-light quark yield, D_{inc}, where $D_{inc} = D^0 + \bar{D}^0 + D^+ + D^-$ and $D_s = D_s^+ + D_s^-$. STAR has measured the D^0 yield in Au+Au collisions at $\sqrt{s_{NN}} = 200$ GeV using an mixed event technique [27].

Since RHIC has not yet published a $D^{\pm}$ spectra, the ratio of D/D^0 is extracted from Pythia or e^+e^- collisions.

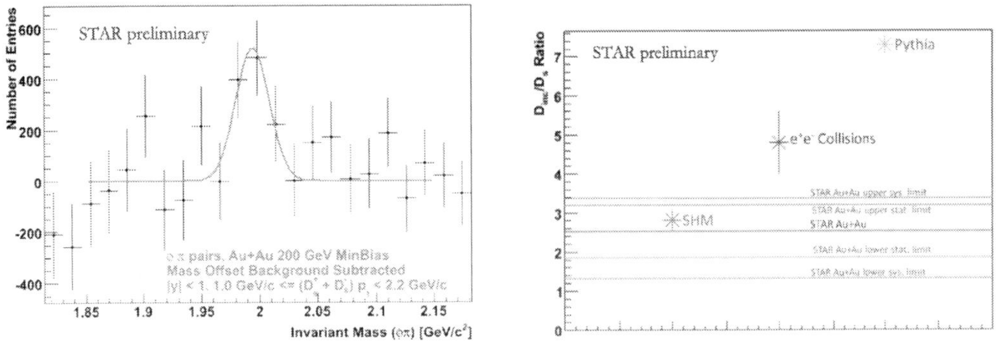

Figure 5. Left: $\phi\pi$ invariant mass after optimized cuts in Au+Au collisions at $\sqrt{s_{NN}} = 200$ GeV. Right: D_{inc}/D_s ratio (black line) compared to estimates from PYTHIA, e^+e^- collisions, and the Statistical Hadronization Model.

Figure 5 (right) shows the D_{inc}/D_s ratio predicted using PYTHIA, e^+e^- collider data, and STAR's measurement from Au+Au collisions. The value from STAR is consistent with SHM prediction of 2.8 at the conditions at RHIC, with s/S $\approx$ 0.03 and $S^Q = S^H$, where s/S is the strangeness to entropy ratio [18]. This result suggests an enhancement of D_s in the QGP.

4. Conclusions

We summarized the STAR collaboration measurements of open heavy flavor production in heavy ion and proton-proton collisions at RHIC. Results of non-photonic electrons from heavy flavor decays and D meson reconstruction from the hadronic channel were shown. Heavy flavor electrons exhibit a similar suppression to that of the light hadrons at intermediate p_T . Currently, most theoretical models underpredict the observed suppression. One model that uses radiative and collisional energy loss, but for charm only, describes the data well. However, e-h and e-D^0 azimuthal correlation measurements show that at a p_T of 5 GeV/c and higher the contribution to the electron spectrum from beauty is comparable to that of charm.

In addition, D mesons can be measured directly. The momentum distribution of D^0 in Cu+Cu collisions fit with a blast wave function indicates that the D mesons do not have as strong a radial flow as the lighter quark hadrons. This suggests that the charmed hadrons interact with and decouple from the QCD matter differently than the light hadrons. We also show the first D meson measurement in heavy ion collisions, using a secondary vertexing technique. The direct reconstruction of D^0 will further constrain the D and B contribution to the non-photonic electron spectrum and offer insight into the charm's interaction with the medium. Finally, a preliminary $D_s^{\pm}$ measurement using secondary vertexing was shown. Its yield, compared to the inclusive D, hints at statistical hadronization. It is important to recognize that significant uncertainties are present in the data, as well as in current models. However, the measurements do provide constraints on theory, which will aid in a greater understanding of heavy flavors interaction with the hot and dense matter created in heavy ion collisions.

References

[1] E. Shuryak, arXiv:hep-ph/0312227 (2004)
[2] T. Hirano and M. Gyulassy, arXiv:nucl-th/0506049 (2006)
[3] C. Adler et al., *Phys. Rev. Lett.* **89** 202301 (2002)
[4] J. Adams et al., *Phys. Rev. Lett.* **91** 172302 (2003)
[5] J. Adams et al., *Phys. Rev. Lett.* **91** 072304 (2002)
[6] Z. Lin and M. Gyulassy, *Phys. Rev. C* **51** 4 2177 (1995)
[7] Yu. Dokshitzer and D.E. Kharzeev, *Phys. Lett. B* **519** 199 (2001)
[8] S.S. Adler et al., *Phys. Rev. Lett.* **98** 172301 (2007)
[9] M. Djordjevic et al., *Phys. Lett. B* **632** 81 (2006)
[10] N. Armesto et al., *Phys. Lett. B* **637** 362 (2006)
[11] S. Wicks et al., *Nucl. Phys. A* **784** 426 (2007)
[12] H. van Hees et al., *Phys. Rev. C* **73** 034913 (2006)
[13] N. Armesto et al., *Phys. Lett. B* **637** 362 (2006)
[14] A. Adil and I. Vitev, arXiv:hep-ph/0611109v2 (2006)
[15] M. Cacciari, P. Nason and R. Vogt, *Phys.Rev. Lett.* **95** 122001 (2005)
[16] E. Schnedermann, E. Sollfrank, U. Heinz *Phys. Rev. C* **48** 2462 (1993)
[17] A. Iordanove et al., *J. Phys. G.* **35** 044008 (2008)
[18] I. Kuznetsova and J. Rafelski, *Eur. Phys. J. C* **51** 113 (2007)
[19] A. Mischke,*Phys. Lett. B* **671** 361 (2009)
[20] C. Amsler et al. (Particle Data Group), *Phys. Rev. Lett. B* **667** 1 (2008)
[21] Y. Fisyak et al., *J. Phys.: Conf. Ser.* **119** 032017 (2008)
[22] S. Sakai et al. (STAR Collaboration), proceedings for poster presentation at QM08, submitted to Ind. J. Phys.
[23] A. Mischke et al., *Eur. Phys. J. C* **61** 807 (2009)
[24] G. Wang et al., *J. Phys. G: Nucl. Part. Phys.* **35** 104107 (2008)
[25] S. Baumgart, *A Study of Open Charm Production in Heavy Ion Collisions of Center-of-Mass Energy 200 GeV per Nucleon.* PhD thesis. Yale University (2009)
[26] S. LaPointe, arXiv:nucl-ex/0907.4730 (2009)
[27] H. Zhang et al., *Nucl. Phys. A.* **774** 701(2006)

Open charm meson analysis in proton-proton collisions at the LHC with ALICE

G Ortonal[1]
1 INFN Torino & Università Degli Studi di Torino
E-mail: ortona@to.infn.it

Abstract. The extremely high energies that will be reached with the Large Hadron Collider (LHC) at CERN will allow studying the production of open charm with high statistics in both proton-proton and Pb-Pb collisions. The study of open charm (D) mesons in Pb-Pb collisions will be a powerful tool to investigate the production of heavy flavours and their interaction with the medium produced in such collisions (QGP). Heavy flavour yields will provide also a normalization for quarkonia production. We will present a general overview of the ALICE collaboration heavy flavour program, then we will focus on the analysis and reconstruction strategies developed for the study of the charmed (D) mesons by the ALICE collaboration for proton-proton collisions, with special emphasis on the charged D mesons. Finally, some expected results obtained with MonteCarlo production will be shown.

1. Introduction

At the Large Hadron Collider (LHC) at CERN energies as high as $\sqrt{s_{NN}} = 5.5$ TeV for Pb-Pb collisions and $\sqrt{s_{NN}} = 14$ TeV for proton-proton collisions will be reached, opening up a new domain for the study of strongly interacting matter in conditions of high temperature (3 to 5 times higher than the critical temperature) and high energy density (15 to 60 GeV/fm^3).
QCD calculations[1] at high temperature and high energy density predict in such conditions the formation of a deconfined state of matter, called Quark-Gluon Plasma. In the study of the properties of the produced (deconfined) state, heavy quarks play a crucial role. Heavy quarks and hard partons, abundantly produced at LHC energies in the initial hard scattering processes, are sensitive probes of the medium formed in the collisions as they may lose energy by gluon radiation while propagating through the medium itself.
At LHC energy, charm and beauty will be abundantly produced. The cross-sections at LHC are expected to increase by about a factor of 10 for charm and 100 for beauty with respect to RHIC. The baseline production cross-section of $Q\bar{Q}$ pairs for ALICE simulation studies has been calculated in the framework of collinear factorization and pQCD[2], including the nuclear modification of the parton distribution functions (PDFs)[3]. The expected $c\bar{c}$ and $b\bar{b}$ production yields for pp collisions at the maximum energy available at the LHC ($\sqrt{s_{NN}} = 14$ TeV) are 0.16 and 0.0072, respectively[4]. For the 5 % most central Pb-Pb collisions at $\sqrt{s_{NN}} = 5.5$ TeV, the expected yields are 115 and 4.6 respectively, but it has to be noted that these predictions have large uncertainties, of about a factor 2 to 3, depending on the choice of the quark masses and QCD scales.

[1] For the ALICE collaboration

Meson	Decay channel	$c\tau$	BR
D^0	$D^0 \to K^-\pi^+$	$122.9\mu m$	$(3.91 \pm 0.05)\%$
D^0	$D^0 \to K\pi\pi\pi$	$122.9\mu m$	$(8.14 \pm 0.20)\%$
D^+	$D^+ \to K^-\pi^+\pi^+$	$311.8\mu m$	$(9.2 \pm 0.25)\%$
D_s^+	$D_s^+ \to K^+K^-\pi^+$	$149.9\mu m$	$(5.50 \pm 0.28)\%$
D^{*+}	$D^{*+} \to D^0\pi^+$		$(61.9 \pm 2.9)\%$

Table 1. Golden Channels for the reconstruction of open charm mesons[5]

2. The ALICE experiment at the LHC

The ALICE experiment, described in detail in[6], has very good performances for heavy flavour measurements[7].
Experimentally, the two key elements for a rich heavy-flavour program are: good tracking/vertexing and good particle identification.In ALICE, Particle tracking relies on the six concentric layers of high resolution silicon detectors of the Inner Tracking System (ITS)[2], a large volume of time projection chamber (TPC)[3] , and a high granularity transition-radiation detector (TRD). The ALICE detection strategy for charm and beauty hadrons relies on resolving secondary detached vertices reconstructed from tracks with large impact parameters (d0)[4]. The precision on impact parameter measurement is mainly provided by the two innermost ITS layers (SPD) in the bending plane (rϕ) and the two intermediate ITS layer (SDD) for z-coordinate. A resolution σ_{d0} better than $60\mu m$ in the bending plane is achieved for tracks with $p_T > 1.5GeV/c$. particle identification is performed, in the central rapidity region, over the full azimuth by a dE/dx measurement in the tracking detectors (TPC and ITS), via time of flight measurement using Time of Flight detector(TOF) and transition radiation measurement in the Transition Radiation Detector (TRD). Muons are identified in the Forward Muon Spectrometer in the pseudorapidity range $-4 < \eta < -2.5$

3. Open Charm measurement

Open charm meson will be measured through their hadronic decays in fully reconstructed topologies. The strategy to reduce the large combinatorial background is based on the selection of displaced-vertex topologies, i.e. large separation between the primary and the secondary vertex, good alignment between the reconstructed D meson momentum and the flight line, large impact parameter. The golden channels that will be studied are listed in table 1.

After the reconstruction, an invariant mass analysis is used to extract the raw signal yield, and then corrections for the detector acceptance, for selection efficiency and reconstruction efficiency are applied.

D^0 reconstruction

The most promising channel for the study of D^0[7, 8] is the channel $D^0 \to K^-\pi^+$, although also the 4-prongs decay $D^0 \to K\pi\pi\pi$ will be studied. The selection strategy is based on the application of 2 different cuts: the product of the impact parameter of the two tracks ($d_0^K \times d_0^\pi$)

[2] Two innermost layers equipped with silicon pixel (SPD) plus two middle layers of silicon drift detectors (SDD), and two outermost layers of silicon strip detectors (SSD)

[3] see J. Alme et al, Nucl. Instr. Meth. A (in print), arXiv:1001.1950

[4] the impact parameter being the distance of closest approach of a particle trajectory to the primary vertex

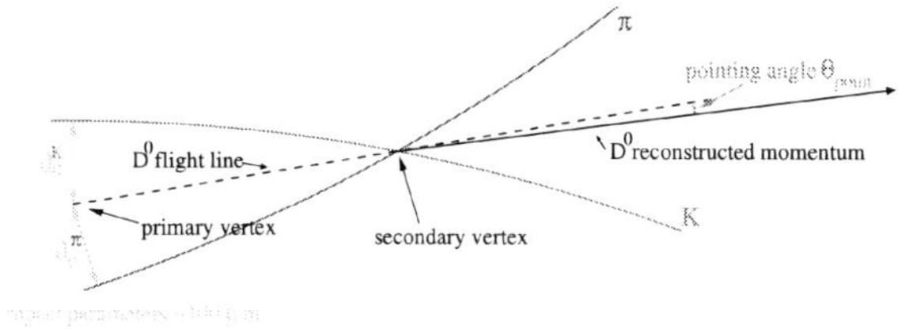

Figure 1. Schematic representation of the decay $D^0 \rightarrow K^-\pi^+$

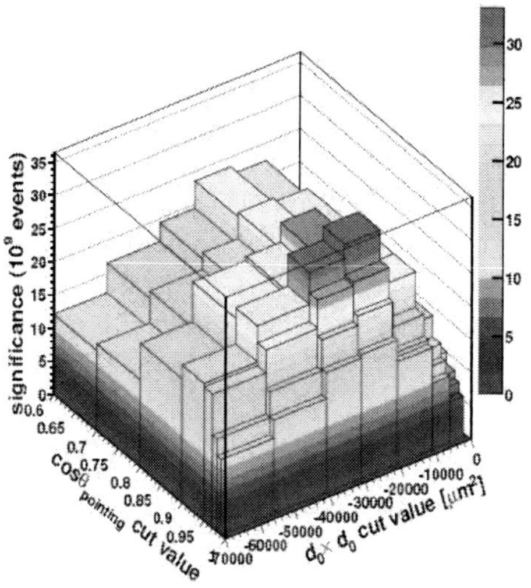

Figure 2. D^0 significance normalized to 10^9 events (1 year of data taking at 14 TeV), in function of cut variables

and the cosinus of the pointing angle[5] of the reconstructed D^0 meson. Figure 1 shows a sketch of this decay topology. The goal of this selection is to increase the significance (the ratio $\frac{S}{\sqrt{S+B}}$, where S is the signal and B the background). The effect of this cuts on the significance is shown in figure 2. As shown in Figure 3, the statistical error on D^0 is expected to be smaller than 20% after the first year of data taking at 14 TeV[10].

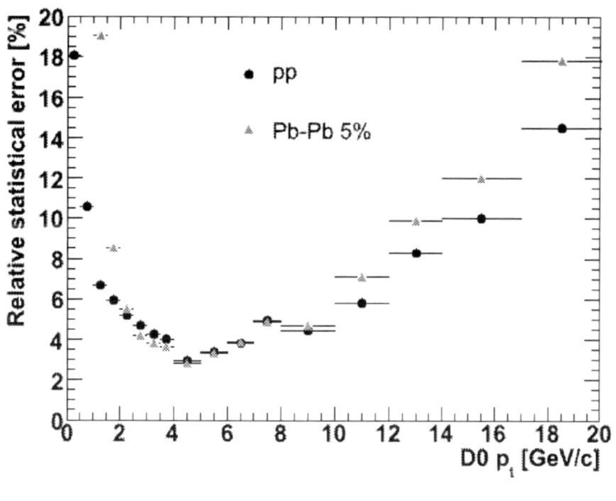

Figure 3. Relative statistical error for D^0 after 1 year of data taking in pp (14 TeV) collisions and Pb-Pb (5.5 ATev) collisions

[5] If the reconstructed particle is a real D meson, then $\cos\theta_{point} \approx 1$, being θ_{point} the angle between the momentum of the reconstructed particle and the flight line towards the primary vertex

D^+ reconstruction
The D^+ meson will be reconstructed in the channel $D^+ \rightarrow K^-\pi^+\pi^+$[9]. With respect to the decay $D^0 \rightarrow K^-\pi^+$ this topology is affected by higher combinatorial background. On the other hand, as shown in table 1, the branching ratio of this decay is higher and the D^+ has longer life time ($c\tau \sim 310\mu m$). Thus we can use a cut on the distance between primary and secondary vertex to remove combinatorial background. Triplets with the correct charge sign combination will be built, then the main cuts for a selection strategy to separate the signal from the huge combinatorial background for the D^+ are therefore the distance between the primary and secondary vertices and the cosinus of the pointing angle of the reconstructed meson. To perform this selection, an extremely good vertexing resolution, both on the primary and secondary vertex must be reached. The expected performance on vertex resolution of ALICE is shown in figure 5. Figure 4 shows the expected statistical error after 1 year of data taking. The significance for the D^+ is shown in figure 6 for both pp and Pb-Pb collisions.

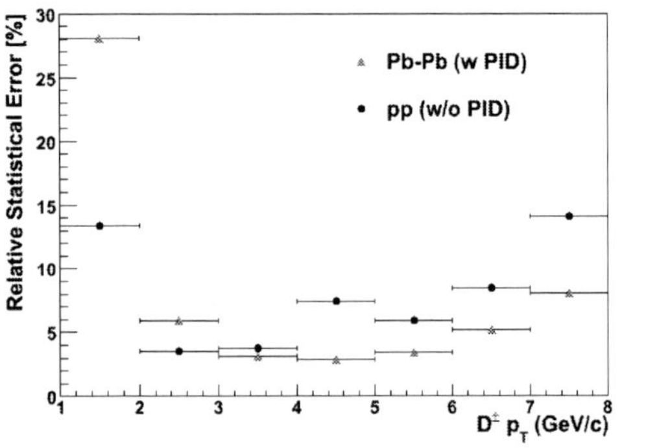

Figure 4. Relative statistical error for D^+ after 1 year of data taking in pp (14 TeV) collisions and Pb-Pb (5.5 ATev) collisions

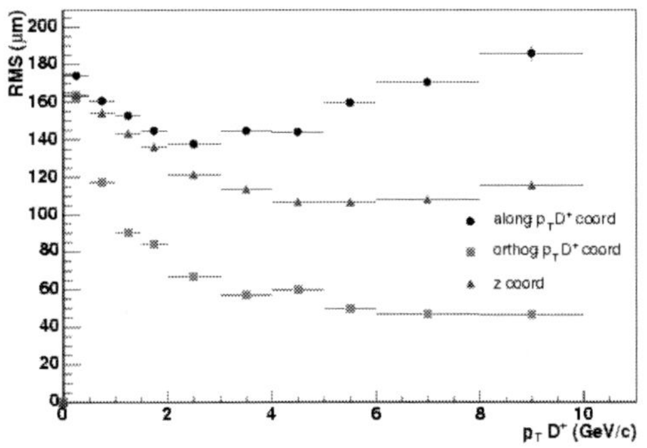

Figure 5. Resolution on the D^+ decay vertex

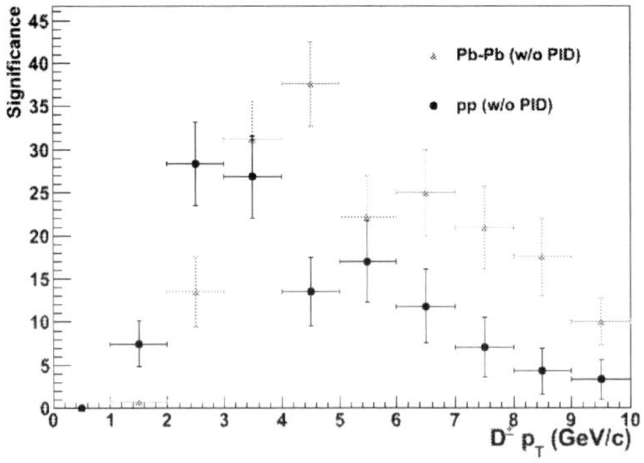

Figure 6. D^+ significance after 1 year of data taking in pp (14 TeV) collisions and Pb-Pb (5.5 ATev) collisions

D^* and D_s reconstruction

D_s reconstruction is very challenging, as the yield of D_s against the huge combinatorial background is low. But D_s is likely to decay in a final state $K^-K^+\pi^+$ through an intermediate resonant states, so the analysis can take advantage from this to separate signal from background. The analysis will be performed in the channel $D_s \to \phi\pi^+ \to K^-K^+\pi^+$, where simulations show[11] that reconstruction of D_s is feasible, even without PID, for reconstructed p_T as low as 3-4 GeV/c.

D^{*+} has a large branching ratio ($\sim$ 67%) for the channel $D^{*+} \to D^0\pi^+_{soft} \to K^-\pi^+\pi^+_{soft}$. The analysis technique is based on the reconstruction of the D^0 and on the finding of a suitable π track of low momentum (π_{soft}). An estimation of the significance for the D^* after one year of data taking is provided in figure 7.

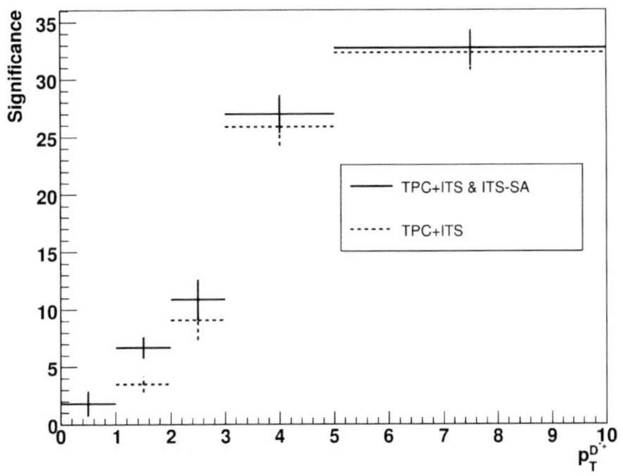

Figure 7. D^* significance after 1 year of data taking in pp (14 TeV) collisions, ($\sim 10^9$ events). Solid: PID performed with both TPC and ITS, dotted: ITS only.

4. Energy Loss

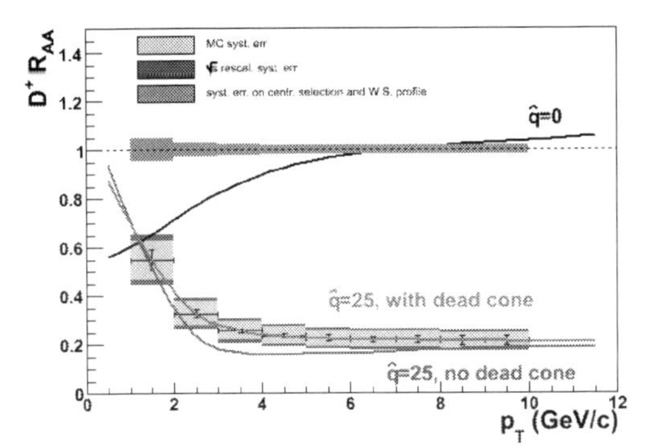

Figure 8. Nuclear modification factor for D^+ meson, 1 year of data taking, energy loss calculation from[12]

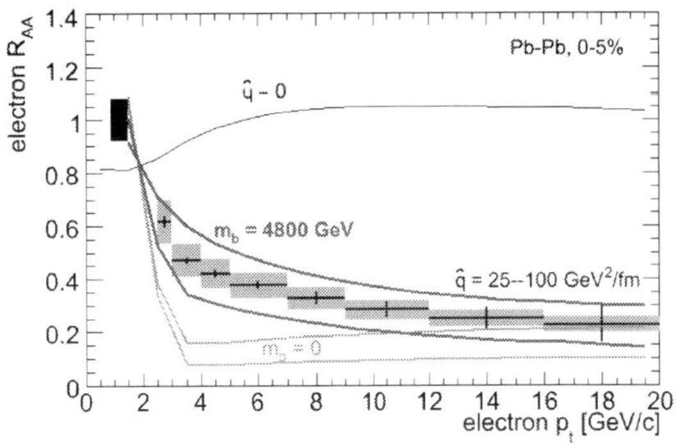

Figure 9. Nuclear modification factor for B meson, 1 year of data taking, energy loss calculation from[12]

Parton energy loss is expected to depend strongly on the properties of the medium (gluon density, volume) and of the probe (colour charge, mass) travelling through it. Charm and beauty quarks are expected to be qualitativerly different probes with respect to light partons, as their energy loss is reduced due to mass effects. In figures 8 and 9 is shown the expected measure of the nuclear modification factor after 1 year of data taking for the D^+ and the electrons coming from B decay. The ratio $R_{Dh}(Pt) = \frac{R^D_{AA}(Pt)}{R^h_{AA}(Pt)}$, shown in figure 10 is an interesting probes of the colour charge dependance of the energy loss, as ligth hadrons mainly come from gluons, carrying a stronger colour charge. Also, the ratio $R_{BD}(Pt) = \frac{R^B_{AA}(Pt)}{R^D_{AA}(Pt)}$ (figure 11) is a probe of of the mass dependance of the energy loss, as B and D come from objects with the same colour charge but different mass.

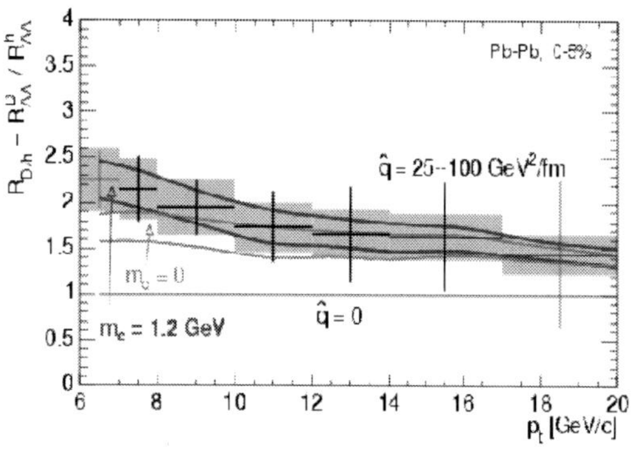

Figure 10. Heavy-to-light ratio, 1 year of data taking, energy loss calculation from[12]

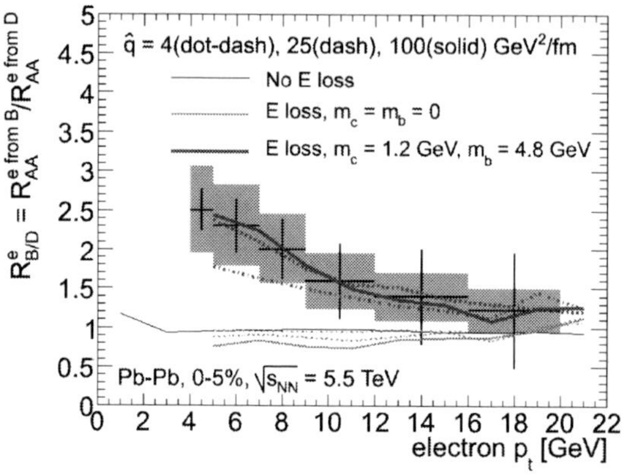

Figure 11. R_{BD}, 1 year of data taking, energy loss calculation from[12]

5. Charm Flow

Flow, and expecially the elliptic flow[13, 14],is an important probe which can provide information about the thermalization of the medium (QGP) produced in a non-central high energy collision. If thermalization is achieved, then collective behaviour should appear. Heavy flavour, due to their higher mass, are less likely to thermalize. Thus, the measurement of the flow of heavy flavours is interesting to study the properties of the QGP.

In figures 12 and 13 is shown the expected measurement of the elliptic flow of D^+ mesons for 2×10^7 Pb-Pb events for two different cases: minimum bias trigger (in the left panel) and semi-peripherical trigger (in the right panel) in the two extreme conditions of no charm flow (meaning that charm doesn't thermalize in the medium) and of charm flowing in the same way as light partons. Statistical error bars are strongly reduced when we move from minimum bias to semi-peripheral ($6 < b < 9$ fm) events, thus $v_2(p_T)$ should be measurable down to $p_T \sim 1GeV/c$.

6. conclusions

With its excellent tracking, vertexing and particle identification capabilities, ALICE has very promising prospectives for the studies of heavy flavour production. It has been shown there is good feasibility of reconstruction of D and B mesons with a good significance in one year of data taking at nominal luminosity. This will allow to measure with good precision the energy loss and the azimuthal anisotropy of those particles.

[1] Karsch F 2001 hep-lat/0106019 and Cheng M *et al.*, 2008 *Phys. Rev. D* **77** 014511.
[2] Mangano M L, Nason P and Radolfi G 1992 *Nucl. Phys. B* **373** 295.
[3] Eskola K J, Kolhinen V J and Salgado C A 1999 *Eur. Phys. J. C* **9** 61.
[4] Armesto N *et al.* 2008 *J. Phys. G* **35** 054001 arXiv:0711.0974

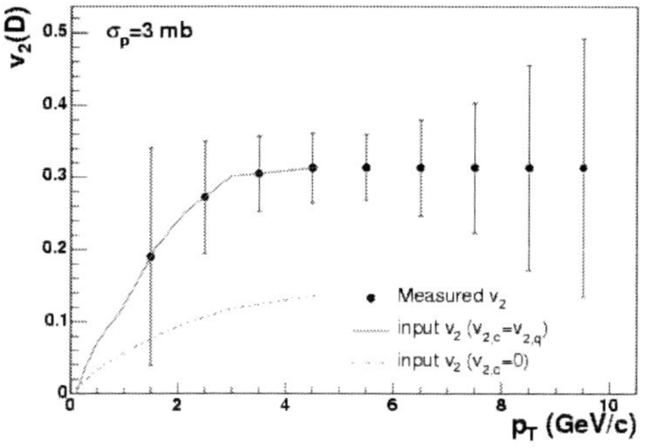

Figure 12. Expected D^+ elliptic flow after 2×10^7 Pb-Pb minimum bias events. Solid: $v_2^D = v_2^q$, dotted: $v_2^D = 0$

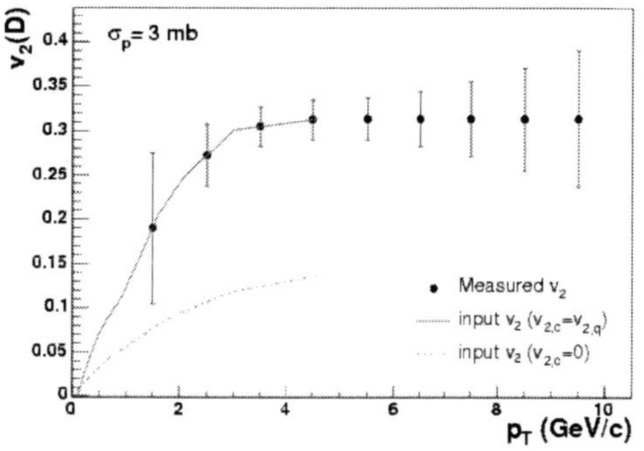

Figure 13. Expected D^+ elliptic flow after 2×10^7 Pb-Pb semi-peripheral events ($6 < b < 9$ fm). Solid: $v_2^D = v_2^q$, dotted: $v_2^D = 0$

[5] Amsler C *et al.* (Particle Data Group), PL **B667**. 1 (2008) and 2009 partial update for 2010 edition
[6] ALICE Collaboration *The Alice Experiment at CERN LHC* JINST 0803: S08002, 2008.
[7] ALICE Collaboration, *Physics Performance Report vol. II*, CERN/LHCC 2005-030 and *J. Phys. G***32** 1295 (2006).
[8] Dainese A, Ph. D. Thesis and arxiv:nucl-ex/0311004.
[9] Bruna E, Ph. D. Thesis and arxiv:nucl-ex/0703005v2.
[10] Prino F, arxiv:nucl-ex/08103086v1.
[11] Senyukov S, Ph. D. Thesis, University of Torino.
[12] Armesto N *et al.* 2005 *Phys. Rev. D* **71** 054027.
[13] Ko C M, Chen L W and Zhang B W 2007 *Braz J. Phys.* **37** 969.
[14] Masera M, Ortona G, Poghosyan M and Prino F 2009 *Phys. Rev. C* **79** 064909.

Journal of Physics: Conference Series **230** (2010) 012008 doi:10.1088/1742-6596/230/1/012008

Reconstructed Jets at RHIC

Sevil Salur

Department of Physics, University of California, One Shields Avenue, Davis, CA 95616

E-mail: salur@physics.ucdavis.edu

Abstract. To precisely measure jets over a large background such as pile up in high luminosity p+p collisions at LHC, a new generation of jet reconstruction algorithms is developed. These algorithms are also applicable to reconstruct jets in the heavy ion environment where large event multiplicities are produced. Energy loss in the medium created in heavy ion collisions are already observed indirectly via inclusive hadron distributions and di-hadron correlations. Jets can be used to study this energy loss in detail with reduced biases. We review the latest results on jet-medium interactions as seen in A+A collisions at RHIC, focusing on the recent progress on jet reconstruction in heavy ion collisions.

1. Introduction

Inclusive hadron distributions and di-hadron correlations at high transverse momentum were utilized to measure jet quenching at RHIC. While these measurements are used to avoid the complex backgrounds of high multiplicity heavy ion environments they are biased towards the jet fragmentation particles that has the least interaction with the medium. Full jet reconstruction in A+A collisions can overcome theses geometric biases as the energy flow is measured independently of the fragmentation details. New observables such as energy flow, jet substructure and fragmentation functions that can be measured in multiple channels (inclusive, di-jets, h-jets and gamma-jets) and these measurements is expected to significantly reduce the effect of the geometric biases. In this article we review the current status of the new techniques developed for full jet reconstruction in heavy ion collisions at RHIC. Further experimental details of jet reconstruction measurements utilizing STAR and PHENIX experiments can be found in these proceedings [1, 2, 3] and from the earlier publications [4, 5, 6, 7, 8, 9, 10, 11, 12, 13, 14, 15, 16].

2. Jet Measurements

Jet reconstruction algorithms are constantly improved during the last 20 years. Experimentally and theoretically consistent jet definitions are used to search for jets in leptonic and hadronic colliders and to calculate their expected cross-sections. Cone, sequential recombination and gaussian filtering algorithms are explored at RHIC. For an overview of jet algorithms in high energy collisions, see [17, 18, 19, 20, 21, 22, 23, 24, 25] and the references therein.

Figures 1 and 2 show the inclusive differential cross sections for $p+p \rightarrow jet+X$ at $\sqrt{s}$ =200 GeV versus jet p_T for both the PHENIX and STAR experiments. Jets that are shown in Figure 1 with a wide p_T range ($5 < p_T < 65$ GeV) are reconstructed via a gaussian filtering algorithm with a parameter of $\sigma = 0.3$ at the PHENIX detector. This cross-section is compared to jets reconstructed with a cone radius of 0.4 from STAR at the same center of mass energy shown as magenta circles [26]. Both jet cross-sections agree well with each other within their statistical and systematic uncertainties. The STAR cone jets were compared to NLO pQCD cross-section using the CTEQ6M parton distributions [26]. These calculations (also shown in Figure 1 as the solid line) show a satisfactory agreement for cross-sections over 7 orders of magnitude [26, 27]. New measurements from STAR experiment for the sequential recombination algorithm jets using FastJet suite of algorithms are presented in Figure [12, 28]. Resolution parameter of 0.4 is used for both k_T and anti-k_T algorithms and these cross-sections also agree well within the uncertainties with the previously published cone jets shown as stars in Figure 2 reconstructed with the STAR experiment [26].

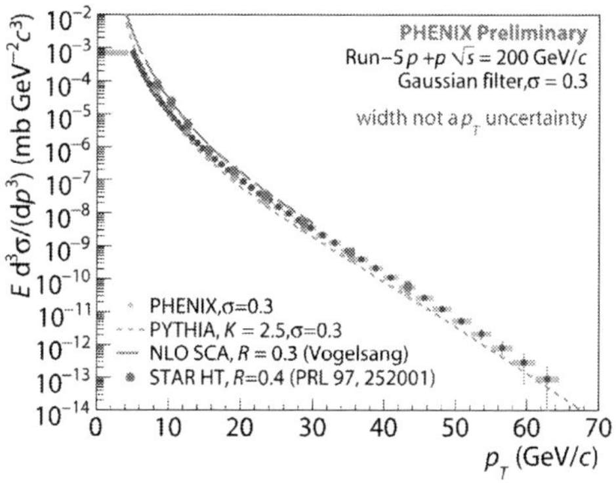

Figure 1. Jet cross-section in p+p collisions at $\sqrt{s} = 200$ GeV in the PHENIX detector compared with the jet measurement by the STAR detector [26, 13]. Pythia jets and next to the leading order calculations are shown in dashed and solid lines [29, 27].

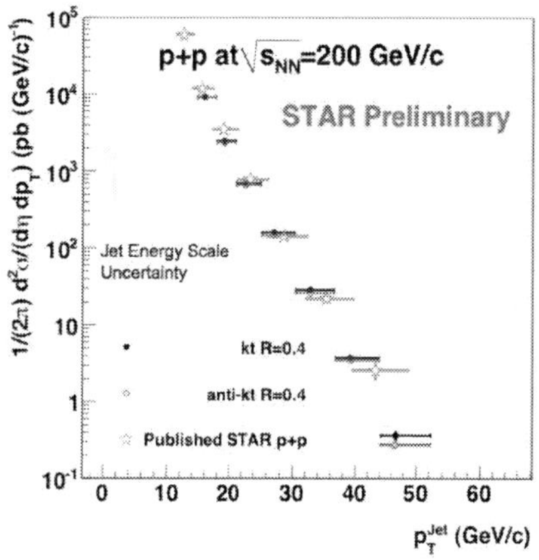

Figure 2. Comparison of the jet cross-sections for the p+p collisions obtained by the sequential recombination (k_T and anti-k_T shown as circles) algorithms and the cone algorithm shown as stars from STAR [12, 26].

Jet measurements require various levels of corrections depending on the detector subsystems and jet algorithms response before they can be compared to theoretical calculations. Many of the preliminary estimation of these corrections are applied to the jets reconstructed in p+p collisions shown in Figures 1 and 2. In the following sub-sections we review the main corrections for the jet measurements. We will concentrate this discussion with the STAR detector system only.

2.1. Particle and Jet Level Corrections

Jets are reconstructed using cone and sequential recombination algorithms that cluster reconstructed time projection chamber (TPC) tracks and barrel electromagnetic calorimeter (BEMC) energy deposits. Corrections for double-counting of energy due to hadronic energy deposition in the BEMC of minimum ionizing particles and electrons is required at the particle level before running the jet finders to estimate both the jet energy and the underlying diffuse background of events correctly [30]. Limited efficiency of tracking and unobserved neutral energy such as K_L^0 mesons and neutrons will result into an energy shift and is required to be corrected at the jet level with Monte-Carlo studies and with data driven ways for example by utilizing the measured protons and charged kaons. Similarly BEMC calibration scale is required to be assessed and corrected as it also results into an energy shift in the jet spectrum. It is possible to deconvolve the resulting gaussian smearing purely due to detector effects such as TPC momentum and BEMC energy resolution from the jet distributions [31]. Fake jets that are purely due to random clustering of event fluctuations have to be identified and removed from jet spectrum statistically. Detailed studies to estimate the fake jets with randomization of particles in an event and di-jets leads to similar results with the major contribution of fake jets for jets with a p_T below 15 GeV in the heavy-ion collisions [2, 12].

2.2. *Underlying Heavy Ion Event Background*

The complex heavy ion background makes the full jet reconstruction a challenging task. Assumptions required by the full jet reconstruction in these complex environments require further investigation of the algorithmic responses of full jet reconstruction by utilizing various reconstruction algorithms. A fundamental requirement that the signal and the background are two separable components can be violated by the presence of the jets and their effect on the background estimation. For example the initial state radiation, even though expected to be small compared to jet energy, might be different in A+A relative to p+p.

Initial state processes resulting in the enhancement of the multiplicity of the underlying events appears to be distributed uniformly in the simpler p+p case [10] and therefore it is fully accounted for the estimation of the background under jets [19, 32]. However in the A+A collisions, "the p+p correspondent" underlying event may be modified, possibly generating non-uniform structures. These non-uniformities in the background might be even larger due to the final state processes. Energy loss of the jet in matter might modify the event shape, resulting in non-uniform structures such as the ridge.

Azimuthal and longitudinal anisotropy of heavy ion events will also result into non-uniform backgrounds. Some of these sources of correlated backgrounds can be brought under quantitative control by using different collision systems. On the other hand, other observed effects might help us to understand details of the jet interactions with the heavy ion environment and may give further insight into the structures that are observed in di-hadron correlations and their origins.

2.3. *Biases or Physics Observables*

The ultimate goal of full jet reconstruction is to investigate the jet quenching in heavy ion collisions at the partonic level, without any ambiguities being introduced by hadronization and geometric biases of the inclusive spectrum and di-hadron measurements. However, it is possible that new biases can be introduced when reconstructing jets. For example, all jet algorithms have various parameters for searching and defining jets, and the effects of varying these parameters need to be explored in detail for a full understanding of jet reconstruction and for a systematic study of jet broadening effects.

A bias will be introduced while trying to reduce the effect of the background fluctuations in heavy ion collisions with the minimum threshold cuts on the track momenta and BEMC tower energies (p_T^{cut}). Comparison of the jet spectra with a variation of p_T^{min} threshold cuts in Au+Au and the N_{Binary} scaled p+p collisions show a good agreement between Au+Au and N_{Binary} scaled p+p jet measurements for the lowest value of the p_T^{min} cut but a much poorer one with the larger p_T threshold cut [5, 6]. This suggests that the threshold cuts introduce biases which are not fully corrected by the procedures that use fragmentation models that are developed for $e^+ + e^-$ and p+p collisions.

The resolution parameter or cone size, which restricts the area of the jet and thereby the amount of energy flow, can be a harder parameter to calculate hence interpret in heavy ion collisions than in p+p collisions. If the jets are broader in the heavy ion environment, the same resolution parameter might not be sufficient to recover the same fraction of jet energy in comparison to p+p jets. This bias needs to be investigated by varying the resolution parameter and by looking into the jet profile of these jet definitions. Variation of these biases introduced by both resolution parameter and p_T^{min} cut might serve useful to study the evaluation of the quenching on jets to determine the distribution of the lost energy [33].

Another bias is introduced with the event selection. For example to enhance the recorded rate of high p_T particles and jets, events above some threshold in the BEMC are collected. (This threshold is 5.4 GeV for the STAR experiment during 2006 and 2007 runs.) This is very similar to the case of jets that are reconstructed with seeded infrared unsafe algorithms. The online BEMC tower triggers introduce a strong bias of reconstructed jets that are fragmenting hard in comparison to the jets that are reconstructed without a seed. New observables such as intra-jet energy distributions, away side of the tower triggered di-jets and hadron-jet correlation studies can be used to further elucidate the effect of these biases.

3. Jet Results in Heavy Ion Collisions

Uncorrected jet spectra reconstructed via a gaussian filtering algorithm in four centrality classes of Cu+Cu collisions at $\sqrt{s_{NN}} = 200$ GeV collected by the PHENIX experiment is shown in Figure 3 [13]. Unfolded and corrected jet results are presented later in [14]. The Figure 4 shows the comparison of the inclusive jet spectra reconstructed by $\mathrm{k_T}$ and anti-$\mathrm{k_T}$ sequential recombination algorithms for the most central Au+Au collisions collected by the STAR experiment [12]. Systematic uncertainties due to the unfolding procedure are shown as the envelopes in red and black and the jet energy resolution as the yellow bar. Jet spectra are consistent between the two sequential reconstruction algorithms extending to 50 GeV kinematic reach.

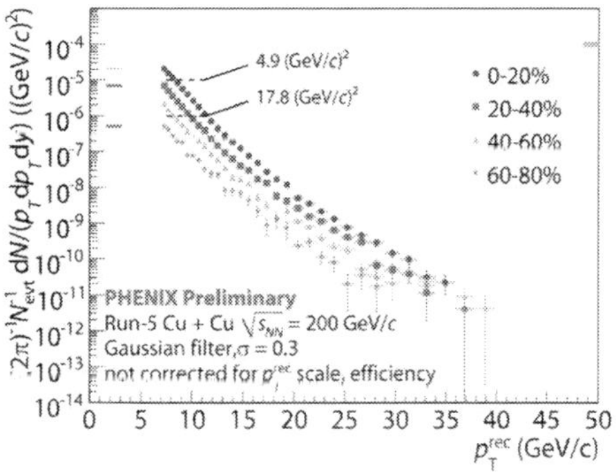

Figure 3. Raw jet yield vs jet p_T obtained by the gaussian filtering algorithm after fake jet correction for $0-20\%$, $20-40\%$, $40-60\%$ and $60-80\%$ centralities of Cu+Cu collisions [13].

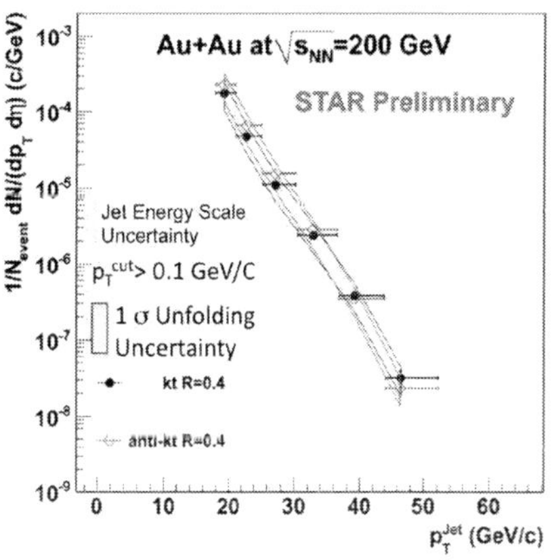

Figure 4. Jet yield per event vs transverse jet energy for the central Au+Au collisions obtained by the sequential recombination ($\mathrm{k_T}$ and anti-$\mathrm{k_T}$) algorithms from STAR .

For the full jet reconstruction in heavy ion collisions, $\mathrm{N_{Binary}}$ scaling as calculated by a Glauber model [34] is expected if the reconstruction is unbiased, i.e. if the full jet energy is recovered independent of the fragmentation details, even in the presence of strong jet quenching. This scaling is analogous to the cross section scaling of high p_T direct photon production in heavy ion collisions observed by the PHENIX experiment [35].

The nuclear modification factor ($\mathrm{R_{AA}}$) for the jets defined with a gaussian filtering algorithm in the smaller system sizes such as Cu+Cu collisions are presented in Figure 5. While the experimental uncertainties are large, a trend towards a larger suppression reaching that of single particle suppression for higher centralities is observed. This much suppression implies that the gaussian filtering algorithm with $\sigma = 0.3$ seems to be geometrically biased like the di-hadron correlations and is only recovering a small fraction of the jet energy rather than the full jet energy.

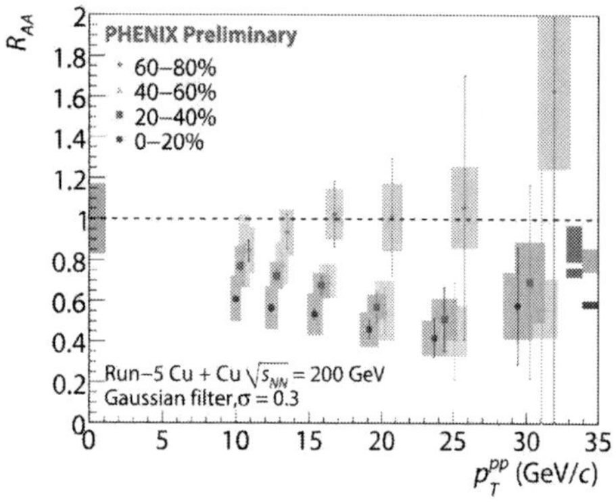

Figure 5. Centrality dependence of the nuclear modification factors of reconstructed jets (Cu+Cu collisions divided by N_{Binary} scaled p+p collisions at $\sqrt{s_{NN}} = 200$ GeV) measured with the PHENIX detector [14].

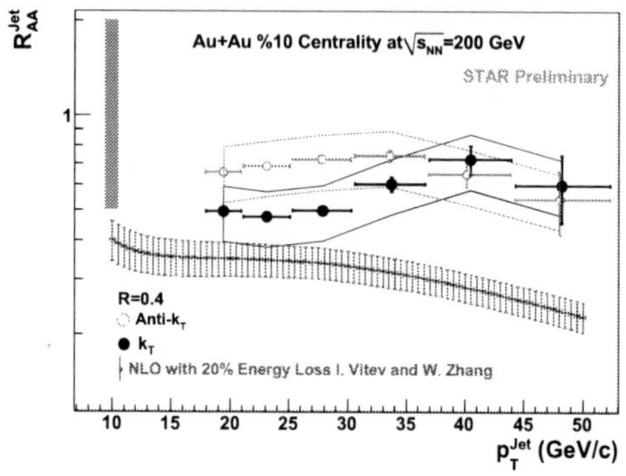

Figure 6. Momentum dependence of the nuclear modification factor of jet spectra reconstructed with $\mathrm{k_T}$ and anti-$\mathrm{k_T}$ algorithms (0-10% most central Au+Au divided by $\mathrm{N_{Binary}}$ scaled p+p collisions) at the STAR detector and compared with a partonic level NLO calculation [12, 36].

The nuclear modification factor for the reconstructed jet spectra with the STAR detector for the most central Au+Au collisions at $\sqrt{s_{NN}} = 200$ GeV can be found in Figure 6. The envelopes shown around the data points for the resolution parameter of 0.4 from $\mathrm{k_T}$ and anti-$\mathrm{k_T}$ jets represent the one sigma uncertainty of the deconvolution of the heavy ion background fluctuations. The total systematic uncertainty on the jet energy scale is around 50%, shown as the gray band. Since the $\mathrm{R_{AA}}$ measurements are from two separate years main uncertainty due to the BEMC calibration do not cancel as naively expected from the ratio. How to improve the jet energy resolution at STAR is currently being explored. The momentum dependence of the nuclear modification factor is observed to be larger than the amount of suppression of π meson $\mathrm{R_{AA}}$. A hint of a suppression of jet $\mathrm{R_{AA}}$ below 1 for jets with p_T above 30 GeV can also be observed with the large uncertainties in mind. It is expected that for a smaller resolution parameter in the sequential recombination algorithms, this suppression should reach to single particle suppression at large momentum.

The systematic variation of the size of the resolution parameter and the suppression amounts in nuclear modification factors can be used to evaluate the quenching effects and the expected amount of jet broadening. The jet $\mathrm{R_{AA}}$ of STAR is compared with next-to-leading order calculation by I. Vitev and B.W. Zhang as presented by the blue histogram in Figure [36]. Error bars around the theoretical data points are due to variation of the energy loss by 20%. Details of these partonic level calculations and system size and resolution parameter dependences can be found in [36]. While the hadronization effect is expected to be small in a nuclear modification factor, STAR data points are systematically higher than that of the calculation. Improvements in the jet energy scale from the experimental side and the addition of hadronization effects from the theoretical side is required to investigate further if there is a real discrepancy in data and model calculation due to other unexplored quenching effects.

4. Conclusions

Large momenta reaching 50 GeV can be studied in heavy ion collisions for the first time with the full jet reconstruction. New physics effects should be considered when interpreting the results at large momentum. For example the momentum dependence of the relative contributions of quark and gluon sub-processes to inclusive jet production might be altered. The quark and gluon contributions already vary with respect to the jet momentum in elementary p+p collisions [37]. The relative contributions might be even more different in a heavy ion environment when a quark gluon plasma is produced, affecting the expected shape of the jet spectra and therefore of the nuclear modification factors. Another possibility is that at large momentum fraction x, initial state effects (such as the EMC effect which is the deviation between structure functions of heavy ions to light ions) are observed to be as large as 15% [38]. Some other contributions like the EMC effect might be playing a major role in the relative suppression or enhancement of nuclear modification factors at large momentum.

The new Monte-Carlo based simulations of jet quenching in medium such as Jewel [39], Q-Pythia [40] and YaJEM [41] and complementary analytic calculations [33, 36, 42, 43] recently became available to pursue a quantitative analysis of jet quenching as observed in heavy ion collisions. However there are many uncertainties (e.g., how hadronization is treated) in the predictions of these models and calculations. To confront the calculations with data, new robust QCD jet observables that are unaffected by the p_T cuts and hadronization need to be explored experimentally. For example the subjet observable is infrared safe and insensitive to hadronization and will be used to study the jet quenching [39].

The studies shown here indicate that reconstruction of jets with a uniquely large kinematic limit is possible in heavy ion events. Jet reconstruction in heavy ion collisions is not yet free of biases due to selection of particles such as p_T cuts to reduce the fluctuations of heavy ion background, requirement of algorithmic parameters such as cone size, gaussian width or resolution parameter, and the collection of events with thresholds to enhance the jet rates. Multiple channels like di-jets, h-jets, gamma-jets to measure qualitatively new observables such as energy flow, jet substructure and fragmentation functions will help to to assess fully the systematic uncertainties of jet measurements and the scale of jet broadening.

Acknowledgments

The author wishes to thank Ivan Vitev for the valuable discussions and providing the theoretical calculations and the organizers for the invitation and for the fruitful meeting.

[1] Y. Shi for the PHENIX Collaboration, these proceedings.
[2] E. Bruna for the STAR Collaboration, these proceedings.
[3] A. Timmins for the STAR Collaboration, these proceedings.
[4] S. Salur, "Searching for Jets in Ultra-Relativistic Heavy Ion collisions", 24th Winter Workshop on Nuclear Dynamics Proceedings, South Padre, TX (2008), arXiv:0809.1606.
[5] S. Salur, "Jet Reconstruction in Heavy Ion Collisions", 25th Winter Workshop on Nuclear Dynamics Proceedings, Big Sky, Montana, (2009), arXiv:0905.1917.
[6] S. Salur for the STAR Collaboration, Eur. Phys. J. C61:761-767 (2009), arXiv:0809.1609; S. Salur for the STAR Collaboration, Eur. Phys. J. C62:119-125 (2009), arXiv:0810.0500.
[7] J. Putschke for the STAR Collaboration, Eur. Phys. J. C61:629-635 (2009), arXiv:0809.1419.
[8] E. Bruna for the STAR Collaboration, *Nucl. Phys.* A830:267C-270C (2009), arXiv:0907.4788.
[9] E. Bruna for the STAR Collaboration, "Jet fragmentation in STAR going from p+p to Au+Au", 25th Winter Workshop on Nuclear Dynamics Proceedings, Big Sky, Montana, (2009), arXiv:0905.4763.
[10] H. Caines for the STAR Collaboration, *Nucl. Phys.* A830:263C-266C (2009), arXiv:0907.3460.
[11] J. Kapitan for the STAR Collaboration, *Nucl. Phys.* A830:619C-620C (2009), arXiv:0907.3830.
[12] M. Ploskon for the STAR Collaboration, *Nucl. Phys.* A830:255C-258C (2009), arXiv:0908.1799
[13] Y. Shi for the PHENIX Collaboration, *Nucl. Phys.* A830:251C-254C (2009), arXiv:0907.4725.

[14] Y. Shi for the PHENIX Collaboration, "Direct jet reconstruction in p + p and Cu + Cu at PHENIX", e-proceedings of DPF 2009, Detroit, MI, arXiv:0911.3399 (2009).
[15] S. Salur, *Nucl. Phys.* A830:139C-146C (2009), arXiv:0907.4536.
[16] S. Salur, "Full Jet Reconstruction in Heavy Ion Collisions: Prospects and Perils", e-proceedings of DPF 2009, Detroit, MI, arXiv:0910.2081 (2009).
[17] S.D Drell, D. J. Levy and T.M. Yan, *Phys. Rev.* 187, 2159 (1969); N. Cabibbo, G. Parisi and M. Testa, Lett. Nuovo Cimento 4,35 (1970); J.D. Bjorken and S.D. Brodsky, *Phys. Rev.* D 1, 1416 (1970); G. Sterman and S. Weinberg, *Phys. Rev. Lett.* 39, 1436 (1977) .
[18] M. H. Seymour, CERN-TH/95-176, hep-ph/9506421 (1995).
[19] M. Cacciari, G. P. Salam, G. Soyez, *JHEP* 0804:005 (2008), arXiv:0802.1188; and references therein.
[20] B. Flaugher and K. Meier. FERMILAB-CONF-90-248-E, Dec (1990).
[21] N. Brown, W. J. Stirling *Z.Phys.* C 53:629-636 (1992).
[22] S. Catani, Y. L. Dokshitzer, M. Olsson, G. Turnock and B. R. Webber, *Phys. Lett.* B 269, 432 (1991); S. Catani, Y. L. Dokshitzer, M. H. Seymour and B. R. Webber, *Nucl. Phys.* B 406, 187 (1993); S. D. Ellis and D. E. Soper, *Phys. Rev.* D 48, 3160 (1993).
[23] G. C. Blazey et al. FERMILAB-CONF-00-092-E, hep-ex/0005012 (2000).
[24] C. F. Berger, T. Kucs, and G. Sterman, *Phys. Rev.* D 68, 014012 (2003).
[25] Yue-Shi Lai and Brian A. Cole, arXiv:0806.1499 (2008).
[26] B. I. Abelev, et. al. STAR Collaboration *Phys. Rev. Lett.* 97 252001 (2006).
[27] B. Jager, M. Stratmann, and W. Vogelsang, *Phys. Rev.* D 70, 034010 (2004).
[28] M. Cacciari, G. P. Salam, G. Soyez, http://www.lpthe.jussieu.fr/~salam/fastjet.
[29] T. Sjostrand, S. Mrenna, P. Skands, Comput. Phys. Commun. 178, 852-867 (2008); T. Sjostrand, S. Mrenna, P. Skands hep-ph:0710.3820 (2007); T. Sjostrand, P. Eden, C. Friberg, L. Lonnblad, G. Miu, S. Mrenna, E. Norrbin *Comput. Phys. Commun.* 135, 238-259 (2001).
[30] J. Adams et. al. STAR Collaboration *Phys. Rev.* C70:054907 (2004), nucl-ex/0407003.
[31] A. Bhapet. al.Nucl. Instrum. Meth. A566:375-412 (2006)
[32] M. Cacciari, G. P. Salam, S. Sapeta, arXiv:0912.4926 (2009).
[33] I. Vitev, B. W. Zhang, S. Wicks, Eur. Phys. J. C 62:139-144 (2009), arXiv:0810.3052.
[34] R. J. Glauber, *Phys. Rev.* 100, 242 (1955).
[35] PHENIX Collaboration, S. S Adler et al. *Phys. Rev. Lett* 96 202301 (2006), nucl-ex/0601037.
[36] I. Vitev, B. W. Zhang, *Phys. Rev. Lett.* 104:132001 (2010), arXiv:0910.1090.
[37] W. Vogelsang Private Communication
[38] J. Gomez et al., SLACPUB5813 June 7, (2001); D.F. Geesaman et al., Ann. Rev. Nucl. Part. Sci. 45, 337 (1995); B. A. Cole. et al, hep-ph/0702101 (2007).
[39] K. Zapp, G. Ingelman, J. Rathsman, J. Stachel, U. A. Wiedemann arXiv:0805.4759 (2008); K. Zapp, G. Ingelman, J. Rathsman, J. Stachel, U. A. Wiedmann, Eur. Phys. J. C60:617-632 (2009), arXiv:0804.3568.
[40] N. Armesto, L. Cunqueiro and C. A. Salgado Eur. Phys. J. C 61:775-778 (2009), arXiv:0809.4433; N. Armesto, L. Cunqueiro, C.A. Salgado, arXiv:0907.1014 (2009).
[41] T. Renk, *Phys. Rev.* C80:044904 (2009), arXiv:0906.3397; T. Renk, arXiv:0808.1803 (2008).
[42] I. Vitev, S. Wicks, B. W. Zhang *JHEP* 0811:093 (2008), arXiv:0810.2807.
[43] N. Borghini *Phys. Lett.* B 678:82-85, (2009), arXiv:0902.2951.

Journal of Physics: Conference Series **230** (2010) 012009 doi:10.1088/1742-6596/230/1/012009

Overview on jet results from STAR

Elena Bruna, for the STAR Collaboration
Physics Department, Yale University
272 Whitney Avenue, New Haven, CT-06511 USA
E-mail: elena.bruna@yale.edu

Abstract. Full jet reconstruction allows access to the parton kinematics over a large energy domain and can be used to constrain the mechanisms of energy loss in heavy-ion collisions. Such measurements are challenging at RHIC, due to the high-multiplicity environments created in heavy-ion collisions. In these proceedings, we report an overview of the results on full jet reconstruction obtained by the STAR experiment. Jet measurements in 200 GeV p+p show that jets are calibrated pQCD probes and provide a baseline for jet measurements in Au+Au collisions. Inclusive differential jet production cross sections and ratios are reported for central 200 GeV Au+Au collisions and compared to p+p. We also present measurements of fully reconstructed di-jets at mid-rapidity, and compare spectra and fragmentation functions in p+p and central Au+Au collisions.

1. Introduction

Jet quenching at RHIC was observed via single hadron and di-hadron suppression at high p_T in Au+Au collisions relative to p+p and d+Au [1]. Although these measurements showed evidence for parton energy loss in the medium, they were limited by the inability to constrain the kinematics of the hard-scattered parton and were biased to select jets coming from the surface of the medium [2] (geometrical bias).

In the last couple of years, a step forward was made toward full jet reconstruction in heavy-ion collisions, enabled by the modern jet-finding techniques well suited for high-multiplicity environment. In these proceedings, we report an overview of the results from full jet reconstruction in STAR [3, 4, 5, 6].

Unbiased full jet reconstruction allows a measurement of the energy of the hard scattering, even in the presence of the underlying heavy-ion collision. Assuming that jets are a well calibrated pQCD probe (i.e. the predicted pQCD energy distribution agrees with reference measurements such as in p+p collisions), the inclusive jet production cross section should follow binary collision scaling from p+p to Au+Au collisions. Unbiased reconstruction of jets should also allow the measurement of the softening fragmentation functions (FF) in Au+Au compared to baseline p+p jets of the same energy, as predicted by theoretical models [7].

2. Experimental techniques

The STAR experiment is well suited for full jet reconstruction analysis thanks to the complete azimuthal coverage and pseudo-rapidity acceptance $|\eta| < 1$ of the Time Projection Chamber (TPC) and the Barrel Electromagnetic Calorimeter (BEMC) that allow the reconstruction of charged particles and neutral energy, respectively.

The results presented in these proceedings are based on p+p year 2006 and central Au+Au year 2007 events.

Two online triggers are used for the p+p data set: the High Tower (HT) and the Jet Patch (JP) trigger. The HT trigger requires, in addition to the minimum bias condition, one tower in the BEMC above a transverse-energy threshold E_t of 5.4 GeV. In the JP trigger the total energy of a larger area of $\Delta\eta \times \Delta\phi = 1 \times 1$ has to be above a threshold of 8 GeV. The two triggers used for the Au+Au data set are the minimum bias trigger (MB) and the HT trigger in the BEMC. For Au+Au, the E_t threshold for HT is 5.75 GeV.

Jets are reconstructed using the "k_T" and "anti-k_T" recombination algorithms that are part of the FastJet package [8, 9] and are collinear and infrared safe. In the "k_T" algorithm, pairs of particles are clustered according to their spatial separation with a distance scale R (resolution parameter of the jet algorithm) and a weight given by the p_T of the particles. In "anti-k_T", the weight is given by the inverse p_T of the particles. Therefore, hard particles are clustered before soft particles. As a consequence, "anti-k_T" jets have a more circular shape and their radius is defined by the R parameter.

A data-driven correction scheme for detector resolution and background effects was developed, as well as a first attempt to address the systematic uncertainties in the measurements [3, 5, 6].

3. Background in jet analyses in Au+Au

The background characterization is a challenge in full jet reconstruction because of the presence of the underlying event in heavy-ion collisions. The p_T of the reconstructed jet is conceptually defined in FastJet as:

$$p_{T,rec}^{jet} = p_{T,true}^{jet} + \rho A \pm \sigma\sqrt{A}$$

Fastjet provides an estimate of the background p_T per unit area ρ, that is subtracted to get the jet component $p_{T,rec}^{jet}$. The fluctuations of the background in a central Au+Au event are approximated by a Gaussian function. The width σ is expected to scale with R and is of the order of 6-7 GeV in case of R=0.4. Non-statistical background fluctuations [10] remain to be included. A resolution parameter $R = 0.4$ suppresses the background without losing a significant fraction of the jet energy (~80% of the jet energy lies within R=0.4 for 20 GeV p+p jets [3]). Upward fluctuations of uncorrelated background particles can be clustered into "fake jets". The fake jet spectrum is estimated using different procedures in the MB and HT Au+Au triggers. In the MB Au+Au data, the fake jet spectrum is estimated by removing the leading particles from the jets and running the jet-finding algorithms after randomizing in azimuth the remaining particles (tracks and towers). The background jet spectrum in di-jet analysis (HT Au+Au data) is given by fake jets and jets originating from additional hard scatterings not associated with the HT trigger. The rate of background jets is estimated from the jet spectrum at 90° in ϕ from the trigger jet.

The background fluctuations are corrected via an unfolding procedure after removing the background jet contribution from the measured jet spectrum, both in MB and HT Au+Au data sets.

4. Jet measurements in p+p and d+Au

The study of jet properties in p+p collisions is important to understand the production mechanism in QCD as well as the hadronization process at RHIC energies. In addition, jet measurements in p+p and d+Au are needed as references to interpret the medium effects in Au+Au. The very good agreement between the measured jet cross section in p+p and the NLO calculation over several orders of magnitude, as shown in Fig. 1 [11], indicates that jets in p+p may be considered as a calibrated probe and can be used as a baseline for comparison to Au+Au results. STAR also reported fragmentation functions (FF) in p+p [3]. The FF as a function of

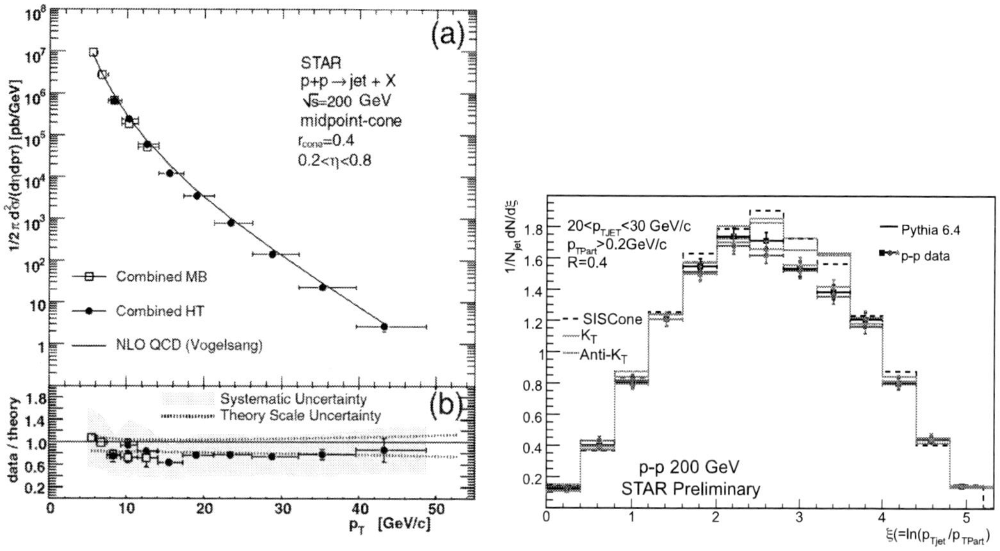

Figure 1. Left: (a) Inclusive differential jet cross sections for MB (open squares) and HT (filled circles) p+p data from the years 2003 and 2004. The curve shows a NLO calculation. (b) Ratio between data and theory. Right: ξ distributions for jets reconstructed in p+p with $20 < p_{T,rec}^{jet} < 30$ GeV/c and R=0.4, compared to PYTHIA for three different jet finding algorithms, k_T (red triangles), anti-k_T (blue circles) and SISCone (black squares).

$\xi = \ln(p_{T,rec}^{jet}/p_{T,hadr})$ for jets with reconstructed p_T in the range 20 - 30 GeV are shown in Fig. 1 (right) for R=0.4, for different jet finding algorithms applied to data and PYTHIA simulations. There is a reasonable agreement between data and PYTHIA. The FF are not corrected yet at the particle level, therefore the comparison is made by passing PYTHIA simulations through STAR's simulation and reconstruction algorithms.

Full jet reconstruction in d+Au allows exploration of possible cold nuclear matter effects, such as an additional k_t broadening. In particular, STAR performed a measurement of $k_{t,raw} = p_{T,1}\sin(\Delta\phi)$ via di-jets [4], with $p_{T,1} > p_{T,2}$. A study based on PYTHIA jets embedded in real d+Au events suggests that the contribution of the jet energy resolution and possible underlying background to the measured uncorrected $k_{t,raw}$ is small. Fig. 2 shows the $k_{t,raw}$ distributions for p+p and d+Au. The similar widths indicate that nuclear effects are rather small.

5. Inclusive jet measurements

The R_{AA}^{jet}, reported in Fig. 3 (left) [5], is the ratio of the jet yield in 0-10% most central MB Au+Au events to the jet yield in p+p scaled by the number of binary collisions. k_T and anti-k_T algorithms were used with two choices of R parameter, 0.4 and 0.2. The kinematic range extends up to about 50 GeV/c, significantly increased as compared to inclusive single particle measurements. The measurements are corrected for detector resolution and background fluctuations. The systematic uncertainty in the jet energy scale is dominated by the BEMC calibration (5%), the unobserved neutral energy (3%) and the momentum resolution of the charged particles (2%).

In the case of an unbiased jet reconstruction, R_{AA}^{jet} is expected to be close to unity because the jet production is a hard process and should scale as N_{bin}. It can be noticed that the R_{AA}^{jet} for R=0.4 recovers more jets compared to single particle R_{AA} and R=0.2. However, even

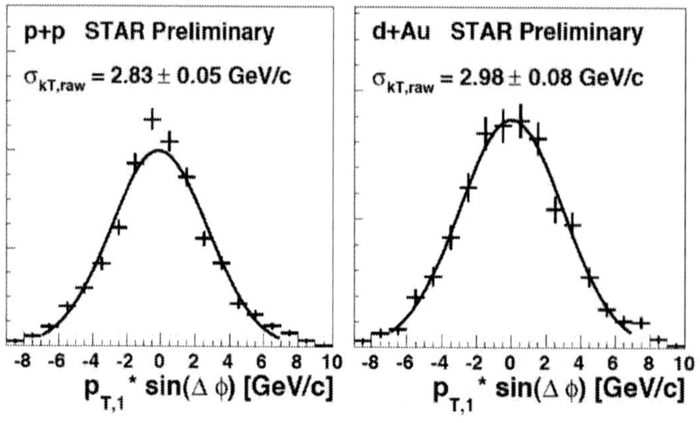

Figure 2. Uncorrected $k_{t,raw}$ for p+p (left) and d+Au (right) collisions, for $10 < p_{T,2} < 20$ GeV/c.

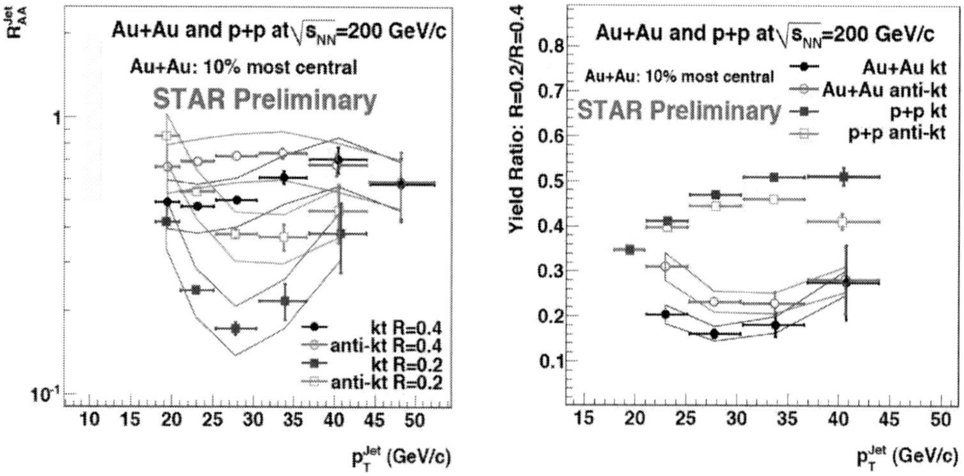

Figure 3. Left: R_{AA}^{jet} for 0-10% most central MB Au+Au events. k_T and anti-k_T algorithms are used with R=0.4 and 0.2. The systematic uncertainty is shown as the vertical yellow band. Solid lines represent systematic uncertainties due to the unfolding of the background fluctuations. Right: Ratio of jet yields R=0.2/R=0.4, using both k_T and anti-k_T, for each collision system.

though the systematic uncertainties are quite large, there are indications that unbiased jet reconstruction is not achieved even for R=0.4 because R_{AA}^{jet} does not approach 1 at large p_T^{jet}. Possible explanations could be an overall absorption or that significant broadening of the jet energy profile beyond R=0.4 occurs in Au+Au, which would cause an underestimation of the reconstructed jet energy in Au+Au relative to p+p. Measurements of the jet energy profile could help to distinguish between the two phenomena. Fig. 3 (right) shows the ratio of the jet yields for R=0.2 over R=0.4 for p+p and Au+Au. The ratio increases with the jet p_T in p+p, consistent with PYTHIA calculations which predict that jets become more collimated as their energy increases. The ratio is significantly suppressed for Au+Au, indicating a broadening of the jet energy profile going from R=0.2 to R=0.4. Assuming a continuous evolution, it is expected that the broadening mechanism extends beyond R=0.4.

Journal of Physics: Conference Series **230** (2010) 012009 doi:10.1088/1742-6596/230/1/012009

6. Di-jet analysis

STAR performed a measurement of di-jets in HT p+p and 0-20% HT Au+Au events [6], using the anti-k_T algorithm with R=0.4. The analysis searched for a "trigger" jet matching the online triggered tower in the BEMC and a "recoil" jet on the away side of the trigger jet (i.e. $\Delta\phi \sim \pi$). For the trigger jet a $p_t^{cut} = 2$ GeV/c was applied to tracks and towers in order to have the same energy scale in p+p and Au+Au. The p_t^{cut} introduces a strong bias in the jet population in Au+Au, causing the trigger jets to more likely come from the surface of the medium. In order to avoid a similar surface bias on the recoil jets, the latter were reconstructed with a minimal $p_t^{cut} = 0.15$ GeV/c given by the TPC acceptance. The requirement of $p_{t,rec}^{jet}(trigger) > 10$ GeV/c for the reconstructed trigger jet minimizes the contribution of "fake" jets. The results of this analysis are not corrected for the difference in tracking efficiency between p+p and Au+Au. An absolute correction to get the parton energy will be performed in the future. However, we applied a data-driven unfolding of the background fluctuations from the Au+Au di-jet spectra and FF of recoil jets to directly compare p+p and Au+Au. The ratio of recoil jet spectra normalized to the number of trigger jets in Au+Au over p+p is shown in Fig. 4 (left). A significant suppression of recoil jets is observed in Au+Au relative to p+p, in contrast with the expected ratio of 1 for unbiased jet reconstruction. This could be explained by a broadening of the jet structure in an area beyond R=0.4 (see Sec. 5), resulting an underestimation of the jet energy within R=0.4 in Au+Au relative to p+p. Such an effect can be attributed to the trigger bias that forces the recoil jets to have large in-medium path-length and hence suffer a maximum energy loss.

STAR also measured the FF of recoil jets in p+p and Au+Au. The FF are measured for charged hadrons within R=0.7 around the jet axis, whereas the reconstructed jet energy is defined for R=0.4. This method includes additional soft fragmentation particles. The charged particles included in the FF in Au+Au contain the jet fragments plus background particles., where the latter must be subtracted in order to measure the jet fragmentation. The background component is estimated on an event-by-event basis from the charged particle spectrum in the area outside the two jets with the highest reconstructed energy. The FF are measured in Au+Au for $p_{t,rec}^{recoil}(AuAu) > 25$ GeV/c. The p+p jets corresponding to $p_{t,rec}^{recoil}(AuAu) > 25$ GeV/c are selected taking the background fluctuations into account, assuming a Gaussian parameterization with $\sigma = 6$ GeV. The mean jet p_t of the selected p+p jets is $p_{t,rec}^{recoil}(pp) \simeq 25$ GeV/c. The ratio of the $z = p_{t,hadr}/p_{t,jet}^{recoil}$ distributions of Au+Au to p+p is shown in Fig. 4 (right).

No significant modification of the fragmentation functions is observed for recoil jets in Au+Au relative to p+p, for $z > 0.2$. This is in contrast with the theoretical expectation of a softening of the FF for an unbiased jet population and with the measured suppression of high-p_t hadrons at RHIC ($R_{AA} \sim 0.2$ [1]), both of which are related to parton energy loss in the medium. The low-z part of the fragmentation function ($z < 0.1$) is still under investigation because it might be affected by a large uncertainty due to background subtraction.

7. Summary

We presented in these proceedings an overview of the STAR results on jet studies. Full jet reconstruction extends the kinematic reach of studies of parton energy loss mechanisms in the dense medium at RHIC. Jet measurements in p+p show that full reconstructed jets are well calibrated tools to probe the medium. Preliminary results from d+Au suggest that there are no major cold nuclear matter effects which would cause additional k_t broadening in di-jet measurements.

A suppression of R_{AA}^{jet} was observed in inclusive jet measurements for $p_{t,jet} > 20$ GeV/c, suggesting that the jet energy is not fully recovered within R=0.4 and R=0.2. This is consistent with the observed decrease of jets with R=0.2 relative to R=0.4 in Au+Au with respect to p+p, indicating a broadening of the jet profile as observed in going from R=0.2 to R=0.4.

In di-jet analysis, the online trigger maximizes the in-medium path of the recoil jets making

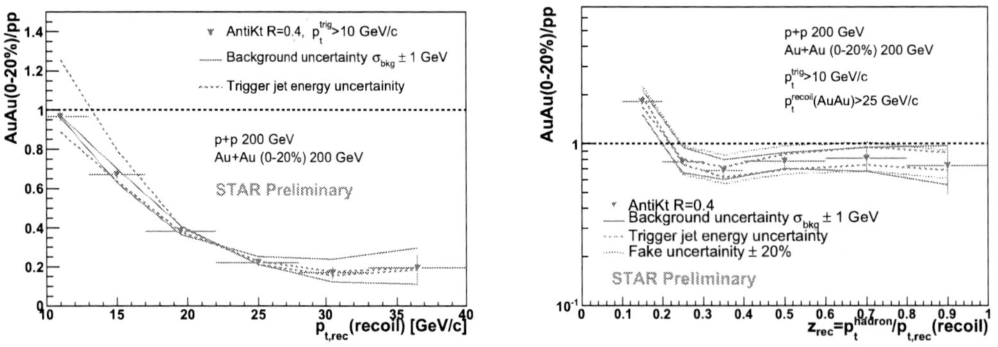

Figure 4. Left: ratio of p_t spectra of recoil jets in Au+Au (corrected for fake jets and background fluctuations) to p+p. The curves indicate the systematic uncertainties in the estimation of the background fluctuations (solid) and of the p_t of the trigger jet assuming a jet-p_t resolution of 25% (dotted) [3]. Right: ratio of the z distributions measured in 0-20% central Au+Au events to p+p collisions for $p_{t,rec}^{recoil}(AuAu) > 25$ GeV/c. The curves indicate the systematic uncertainties in the estimation of the background fluctuations (solid) and of the trigger jet energy (dotted). The uncertainty due to the estimation of fake jets (small dotted curves) was assumed to only affect the normalization of the z distributions.

them more susceptible to energy loss. Given this selection bias, a suppression of the recoil jets in Au+Au relative to p+p was observed. This suppression can be explained with a scenario where the jet is broadened and its energy is not fully recovered in $R = 0.4$ with respect to p+p with the current jet-finding algorithms, causing a shift of the recoil spectrum in Au+Au towards smaller energies. The absence of a strong modification of the FF in Au+Au with respect to p+p could be due to an underestimation of the jet energy within R=0.4 owing to the preceding jet broadening picture.

Future measurements will attempt to determine wether or not jet broadening occurs beyond R=0.4, as suggested by the R_{AA}^{jet} and di-jet spectra measurements. The jet-energy profile will also be investigated in order to improve our understanding of the mechanism of jet broadening in the medium. Further systematic studies, including a detailed background characterization, are being conducted in order to access the maximum possible kinematic range of the hard scattering.

8. Acknowledgments

The author wishes to thank the Bulldog facility at the Yale University.

[1] J Adams et al., STAR Collaboration, Nucl. Phys. A 757 (2005) 102.
[2] T. Renk, K. Eskola, Phys. Rev. C75 (2007) 054910, arXiv:hep-ph/0610059.
[3] H. Caines for the STAR Collaboration, Quark Matter 2009 Proceedings, Nucl. Phys. A830 (2009) 263c, arXiv:0907.3460.
[4] J. Kapitan for the STAR Collaboration, Proceedings EPS 2009 Conference, PoS(EPS-HEP 2009) 041, arXiv:0909.2951.
[5] M. Ploskon for the STAR Collaboration, Quark Matter 2009 Proceedings, Nucl. Phys. A830 (2009) 255c, arXiv:0908.1799.
[6] E. Bruna for the STAR Collaboration, Quark Matter 2009 Proceedings, Nucl. Phys. A830 (2009) 267c, arXiv:0907.4788.
[7] S. Sapeta and U. A. Wiedemann, Eur. Phys. J. C 55, 293 (2008), arXiv:0707.3494.
[8] M. Cacciari, G. Salam, G. Soyez, JHEP 0804 (2008) 005, arXiv:0802.1188.
[9] M. Cacciari, G. Salam, G. Soyez, JHEP 0804 (2008) 063, arXiv:0802.1189.
[10] J. Adams et al., STAR Collaboration, J. Phys. G 32, (2006) L37-L48, arXiv:nucl-ex/0509030v3.
[11] B. Abelev et al., STAR Collaboration, Phys. Rev. Lett. 97 (2006) 252001.

Measurement of jet fragmentation in $p + p$ collision at $\sqrt{s} = 200$ GeV with the PHENIX detector

Yue-shi Lai (for the PHENIX Collaboration)

Columbia University, New York, NY 10027-7061 and Nevis Laboratories, Irvington, NY 10533-2508, USA

E-mail: ylai@phys.columbia.edu

Abstract. Measurement of jet fragmentation property in $p + p$ collisions provides a crucial baseline for the study of a possible, modified fragmentation behavior at the presence of the quark-gluon plasma. We describe the first measurement of jet fragmentation functions in $\sqrt{s} = 200$ GeV $p + p$ collisions at RHIC. This is facilitated by extending the linear least square unfolding technique employed by high energy and nuclear experiments to multidimensional spectra, which allows us to correct the inefficiencies in the jet energy response encountered at detectors such as PHENIX.

1. Introduction

Hadronization of jet production into final state particles, together with the production of hard scattering, are complementary parts of the QCD description of particle production. Fragmentation function in $p + p$ collisions further provides a baseline for any measurement of the fragmentation property of reconstructed jets in heavy ion collisions. The narrow detector aperture and the need to operate within the large multiplicity heavy ion background has traditionally limited the application of direct jet reconstruction at PHENIX. Fragmentation properties of jet production therefore been only measured via two particle correlations by PHENIX [1, 2]. The Gaussian filter jet reconstruction algorithm [3] provides a cone-like algorithm that at the same time reduces the sensitivity to large angle fragments (or the lack of such, due to acceptance limit), and is therefore well-suited for both $p + p$ and heavy ion measurement at PHENIX.

Measurement of fragmentation function using a detector with inefficiency due to the lack of hadronic calorimetry is difficult, since the full jet energy is not directly accessible, but has to be reconstructed from combining track momenta with the electromagnetic cluster energies. While the energy carried by long-lived neutral hadrons such as n, K^0_L are fully lost, the inherent momentum resolution and background in a typical tracking system limits the ability to accurately measure tracks with $p \gg 20$ GeV/c. The resulting large difference in the true and measured jet energy scale is difficult to correct multiplicatively, but can be addressed by using unfolding the measured spectra with the true-to-detector-level jet energy transfer function. This is particularly difficult for spectra that contain additional dimensions than the jet energy, such as the fragmentation function. As such, none of the RHIC experiment, which all lacks hadronic calorimetry, has attempted the measurement of fragmentation functions so far.

Jet reconstruction using the $\sigma_{\text{filter}} = 0.3$ Gaussian filter has been applied to PHENIX for the measurement of the inclusive jet cross section in $p + p$ and jet yield in Cu + Cu [4, 5]. In this proceeding, we report the development of a multidimensional unfolding technique for the energy scale correction of fragmentation functions, and the measurement of $p + p$ fragmentation functions for charged particles and neutral electromagnetic clusters using the PHENIX detector.

2. Jet reconstruction by Gaussian filtering

The iterative cone algorithm with the size parameter R is known to be equivalent to a local maximization of a filter output in (η, ϕ) with a special choice of the angular weight function to be the flat $k(r^2) = \theta(R^2 - r^2)$ and $r^2 = \eta^2 + \phi^2$ (note that unlike $k(r^2)$, the filter kernel is $h(r^2) \propto -\int dr^2 k(r^2) \propto \max(0, 1 - r^2/R^2)$, and not flat) [6, 7]. Similarly, the Gaussian filter uses a Gaussian distributed weighting. The Gaussian weighting takes advantage the property of high-p_T jets being collimated emissions of particles, and enhances the center signal to the periphery, which is more likely to be contaminated by the event background. By avoiding an sharp radial cut-off, the algorithm also becomes analytically collinear and infrared safe (we further verified the practical infrared safety using a procedure analogously to [8]).

In $p + p$ events and without a large event background, the event transverse momentum density is the sum of point-like final state particles $p_{T,i}$

$$p_T(\eta, \phi) = \sum_{i \in F} p_{T,i} \delta(\eta - \eta_i) \delta(\phi - \phi_i). \tag{1}$$

The Gaussian filtering of p_T is the linear-circular convolution of $p_T(\eta, \phi)$ with a Gaussian distribution

$$p_T^{\text{filt}}(\eta, \phi) = \iint d\eta' d\phi' p_T(\eta', \phi') \exp\left(-\frac{(\eta - \eta')^2 + (\phi - \phi')^2}{2\sigma_{\text{filter}}}\right). \tag{2}$$

The output of the filter for a given (η, ϕ) position is the Gaussian weighted transverse momentum in the said event. The local maxima in $p_T^{\text{filt}}(\eta, \phi)$ are the reconstructed jets using the Gaussian filter. More detailed discussion of the property of Gaussian filter for jet reconstruction can be found in [3, 5].

3. Multidimensional unfolding

The true particle level and the measured energy scale in the PHENIX central arm differ due to inefficiencies, the lack of hadronic calorimetry, and the loss of out-of-acceptance fragments due to the $\Delta\eta = 0.7$ pseudorapidity coverage and also the $\Delta\phi = \pi$ partial azimuthal coverage. This difference means that a full unfolding is needed to extract the true energy instead of a first order multiplicative correction.

Linear least square unfolding with the Phillips–Tikhonov regularization [9, 10] has been widely used by e.g. GURU [11] for the unfolding of spectra. Binned spectra can be mathematically regarded as a vector of individual bins, and the linear bin-to-bin migration due to differing energy scales is then conveniently expressed as a matrix. For a typical transfer matrix **A** generated using Monte Carlo detector simulation, if the measured spectrum **b** with the covariance matrix **C** is unfolded using the simple linear least square relation

$$\underset{\mathbf{x}}{\text{minimize}} \quad \|\mathbf{A}\mathbf{x} - \mathbf{b}\|_{\mathbf{C}}^2, \tag{3}$$

($\|\mathbf{u}\|_{\mathbf{C}} = \mathbf{u}^T \mathbf{C}^{-1} \mathbf{u}$ denotes the Mahalanobis distance) the propagation of fluctuations in **A** would result in large, non-statistical fluctuation in the unfolding result **x**. The regularized unfolding determines the unfolded spectrum **x** by solving the minimization problem

$$\underset{\mathbf{x}}{\text{minimize}} \quad \|\mathbf{A}\mathbf{x} - \mathbf{b}\|_{\mathbf{C}}^2 + \tau \|\mathbf{L}\mathbf{x}\|^2, \tag{4}$$

where τ is the regularization parameter and **L** is a linear operator that describes the amount of discontinuity or "noisiness" of the unfolding result.

Spectra in particle and nuclear physics can cross multiple orders of magnitude and a homogeneous **L** would cause the large magnitude part of the spectra to dominate the singular vectors, thus exhausting most degrees of freedom purely to reproduce the variation in magnitudes. This can be solved by either scaling **L** to approximately match the variations in **x**, or to prescale **x**. We follow the approach used by GURU to prescale the unknowns by

$$x_i \mapsto \hat{x}_i = \frac{x_i}{x_i^{\text{ini}}}, \tag{5}$$

Journal of Physics: Conference Series **230** (2010) 012010 doi:10.1088/1742-6596/230/1/012010

and the linear system $\mathbf{A}$ is therefore inversely scaled by

$$a_{ij} \mapsto \hat{a}_{ij} = a_{ij} x_i^{\text{ini}} \tag{6}$$

(column-wise scaling), with the purpose of preventing exceedingly large or small $\hat{x}_i$ to appear numerically.

The typical choice for $\mathbf{L}$ is a second order finite-difference matrix

$$\mathbf{L}^{(2)} = \begin{pmatrix} -1 & 1 & 0 & \cdots & 0 \\ 1 & -2 & 1 & & \vdots \\ 0 & 1 & -2 & & 0 \\ \vdots & & & \ddots & 1 \\ 0 & \cdots & 0 & 1 & -1 \end{pmatrix} \tag{7}$$

The minimization with respect to the second order derivative describes a continuity constraint that restricts the shape of $\hat{\mathbf{x}}$ to be cubic spline-like. The simplest D-dimensional generalization of $\mathbf{L}^{(2)}$ are the isotropic axial derivatives $\partial^2/\partial x_k^2$, $k = 1, \dots, D$, and the minimization with respect to it behaves similarly to the D-dimensional cubic tensor splines. We found this type of regularization to be sufficient for the purpose of unfolding multidimensional spectra with Poisson statistics, other choices used e.g. in image restoration are reviewed in [12].

The solution to (4) can be written as a least square problem with the matrix pencil $(\hat{\mathbf{A}} + \sqrt{\tau}\mathbf{L})$, and can be solved by a SVD of $(\mathbf{\Lambda}^{-1}\hat{\mathbf{A}})\mathbf{L}^{-1} = \mathbf{U\Sigma V}^T$, where $\mathbf{C} = \mathbf{\Lambda\Lambda}^T$ is the Cholesky decomposition of the measurement covariance, and provided $\mathbf{L}$ is nonsingular. The software package GURU provides an implementation of this method for diagonal $\mathbf{C}$ and using $\mathbf{L}^{(2)}$ (which is strictly speaking singular, but the rectangular nature allows it to be approximately inverted by perturbing $\mathbf{L}^{(2)} \mapsto \mathbf{L}^{(2)} + \epsilon\mathbf{I}$, $|\epsilon| \ll 1$). However, a D-dimensional continuity constraint for N measurements consists of at least DN axial derivatives and therefore $\mathbf{L}$ becomes a $DN \times N$ matrix, it is therefore generally not possible to invert $\mathbf{L}$ and solve the multidimensional unfolding in the fashion of GURU.

The least square problem with a matrix pencil $(\mathbf{\Lambda}^{-1}\hat{\mathbf{A}} + \sqrt{\tau}\mathbf{L})$ can also be solved by the generalized SVD (GSVD) [13] of a matrix pair $\text{GSVD}(\mathbf{\Lambda}^{-1}\hat{\mathbf{A}}, \mathbf{L})$, which simultaneously decomposes both matrices into

$$\begin{aligned} \mathbf{\Lambda}^{-1}\hat{\mathbf{A}} &= \mathbf{U}\begin{pmatrix}0 & \mathbf{\Sigma}_1\end{pmatrix}\mathbf{X}^{-1} \\ \mathbf{L} &= \mathbf{V}\begin{pmatrix}0 & \mathbf{\Sigma}_2\end{pmatrix}\mathbf{X}^{-1}. \end{aligned} \tag{8}$$

$\text{GSVD}(\mathbf{\Lambda}^{-1}\hat{\mathbf{A}}, \mathbf{L})$ is therefore closely related to $\text{SVD}((\mathbf{\Lambda}^{-1}\hat{\mathbf{A}})\mathbf{L}^{-1})$, which can be immediately seen by comparing the GSVD of the quotient matrix $(\mathbf{\Lambda}^{-1}\hat{\mathbf{A}})\mathbf{L}^\dagger = \mathbf{U}(\mathbf{\Sigma}_1\mathbf{\Sigma}_2^{-1})\mathbf{V}^T$ ($\mathbf{L}^\dagger$ being the $(\mathbf{\Lambda}^{-1}\hat{\mathbf{A}})$-weighted pseudoinverse of $\mathbf{L}$ [14]) to the ordinary SVD form $(\mathbf{\Lambda}^{-1}\hat{\mathbf{A}})\mathbf{L}^{-1} = \mathbf{U\Sigma V}^T$.

With the generalized singular value pairs $\mathbf{\Sigma}_1 = \text{diag}(\alpha_i)$, $\mathbf{\Sigma}_2 = \text{diag}(\beta_i)$, the SVD singular values of $(\mathbf{\Lambda}^{-1}\hat{\mathbf{A}})\mathbf{L}^\dagger$ are $\gamma_i = \alpha_i/\beta_i$. The solution then can be written as the Tikhonov filter for the singular values

$$\mathbf{\Sigma}'^{-1} = \text{diag}\left(\frac{\gamma_i^2}{\gamma_i^2 + \tau}\alpha_i^{-1}\right), \tag{9}$$

and the mean and covariance matrix of the unfolding solution is

$$\begin{aligned} \hat{\mathbf{x}} &= \mathbf{X\Sigma}'^{-1}\mathbf{U}^T(\mathbf{\Lambda}^{-1}\mathbf{b}) \\ \text{Cov}\,\hat{\mathbf{x}} &= \mathbf{X\Sigma}'^{-1}(\mathbf{\Sigma}'^{-1})^T\mathbf{X}^T. \end{aligned} \tag{10}$$

We implemented this multidimensional generalization to GURU's regularized linear least square unfolding based on LAPACK's DGGSVD [15], which implements the dense GSVD algorithm by Bai, Demmel & Zha [16, 17].

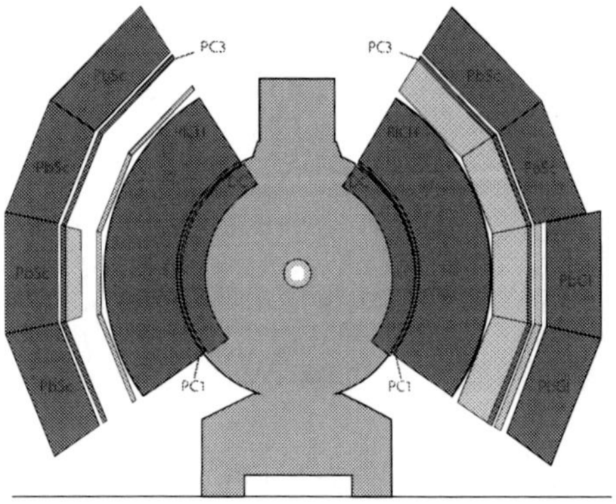

Figure 1. The PHENIX central arm detectors for RHIC Run-5 (year 2004/2005), viewed along the beam axis from the south towards north. Dark regions indicate detectors used for the jet reconstruction: The drift chamber (DC), the pad chamber layers 1 and 3 (PC1/PC3), the ring-imaging Čerenkov detector (RICH), and the Pb scintillator (PbSc) and Pb glass (PbGl) electromagnetic calorimeters.

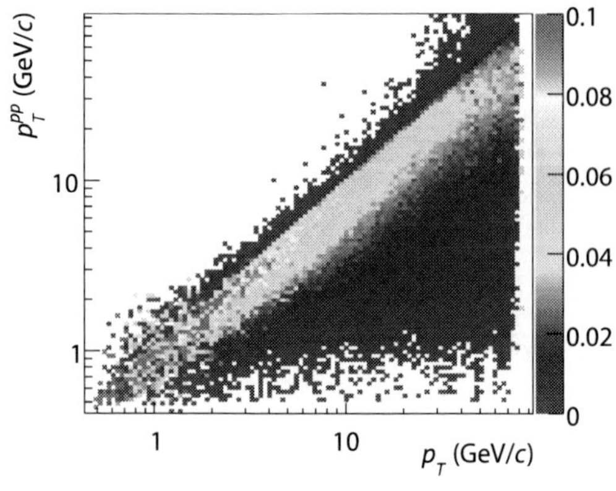

Figure 2. The PHENIX jet $P(p_T^{pp}|p_T)$ transfer matrix for $\sqrt{s} = 200$ GeV and $\sigma = 0.3$ Gaussian filter, derived from the GEANT simulation of $\approx 1.6 \times 10^7$ PYTHIA events. The $p_T^{pp} < p_T$ region is dominated by n, K_L^0 energy loss.

4. Experimental setup

Figure 1 shows the Run-5 PHENIX "central arm" configuration for RHIC Run-5 (year 2004/2005). PHENIX has two mid-rapidity spectrometers with an aperture of $|\eta| < 0.35$ and $\Delta\phi = \pi/2$ each. The components used for jet reconstruction are the drift chamber (DC), the pixel pad chamber layers 1 and 3 (PC1/PC3), the ring-imaging Čerenkov detector (RICH), and the electromagnetic calorimeters (EMCal). For the data presented in this paper, DC/PC1/PC3 provide momentum measurement for charged particles, and the EMCal the energy for photons.

Pattern recognition and momentum reconstruction in the tracking system formed by DC and PC1/3 is performed using the combinatorial Hough transform, with the momentum scale determined by the time-of-flight measurement of identified $\pi^{\pm}$, $K^{\pm}$, and $p/\bar{p}$. The momentum resolution of the tracking system is determined as $\delta p/p = 0.7\% \oplus 1.0\% p/(\text{GeV}/c)$. Two calorimeter technologies were used, six of the total eight sectors are covered by Pb-scintillator (PbSc) calorimeters with resolution of $\sigma_E/E = 8.1\%/\sqrt{E} \oplus 2.1\%$ and a granularity of $\Delta\eta \times \Delta\phi \approx 0.01 \times 0.01$, and two sectors by Pb-glass (PbGl) calorimeters with $\sigma_E/E = 5.9\%/\sqrt{E} \oplus 0.8\%$ and $\Delta\eta \times \Delta\phi \approx 0.008 \times 0.008$. The intrinsic timing resolution for 1 GeV $\pi^{\pm}$ are about 200–300 ps for both technologies.

Since PHENIX currently lacks an in-field tracking capability, conversion electrons in the DC can produce a displaced track that has the appearance a high p_T track originating from the event vertex. We use the information from the RICH and dE/dx measurement to identify and remove these tracks. To provide additional suppression at the cross section level of jets with $p_T > 20$ GeV, we use the fact that conversion electrons are geometrically unlikely to coincide with the direction of the jet production, and require the reconstructed jet to have a minimum multiplicity of 3 particles measured within a radial angle of 60°, and the charged fraction of the jet p_T to be below 0.9 to remove single track events, or when the jet is dominated by a large p_T track.

The absolute energy scale of the calorimeter clusters are determined using both the reconstructed π^0 masses from the observed $\pi^0 \to \gamma\gamma$ decays, and checked using the E/p from energies from cluster matching RICH identified $e^{\pm}$ tracks. Shower shape cuts are used to remove clusters. The residual

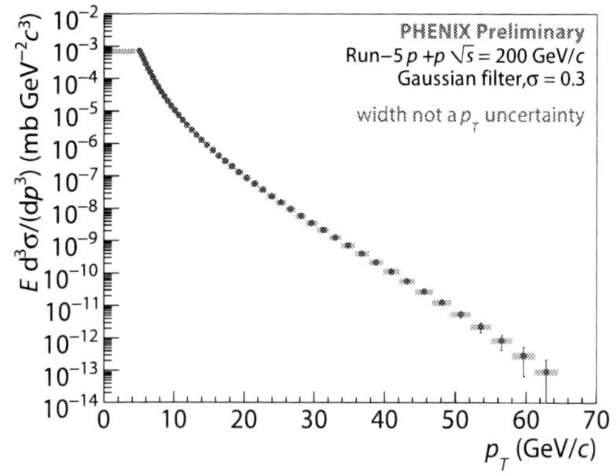

Figure 3. PHENIX Run-5 $p+p$ at $\sqrt{s} = 200$ GeV invariant jet cross section spectrum as a function of p_T. The shaded box to the left indicates the overall normalization systematic uncertainty, shaded boxes associated with data points indicate point-to-point systematic uncertainties, and error bars indicate statistical uncertainties.

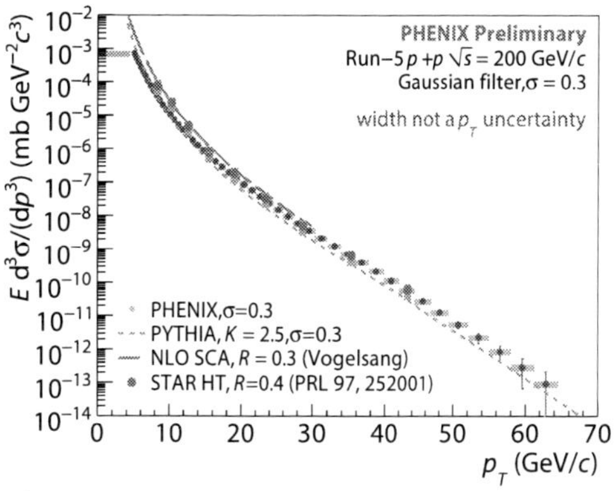

Figure 4. PHENIX Run-5 $p+p$ at $\sqrt{s} = 200$ GeV invariant jet cross section spectrum as a function of p_T, with comparison to [18], next-to-leading order calculation from [19], and PYTHIA assuming $K = 2.5$. The shaded box to the left indicates the overall normalization systematic uncertainty, shaded boxes associated with data points indicate point-to-point systematic uncertainties, and error bars indicate statistical uncertainties.

uncertainty in the energy scale is $\pm 3\%$ (syst.). Since the measurement extends to very low cross sections, processes such as upward beam interaction can deposit energy in the EMCal that randomly coincides with an event. This is suppressed by measuring the time-of-flight of the clusters and rejecting those that are out of synchronization with the collision.

The PHENIX minimum bias (MB) trigger is defined by the coincident firing of the two beam-beam counters (BBC) located at $3.0 < \eta < 3.9$. The Van de Meer/vernier scan method is used to measure the BBC cross section, with $\sigma_{\mathrm{BBC}} = 22.9 \pm 2.3$ mb (syst.). The efficiency of BBC MB trigger on an event containing a jet with $p_T^{\mathrm{rec}} > 2$ GeV/c is $\epsilon_{\mathrm{BBC}} = 0.86 \pm 0.05$ (syst.) and constant with respect to p_T^{rec} within that uncertainty. We require the collision vertex to be within $|z| < 25$ cm along the beam axis, derived from the timing difference between the firing of the two BBC.

5. Inclusive jet cross section in $p+p$

The data presented in the following sections were obtained from the PHENIX $p+p$ dataset from the RHIC Run-5 (year 2004/2005). After removal of bad quality runs, a total of 1.47×10^9 minimum bias $p+p$ and 1.16×10^9 triggered $p+p$ events are being used.

The measurement of inclusive jet cross section in Run-5 $p+p$ collisions uses the combined minimum bias and triggered data set, and is based on the regularized unfolding of the jet spectrum $dN_{\mathrm{jet}}/dp_T^{\mathrm{jet,rec}}$ by the jet energy scale transfer function $P(p_T^{\mathrm{jet,rec}}|p_T^{\mathrm{jet}})$, which is evaluated by GEANT simulation using PYTHIA 6.4.20 events. A total of 1.6×10^7 events were simulated with 14 different minimum Q^2 settings varying between $\sqrt{Q^2} > 0.5$ GeV/c and $\sqrt{Q^2} > 64$ GeV/c. The resulting transfer matrix $P(p_T^{pp}|p_T)$ is shown in figure 2.

The unfolding is equivalent to the inversion of the 1D Fredholm equation

$$\frac{dN_{\mathrm{jet}}}{dp_T^{\mathrm{jet,rec}}} = \int dp_T^{\mathrm{jet}}\, P(p_T^{\mathrm{jet,rec}}|p_T^{\mathrm{jet}}) \frac{dN}{dp_T^{\mathrm{jet}}}. \tag{11}$$

Journal of Physics: Conference Series **230** (2010) 012010 doi:10.1088/1742-6596/230/1/012010

Then the invariant cross section can be evaluated as

$$\frac{Ed^3\sigma^{\text{jet}}}{dp^3} = \frac{1}{2\pi p_T}\frac{d^2\sigma^{\text{jet}}}{dp_T^{\text{jet}}dy} = \frac{\sigma_{\text{BBC}}}{A\,\epsilon_{\text{BBC}}}\frac{1}{p_T^{\text{jet}}}\frac{1}{N_{\text{evt}}}\frac{dN^{\text{jet}}}{dp_T^{\text{jet}}} \tag{12}$$

where dN/dp_T is the unfolding result from (11) and

$$A = 2(\Delta\eta - 2d)(\Delta\phi/2 - 2d) \tag{13}$$

the fiducially reduced PHENIX central arm acceptance area. This measurement is detailed in [5].

The choice of regularization parameter τ can translate into an uncertainty on the low frequency, global shape of a spectrum. We address this by evaluating the systematic uncertainty from varying τ over the entire meaningful range between $\approx 4^D$ degrees of freedom up to the Nyquist frequency, D being the unfolding dimension.

Figure 3 shows the PHENIX preliminary $p+p$ jet spectrum measured using the Gaussian filter, plotted in invariant cross sections. The shaded box to the left indicates the overall normalization systematic uncertainty, shaded boxes associated with data points indicate point-to-point systematic uncertainties, and error bars indicate statistical uncertainties. We show the unfolded spectrum out to the p_T bin where the nominal yield for the number of sampled events reaches the level of 1 jet, namely 60 GeV/c.

Figure 4 shows the same spectrum as in Figure 3, compared against the spectrum from [18], the next-to-leading order (NLO) calculation using the small cone approximation (SCA) [19], and the leading order PYTHIA spectrum assuming $K = 2.5$. The comparison to [18] and NLO SCA involve different jet definitions, a residual difference should be expected, even though for $p_T > 15$ GeV/c it appears to be small between filter and cone jets for the Gaussian size $\sigma = 0.3$ used in this analysis. Our spectrum is close to [18] within its p_T reach. The spectrum also follows approximately the shape of the NLO SCA calculation, and the leading order PYTHIA spectrum, if $K = 2.5$ is assumed. However, a more appropriate comparison would involve Gaussian filter based NLO calculations, which we plan to perform in the future.

6. Fragmentation functions in $p+p$

We use the Run-5 $p+p$ minimum bias data set for the measurement of the fragmentation function. For reconstructed jets, we measure the longitudinal contribution of the fragment momentum to the jet as

$$z = w_{\text{filter}}(\Delta R)\frac{p_\parallel}{p^{\text{jet}}} = \exp\left(-\frac{\Delta\eta^2 + \Delta\phi^2}{2\sigma_{\text{filter}}}\right)\frac{\mathbf{p}\cdot\mathbf{p}^{\text{jet}}}{(p^{\text{jet}})^2}. \tag{14}$$

This choice preserves the identity $\sum z = 1$ for a weighted jet definition, compared to traditional algorithms that would assign zero or unitary weights for nonconstituent and constituent fragments. For practical purpose, both $p_\parallel$ and p_T^{jet} are binned logarithmically in the $dN/(dp_\parallel dp_T^{\text{jet}})$ distribution, which allows the $p_\parallel/p_T$ division to be performed as bin shifts. For the narrow mid-pseudorapidity acceptance of the PHENIX central arms, the difference in $z = p_\parallel/p^{\text{jet}}$ and $z = p_\parallel/p_T^{\text{jet}}$ division is $\leq 6\%$ and negligible within our uncertainties.

We use RICH to reject electrons for the measurement of charged fragmentation function, since in PHENIX, they are produced predominantly from γ conversions in the beam pipe and Dalitz decays, therefore not associated with the charged fragmentation of jets. At the moment, we also do not reconstruct γ pairs from π^0 decays, although we envision to provide identified neutral fragmentation function measurements in the future.

Since the PHENIX detector resolution for single particle far exceeds the resolution for jet energy, we assume a diagonal transfer function for particle energies in the transfer function

$$P(p_T^{\text{jet,rec}}, p_\parallel^{\text{rec}} | p_T^{\text{jet}}, p_\parallel) = P(p_T^{\text{jet,rec}} | p_T^{\text{jet}}, p_\parallel)\delta(p_\parallel^{\text{rec}} - p_\parallel), \tag{15}$$

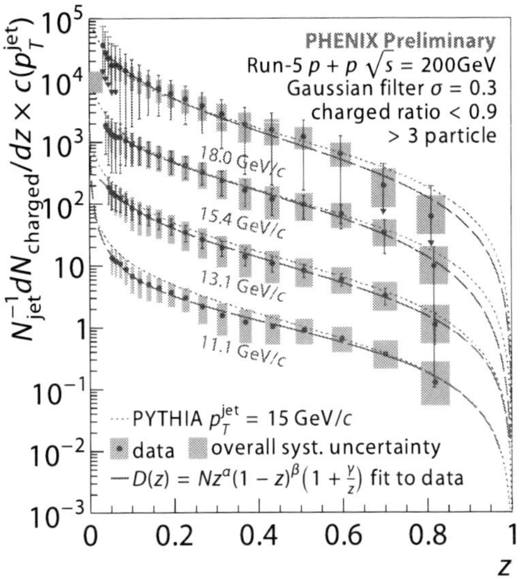

Figure 5. PHENIX Run-5 $p+p$ at $\sqrt{s} = 200$ GeV charged (with electron rejection) jet fragmentation function with respect to $z = w_{\text{filter}} p_{\parallel}/p^{\text{jet}}$ and vertically scaled by $c(p_T^{\text{jet}}) = 10^k, k = 0, \ldots, 3$. The shaded box to the left indicates the overall normalization systematic uncertainty, shaded boxes associated with data points indicate point-to-point systematic uncertainties, and error bars indicate statistical uncertainties. Biases from the jet level cuts are fully quantified in the systematic uncertainties. The blue curve indicates the $D(z) = Nz^{\alpha}(1-z)^{\beta}\left(1+\frac{\gamma}{z}\right)$ fit to data, while the red curve shows the same fit from PYTHIA at $p_T^{\text{jet}} = 15$ GeV/c.

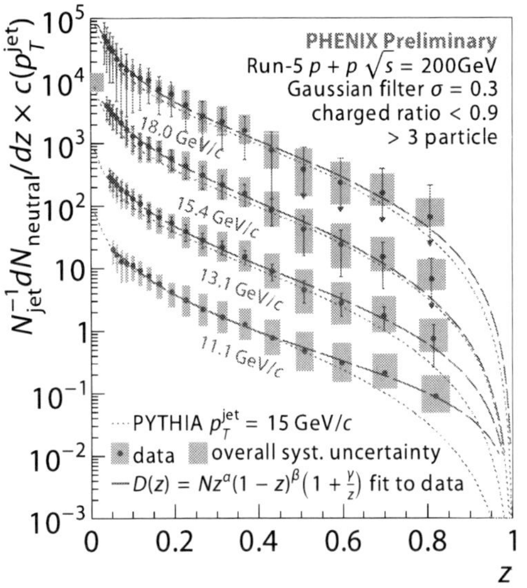

Figure 6. PHENIX Run-5 $p+p$ at $\sqrt{s} = 200$ GeV (electromagnetic) neutral jet fragmentation function with respect to $z = w_{\text{filter}} p_{\parallel}/p^{\text{jet}}$ and vertically scaled by $c(p_T^{\text{jet}}) = 10^k, k = 0, \ldots, 3$. The shaded box to the left indicates the overall normalization systematic uncertainty, shaded boxes associated with data points indicate point-to-point systematic uncertainties, and error bars indicate statistical uncertainties. Biases from the jet level cuts are fully quantified in the systematic uncertainties. The blue curve indicates the $D(z) = Nz^{\alpha}(1-z)^{\beta}\left(1+\frac{\gamma}{z}\right)$ fit to data, while the red curve shows the same fit from PYTHIA at $p_T^{\text{jet}} = 15$ GeV/c.

although the use of a full transfer function is planned for the future. The same GEANT simulation result as for the jet spectrum is used to evaluated the so reduced conditional transfer function $P(p_T^{\text{jet,rec}} | p_T^{\text{jet}}, p_{\parallel})$.

The $(p_T^{\text{jet}}, p_{\parallel})$ distribution is then unfolded by the regularized inversion of the 2D equation

$$\frac{dN}{dp_T^{\text{jet,rec}}\, dp_{\parallel}^{\text{rec}}} = \iint dp_T^{\text{jet}}\, dp_{\parallel}\, P(p_T^{\text{jet,rec}}, p_{\parallel}^{\text{rec}} | p_T^{\text{jet}}, p_{\parallel}) \frac{dN}{dp_T^{\text{jet}}\, dp_{\parallel}}. \tag{16}$$

Similarly, the jet spectrum is unfolded by inverting (11). The per-jet normalized fragmentation function is then extracted by dividing the unfolded $(p_T^{\text{jet}}, p_{\parallel})$ distribution by the unfolded jet spectrum:

$$D(z) = \frac{1}{\epsilon(z, p_T^{\text{jet}})} \left(\frac{dN_{\text{jet}}}{dp_T^{\text{jet}}}\right)^{-1} \frac{dN}{dp_T^{\text{jet}}\, dz} = \frac{1}{\epsilon(z, p_T^{\text{jet}})} \left(\frac{dN_{\text{jet}}}{dp_T^{\text{jet}}}\right)^{-1} p^{\text{jet}} \left.\frac{dN}{dp_T^{\text{jet}}\, dp_{\parallel}}\right|_{z=p_{\parallel}/p^{\text{jet}}} \tag{17}$$

where $\epsilon(z, p_T^{\text{jet}})$ is the single particle efficiency evaluated by GEANT simulation. While the efficiency for neutral clusters are constant at $\epsilon = 0.81 \pm 0.04$ (syst.), the efficiency for charged tracks exhibits a strong

z-dependence both due to the magnetic field causing low p_T tracks to be bend out of the azimuthal acceptance, and tracking cuts resulting in the rejection of high p_T $\pi^\pm$ misidentified as electrons. For $p_T^{\text{jet}} = 18\,\text{GeV}/c$, the efficiency for charged tracks saturates for $z > 0.3$ to $\epsilon = 0.58 \pm 0.06$ (syst.).

Figure 5 shows the $p+p$ at $\sqrt{s} = 200$ GeV charged (with electron rejection) jet fragmentation function, vertically scaled by $c(p_T^{\text{jet}}) = 10^k, k = 0,\ldots,3$. The shaded box to the left indicates the overall normalization systematic uncertainty, shaded boxes associated with data points indicate point-to-point systematic uncertainties, and error bars indicate statistical uncertainties. Biases from the jet level cuts are fully quantified in the systematic uncertainties. The blue curve indicates the $D(z) = Nz^\alpha(1-z)^\beta\left(1 + \frac{\gamma}{z}\right)$ fit to data, while the red curve shows the same fit from PYTHIA at $p_T^{\text{jet}} = 15\,\text{GeV}/c$.

Figure 6 shows the same plot, for the $p+p$ at $\sqrt{s} = 200\,\text{GeV}$ (electromagnetic) neutral jet fragmentation function. In both set of fragmentation measurements, the fragmentation function for $p_T^{\text{jet}} \geq 15\,\text{GeV}/c$ agree well within our current uncertainty with PYTHIA. We could reach a maximum $z \approx 0.81$ with our measurement.

7. Summary

We presented the first measurement of jet fragmentation function in $p+p$ collisions performed at RHIC. It demonstrates for the ability of PHENIX to access high-z jet fragmentation property for both charged particles and photons. Techniques to unfold multidimensional spectra such as the fragmentation function has been discussed, which significantly facilitated us to make such a measurement.

Restricting the data to the minimum bias set currently constrains our statistics and ability to lower the unfolding systematic uncertainty. Our current development of the trigger efficiency correction for fragmentation function measurement would allow us to address this problem in the future.

References

[1] Adler S S *et al.* (PHENIX) 2006 *Phys. Rev.* D **74** 072002
[2] Adare A *et al.* (PHENIX) 2009 *Phys. Rev.* C **80** 024908
[3] Lai Y s and Cole B A 2008 (*Preprint* `arXiv:0806.1499`)
[4] Lai Y s (PHENIX) 2009 *Nucl. Phys.* A **830** 251c–254c
[5] Lai Y s (PHENIX) 2009 (*Preprint* `arXiv:0911.3399`)
[6] Cheng Y z 1995 *IEEE Trans. Pattern Anal. Mach. Intell.* **17** 790–799
[7] Fashing M and Tomasi C 2005 *IEEE Trans. Pattern Anal. Mach. Intell.* **27** 471–474
[8] Salam G P and Soyez G 2007 *J. High Energy Phys.* **2007** JHEP05(2007)086
[9] Phillips D L 1962 *J. ACM* **9** 84–97
[10] Tychonoff A N 1963 *Dokl. Akad. Nauk SSSR* **151** 501–504
[11] Höcker A and Kartvelishvili V 1996 *Nucl. Instrum. Methods* A **372** 469–481
[12] Karayiannis N B and Venetsanopoulos A N 1990 *IEEE Trans. Acoust. Speech Signal Processing* **38** 1155–1179
[13] Van Loan C F 1976 *SIAM J. Numer. Anal.* **13** 76–83
[14] Eldén L 1982 *BIT Numer. Math.* **22** 487–502
[15] Anderson E *et al.* 1999 LAPACK *Users' Guide* 3rd ed (Philadelphia, PA: SIAM)
[16] Bai Z j and Zha H y 1993 *SIAM J. Sci. Comput.* **14** 1007–1012
[17] Bai Z j and Demmel J W 1993 *SIAM J. Sci. Comput.* **14** 1464–1486
[18] Abelev B I *et al.* 2006 *Phys. Rev. Lett.* **97** 252001
[19] Jäger B, Stratmann M and Vogelsang W 2004 *Phys. Rev.* D **70** 034010

Journal of Physics: Conference Series **230** (2010) 012011 doi:10.1088/1742-6596/230/1/012011

Strangeness production in jets from p+p $\sqrt{s} = 200$ GeV collisions

Anthony R. Timmins for the STAR Collaboration

Department of Physics and Astronomy, Wayne State University, 666 W. Hancock, Detroit, MI 48201, USA

E-mail: tone421@rcf.rhic.bnl.gov

Abstract.
Measurements of strangeness production in jets help illuminate the QCD mechanisms in fragmentation. Furthermore, they provide a crucial baseline for heavy-ion studies where modifications in jet chemistry have recently been predicted. We present new results on strange particle production in jets from p+p $\sqrt{s} = 200$ GeV collisions measured by the STAR experiment. The momentum distributions of the Λ , $\bar{\Lambda}$ and K^0_S particles are obtained using various jet finding algorithms, and then compared to various models. Strange particle ratios in jets are obtained and compared to values obtained from the inclusive spectra. Finally, we show jets tagged with leading strange baryons and mesons, in order to investigate whether gluon or quark jets can be isolated in this way.

1. Introduction

We report measurements of strange particle momentum distributions in jets from p+p $\sqrt{s} = 200$ collisions measured by the STAR experiment. These results provide additional tests to QCD inspired models of fragmentation such as PYTHIA [1] or the Modified Leading Log Approximation (MLLA) [2, 3, 4, 5], beyond the usual comparisons to charged hadron momentum distributions. Furthermore, predictions for the modification of jet chemistry in heavy-ion collisions relative to p+p have been made [6]. These predictions show the K/π and p/π ratios in jets are expected to be higher in the presence of a medium. It is therefore essential we test theoretical descriptions of identified particle momentum distributions in p+p jets, if predictions in the modification of heavy-ion jet chemistry are to be made. Finally, since jets are from hard processes, comparisons of particles yields in jets to inclusive spectrum measurements may help elucidate the contributions from hard processes to those measurements as a function of p_T.

2. Analysis

The data presented are from $\sim$ 8 million jet-patch triggered p+p $\sqrt{s_{NN}} = 200$ GeV events recorded in 2006 by the STAR experiment. The jet-patch trigger is designed to select events with jets, and thus requires energies above 8 GeV to be deposited in an area $\Delta\eta \times \Delta\phi = 1 \times 1$ within the electromagnetic calorimeter [7]. The Time Projection Chamber (TPC) is used to detect the charged particles [8], while the barrel electromagnetic calorimeter, BEMC, is used for neutral particles. Three jet finders are employed from the fast-jet package [9]: k_t, anti-k_t and SIS. We select recoil jets with respect to the triggered jets to measure fragmentation functions.

This is because the triggered jets have a bias towards the neutral jet component. The jets are found within a resolution parameter, R = 0.4 and a jet axis η cut < 0.6 to ensure the jet area lies within the TPC/BEMC acceptance. The resolution parameter can be thought of as an effective jet radius in the η, ϕ plane. The minimum p_T cut for a charged particle is 200 MeV/c, while the minimum calorimeter tower energy is 200 MeV. Towers which match charged particles are removed from the jet to avoid over counting of electrons and Minimum Ionising Particles (MIPs). No corrections are applied to the jet energy scale which means the reconstructed jet p_T is likely to be less than the actual jet p_T due to TPC/BEMC detection inefficiencies, and missing neutral energy principally from undetected n and K_L^0 particles. Studies are underway to determine a correction for this. Finally, in order to measure the momentum distributions of the Λ, $\bar{\Lambda}$, and K_S^0 particles, V0s are identified within the jet area from TPC tracks. The invariant mass distributions are calculated to extract the yields of these particles via their dominant decay channels: $\Lambda \rightarrow p + \pi^-$, $\bar{\Lambda} \rightarrow \bar{p} + \pi^+$, $K_S^0 \rightarrow \pi^+ + \pi^-$. In conjunction with a minimum p_T cut of 1 GeV/c, a series of topological cuts are placed on the V0s to minimise the relative background contribution, and ensure the signal to background ratios are approximately the same for each particle. The residual background after all cuts is not yet removed from the fragment yields.

3. Results

Strange particle momentum distributions as a function of the reconstructed jet p_T are shown in figure 1. The momentum distribution is expressed in $\xi = log(jp_T/fp_T)$, where jp_T and fp_T are the reconstructed jet and fragment p_T respectively. For a particular particle species, the integral of the ξ function gives its mean multiplicity per jet. The uncertainties are statistical added in quadrature to the small differences obtained by comparing the three different jet finders. A data point is the mid point value between the lowest and highest values the three jet finders give.

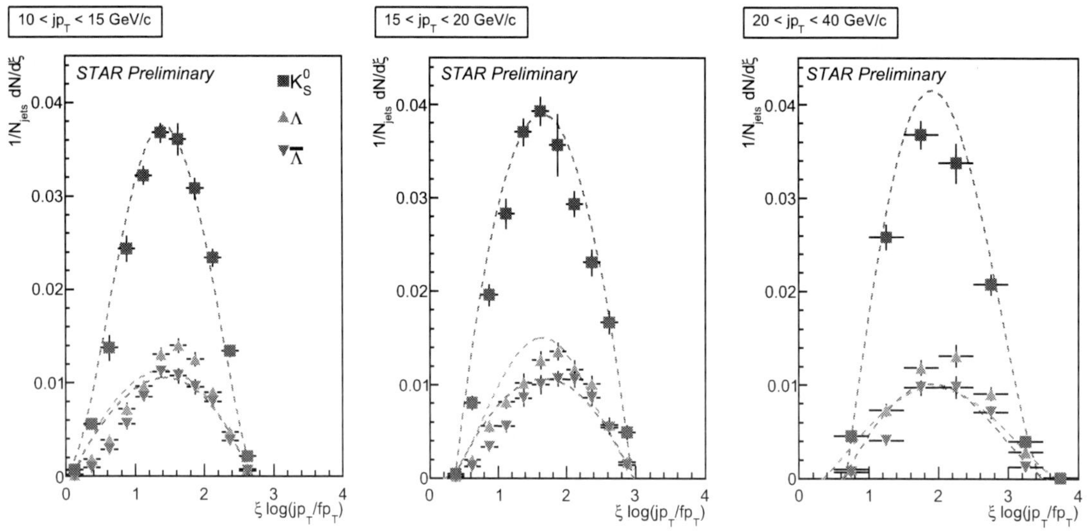

Figure 1. Data points show strange particle momentum distributions for three reconstructed jet p_T ranges in p+p $\sqrt{s_{NN}} = 200$ GeV collisions. The strange particle yields are not corrected for tracking efficiency. The symbols jp_T and fp_T are the reconstructed jet and fragment p_T respectively. The curves show polynomial fits to PYTHIA 6.4 predictions. These predictions are obtained from PYTHIA events ran through the STAR detector simulator. The same applies to all other figures.

The measured V0 yields are not yet corrected for acceptance and TPC tracking inefficiencies. The PYTHIA predictions are obtained by processing PYTHIA events with STAR's detector simulation, and running the track, jet, and V0 reconstruction software on the simulated output in the same way as the real data. This ensures the PYTHIA events are subject to the same acceptance restrictions/detector inefficiencies as the real data. All PYTHIA predictions shown in these proceedings are from version 6.4 with Tune A. We find that PYTHIA gives a reasonable description of the K^0_S data. JETSET (the jet production scheme in PYTHIA) has also been shown to describe $K^{\pm}$ fragmentation functions in $e^+ + e^-$ collisions well, where the jet energies range $5 < E_{jet} < 46$ GeV [10]. Furthermore, PYTHIA has been shown to describe charged hadron momentum distributions in p+p collisions at the same center of mass energy [11]. On the other hand, we note that the PYTHIA description of Λ and $\bar{\Lambda}$ momentum distributions are less satisfactory. Although the predictions appear to predict the correct yield over all ξ, they tend to over predict the yields at low ξ and under predict the yields at intermediate ξ. Whether this is due to a mismatch in the reconstruction of PYTHIA events and real data, or due to physics, is under investigation.

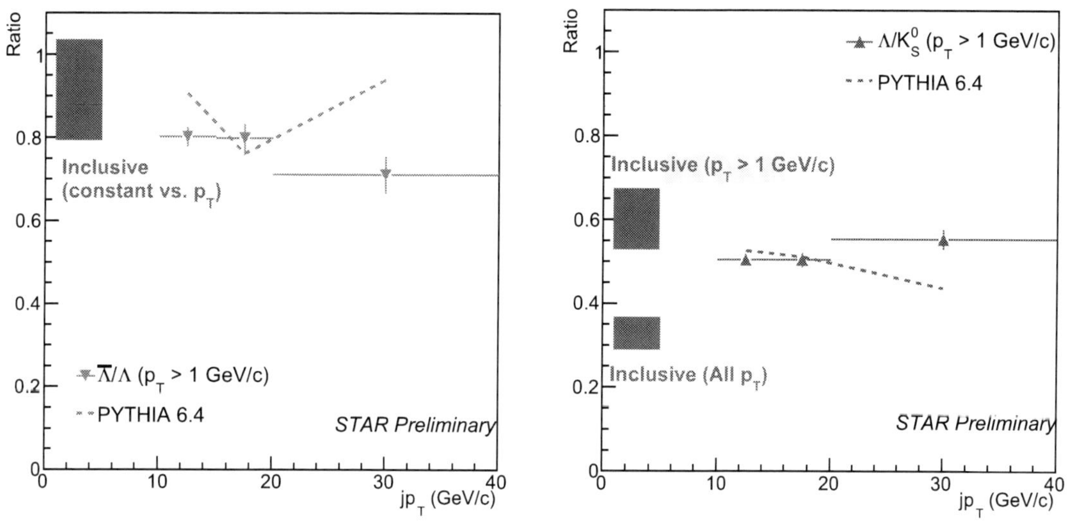

Figure 2. $\bar{\Lambda}/\Lambda$ and Λ/K^0_S ratios in jets for three reconstructed jet p_T ranges in p+p $\sqrt{s_{NN}} = 200$ GeV collisions. In each case, the minimum particle p_T is 1 GeV/c and ratios are corrected for TPC efficiency and acceptance. The uncertainties are statistical added in quadrature to the small differences obtained by comparing the three different jet finders. Inclusive refers to values obtained from inclusive spectra measurements [12]. The vertical size of shaded boxes reflects statistical and systematic uncertainties added in quadrature.

In figure 2, we show ratios of integrated yields $\bar{\Lambda}/\Lambda$ and Λ/K^0_S for $p_T > 1$ GeV/c as a function of reconstructed jet p_T. The analysis shows the $\bar{\Lambda}/\Lambda$ ratio is below 1 showing the baryon asymmetry in the colliding system (p+p) is transferred to jets over all jet energies measured. We also find the $\bar{\Lambda}/\Lambda$ ratio is consistent with the value obtained from inclusive spectra measurements [12]. PYTHIA also seems to reproduce the magnitude of the ratio. The jaggedness of the predictions is due statistical fluctuations which are $\sim 10\%$ and not shown. Regarding the Λ/K^0_S ratios in jets, these are higher than values from inclusive spectrum measurements over all p_T. The inclusive spectrum measurement will be dominated by production below 1 GeV/c. When we compare the ratios in jets to the inclusive spectrum measurement above 1 GeV/c, we find they are generally consistent. This may mean that Λ and K_S^0 spectrum measurements with $p_T > 1$ GeV/c have a dominant contribution from hard processes i.e. jet production. However, measurements of the same ratio from the underlying event are needed to further qualify this observation. We also find PYTHIA again reproduces the magnitude of the ratio.

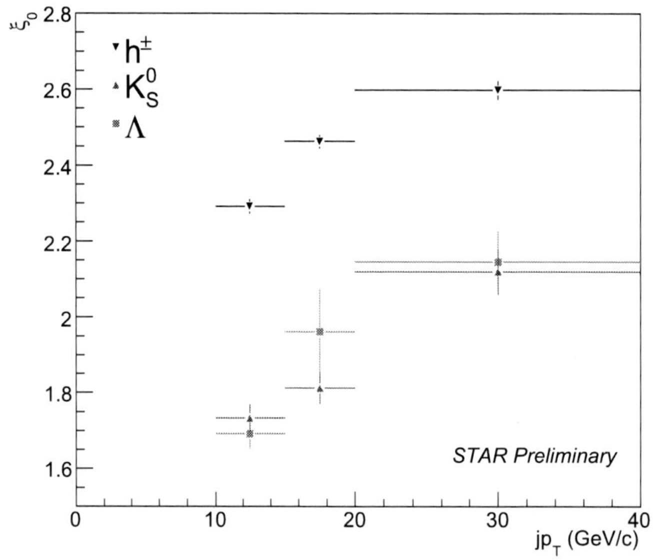

Figure 3. Peak position of ξ (ξ_0) for charged hadrons, Λ and K_S^0 particles for three reconstructed jet p_T ranges in p+p $\sqrt{s_{NN}} = 200$ GeV collisions. The uncertainties are statistical added in quadrature to the small differences obtained by comparing the three different jet finders.

In figure 3, we show the position of the maximum of the ξ distribution, ξ_0, for the various particles as a function of the reconstructed jet p_T. Since the p_T dependent tracking inefficiency/acceptance will distort the shape of the ξ distribution, we correct the distributions using embedding which determines the V0 reconstruction efficiency. A Gaussian is fitted in the peak region to extract ξ_0. The fit range is chosen to avoid the region in ξ affected by the minimum p_T cut on the fragments. In the MLLA scheme, the position of the maximum can be described by the following relation [5];

$$\xi_0 = Y + \sqrt{cY} - c \tag{1}$$

where c is a constant and Y is:

$$Y = log\left(\frac{E_{jet}sin(\theta_0)}{Q_0}\right) \tag{2}$$

where θ_0 is the jet opening angle, and Q_0 is the effective transverse momentum (relative to the jet axis) where a parton ceases to branch in the parton shower. MLLA predictions for momentum distributions of hadrons with mass M_0 typically assume $Q_0 \simeq M_0$ [6]. Equation 2 then implies a scaling behaviour of ξ_0 with the mass of the hadron i.e. ξ_0 should decrease for the higher mass hadrons. The value of ξ_0 should also increase with jet energy for a given hadron which is observed in figure 3 for all particles. However, we find that $\xi_0(K_S^0) \sim \xi_0(\Lambda)$ which is contrary to the expected scaling. On the other hand, $\xi_0(h^\pm) > \xi_0(K_S^0)$ and $\xi_0(\Lambda)$ which is consistent with such a scaling assuming pions dominate particle production in jets.

A similar observation was made for π, K, p fragmentation functions for similar jet energies in $e^+ + e^-$ collisions [13]. Finally, we note in interest that if we extract the mode of the transverse rapidity, y_T, distribution (motivated by this study [14]), we find a mass scaling in y_T space i.e $y_{T,0}(h^{\pm}) > y_{T,0}(K^0_S)) > y_{T,0}(\Lambda)$ for a given reconstructed jet p_T bin. For a given particle species, we also find $y_{T,0}$ increases with jet energy.

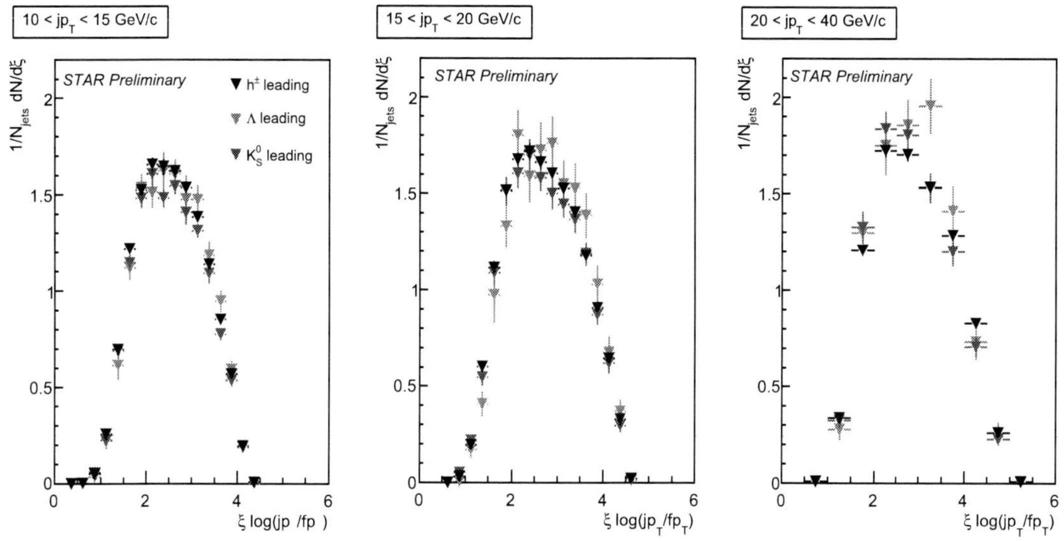

Figure 4. Momentum distributions of non-leading charged hadrons for three reconstructed jet p_T ranges in p+p $\sqrt{s_{NN}} = 200$ GeV collisions. The black points show data for jets where the leading particle is a charged hadron, red corresponds to jets with leading Λ particles, and the blue points correspond to jets with leading K^0_S particles. The charged hadron yields are not corrected for tracking efficiency. jp_T and fp_T are the reconstructed jet and fragment p_T respectively. The uncertainties are statistical added in quadrature to the small differences obtained by comparing the three different jet finders.

In figure 4 we show the momentum distributions of *non-leading* charged hadrons for the cases where the leading particle is a charged hadron, Λ, or K^0_S. We want to investigate whether tagging jets according to the species of the leading hadron, preferentially selects on gluon or quark jets. Both measured data [15] and theoretical predictions show hadron multiplicities are larger in gluon jets compared to quark jets: MLLA gives the ratio 9/4 for gluon/quark jet multiplicities. Thus, if tagging jets with particle A preferentially selected gluon jets, and tagging jets with particle B preferentially selected quark jets, we would expect the jets associated with particle A to have more non-leading charged hadrons compared to jets associated with particle B. However, in figure 4 we observe at given jet energy, the charged hadron multiplicities are the same for each of the tagged jets. This might be expected for the highest jet p_T bin where jet production is expected to be dominated by hard scattering of valance quarks, however at lower jet energies there should be mixture of quark and gluon jets [16]. Therefore, our observation suggests quark/gluon jets cannot be tagged in this way. Further studies are underway to confirm this.

4. Summary

In summary, we have shown measurements of strange particle momentum distributions. We have found that PYTHIA describes the K^0_S momentum distributions well, however we observe some deviations for the Λ and $\bar{\Lambda}$ data. We have found that $\bar{\Lambda}/\Lambda$ ratios ($p_T > 1$ GeV/c) in jets are consistent with values obtained from inclusive spectra. Λ/K^0_S ratios ($p_T > 1$ GeV/c) in jets were found to be similar to values obtained from inclusive spectra in the same p_T range. This may suggest that Λ and K^0_S production above 1 GeV/c is dominated by hard processes in p+p $\sqrt{s_{NN}} = 200$ GeV collisions. We investigated a mass scaling of the peak position of ξ (inferred from MLLA), and found although $\xi_0(h^{\pm}) > \xi_0(K^0_S)$ and $\xi_0(\Lambda)$, $\xi_0(K^0_S) \sim \xi_0(\Lambda)$. Finally, we have found tagging jets with leading strange baryons/mesons may not preferentially select on quark/gluon jets.

References

[1] Sjostrand T, Mrenna S, and Skands P 2006 *JHEP* **0605** 026
[2] Dokshitzer Y, Fadin V, and Khoze V 1982 *Phys. Lett. B* **115** 242; Bassetto A, Ciafaloni M, Marchesini G, and Mueller A 1982 *Nucl. Phys. B* **207**
[3] Mueller A 1983 *Nucl. Phys. B* **213** 85
[4] E Malaza E and Webber B (1984) Phys. Lett. B **149** 501
[5] Dokshitzer Y, Mueller A , Khoze V, and Troyan S, *Basics of Perturbative QCD (Editions Frontieres, Paris, 1991)*
[6] Sapeta S and Wiedemann U 2008 *Eur. Phys. J. C* **55** 293
[7] Abelev B *et al.* (STAR Collaboration), 2008 *Phys. Rev. Lett.* **100** 232003
[8] Ackermann K H *et al.*, 2003 *Nucl. Inst. and Meth. A* **499** 659
[9] Salam G and Soyez G 2007 *JHEP* **0705** 086
[10] Anulli F (BABAR Collaboration) *hep-ex/0406017v1*
[11] Caines H (STAR Collaboration) 2009 *Nucl. Phys. A* **830** 263c-266c
[12] Abelev B *et al.* (STAR Collaboration) 2007 *Phys. Rev.* C **75** 064901
[13] Albino S *et al. arXiv:0804.2021v1*
[14] Trainor T and Kettler D 2006 *Phys. Rev. D* **74** 034012
[15] Abreu P (DELPHI Collaboration) 2000 *Eur. Phys. J. C* **13** 573
[16] De Florian D and Vogelsang W 2007 *Phys. Rev. D* **76** 074031

Journal of Physics: Conference Series **230** (2010) 012012 doi:10.1088/1742-6596/230/1/012012

On Fluctuations of Conserved Charges : Lattice Results Versus Hadron Resonance Gas

Pasi Huovinen[1] **and Péter Petreczky**[2]

[1] Institut für Theoretische Physik, Johann Wolfgang Goethe-Universität, Max-von-Laue-Straße 1, 60438 Frankfurt am Main, Germany

[2]Physics Department, Brookhaven National Laboratory, Upton, NY 11973, USA

Abstract. We compare recent lattice results on fluctuation and correlations of strangeness, baryon number and electric charge obtained with p4 improved staggered action with the prediction of hadron resonance gas model. We show that hadron resonance gas can describe these fluctuations reasonably well if the hadron properties are as calculated on the lattice.

1. Introduction

In recent years a lot of progress has been made in understanding the QCD transition at non-zero temperature through lattice calculations (see [1, 2] for reviews). In particular, fluctuations of conserved charges have been studied on the lattice and provided insight on how the relevant degrees of freedom change from hadronic to partonic [3, 4, 5, 6, 7, 8, 9, 10]. At low temperatures it is expected that hadrons are good degrees of freedom and thermodynamic quantities can be well described by hadron resonance gas.

Hadron resonance gas (HRG) turned out to be very successful in describing particle abundances produced in heavy ion collisions [11]. It was also used to estimate QCD transport coefficients [12] as well as chemical equilibration rates [13] close to the transition temperature. Thermodynamic quantities calculated in lattice QCD with rather large quark mass agree well with the HRG model if the masses of the hadrons in the model are tuned appropriately to match the large quark mass used in lattice calculations [14]. Furthermore, the ratios of certain fluctuations calculated on the lattice are also in reasonably good agreement with HRG model predictions at low temperatures [15, 16, 10].

Fluctuations of conserved charges are the most suitable quantities to test the validity of HRG model. They can be defined as derivatives of the pressure with respect to chemical potentials and as such appear in the calculations of thermodynamic quantities at finite baryon density via Taylor expansion. The recent lattice calculations performed with p4 and asqtad actions at or very close to the physical quark mass [6, 7, 10] gave fluctuations that were quite different from the results obtained in HRG model. It was pointed out that lattice discretization effects on the hadron spectrum are responsible for this discrepancy [17]. When taking into account the lattice spacing dependence of the hadron masses in the hadron resonance gas calculations a good agreement between lattice and HRG calculations has been found. Strangeness fluctuations calculated with and stout [8, 9] and HISQ [18] improved staggered fermion actions agree better with the HRG result (see also the discussion in Ref. [19]). This is due to largely reduced discretization effects in the hadron spectrum for these actions. In Ref. [17] we studied baryon number and strangeness fluctuations in HRG, where the hadron masses have been modified to include discretization effects present in lattice calculations. The comparison was done using lattice data obtained with asqtad action on $N_\tau = 6$ and 8 lattices. On the other hand, in the most detailed lattice study of baryon number, strangeness and electric charge fluctuations and correlations up to sixth order was performed with p4 action using lattices with temporal extent $N_\tau = 4$ and $N_\tau = 6$ [10]. In this contribution we are going to extend our calculations in modified HRG model to study different fluctuations and compare them with lattice results obtained with p4 action on $N_\tau = 4$ and 6 lattices. As mentioned above this is also important for constructing realistic equation of state at finite baryon density for hydrodynamic models similarly as this was done for zero baryon density in Ref. [17].

2. Fluctuations of strangeness, baryon number and electric charge

Derivatives of the pressure with respect to chemical potentials of conserved charges, e.g. baryon number (B), electric charge Q and strangeness S at zero chemical potentials can be well calculated in lattice QCD

$$\chi_n^X = T^n \frac{\partial^n p(T,\mu_B,\mu_Q,\mu_S)}{\partial \mu_X^n}|_{\mu_B=\mu_Q=\mu_S=0},\ X = B,Q,S. \tag{1}$$

These are related to quadratic and higher order fluctuations of conserved charges $\chi_2^X = \langle X^2\rangle/(VT^3)$, $\chi_4^X = (\langle N_X^4\rangle - 3\langle N_X^2\rangle^2)/(VT^3)$ etc.[1] Mixed derivatives of the pressure give correlations of conserved charges

$$\chi_{11}^{XY} = T^2 \frac{\partial^2 p(T,\mu_B,\mu_Q,\mu_S)}{\partial \mu_X \partial \mu_Y}|_{\mu_B=\mu_Q=\mu_S=0} = \langle XY\rangle/(VT^3),\ X,Y = B,Q,S. \tag{2}$$

We have calculated these quantities in HRG with hadron spectrum modified to take into account the lattice artifacts. Unfortunately, there is no detailed calculations of the hadron spectrum with p4 action. Therefore the lattice spacing and quark mass dependence of the hadron masses was evaluated using the formulas derived in Ref. [17] based on asqtad calculations. We expect that cutoff dependence of the hadron masses is similar for p4 and asqtad actions, except in the pseudo-scalar meson sector. It is known that the quadratic splitting of non-Goldstone pseudo-scalar mesons is about two times larger for p4 action than for asqtad action [20]. Therefore when evaluating the contributions of kaons and pions to different quantities we simply doubled the quadratic splittings used in Ref. [17]. To get agreement between lattice results and HRG it turned out to be necessary to modify the masses of excited baryons states. Since the cutoff dependence of the excited baryon masses is not known, it was assumed that masses of all excited states up to certain threshold m_{cut}^B are modified the same way as the ground state baryon masses [17]. Values $m_{cut}^B = (1.7 - 2.5)$GeV have been considered in the previous analysis.

In this paper we consider all resonances with mass up to 2.5GeV and use $m_{cut}^B = 1.9$GeV in all calculations. The quadratic fluctuations of strangeness and electric charge calculated in HRG and compared to the p4 lattice results are shown in Fig. 1. The lattice results are well below the HRG curve obtained with physical hadron masses. This is a general feature for all fluctuations. However, taking into account the discretization effects in the hadron spectrum we find a reasonable agreement between HRG and lattice results. In the strange sector discretization effects are slightly overestimated in our approach. We also studied quadratic baryon number fluctuations and baryon number - strangeness correlations and the corresponding results are shown in Fig. 2. The baryon number fluctuations are well described by modified HRG, although the in the low temperature region cutoff effects are under-predicted for $N_\tau = 6$. For

[1] Here we consider the case of zero chemical potential, so $\langle N_X\rangle = 0$.

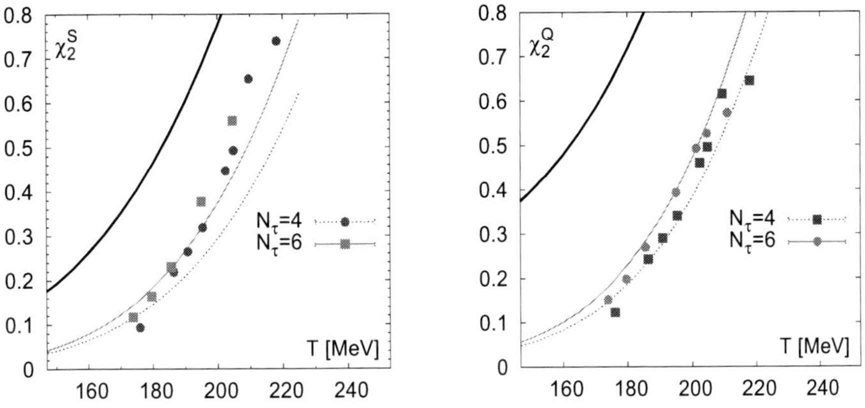

Figure 1. The quadratic fluctuations of strangeness (left) and electric charge (right) calculated on the lattice with p4 action [10] and compared with the HRG calculations in the continuum (solid black line) and on the lattice (blue and red lines).

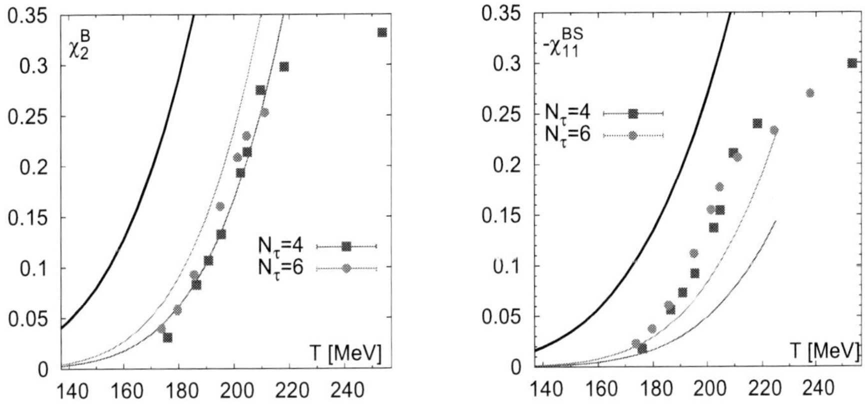

Figure 2. The quadratic fluctuations of baryon number (left) and baryon number - strangeness correlations (right) calculated on the lattice with p4 action [10] and compared with the HRG calculations in the continuum (solid black line) and on the lattice (blue and red lines).

baryon number-strangeness fluctuations the agreement between lattice and modified HRG is not very good. This may imply that the cutoff effects in strange baryon sector are smaller than in non-strange baryon sector and are overestimated in our approach. Interestingly, the new calculations with HISQ action also suggest smaller cutoff effects in baryon number-strangeness fluctuations compared to baryon number fluctuations [19]. Finally we have considered fourth order fluctuations of the electric charge and baryon number. The numerical results are shown in Fig. 3. The HRG calculations with the modified hadron masses agree reasonably well with the p4 lattice data. Note, however, that here the deviations from the resonance gas show up at smaller temperatures.

3. Conclusions

We have studied quadratic and fourth order fluctuations of baryon number, electric charge and strangeness fluctuations in HRG model, where hadron masses have been adjusted to take into

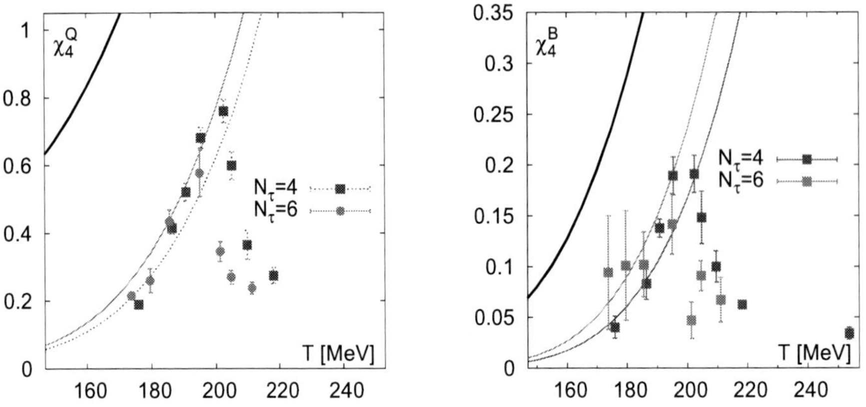

Figure 3. The fourth order fluctuations of the electric charge (left) and baryon number (right) calculated on the lattice with p4 action [10] and compared with the HRG calculations in the continuum (solid black line) and on the lattice (blue and red lines).

account the discretization errors in the hadron spectrum present in the lattice calculations. We have found reasonably good agreement with the lattice calculations of the fluctuations performed using p4 action on $N_\tau = 4$ and 6 lattices. Our calculations explain why all the lattice results fall below the physical HRG result. Our approach does not give a good description of baryon number - strangeness fluctuations. To resolved this issue a more refined treatment of the cutoff effects in the baryon sector is needed.

Acknowledgments
This work was supported by the U.S. Department of Energy under contract DE-AC02-98CH1086 and by the ExtreMe Matter Institute (EMMI). P.H. is grateful for support from Center of Analysis and Theory for Heavy Ion Experiment (CATHIE) which enabled him to stay in BNL where large part of this work was finalized.

[1] C. E. DeTar, PoS **LATTICE2008**, 001 (2008)
[2] P. Petreczky, Nucl. Phys. A **830**, 11C (2009); P. Petreczky, Nucl. Phys. Proc. Suppl. **140**, 78 (2005)
[3] R. V. Gavai and S. Gupta, Phys. Rev. D **64**, 074506 (2001).
[4] R. V. Gavai and S. Gupta, Phys. Rev. D **73**, 014004 (2006)
[5] C. Bernard *et al.* [MILC Collaboration], Phys. Rev. D **71**, 034504 (2005)
[6] A. Bazavov *et al.*, Phys. Rev. D **80**, 014504 (2009)
[7] M. Cheng *et al.*, Phys. Rev. D **81**, 054504 (2010)
[8] Y. Aoki, Z. Fodor, S. D. Katz and K. K. Szabo, Phys. Lett. B **643**, 46 (2006)
[9] Y. Aoki, S. Borsanyi, S. Durr, Z. Fodor, S. D. Katz, S. Krieg and K. K. Szabo, JHEP **0906**, 088 (2009)
[10] M. Cheng *et al.*, Phys. Rev. D **79**, 074505 (2009)
[11] A. Andronic, P. Braun-Munzinger, K. Redlich and J. Stachel, Phys. Lett. B **571**, 36 (2003)
[12] J. Noronha-Hostler, J. Noronha and C. Greiner, Phys. Rev. Lett. **103**, 172302 (2009)
[13] P. Braun-Munzinger, J. Stachel and C. Wetterich, Phys. Lett. B **596**, 61 (2004)
[14] F. Karsch, K. Redlich and A. Tawfik, Eur. Phys. J. C **29**, 549 (2003)
[15] F. Karsch, K. Redlich and A. Tawfik, Phys. Lett. B **571**, 67 (2003)
[16] S. Ejiri, F. Karsch and K. Redlich, Phys. Lett. B **633**, 275 (2006)
[17] P. Huovinen and P. Petreczky, Nucl. Phys. A **837**, 26 (2010)
[18] A. Bazavov and P. Petreczky [HotQCD Collaboration], PoS **LAT2009**, 163 (2009)
[19] A. Bazavov and P. Petreczky[HotQCD Collaboration] , these proceedings
[20] M. Cheng *et. al.*, Eur. Phys. J. C **51**, 875 (2007)

QCD thermodynamics on the lattice:approaching the continuum limit with physical quark masses

Z. Fodor

Bergische Universität Wuppertal, Gaussstr. 20, D-42119 Wuppertal, Germany
Julich Supercomputing Center, FZJ, D-52425 Jülich, Germany
Institute for Theor. Physics, Eötvös University, H-1117 Budapest, Hungary

E-mail: `fodor (at) physik.uni-wuppertal.de`

Abstract. The T>0 QCD transition is an analytic cross-over. This non-trivial result is shown by means of lattice simulations using physical quark masses and finite size scaling analysis extrapolated to the continuum limit. The transition temperature (T_c) is determined. The discrepancy between our (Wuppertal-Budapest) results and the results of the Bielefeld-Brookhaven-Columbia-Riken Collaboration & hotQCD Collaboration is discussed. Preliminary results on the equation of state are presented.

1. Introduction

During the evolution of the universe there was a transition at $T \approx 200$ MeV. It is related to the spontaneous breaking of the chiral symmetry of QCD. The nature of the QCD transition affects our understanding of the universe's evolution (see e.g. Ref. [1]). Extensive experimental work is currently being done with heavy ion collisions to study the QCD transition (most recently at the Relativistic Heavy Ion Collider, RHIC and at the Large Hadron Collider, LHC). Both for the cosmological transition and for RHIC/LHC, the net baryon densities are quite small, thus the baryonic chemical potentials (μ) are much less than the typical hadron masses (below 50 MeV at RHIC, even smaller at LHC and negligible in the early universe). A calculation at μ=0 is directly applicable for the cosmological transition and most probably also determines the nature and absolute temperature of the transition at RHIC/LHC. Thus we carry out our analysis at μ=0. (For a review on the QCD transition at μ=0 and μ >0 see e.g. [2].)

When we analyze the nature and/or the absolute scale of the $T > 0$ QCD transition for the physically relevant case two ingredients are quite important.

First of all, one should use physical quark masses. The nature of the transition depends on the quark mass, for small or large quark masses it is a first order phase transition, whereas for intermediate quark masses it is an analytic crossover. Since the nature of the transition influences the absolute scale (T_c) of the transition –its value, mass dependence, uniqueness etc.– the use of physical quark masses is also essential for the determination of T_c, too. The absolute scale than goes into all observables, also into the equation of state.

Secondly, the nature of the $T > 0$ QCD transition is known to suffer from discretization errors [3, 4]. The three flavor theory with a large, $a \approx 0.3$ fm lattice spacing and standard action predicts a critical pseudoscalar mass of about 300 MeV. This point separates the first order and cross-over regions. If we took another discretization, with another discretization error,

Journal of Physics: Conference Series **230** (2010) 012013 doi:10.1088/1742-6596/230/1/012013

the critical pseudoscalar mass turns out to be much smaller, well below the physical pion mass of 135 MeV. The only way to determine whether the physical point lies in the crossover or first-order region, is to carry out a careful continuum limit analysis.

In this review the recent results of the Wuppertal-Budapest group are discussed (for the computational setup see [5]). The order and the absolute scale of the transition (T_c) was determined using physical quark masses and a continuum extrapolation. The findings are compared to those of our competitor's (Bielefeld-Brookhaven-Columbia-Riken, which later merged with some part of the MILC Collaboration and formed the hotQCD Collaboration). Preliminary results on the T_c determination for temporal extensions upto 16 and equation of state results for temporal extensions upto 12 are presented.

In Section 2. the most important qualitative features of the QCD transition are summarized. Section 3. presents the basic ingredients of lattice QCD simulations at non-vanishing temperatures. Section 4. discusses the nature of the QCD transition. In Section 5. the absolute scale of the transition is determined. Section 6. discusses some preliminary results on T_c at even finer lattice spacings and the equation of state (EOS). A comparisons based on the hadron resonance gas model is carried out. Finally, Section 7 concludes.

2. The QCD transition

In this section we summarize the qualitative features of the T>0 QCD transition. We use the water-vapor transition as an illustration.

One of the most important piece of information we have is our knowledge about the nature of the transition. Though many take it for granted, it is a highly non-trivial result, that the transition is an analytic one and usually called as a cross-over [6]. In order to show this by means of lattice QCD, physical quark masses were taken, and a finite size scaling analysis was carried out for the continuum extrapolated chiral susceptibilities. (Some details of the analysis will be given in the next section.) This analytic behavior has important consequences for any T_c determination in QCD.

In order to illustrate the most important differences between a real phase transition and an analytic cross-over we recall the water-vapor phase diagram on the temperature versus pressure plane (c.f. [7], Figure 1). We study the transition by fixing the pressure to a given value and than varying the temperature. For smaller pressures (below about 22 MPa) there is a first order phase transition. The density jumps, the heat capacity is infinite, and these singular features appear simultaneously, thus exactly at the same critical temperature. At ≈22.064 MPa pressure and ≈647.096 K temperature there is a critical point with a second order phase transition. This phase transition is also characterized by a singular behavior. (Note, that a real singularity, a phase transition takes place only in infinite size systems. In our example we have a macroscopic amount of water with $O(10^{23})$ molecules. From the practical point of view this is an infinitely large system).

At even larger pressures (p$\gtrsim$22.064 MPa) the water-vapor transition is an analytic one (the behaviors of various observables are analytic, even in the infinite volume limit). As a consequence, in this pressure region if we change the temperature there is no jump in the density only a rapid but continuous change. The inflection point of this density-temperature function (the point with the largest, though finite, derivative) can be used to define the pseudocritical temperature (another usual name for it is "transition temperature") related to the density. Similarly, the heat capacity is always finite, but it has a pronounced peak as we increase the temperature. The position of this peak can be used to define the pseudocritical temperature related to the heat capacity. Despite the fact that there is no singularity the inflection point and peak position are well defined.

The most important message here is that the various transition temperatures (e.g. those related to the density or heat capacity) behave differently depending on whether we are in the

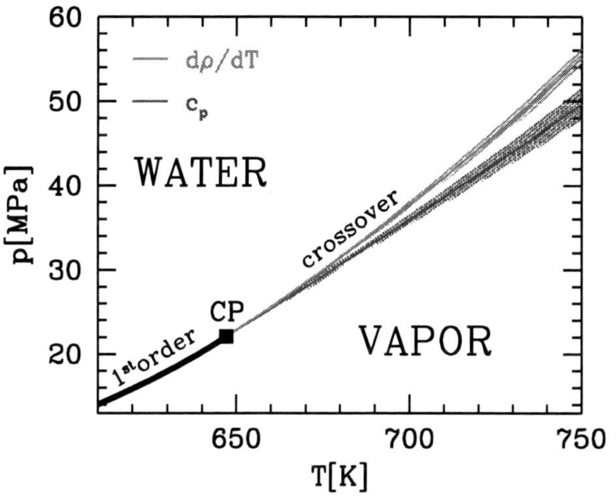

Figure 1. The phase diagram of water around its critical point (CP). For pressures below the critical value (p_c) the transition is first order, for $p > p_c$ values there is a rapid crossover. In the crossover region the critical temperatures defined from different quantities are not necessarily equal. This can be seen for the temperature derivative of the density ($d\rho/dT$) and the specific heat (c_p). The bands show the experimental uncertainties (see [8]).

singular (real phase transition) or non-singular (analytic cross over) region. As it is indicated on the figure for a real phase transition these critical temperatures coincide, whereas in the non-singular region (for pressures above 22.064 MPa) the pseudocritical temperatures can differ considerably. The fast change (though no jump) in the density is at a lower temperature than the peak in the heat capacity. The transition is a broad cross-over. The pseudocritical temperatures, related to various observables, are separated, but both of them are in the broad transition temperature region. This separation does not mean that we have two transitions (one for the density and one for the heat capacity), it merely reflects the broadness of the transition.

It is easy to see that different observables can give different pseudocritical temperatures. Let us study an observable X, which characterizes the transition as a function of the temperature $X(T)$. For a real phase transition its singular behavior appears at the same temperature even if we multiplied it by T (an infinitely high peak keeps its position). For an analytic cross-over we have a peak with a finite height and a finite width. Multiplying it by T (thus constructing a new observable $T \cdot X$, which is also sensitive to the transition) shifts the peak position to larger temperature values. The value of T_c is shifted. The pseudocritical temperature is well defined for any definition, but it is not unique. Though a T_c related to some observable is informative a more complete description is given by the whole temperature dependence of $X(T)$.

The determination of such curves is the main goal of any study on the QCD transition (c.f. our earlier studies [7, 9]). Since the QCD transition (at vanishing chemical potential) is an analytic cross-over one wants to obtain these smooth curves for several observables. Though the characteristic points of such curves contain obviously less informations than the curves themselves, their knowledge tells a lot about the transition, too.

Before we list the observables we study in detail, it is worth to mention that the crossover nature of the QCD transition is related to the specific values of the quark masses we have in nature. For two or three flavor QCD with vanishing quark masses or with infinitely massive quarks one has real phase transitions. There are order parameters (in the former case the chiral susceptibility/condensate signaling the chiral phase transition; in the latter case the Polyakov line signaling the deconfinement phase transition) which show a non-analytic behavior as we change the temperature. As we pointed out earlier, the highly non-trivial result about the analytic nature of the QCD transition with physical quark masses implies, that no observable can be treated as an order parameter. All of the observables show analytic temperature dependence. There is neither a chiral nor a deconfinement phase transition. Note however, that similarly to the density or to the heat capacity in the water-vapor cross-over transition, the observables chiral susceptibility/condensate and the Polyakov line can develop a pronounced peak or show a rapid change. The peak position or the inflection points for such a cross-over are usually expected to be at different temperatures. Again, we do not say [7, 9] that there are two phase transitions and one of them is at a lower temperature than the other. The separation of the pseudocritical temperature is merely a sign of the broad analytic transition [6].

Since the chiral susceptibility/condensate and the Polyakov loop are not order parameters, they are just used to signal the cross-over. In principle any other quantity showing rapid changes or developing a peak in the transition region can be studied. The temperature dependences of these observables can be compared with the predictions of other lattice results or model calculations.

The reason for calculating the temperature dependence of these many observables is obvious. The more observables we study the broader picture we have on the QCD transition. To be more specific, the chiral susceptibility/condensate and the Polyakov loop are remnants of the real phase transition order parameters (for other mass regions of the phase plane). The strange quark number susceptibility is a particularly attractive quantity from the theoretical point of view. It is related to a conserved current, thus no renormalization ambiguities appear, which makes direct comparisons particularly easy.

The various observables (listed in the previous paragraph) lead to different transition temperatures, they are typically between 150 and 170 MeV, thus well within the broadness of the transition. Let us emphasize again, the difference between the T_c values does not mean that one of the phase transition happens at a lower temperature than the other, quite the contrary: no phase transition happens at all. Our new results confirm our earlier findings and their interpretation by all means: the transition temperatures scatter within the broad temperature interval, characteristic to the cross-over.

3. Lattice formulation

Thermodynamical quantities can be obtained from the partition function which can be given by a Euclidean path-integral: $Z = \int dU d\bar{\Psi} d\Psi \exp -S_E(U, \bar{\Psi}, \Psi)$, where U and $\bar{\Psi}, \Psi$ are the gauge and fermionic fields and S_E is the Euclidean action. The lattice regularization of this action is not unique. There are several possibilities to use improved actions which have the same continuum limit as the straightforward unimproved ones. The advantage of improved actions is that the discretization errors are reduced and therefore a reliable continuum extrapolation is possible already from larger lattice spacings. On the other hand, calculations with improved actions are usually more expensive than with the unimproved one.

Our choice for S_E is to use a tree level Symanzik improved gauge action and a stout improved staggered fermion action [5]. This choice is motivated by at least two reasons. It is computationally relatively cheap (comparable to the unimproved actions) and the discretization effects coming from $T = 0$ and $T > 0$ simulations are balanced.

In order to carry out $T > 0$ simulations we have to fix the parameters of the action, the gauge coupling and the quark masses. This is usually done at zero temperature, where the results of lattice computations can be compared to experiments. In order to fix three parameters, three quantities are needed. We chose to use the pion and kaon masses (m_π, m_K) and the leptonic decay constant of the kaon (f_K). These can all be determined from $T = 0$ lattice simulations. Lattice calculations can only yield dimensionless quantities, i.e. am_π, am_K and af_K, where a is the lattice spacing. For any value of the coupling, the mass parameters can be tuned to obtain the correct ratios for m_π/f_K and m_K/f_K. These $m_{ud}(\beta)$ and $m_s(\beta)$ functions are called the line of constant physics (LCP). All finite T simulations have to be carried out along the LCP.

A very important step of any lattice analysis is the continuum extrapolation. Several simulations have to be carried out at different lattice spacings and the results have to be extrapolated to the continuum. Since $T = 1/(N_t a)$, where N_t is the number of lattice sites in the temporal direction, if we are interested in a given temperature range (around the transition) then a decreasing lattice spacing corresponds to increasing N_t.

4. The order of the transition

In this section the nature of the QCD transition is discussed. The details of the calculations can be found in [6]. In order to determine the nature of the transition one should apply finite size scaling techniques for the chiral susceptibility $\chi = (T/V) \cdot (\partial^2 \log Z/\partial m_{ud}^2)$. This quantity shows a pronounced peak as a function of the temperature. For a first order phase transition, such as in the pure gauge theory, the peak of the analogous Polyakov susceptibility gets more and more singular as we increase the volume (V). The width scales with 1/V the height scales with volume. A second order transition shows a similar singular behavior with critical indices. For an analytic transition (what we call a cross-over) the peak width and height saturates to a constant value.

We carried out a finite size scaling analysis with the continuum extrapolated height of the renormalized susceptibility. The renormalization of the chiral susceptibility can be done by taking the second derivative of the free energy density (f) with respect to the renormalized mass (m_r). The logarithm of the partition function contains quartic divergences. These can be removed by subtracting the free energy at $T = 0$: $f/T^4 = -N_t^4 \cdot [\log Z(N_s, N_t)/(N_t N_s^3) - \log Z(N_{s0}, N_{t0})/(N_{t0} N_{s0}^3)]$. This quantity has a correct continuum limit. The subtraction term is obtained at T=0, for which simulations are carried out on lattices with N_{s0}, N_{t0} spatial and temporal extensions (otherwise at the same parameters of the action). The bare light quark mass (m_{ud}) is related to m_r by the mass renormalization constant $m_r = Z_m \cdot m_{ud}$. Note that Z_m falls out of the combination $m_r^2 \partial^2/\partial m_r^2 = m_{ud}^2 \partial^2/\partial m_{ud}^2$. Thus, $m_{ud}^2 [\chi(N_s, N_t) - \chi(N_{s0}, N_{t0})]$ also has a continuum limit (for its maximum values for different N_t, and in the continuum limit we use the shorthand notation $m^2 \Delta\chi$).

In order to carry out the finite volume scaling in the continuum limit we took three different physical volumes. For these volumes we calculated the dimensionless combination $T^4/m^2\Delta\chi$ at 4 different lattice spacings: 0.3 fm was always off, otherwise the continuum extrapolations could be carried out. The volume dependence of the continuum extrapolated inverse susceptibilites is shown on Figure 2.

Our result is consistent with an approximately constant behaviour, despite the fact that we had a factor of 5 difference in the volume. The chance probabilities, that statistical fluctuations changed the dominant behaviour of the volume dependence are negligible. As a conclusion we can say that the staggered QCD transition at $\mu = 0$ is a cross-over.

5. The transition temperature

An analytic cross-over, like the QCD transition has no unique T_c, different observables and/or definitions may lead to different temperatures.

In ref. [7] we considered three quantities to locate the transition point: the chiral susceptibility, the strange quark number susceptibility and the Polyakov-loop. The obtained temperatures were

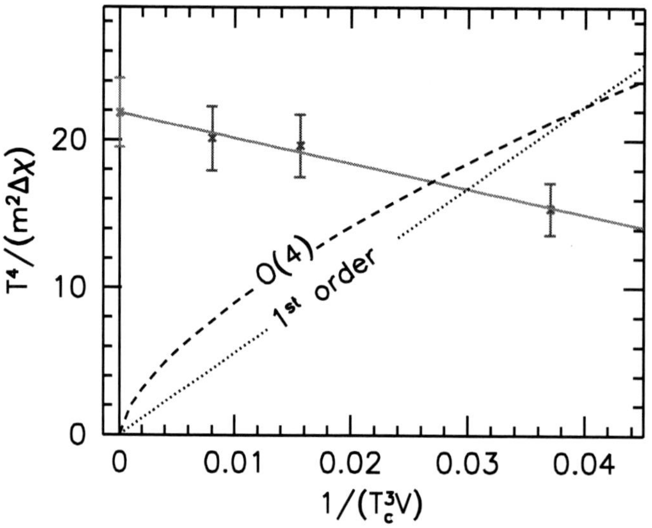

Figure 2. Continuum extrapolated susceptibilities $T^4/(m^2\Delta\chi)$ as a function of $1/(T_c^3V)$. For true phase transitions the infinite volume extrapolation should be consistent with zero, whereas for an analytic crossover the infinite volume extrapolation gives a non-vanishing value. The continuum-extrapolated susceptibilities show no phase-transition-like volume dependence, though the volume changes by a factor of five. The $V\to\infty$ extrapolated value is 22(2) which is 11σ away from zero. For illustration, we fit the expected asymptotic behaviour for first-order and O(4) (second order) phase transitions shown by dotted and dashed lines, which results in chance probabilities of 10^{-19} (7×10^{-13}), respectively.

151(3)(3) MeV, 175(2)(4) MeV and 176(3)(4) MeV, respectively. The first errors come from the finite temperature analysis, the second ones from the $T=0$ scale setting. We used physical quark masses and four lattice spacings corresponding to $N_t=4,6,8$ and 10. These results, especially the one coming from the chiral susceptibilty, are in contradiction with those of ref. [10]. They obtain $T_c=192(7)(4)$ MeV from both the chiral susceptibility and Polyakov loop susceptibility. They used $N_t=4$ and 6 lattices with a p4 improved action.

Since then both collaborations improved on their data. We have used physical quark masses also for the $T=0$ simulations and took even finer lattice spacings ($N_t=12$ and $N_t=16$ at one point) at finite temperature [9]. The HotQCD collaboration uses two lattice actions (asqtad and p4) and they extended their simulations to $N_t=8$ lattices [11]. In the following we compare the latest results of the two groups to each other.

A possible reason for the discrepancy could be the uncertainty coming from the scale setting. Different quantities can be used to set the lattice spacing and the results should not depend on this choice. Figure 3 shows the masses of the Ω baryon, $\phi(1020)$ meson and $K^*(892)$ meson as well as the ratio of the quark masses and f_K/f_π obtained from our $T=0$ simulations [9]. The agreement to the experimental values shows that using any of the hadron masses to set the lattice spacing would give the same results.

The light quark chiral condensate ($\langle\bar\psi\psi\rangle$) is minus one times the first derivative of the free energy density with respect to the light quark mass. It is ultraviolet divergent, a possible way of removing divergences was proposed in [12]. *If one assumes* that the additive divergences of the free energy density depend on the quark masses only through the combination $m_{ud}^2+m_s^2$, then one can get rid of the additive divergences in $\langle\bar\psi\psi\rangle$ by using the strange quark condensate

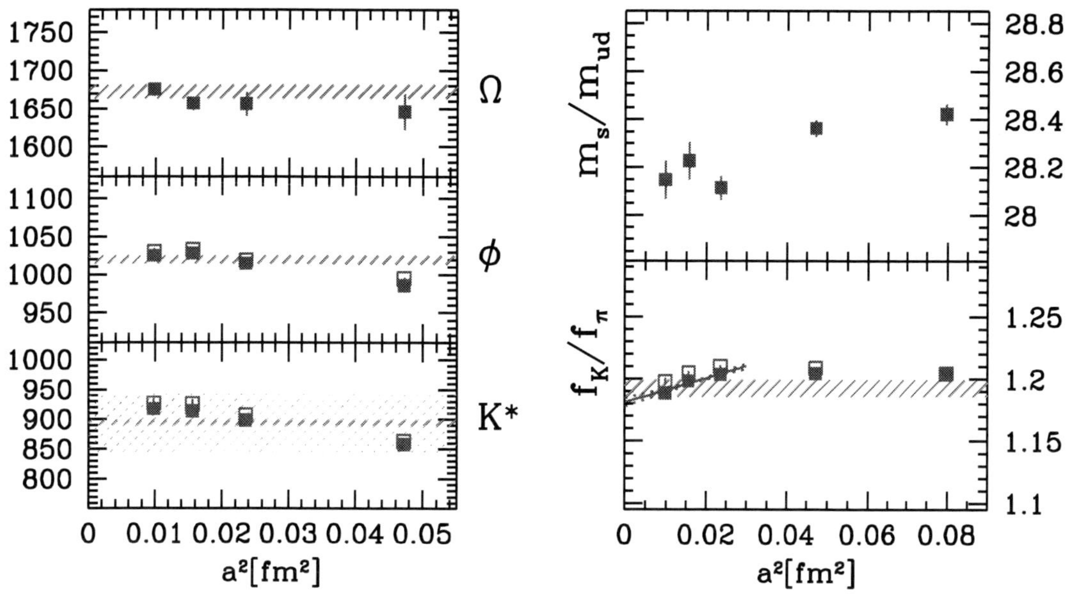

Figure 3. Left panel: masses of Ω baryon, $\phi(1020)$ meson and $K^*(892)$ meson in MeV on our four finest lattices as a function of the lattice spacing squared. Right panel: quark mass ratio and f_K/f_π for all five ensembles. See text for a detailed explanation.

($\langle\bar{s}s\rangle$):

$$\Delta_{l,s} = \langle\bar{\psi}\psi\rangle - \frac{2m_{ud}}{m_s}\langle\bar{s}s\rangle. \tag{1}$$

The remaining multiplicative divergences can be removed by dividing with the same quantity at zero temperature:

$$\Delta_{l,s} \to \frac{\Delta_{l,s}(T)}{\Delta_{l,s}(T=0)}. \tag{2}$$

On Figure 4(left) we plot this quantity as a function of the temperature. There is no significant lattice spacing or volume dependence for lattices of $N_t = 8, 10$ and 12 and for aspect ratios 3-4. For comparison we take the $N_t = 8$ data of the 'hotQCD' collaboration from [13]. We can see a huge disagreement between the curves in the transition regime. The shift between the curves of the different groups is in the order of 35 MeV.

The strange quark number susceptibility (χ_s) is defined as minus one times the derivative of the free energy density with respect to the square of the strange quark chemical potential. It is conveniently normalized by T^2, by which it will asymptotically reach one as the temperature is increased to infinity (Stefan-Boltzmann limit).

Our results on $N_t = 8, 10$ and 12 are shown in Figure 4(right). We also have an additional point on a very fine lattice ($N_t = 16$) at a high temperature. The comparison with the results of the 'hotQCD' collaboration (see Reference [14]) brings us to a similar conclusion as for the chiral condensate. Around the transition point there is an approximately 20 MeV shift between the results of the two groups. For larger than ~ 230 MeV temperatures our finer lattices are in good agreement with the 'hotQCD' results.

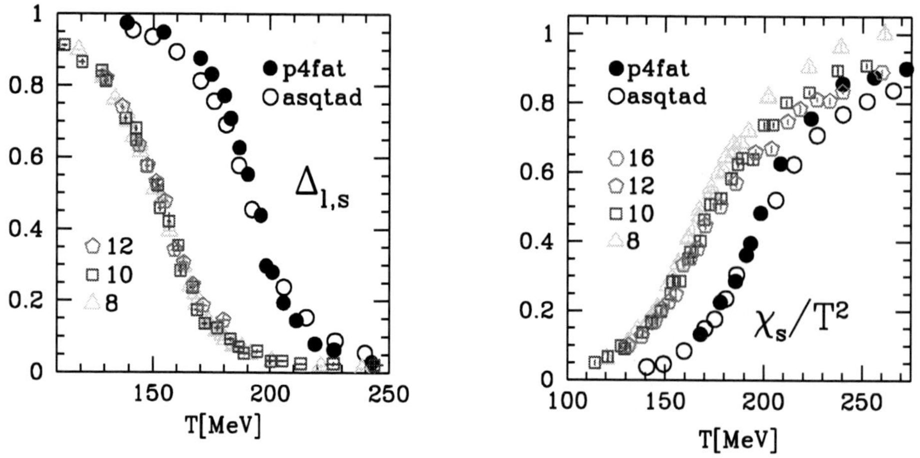

Figure 4. Left panel: renormalized chiral condensate as a function of the temperature. For comparison the 'HotQCD' results are also shown (see text). Right panel: strange quark number susceptibility.

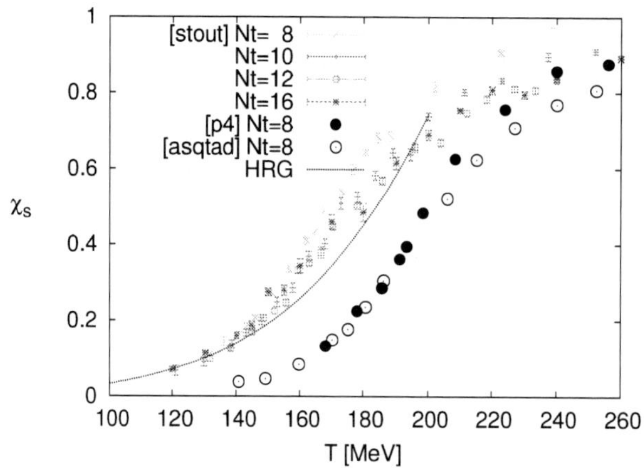

Figure 5. Preliminary results for the strange susceptibility on lattices upto N_t=16 and a comparison with the hadron resonance gas model (HRG). In our (Wuppertal-Budapest) analysis a stout improved action is used. A good agreement is found with the hadron resonance gas model using physical hadron masses. The results of the hotQCD Collaboration (asqtad and p4 actions) can be described only by using a non-physical distorted spectrum, characteristic to their choice of action and lattice spacings [15, 16].

6. Hadron resonance gas model and equation of state

In this section some preliminary results are presented. For the transition temperature determination even finer lattices are used, the temporal extension of the lattice is increased to 16. Earlier findings are confirmed. The results are compared with the predictions of the hadron resonance gas model (c.f. the transition curve for the strange quark number susceptibility on Figure 5). The physical hadron spectrum is applied for the HRG model. A very good agreement is found upto 150-170 MeV, above which the QCD transition takes place and free hadrons are not the dominant degrees of freedom any more. Note, that results of the hotQCD Collaboration can

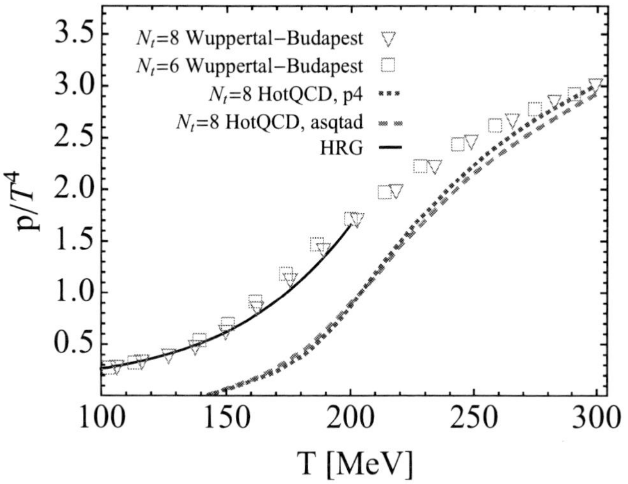

Figure 6. Preliminary results for the pressure as a function of the temperature with our (Wuppertal-Budapest) stout improved action. As it can be seen the hotQCD collaboration (asqtad and p4 actions) has a larger characteristic temperature, and a steeper increase than the Wuppertal-Budapest result. The difference can not be explained by a 5 MeV shift to the left expected by the cut-off dependence between N_t=6 and 8 [11]. The hotQCD result can only be described by using a hadron resonance gas model (HRG) with a non-physical distorted spectrum, characteristic to their choice of action and lattice spacings [15, 16]. The Wuppertal-Budapest result is in complete agreement with the hadron resonance gas model with physical input.

not be reproduced by a physical spectrum. Their results has a relatively large taste symmetry violation and a mass spectrum, which is distorted to larger masses. Incorporating this non-physical distorted spectrum [15, 16] into the hadron resonance gas model one can successfully reproduce the result of the hotQCD Collaboration.

Preliminary results of our Wuppertal-Budapest Collaboration for the equation of state calculation are also presented on Figure 6. Temporal extensions of 6 and 8 are used. In order to verify the correctness of the result in the continuum limit the calculations were repeated at various temperatures on N_t=10 and 12 lattices. Since there is a large discrepancy between the absolute scales of the Wuppertal-Budapest result and that of the hotQCD Collaboration one expects a large discrepancy in the equation of state, too. Indeed, there is a remarkable difference between the two results. The message is the same as that of the previous paragraph. The Wuppertal-Budapest results can be described by the hadron resonance gas model using physical mass parameters (c.f. Figure 6). Again, the agreement is valid upto 150-170 MeV, which signals the onset of the QCD transition. In contrast to that, the hotQCD results are reproduced only by using a non-physical distorted spectrum [15, 16].

More details on these results will be presented in a forthcoming publication [17].

7. Conclusions

Recent results of the Wuppertal-Budapest group were reviewed. The order of the cosmological is found to be an analytic crossover. Due to the crossover nature of the transition different definitions may lead to different transition temperatures. The T_c values published by the different groups are given in Table 1. Our latest results are consistent with the ones from 2006. The small difference comes from the fact that the experimental value of f_K has changed slightly since 2006. The discrepancy with the 'HotQCD' results is still present and has to be resolved

	$\Delta\chi_{\bar{\psi}\psi}/T^4$	$\Delta\chi_{\bar{\psi}\psi}/T^2$	$\Delta\chi_{\bar{\psi}\psi}$	$\Delta_{l,s}$	L	χ_s
Wuppertal-Budapest '09	146(2)(3)	152(3)(3)	157(3)(3)	155(2)(3)	170(4)(3)	169(3)(3)
Wuppertal-Budapest '06	151(3)(3)	-	-	-	176(3)(4)	175(2)(4)
RBCBC (ref. [10])	-	192(4)(7)	-	-	192(4)(7)	-

Table 1. Continuum extrapolated transition temperatures at the physical point for different observables and in different works. The first three columns give T_c obtained from the chiral susceptibility using different normalizations. The other three columns give T_c from the renormalized chiral condensate, renormalized Polyakov-loop and the strange quark number susceptibility.

by future work.

Preliminary results were presented for the transition temperature and for the equation of state on even finer lattices. All results are in agreement with our earlier findings. The results of the Wuppertal-Budapest collaboration can be reproduced by using a hadron resonance gas model with physical hadron spectrum, whereas the hotQCD results can only be described by using a hadron resonance gas model with a non-physical distorted spectrum, characteristic to their choice of action and lattice spacings [15, 16].

Acknowledgments
Computations were performed on the BlueGene at FZ Jülich and on clusters at Wuppertal and Budapest equipped with graphics cards. This work is supported in part by DFG grants SFB-TR 55, FO 502/1-2, EU grant (FP7/2007-2013)/ERC n°208740. Special thanks go to S.D. Katz, for the fruitful work and for his help in prepairing this manuscript. The author thanks Y. Aoki, S. Borsanyi, S. Durr, A. Jakovac, S. Krieg, C. Ratti and K.K. Szabo for the pleasant collaboration.

References

[1] D. J. Schwarz, Annalen Phys. **12**, 220 (2003) [arXiv:astro-ph/0303574].
[2] Z. Fodor and S. D. Katz, arXiv:0908.3341 [hep-ph].
[3] P. de Forcrand, S. Kim and O. Philipsen, PoS **LAT2007**, 178 (2007) [arXiv:0711.0262 [hep-lat]].
[4] G. Endrodi, Z. Fodor, S. D. Katz and K. K. Szabo, PoS **LAT2007**, 182 (2007) [arXiv:0710.0998 [hep-lat]].
[5] Y. Aoki, Z. Fodor, S. D. Katz and K. K. Szabo, JHEP **0601**, 089 (2006) [arXiv:hep-lat/0510084].
[6] Y. Aoki, G. Endrodi, Z. Fodor, S. D. Katz and K. K. Szabo, Nature **443**, 675 (2006) [arXiv:hep-lat/0611014].
[7] Y. Aoki, Z. Fodor, S. D. Katz and K. K. Szabo, Phys. Lett. B **643**, 46 (2006) [arXiv:hep-lat/0609068].
[8] B. Spang, http://www.cheresources.com/iapwsif97.shtml
[9] Y. Aoki, S. Borsanyi, S. Durr, Z. Fodor, S. D. Katz, S. Krieg and K. K. Szabo, JHEP **0906**, 088 (2009) [arXiv:0903.4155 [hep-lat]].
[10] M. Cheng *et al.*, Phys. Rev. D **74**, 054507 (2006) [arXiv:hep-lat/0608013].
[11] A. Bazavov *et al.*, arXiv:0903.4379 [hep-lat].
[12] M. Cheng *et al.*, Phys. Rev. D **77**, 014511 (2008) [arXiv:0710.0354 [hep-lat]].
[13] F. Karsch [RBC Collaboration and HotQCD Collaboration], J. Phys. G **35**, 104096 (2008) [arXiv:0804.4148 [hep-lat]].
[14] F. Karsch, PoS C **POD07**, 026 (2007) [arXiv:0711.0656 [hep-lat]].
[15] P. Huovinen and P. Petreczky, Nucl. Phys. A **837**, 26 (2010) [arXiv:0912.2541 [hep-ph]].
[16] P. Huovinen and P. Petreczky, arXiv:1005.0324 [hep-ph].
[17] S. Borsanyi et al., work in preparation.

Deconfinement and chiral transition with the highly improved staggered quark (HISQ) action

Alexei Bazavov[1] **and Peter Petreczky**[2] **[The HotQCD collaboration]**[1]
[1]Department of Physics, University of Arizona, Tucson, AZ 85721, USA
[2]Physics Department, Brookhaven National Laboratory, Upton, NY 11973, USA

Abstract.
We report on investigations of the chiral and deconfinement aspects of the finite temperature transition in 2+1 flavor QCD using the Highly Improved Staggered Quark (HISQ) action on lattices with temporal extent $N_\tau = 6$ and $N_\tau = 8$. We have performed the calculations for physical values of the strange quark mass m_s and the light quark masses $m_l = 0.2m_s$ and $0.05m_s$. Several finite temperature observables, including the renormalized Polyakov loop, the renormalized chiral condensate and the chiral susceptibility have been calculated. We also study the fluctuations and correlations of different conserved charges as well as the trace anomaly at finite temperature. We compare our findings with previous calculations that use different improved staggered fermion formulations: asqtad, p4 and stout.

1. Introduction

Quantum Chromodynamics (QCD) is the theory of strong interactions that successfully describes a wealth of experimentally observed phenomena. At high energies the coupling of the theory runs to zero, making interactions between the constituents of the theory (quarks and gluons) weaker, eventually approaching the limit of non-interacting ideal gas.

In the region of low energies (up to a few hundred MeV) the coupling of the theory is of $O(1)$ (thus justifying the name strong interactions) and the perturbation theory is not reliable. In this phase of the theory the degrees of freedom are hadrons, while quarks and gluons are confined into objects with zero net color charge.

The outlined picture suggests that at some temperature a transition from the confined hadronic phase to the deconfined quark-gluon plasma (QGP) phase takes place. Creating conditions for achieving QGP has been the subject of recent experiments at the Relativistic Heavy-Ion Collider (RHIC) at the Brookhaven National Laboratory and is going to be the goal of the future heavy-ion program at the Large Hadron Collider (LHC) at CERN.

Since the physics of the hadronic phase and the transition to QGP is essentially non-perturbative, lattice QCD techniques have been extensively applied to its studies. Putting fermions on the lattice leads to the infamous fermion doubling problem, and several lattice fermion formulations deal with it in different ways. Improved staggered fermion formulations

[1] The HotQCD Collaboration members are: A. Bazavov, T. Bhattacharya, M. Cheng, N.H. Christ, C. DeTar, S. Gottlieb, R. Gupta, U.M. Heller, C. Jung, F. Karsch, E. Laermann, L. Levkova, C. Miao, R.D. Mawhinney, S. Mukherjee, P. Petreczky, D. Renfrew, C. Schmidt, R.A. Soltz, W. Soeldner, R. Sugar, D. Toussaint, W. Unger and P. Vranas

are widely used to study QCD at non-zero temperatures and densities, see e.g. Ref. [1, 2] for recent reviews, for, at least, two reasons: they preserve a part of the chiral symmetry of the continuum QCD which allows one to study the chiral aspects of the finite temperature transition, and are relatively inexpensive to simulate numerically because due to absence of an additive mass renormalization the Dirac operator is bounded from below. However, there are at least two problems with staggered fermion formulation. The first one is the validity of the rooting procedure, i.e. the way to reduce the number of tastes (unwanted unphysical degrees of freedom that result from fermion doubling) from four to one, and the other is breaking of the taste symmetry at finite lattice spacing. The discussion of the validity of rooted staggered fermions is presented in Refs. [3, 4]. To reduce the taste violations smeared links, i.e. weighted averages of different paths on the lattice that connect neighboring points, are used in the staggered Dirac operator and several improved staggered formulations, like p4, asqtad, stout and HISQ differ in the choice of the smeared gauge links. The ones in the p4 and asqtad actions are linear combinations of single links and different staples [5, 6] and therefore are not elements of the SU(3) group. It is known that projecting the smeared gauge fields onto the SU(3) group greatly improves the taste symmetry [7]. The stout action [8] and the HISQ action implement the projection of the gauge field onto SU(3) (or simply U(3)) group and thus achieve better taste symmetry at a given lattice spacing. For studying QCD at high temperature it is important to use discretization schemes which improve the quark dispersion relation, thus eliminating the tree level $\mathcal{O}(a^2)$ lattice artifacts in thermodynamic quantities. The p4 and asqtad actions implement this improvement by introducing 3-link terms in the staggered Dirac operator. In this paper we report on exploratory studies of QCD thermodynamics with the HISQ action which combines the removal of tree level $\mathcal{O}(a^2)$ lattice artifacts with the addition of projected smeared links that greatly improve the taste symmetry. We also compare our results with the previous ones obtained with the asqtad, p4 and stout actions [9, 10, 11, 12, 13, 14] at comparable quarks masses and lattice spacings as well as with the asqtad action at smaller quark masses and larger N_τ [15].

2. Action and run parameters

The Highly Improved Staggered Quark (HISQ) action developed by the HPQCD/UKQCD collaboration [16] reduces taste symmetry breaking and decreases the splitting between different pion tastes by a factor of about three compared to the asqtad action. The net result, as recent scaling studies show [17], [18], is that a HISQ ensemble at lattice spacing a has scaling violations comparable to ones in an asqtad ensemble at lattice spacing $2/3a$.

In this exploratory study we used the HISQ action in the fermion sector and tree-level Symanzik improved gauge action without the tadpole improvement. The strange quark mass m_s was set to its physical value adjusting the quantity $\sqrt{2m_K^2 - m_\pi^2} = m_{\eta_{s\bar{s}}} \simeq \sqrt{2Bm_s}$ to the physical value 686.57 MeV. Two sets of ensembles have been generated along the two lines of constant physics (LCP): $m_l = 0.2m_s$ and $m_l = 0.05m_s$, where m_l is the u and d quark mass. In the first set runs were performed on $16^3 \times 32$ lattices at zero temperature and $16^3 \times 6$ at finite temperature, and in the second set on 32^4 and $32^3 \times 8$ lattices, correspondingly. To estimate the cutoff effects we have also generated a set of high-temperature ensembles on $24^3 \times 6$ lattices along $m_l = 0.05m_s$ LCP. The parameters and statistics of the runs are summarized in Table 1. The molecular dynamics (MD) trajectories have length of 1 time unit (TU) and the measurements were performed every 5 TUs at zero and 10 TUs at finite temperature. Typically, at least 300 TUs were discarded for equilibration at the beginning of the simulations.

The lattice spacing has been determined by measuring the static quark-anti-quark potential. As in previous studies by the MILC collaboration the static potential was calculated fixing the Coulomb gauge and considering temporal Wilson lines of different extent. Forming the ratio of these correlators and fitting them to constant plus linear Ansatz we extracted the static potential

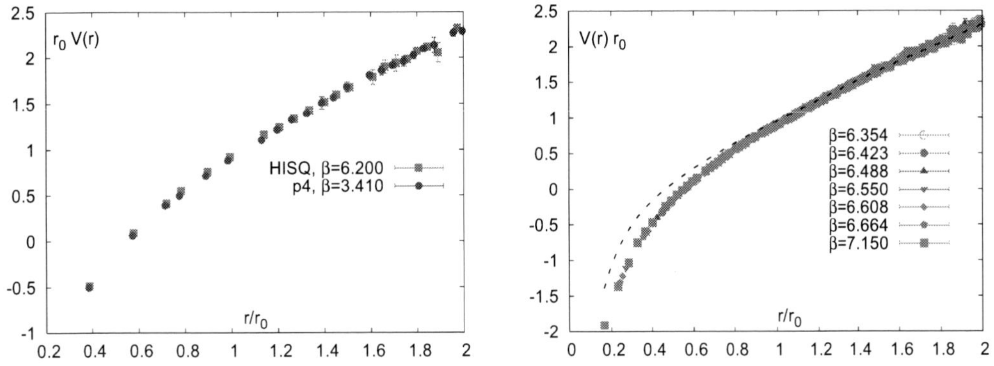

Figure 1. The static potential calculated for $m_l = 0.2m_s$ (left) and $m_l = 0.05m_s$ (right) in units of r_0. In the left figure we compare the HISQ result with the p4 result obtained at similar value of the lattice spacing. The dashed line on the right figure is the string potential (see text).

$V(r)$. We have calculated the Sommer scale r_0 and the related r_1 scale defined as

$$r^2 \frac{dV}{dr}\bigg|_{r=r_0} = 1.65, \qquad r^2 \frac{dV}{dr}\bigg|_{r=r_1} = 1.00, \tag{1}$$

The static potential calculated for $0.2m_s$ and $0.05m_s$ is shown in Fig. 1. The potential has been normalized to the string potential

$$V_{string} = -\frac{\pi}{12r} + \sigma r, \tag{2}$$

at $r = 1.5r_0$ or equivalently to the value $0.91/r_0$ at $r = r_0$. The additive constant determined by this normalization is used to calculate the renormalization constant for the Polyakov loop as will be discussed later. There is no large cutoff dependence visible in the static potential. Furthermore, the static potential calculated with the HISQ action is very similar to that calculated with the p4 action. For the ratio of r_0 and r_1 scales we get

$$r_0/r_1 = 1.459(3),\ m_l = 0.2m_s, \qquad r_0/r_1 = 1.481(6),\ m_l = 0.05m_s \tag{3}$$

To convert from lattice units to physical units we use the value $r_0 = 0.469$ fm determined in [19].

The masses of several hadrons measured on zero-temperature ensembles fall into ranges summarized in Table 2. They give an idea about the lines of constant physics in our calculations. The pion and kaon masses are most sensitive to the choice of the quark mass. We see that the $0.2m_s$ LCP corresponds to the lightest pion mass of about 300 MeV, while the other $0.05m_s$ LCP corresponds to the lightest pion mass of about 160 MeV. The later is very close to the physical value of about 140 MeV. For this reason we refer to $m_l = 0.05m_s$ as the physical quark mass. From the table we see that the systematic error in the choice of the quark masses is about 4% for the $0.2m_s$ LCP and about 3% for the $0.05m_s$ LCP.

As mentioned above the lattice artifacts for the HISQ action are significantly reduced compared to the asqtad action. The taste violations are strongest in the pseudo-scalar meson sector. Therefore we studied the splitting of pseudo-scalar meson masses in the 8 different multiplets for $m_l = 0.2m_s$. The quadratic splittings of the pseudo-scalar meson masses are independent of the quark mass to a very good approximation and therefore it is sufficient to

Journal of Physics: Conference Series **230** (2010) 012014 doi:10.1088/1742-6596/230/1/012014

Table 1. The parameters of the numerical simulations: gauge coupling, strange quark mass and the number of time units (TU), i.e. the number of MD trajectories for each run. Here TU, 0 stands for the number of time units for zero-temperature runs, while TU, T is the number of time units for finite-temperature runs. The last column give the number of time units for $24^3 \times 6$ runs.

$0.2m_s$ LCP runs					$0.05m_s$ LCP runs					
β	a, fm	am_s	TU, 0	TU, T	β	a, fm	am_s	TU, 0	TU, T	TU, T_6
6.000	0.2297	0.115	3,000	6,000	6.195	0.1899	0.0880	2,365	6,110	11,100
6.038	0.2212	0.108	3,000	6,000	6.285	0.1712	0.0790	2,300	6,190	6,750
6.100	0.2082	0.100	3,000	6,000	6.341	0.1612	0.0740	580	7,020	6,590
6.167	0.1954	0.091	3,000	6,000	6.354	0.1595	0.0728	2,295	5,990	5,990
6.200	0.1895	0.087	3,000	6,000	6.423	0.1475	0.0670	2,295	5,990	5,990
6.227	0.1848	0.084	3,000	6,000	6.488	0.1388	0.0620	2,295	5,990	8,790
6.256	0.1800	0.081	3,000	6,000	6.515	0.1352	0.0604	2,045	10,100	10,430
6.285	0.1752	0.079	3,000	6,000	6.550	0.1317	0.0582	2,295	5,990	7,270
6.313	0.1708	0.076	3,000	6,000	6.575	0.1278	0.0564	2,295	14,500	7,330
6.341	0.1665	0.074	3,000	6,000	6.608	0.1241	0.0542	2,295	5,990	6,560
6.369	0.1622	0.072	3,000	6,000	6.664	0.1173	0.0514	2,295	5,990	8,230
6.396	0.1582	0.070	3,000	6,000	6.800	0.1047	0.0448	2,295	5,990	7,000
6.450	0.1505	0.068	3,000	6,000	6.950	0.0921	0.0386	2,295	5,990	7,480
					7.150	0.0770	0.0320	2,295	5,990	4,770

Table 2. Ranges of masses (in MeV) of several hadrons for the two sets of ensembles.

	$0.2m_s$ LCP	$0.05m_s$ LCP
m_π	306-312	158-160
m_K	522-532	496-504
m_ρ	850-883	786-800
m_{K^*}	943-968	910-930
m_ϕ	1035-1061	1032-1057
m_N	1130-1183	1014-1083

study them for larger value of m_l. The results are shown in Fig. 2 and compared to the stout results. In the figure we also show the masses of all the pseudo-scalar mesons as function of the lattice spacing for asqtad and stout actions assuming that the lightest pseudo-scalar mass is fixed to its physical value. For the lattice spacing corresponding to the transition region for temporal extent $N_\tau = 8$ the mass of the heaviest pion is about $400 - 600$ MeV for the stout and asqtad action even for the physical values of the light quark mass. The quadratic splittings in HISQ calculations are 2 to 3 times smaller than in the calculations with the asqtad action and also somewhat smaller than for the stout action. However, the effect of mass splitting in the pseudo-scalar sector is non-negligible even for HISQ.

The smaller taste violations also result in better scaling of other hadron masses. In Fig. 3 the K^*, ρ and ϕ meson masses as well as the nucleon mass calculated for $0.05m_s$ are shown. As one can see there is a good agreement for the vector meson masses calculated on the lattice and the experimental values. The nucleon mass in the HISQ calculations is about 10% larger than

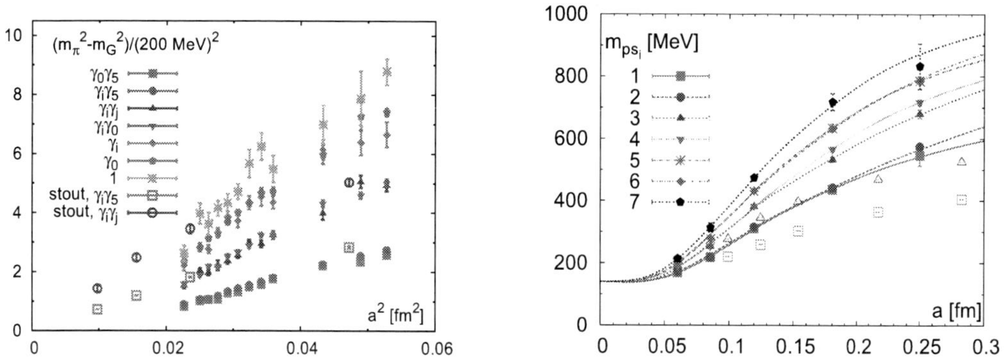

Figure 2. The splitting between different pion multiplets for the HISQ action at $0.2m_s$ and compared to the stout results shown as open symbols (left). In the right figure the non-Goldstone pseudo-scalar meson masses are shown as function of lattices assuming that the lightest (Goldstone) pion mass is fixed to its physical value. The open symbols in the right figure correspond to pseudo-scalar mesons labeled as "1" and "3" (or, equivalently, as $\gamma_i\gamma_5$ and $\gamma_i\gamma_j$) in the stout calculations.

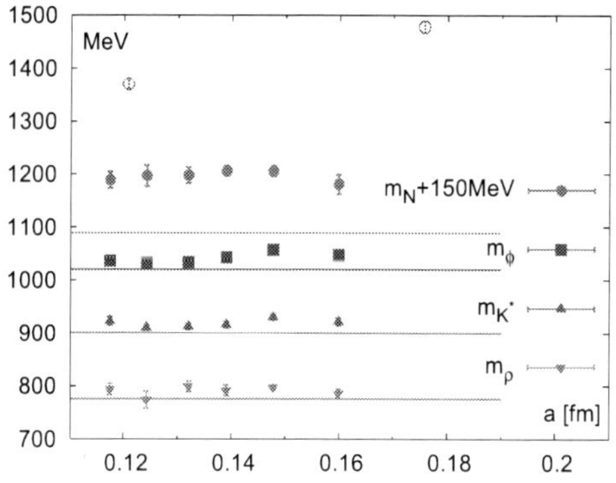

Figure 3. The hadron spectrum for the HISQ action at $m_l = 0.05m_s$ compared with experiment. The open symbols show the nucleon mass for the asqtad action.

the experimental value, but the discrepancy between the lattice calculation and the experiment is largely reduced compared to the asqtad calculations.

3. Deconfinement transition

The deconfinement transition is usually studied using the renormalized Polyakov loop [13, 20, 10, 11]. It is related to the free energy of a static quark anti-quark pair at infinite separation $F_\infty(T)$

$$L_{ren}(T) = \exp(-F_\infty(T)/(2T)), \tag{4}$$

and obtained from the bare Polyakov loop as

$$L_{ren}(T) = z(\beta)^{N_\tau} L_{bare}(\beta) = z(\beta)^{N_\tau} \left\langle \frac{1}{3} \mathrm{Tr} \prod_{x_0=0}^{N_\tau - 1} U_0(x_0, \vec{x}) \right\rangle . \tag{5}$$

Here the multiplicative renormalization constant $z(\beta)$ is related to the additive normalization of the potential $c(\beta)$ as $z(\beta) = \exp(-c(\beta)/2)$ discussed above. To make the comparison with

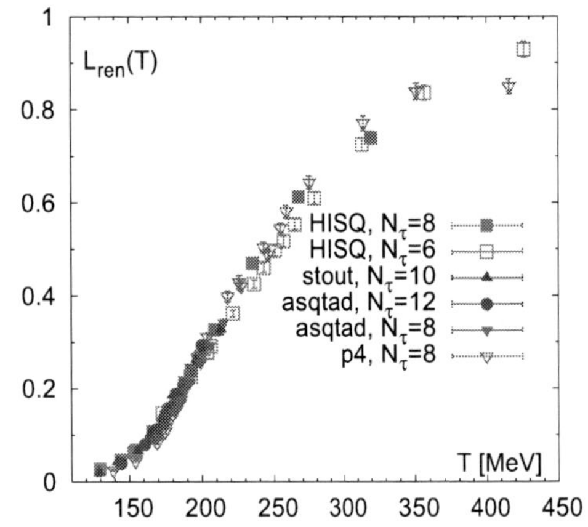

Figure 4. The renormalized Polyakov loop as function of the temperature calculated for the HISQ, stout, asqtad and p4 actions at the physical value of the light quark mass.

stout data the latter have to be multiplied by $\exp(-0.91/(r_0 T))$, as in the stout calculation the potential was normalized to zero at $r = r_0$.

Our results for the quark mass $m_l = 0.05 m_s$ on $32^3 \times 8$ lattices are shown in Fig. 4 and compared to the stout, asqtad and p4 calculations. As one can see the HISQ calculations agree reasonably well with the stout results if the scale is set by r_0 in the stout calculations [2]. On the other hand, the renormalized Polyakov loop calculated with the p4 and asqtad actions is considerably smaller at low temperatures. At temperatures $T > 200$ MeV we see good agreement for different actions. The decrease of $F_\infty(T)$, and thus the increase in the Polyakov loop could be related to onset of screening at high temperatures (e.g. see discussion in Ref. [22]). On the other hand, in the low-temperature region the increase of L_{ren} is related to the fact that there are many static-light meson states that can contribute to the static quark free energy close to the transition temperature, while far away from the transition temperature it is determined by the binding energy of the lowest static-light mesons. The large taste symmetry breaking in the static-light meson sector is probably responsible for the discrepancy between p4 and HISQ results in the low-temperature region. Overall the differences between p4 and HISQ results are the smallest for the renormalized Polyakov loop as we will see in the following sections. This is partly due to the fact that the Polyakov loop is purely a gluonic observable. The taste symmetry breaking associated with light quarks enters only through loops, and the corresponding effect is smaller as for the quark number susceptibilities, for example.

4. The chiral transition

In the limit of zero light quark masses QCD has a chiral symmetry and the finite temperature transition is a true phase transition. The order parameter for this transition is the light chiral condensate $\langle \psi\bar{\psi} \rangle_l$. However, even at finite values of the quark mass the chiral condensate will show a rapid change in the transition region indicating an effective restoration of the chiral symmetry. Since the chiral condensate has an additive ultraviolet renormalization we consider the subtracted chiral condensate [10]

$$\Delta_{l,s}(T) = \frac{\langle \bar{\psi}\psi \rangle_{l,\tau} - \frac{m_l}{m_s} \langle \bar{\psi}\psi \rangle_{s,\tau}}{\langle \bar{\psi}\psi \rangle_{l,0} - \frac{m_l}{m_s} \langle \bar{\psi}\psi \rangle_{s,0}}. \quad (6)$$

[2] In what follows we will show the stout results using the temperature scale set by r_0 instead of the kaon decay constant f_K. We used the published values of r_0 and f_K in Refs. [13, 14] to convert the two scales.

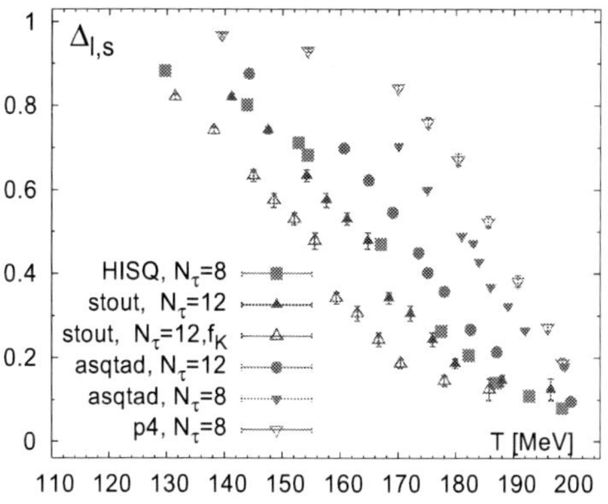

Figure 5. The subtracted chiral condensate for the HISQ action compared with calculations performed for the stout [14], asqtad [15] and p4 actions [12].

Here the subscripts l and s refer to the light and strange quark condensates respectively normalized per single flavor, while the subscripts 0 and τ refer to the zero and finite temperature cases. In Fig. 5 we show the renormalized chiral condensate calculated with the HISQ action and compare it with results obtained with the stout [14], asqtad and p4 [12] results. Our results agree reasonably well with the $N_\tau = 12$ stout results. On the other hand, the subtracted chiral condensate is considerably smaller than for the asqtad and p4 actions on $N_\tau = 8$. This is presumably due to the larger taste violating effects in the asqtad and p4 case. At finer lattice spacings the taste symmetry violations in the asqtad calculations become smaller and therefore the agreement between the asqtad calculations on $N_\tau = 12$ lattice and the HISQ calculations is much better (see Fig. 5). We also note that if f_K is used to set the scale there is no agreement between HISQ and stout calculations.

To study the chiral aspects of the QCD transition we also consider chiral susceptibility defined as the derivative of the light chiral condensate with respect to the quark mass[3]

$$\chi(T) = \frac{\partial \langle \bar{\psi}\psi \rangle_l}{\partial m_l} = \frac{T}{V}\left(\langle (\mathrm{Tr} M_l^{-1})^2 \rangle - \langle \mathrm{Tr} M_l^{-1} \rangle^2 - 2\langle \mathrm{Tr} M_l^{-2} \rangle \right), \tag{7}$$

where $M_l = D + 2m_l$ is the staggered fermion matrix for light quarks. The first term in the above formula describes the fluctuation of the chiral condensate and is also called disconnected chiral susceptibility. The second term is called the connected chiral susceptibility and corresponds to the integrated scalar meson correlation function. For sufficiently small quark masses the chiral susceptibility is dominated by the disconnected part. The transition temperature is defined as a location of the peak of the chiral susceptibility. In Fig. 6 we show the disconnected and connected chiral susceptibility for the HISQ action. We see that both connected and disconnected susceptibilities show a peak-like structure for temperatures around 170 MeV. The disconnected chiral susceptibility has a broad shoulder below the transition temperature. This is due to the effect of the Goldstone modes [21]. In Fig. 7 we compare the HISQ results for the disconnected chiral susceptibility with the corresponding results for asqtad calculations performed at $0.05m_s$ [15]. As one can see there is a good agreement between $N_\tau = 8$ HISQ calculations and $N_\tau = 12$ asqtad calculations.

While the disconnected chiral susceptibility is finite in the continuum limit, the connected part is quadratically divergent for non-vanishing quark masses. To remove this additive ultraviolet divergences the Budapest-Wuppertal group subracted the zero temperature piece from the chiral susceptibility [13]. To remove the multiplicative renormalization they also multiplied the chiral

[3] This is the 2-flavor chiral susceptibility as m_l denotes the light quark mass which is common for u and d quarks.

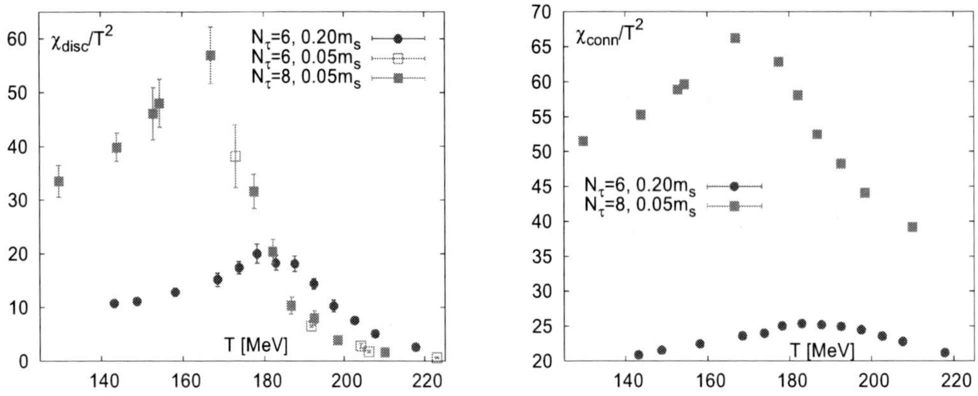

Figure 6. The disconnected (left) and connected (right) chiral susceptibilities calculated for HISQ action for two values of quark masses.

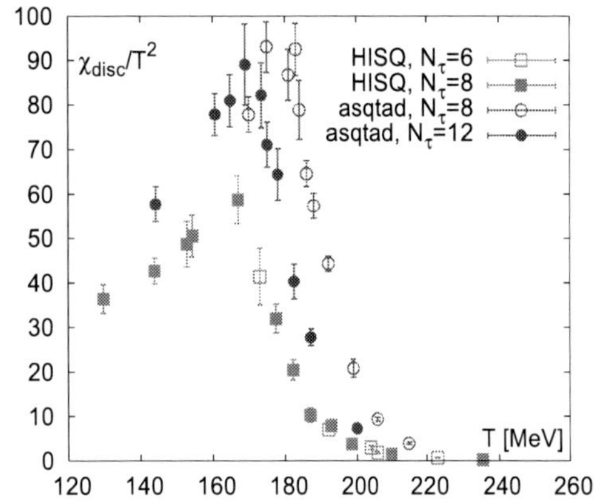

Figure 7. The comparison of the HISQ and asqtad [15] results for the disconnected chiral susceptibility at $m_l = 0.05 m_s$.

condensate by the light quark mass squared [13], i.e. they considered the quantity

$$\chi_r(T) = \frac{m_l^2}{T^4}\chi_R = \left(\frac{m_l^2}{T^4}\chi(T) - \chi(T=0)\right). \tag{8}$$

We calculated the same quantity for the HISQ action and show the comparison in Fig. 8. There is a reasonable agreement between HISQ and stout results if the r_0 is used to fix the scale. However, if the f_K is used to set the scale the stout data shift to smaller temperatures.

5. Fluctuations of conserved charges

Fluctuations and correlations of conserved charges are good probes of deconfinement because they are sensitive to the underlying degrees of freedom, i.e. they can tell whether the relevant degrees of freedom of the system at a given temperature are hadronic or partonic. Here we consider quadratic fluctuations and correlations of conserved charges defined as

$$\frac{\chi_i(T)}{T^2} = \frac{1}{T^3 V}\frac{\partial^2 \ln Z(T,\mu_i)}{\partial(\mu_i/T)^2}\bigg|_{\mu_i=0}, \tag{9}$$

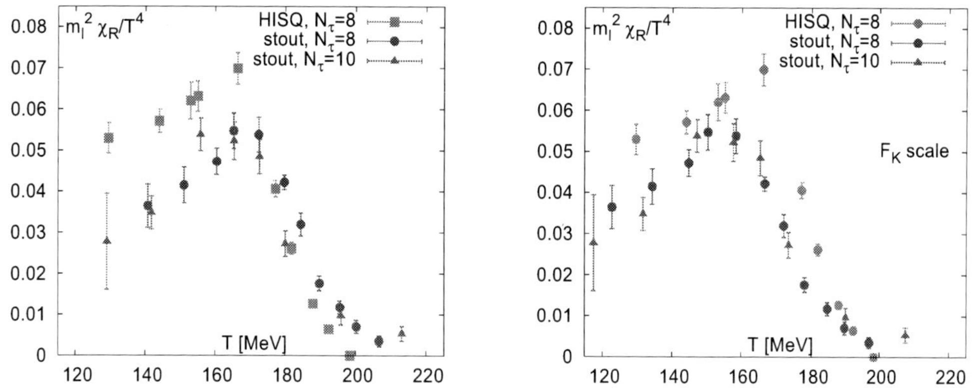

Figure 8. The renormalized chiral susceptibility defined by Eq. (8) for the stout action with lattice spacing fixed by r_0 (left) and f_K (right) [13]. We compared the stout results with the HISQ calculations at $m_l = 0.05m_s$.

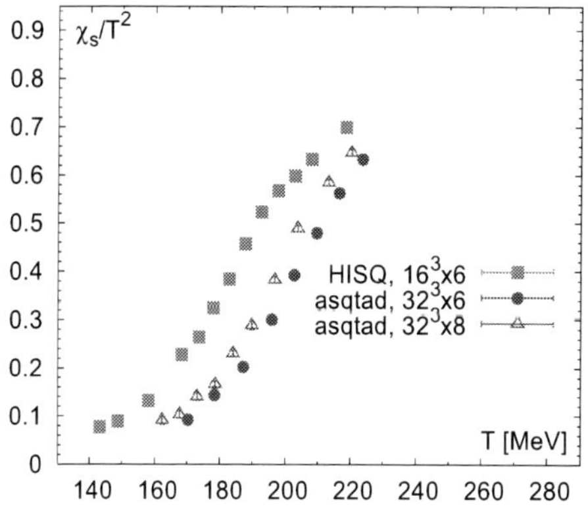

Figure 9. Strangeness fluctuations calculated with the HISQ action and compared to the asqtad action [15] for $m_l = 0.2m_s$.

$$\frac{\chi_{11}^{ij}(T)}{T^2} = \frac{1}{T^3 V} \frac{\partial^2 \ln Z(T, \mu_i, \mu_j)}{\partial(\mu_i/T)\partial(\mu_j/T)}\Bigg|_{\mu_i=\mu_j=0} . \tag{10}$$

Here indeces i and j refer to different conserved charges, like quark number, baryon number, strangeness, isospin etc.

We start our discussion with the strangeness fluctuation, since this quantity is often discussed in the literature in connection with deconfinement [11, 13, 14]. At low temperatures strangeness is carried by massive hadrons and therefore its fluctuations are suppressed. At high temperatures strangeness is carried by quarks and the effect of the non-zero strange quark mass is small. Therefore, in the transition region the strangeness fluctuation rises and eventually reaches a value close to that of an ideal quark gas. For $m_l = 0.2m_s$ our numerical results for strangeness fluctuations are shown in Fig. 9 and compared with asqtad results. As one can see the strangeness fluctuations are larger for the HISQ action at low temperatures, $T < 210$ MeV. In other words, the transition region in the HISQ calculation shifts toward smaller temperatures. This behavior is in fact expected. Due to smaller taste symmetry violation pseudo-scalar masses as well as other hadron masses are smaller and therefore strangeness fluctuations are larger. For the physical quark mass, $m_l = 0.05m_s$ we compare our calculations with the results obtained

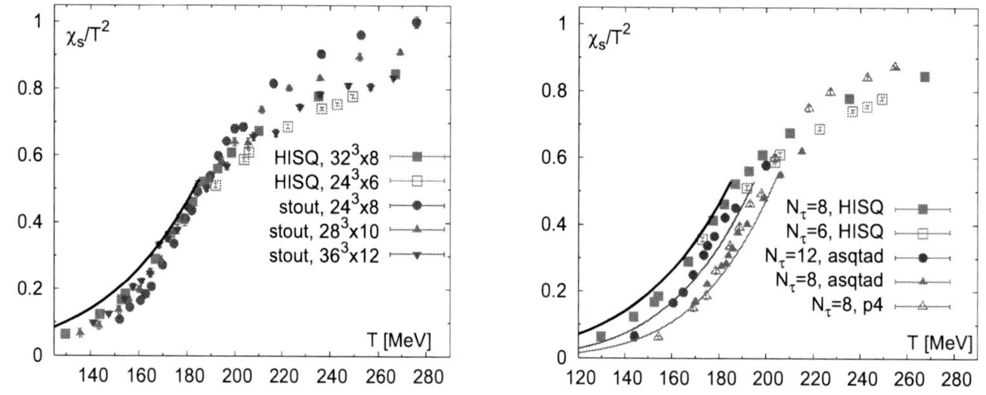

Figure 10. The strangeness fluctuations calculated with the HISQ action for $m_l = 0.05m_s$ and compared to the results obtained with the p4, asqtad and stout actions. The lines show the prediction of the HRG model.

using the p4 action [12] and the stout action [13, 14]. At low temperatures, $T < 200$ MeV the HISQ results are significantly larger than the p4 results but are in good agreement with the stout results. At high temperatures, $T > 200$ MeV the strangeness fluctuations calculated with the HISQ action are in reasonable agreement with the p4 results as well as the $N_\tau = 12$ stout results. We also see that in this temperature region the stout results show some cutoff (N_τ) dependence. This is due to the fact that the tree level $\mathcal{O}(a^2)$ lattice artifacts are not removed in the stout action. In the low temperature region we expect that the Hadron Resonance Gas (HRG) model gives a reasonably good description of the thermodynamic quantities, including strangeness fluctuations. Therefore in Fig. 10 we show the prediction of the HRG model. As one can see all lattice results fall below the HRG model result, although the difference between the lattice data and HRG is the smallest for the HISQ and stout actions. The discrepancy between the HRG model and the lattice data becomes larger at smaller temperatures, although the model is expected to be more reliable there. As explained in Ref. [23] this is due to taste violations leading to the distortion of the hadron spectrum, especially in the pseudo-scalar meson sector. Due to non-negligible splitting in the pseudo-scalar meson mass the contribution of kaons to strangeness fluctuations appears to be smaller than expected in the continuum [23]. As has been discussed in Ref. [23] this is mostly due to the discretization errors in the hadron spectrum. Taking into account the lattice spacing dependence of the hadron masses which enter the HRG calculations it is possible to get good agreement with lattice data [23]. Therefore in the figure we also show the HRG model calculations with hadron masses evaluated at lattice spacing corresponding to $N_\tau = 8$ and $N_\tau = 12$ lattices using the formulas given in Ref. [23]. As one can see from the figure there is a very good agreement between this modified HRG calculation and lattice results. We note again, that for $T < 160$ MeV the splitting in the pseudo-scalar sector is the dominant source of the cutoff effects in the strangeness susceptibility. Finally we note that, when comparing the lattice results with the prediction of HRG we include all resonances up to 2.5 GeV. Including resonances only up to 2.0 GeV will not change the strangeness fluctuations significantly, but further lowering the mass cutoff of the resonances will decrease the fluctuations for $T > 160$ MeV.

We study the quark number and baryon number fluctuations which are sensitive to the light (non-strange) quark or hadron sector. The numerical results are shown in Fig. 11. Due to the much smaller quark mass the taste symmetry breaking effects are much more important in this

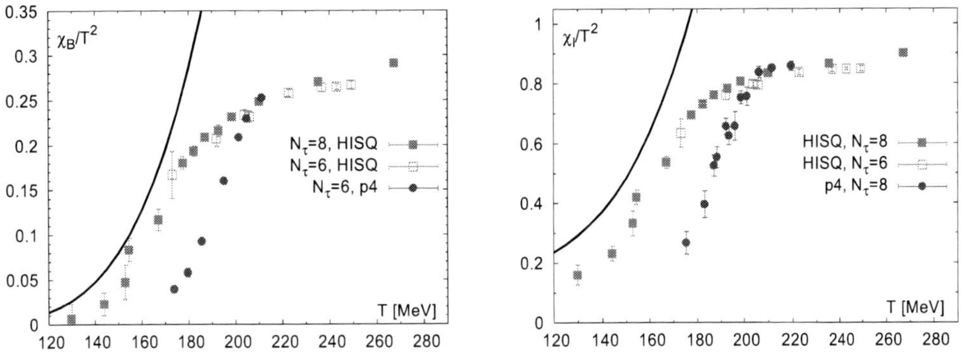

Figure 11. Fluctuation of the baryon number (left) and quark number (right) calculated with the HISQ action. Also shown as the solid line is the prediction of the HRG model. The lattice results are compared with the previous calculations performed with the p4 action [25].

sector. As the consequence the deviations from the HRG result are larger for these quantities and are significant even for HISQ calculations. However, the deviations between HRG and lattice results obtained with the HISQ action are much reduced compared to the previous calculations performed with the p4 action on $N_\tau = 4$ and 6 lattices at $m_l = 0.1 m_s$ [25].

Next let us consider strangeness-baryon number and u- and s-quark number correlations. The numerical results for these quantities are shown in Fig. 12. At low temperatures strangeness-baryon number correlations are due to the presence of strange baryons. At high temperatures strange quarks are the relevant degrees of freedom and carry 1/3 unit of the baryon number and minus one unit of strangeness. As the result strangeness baryon number correlation approaches $-1/3$ at high temperatures. The correlations of and u- and s-quark numbers at low temperatures are due to the presence of strange hadrons. Here both strange mesons and baryons contribute to this quantity. At high temperatures quarks are weakly interacting and the correlation between u- and s-quark numbers is very small. In fact, in perturbation theory it starts at order $g^6 \log(g)$ and numerically gives a result which is 10^4 times smaller than the quark number susceptibility [24]. At low temperatures correlation of the baryon number and strangeness is very well described by HRG. The lattice results for u- and s-quark number correlations obtained with the HISQ action are not incompatible with HRG, but due to the large statistical errors it is difficult to get more precise conclusions. Correlations of conserved charges have been also calculated with the p4 action. In Fig. 12 we also show the corresponding results. As one can see from the figure the p4 results differ significantly from the HISQ results in the low temperature region. This is presumably due to large taste symmetry violations in the p4 calculations.

Finally let us discuss the isospin fluctuations. The numerical results for the isospin fluctuations are shown in Fig. 13 and compared with the light quark number fluctuations. In this figure we compare again the light quark number fluctuations with the strangeness fluctuations. The isospin fluctuations are very similar to the light quark number fluctuations at high temperatures. At low temperatures and in the transition region there are differences between the light quark number and isospin fluctuations. The strangeness and the light quark number fluctuations are also very different at all temperatures $T < 260$ MeV. Only above this temperatures they become similar. One possible reason for the observed difference could be the different sensitivity of these quantities to the singular part of the free energy density. As discussed in Ref. [11] the light quark number fluctuations may be sensitive to the singular part of the free energy density, while strangeness fluctuations are not sensitive to it. Isospin fluctuations

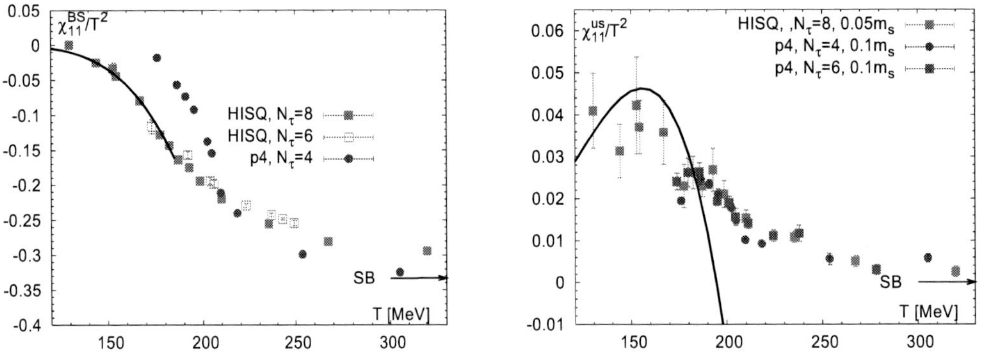

Figure 12. Cross-correlation of the baryon number and strangeness (left) and and u- and s-quark number (right) calculated with the HISQ action. Also shown is the prediction of the HRG model. The lattice results are compared with the previous calculations performed with the p4 action [25].

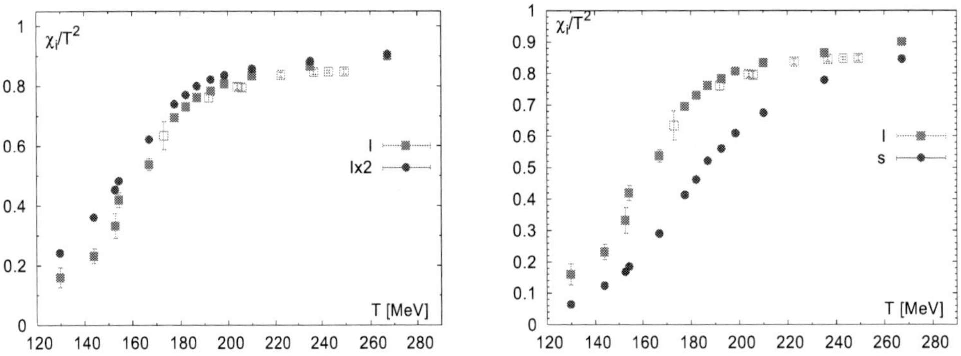

Figure 13. The light quark number susceptibility compared to the isospin susceptibility (left) and to the strangeness susceptibility (right) for $m_l = 0.05m_s$.

are also not sensitive to the singular part of the free energy density as they are not coupled to the sigma field [28].[4] It remains to be seen if the differences between the isospin and light quark number are due to the singular part of the free energy density. Current statistical errors do not allow to make firm conclusions.

Correlations of conserved charges have also been calculated for $m_l = 0.2m_s$. The results are shown in Fig. 14. The results are similar to those obtained for the lighter quark mass. But due to the large quark mass the corresponding hadron masses are larger and the agreement with HRG is not that good. In the figure we also compare our numerical results with the results from previous calculations done with the p4 action [25]. There are significant differences between HISQ and p4 results also for this quark mass. Thus cutoff effects are the dominant source of the difference between p4 and HISQ results. In Fig. 15 we show the light quark number fluctuations in comparison with the isospin and strangeness fluctuations for $m_l = 0.2m_s$. As one can see from the figure the differences between the light quark number and strangeness fluctuations as well as light quark number and isospin fluctuations are smaller for this quark mass. This could be

[4] P.P. thanks F. Karsch and K. Redlich for the discussions on this point.

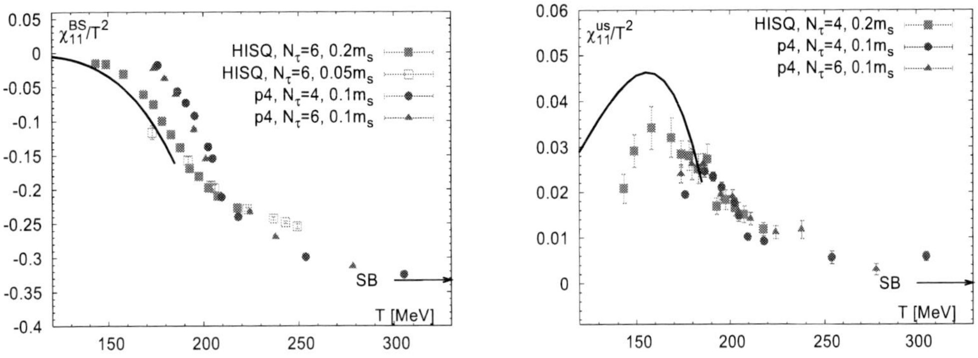

Figure 14. Cross-correlation of the baryon number and strangeness (left) and and u- and s-quark number (right) calculated with the HISQ action for $m_l = 0.2m_s$ and compared with p4 results. Also shown is the prediction of the HRG model.

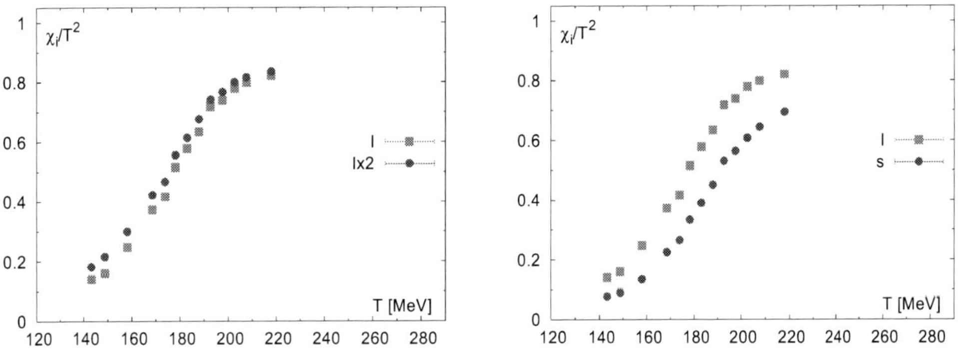

Figure 15. The light quark number susceptibility compared to the isospin susceptibility (left) and to the strangeness susceptibility (right) for $m_l = 0.2m_s$.

due to the fact that the differences in the masses of strange and non-strange hadrons are smaller for $m_l = 0.2m_s$, and also due to the reduced sensitivity of the light quark number fluctuations to the singular part.

To summarize, in this section we have studied fluctuations and correlations of different conserved charges as function of the temperature. We have found that at low temperatures, $T < 170$ MeV, the fluctuations can be well understood in terms of the resonance gas model. To get a quantitative agreement with HRG it may be necessary to consider modifications of the hadron spectrum due to finite lattice spacing in the HRG calculations. This has been shown in some detail for the strangeness fluctuations. At high temperatures, $T > 250$ MeV, the fluctuations and correlations can be understood in terms of weakly interacting quark gas. In the transition region 170 MeV $< T <$ 250 MeV there are no well defined quasi-particles which carry the conserved charges, i.e. it is difficult to define the relevant degrees of freedom. Thus, the deconfinement transition is a very gradual process if defined as the transition from hadronic to partonic degrees of freedom and it is impossible to associate a meaningfull transition temperature with it. Some fluctuations may be sensitive to the singular part of the free energy density. In this case the inflection points will be related to the chiral transition temperature. In the literature

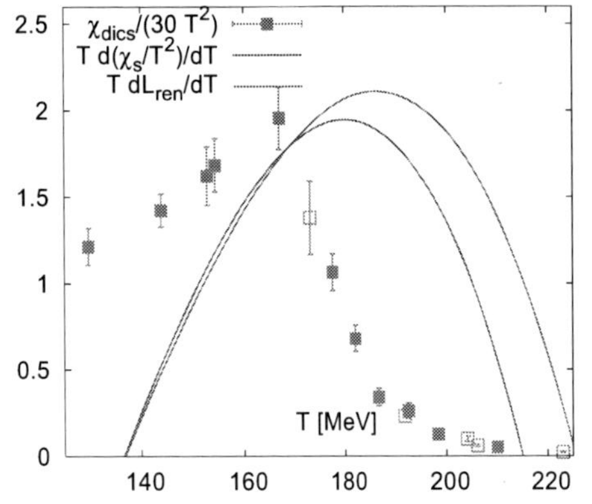

Figure 16. The comparison of the disconnected chiral susceptibility and the derivatives of the Polyakov loop and χ_s.

(see e.g. Refs. [13, 14]) attempts to define the deconfinement transition temperature as inflection points in the strangeness fluctuations and the renormalized Polyakov loop have been made. From the above discussion it is clear that the deconfinement transition temperature defined this way has little to do with the deconfinement temperature in the limit of large or infinite quark mass, which is related to the light modes of the $Z(N)$ center symmetry breaking. Both strangeness fluctuations and the renormalized Polyakov loop are dominated by the regular part of the free energy density. Thus, inflection points of these quantities need not to be related to the chiral transition temperature. In particular, these inflection points could be higher than the chiral transition temperature. This is shown in Fig. 16, where we compare the disconnected chiral susceptibility with the derivatives of the renormalized Polyakov loop and χ_s.

6. Trace anomaly

We also calculated the trace anomaly as function of temperature using the HISQ action for $m_l = 0.2m_s$ and $0.05m_s$. This quantity is often considered in lattice calculations as the pressure and other thermodynamic quantities are obtained by integrating over the trace anomaly to a given temperature. Previous calculations with the p4 and asqtad action gave a trace anomaly which was significantly below the HRG result. This deviation from the resonance gas complicated the use of lattice equation of state in hydrodynamic models since it was not clear how to implement consistent freezout at the hadronic stage of the evolution (see the discussion in Ref. [23]). In Fig. 17 we show our results for the trace anomaly, $\epsilon - 3p$ for the HISQ action and compare them with previous calculations performed with the asqtad and p4 actions. While at high temperatures ($T > 250$ MeV) all lattice data agree, there are noticeable differences in the low-temperature region. The lattice results obtained with the p4 and asqtad actions are lower than the HISQ results. This is due to the smaller breaking of the taste symmetry in the calculations performed with the HISQ action. We also compare the lattice calculations with the parametrization of the trace anomaly obtained from the matching the HRG at low temperature $T < 180$ MeV with the fit to the p4 lattice data at high temperatures for $T > 250$MeV. In the intermediate region the parametrization is constrained by the value of the entropy density at high temperatures. This parametrization gives an entropy density which is 5% below the ideal gas value at temperature of 800 MeV, and is called $s95p$-v1 [23]. This parametrization was used in hydrodynamic model [23]. We also compare the lattice results with the parametrization obtained by Laine and Schroeder [26]. Within the statistical errors there is a good agreement between the lattice results obtained on $N_\tau = 8$ lattice with $m_l = 0.05m_s$ and the $s95p$-v1 parametrization. In the peak region the trace anomaly calculated on $N_\tau = 6$ lattice is significantly below the $N_\tau = 8$ results, while at high temperatures the two calculations agree. Thus, the difference

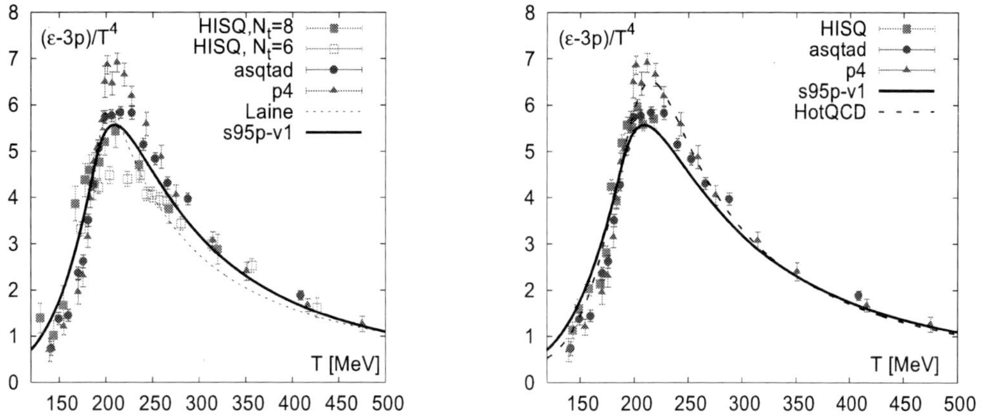

Figure 17. The trace anomaly calculated with the HISQ action for $0.05m_s$ (left) and $0.2m_s$ (right) and compared with p4 and asqtad calculations. The open symbols in the left figure show the $N_\tau = 6$ results. We also compare the lattice results with the parametrizations of the trace anomaly: $s95p$-v1 from Ref. [23] and the one by Laine and Schroeder [26]. In the right figure we also show the HotQCD parametrization of the trace anomaly [11].

between $N_\tau = 6$ and $N_\tau = 8$ calculations follows the expected pattern of the cutoff dependence of the pressure in the free theory, namely at high temperatures the pressure obtained on $N_\tau = 6$ lattice should be below the pressure obtained on $N_\tau = 8$ lattice if the Naik term is used [27]. We also compare the lattice results with the HotQCD parametrization of the trace anomaly [11]. As can be seen in Fig. 17 the HotQCD parametrization agrees with $s95p$-v1 parametrization at high temperatures. In the peak region and in the low-temperature region it is below the $s95p$-v1 parametrization and the HISQ data. This is due to the fact that it is based on $N_\tau = 8$ p4 and asqtad data in the entire temperature region.

7. Conclusions

In this contribution we discussed recent results on QCD thermodynamics with the HISQ action. The use of this action significantly reduces the effect of taste symmetry violation of the staggered fermion formulation. In particular, the quadratic splitting of pseudo-scalar mesons is the smallest for the HISQ action at a given value of the lattice spacing. The reduced lattice artifacts result in more realistic hadron spectrum. In particular, the vector meson masses are close to their continuum values for lattice spacings corresponding to the crossover region on $N_\tau = 8$ lattices. We have calculated different thermodynamic quantities with the HISQ action on $N_\tau = 6$ and $N_\tau = 8$ lattices, including the renormalized Polyakov loop, the chiral condensate, chiral susceptibility, trace anomaly and the fluctuations and correlations of conserved charges. We have found that at low temperatures the Polyakov loop agrees well with the results obtained with the stout action, while at high temperatures it agrees with the p4 and asqtad calculations if the lattice spacing is set by the Sommer scale r_0. The renormalized chiral condensate and chiral susceptibility obtained with the HISQ action agree quite well with the stout calculations but disagree with the results obtained with the p4 and asqtad actions on $N_\tau = 8$ lattices. However, the asqtad results on $N_\tau = 12$ lattices agree well with the ones obtained with the HISQ action on $N_\tau = 8$.

We studied in detail different fluctuations and correlations of conserved charges. When comparing our results with previous ones obtained with the p4 and asqtad actions we have found differences in the low temperature region. Those differences are due to the reduced taste symmetry violation in the HISQ action which gives a more realistic hadron spectrum. We have compared the lattice results in the low temperature region with the prediction of the hadron resonance gas. We have found good agreement between the lattice results and the HRG once the cutoff effects have been taken into account. We found that the agreement between the lattice results and physical HRG is best for the HISQ action. The deconfinement transition can be understood as the transition from hadronic to quark degrees of freedom as seen by the different fluctuations/correlations. We have found that this transition is not abrupt but rather gradual. This implies that it is not possible to define a meaningful deconfinement transition temperature.

Finally we considered the trace anomaly. The calculations performed with the HISQ action give results which agree quite well with the previous results obtained with the p4 and asqtad actions on $N_\tau = 6$ and $N_\tau = 8$ lattices at high temperatures. Differences are seen in the transition and in the low-temperature regions. The HISQ results on the trace anomaly agree well with the HRG in the low temperature region, and overall they agree with the $s95p$-v1 parametrization that has been proposed for use in hydrodynamic models [23].

Acknowledgements
This work has been supported in part by contracts DE-AC02-98CH10886 and DE-FC02-06ER-41439 with the U.S. Department of Energy and contract 0555397 with the National Science Foundation. The numerical calculations have been performed using the USQCD resources at Fermilab as well as the BlueGene/L at the New York Center for Computational Sciences (NYCCS). We thank Z. Fodor and S. Katz for sending us the stout data.

[1] C. E. DeTar, PoS **LATTICE2008**, 001 (2008)
[2] P. Petreczky, Nucl. Phys. A **830**, 11C (2009); P. Petreczky, Nucl. Phys. Proc. Suppl. **140**, 78 (2005)
[3] S. R. Sharpe, PoS **LAT2006**, 022 (2006)
[4] M. Creutz, PoS **LAT2007**, 007 (2007)
[5] F. Karsch, E. Laermann and A. Peikert, Nucl. Phys. B **605**, 579 (2001)
[6] K. Orginos, D. Toussaint and R. L. Sugar [MILC Collaboration], Phys. Rev. D **60**, 054503 (1999)
[7] A. Hasenfratz, Nucl. Phys. Proc. Suppl. **119**, 131 (2003)
[8] Y. Aoki, Z. Fodor, S. D. Katz and K. K. Szabo, JHEP **0601**, 089 (2006)
[9] C. Bernard *et al.* [MILC Collaboration], Phys. Rev. D **71**, 034504 (2005)
[10] M. Cheng *et al.*, Phys. Rev. D **77**, 014511 (2008)
[11] A. Bazavov *et al.*, Phys. Rev. D **80**, 014504 (2009)
[12] M. Cheng *et al.*, Phys. Rev. D **81**, 054504 (2010)
[13] Y. Aoki, Z. Fodor, S. D. Katz and K. K. Szabo, Phys. Lett. B **643**, 46 (2006)
[14] Y. Aoki, S. Borsanyi, S. Durr, Z. Fodor, S. D. Katz, S. Krieg and K. K. Szabo, JHEP **0906**, 088 (2009)
[15] HotQCD Collaboration, work in progress
[16] E. Follana *et al.* [HPQCD collaboration and UKQCD collaboration], Phys. Rev. D **75** (2007) 054502
[17] A. Bazavov *et al.* [MILC Collaboration], PoS **LAT2009**, 123 (2009)
[18] A. Bazavov *et al.* [MILC collaboration], arXiv:1004.0342 [hep-lat].
[19] A. Gray *et al.*, Phys. Rev. D **72**, 094507 (2005)
[20] O. Kaczmarek, F. Karsch, P. Petreczky and F. Zantow, Phys. Lett. B **543**, 41 (2002)
[21] S. Ejiri *et al.*, Phys. Rev. D **80**, 094505 (2009)
[22] P. Petreczky, Eur. Phys. J. C **43**, 51 (2005)
[23] P. Huovinen and P. Petreczky, Nucl. Phys. A **837**, 26 (2010)
[24] J. P. Blaizot, E. Iancu and A. Rebhan, arXiv:hep-ph/0303185.
[25] M. Cheng *et al.*, Phys. Rev. D **79**, 074505 (2009) [arXiv:0811.1006 [hep-lat]].
[26] M. Laine and Y. Schroder, Phys. Rev. D **73**, 085009 (2006)
[27] U. M. Heller, F. Karsch and B. Sturm, Phys. Rev. D **60**, 114502 (1999)
[28] C. R. Allton *et al.*, Phys. Rev. D **71**, 054508 (2005)

Journal of Physics: Conference Series **230** (2010) 012015 doi:10.1088/1742-6596/230/1/012015

Lattice QCD and Hydro/Cascade Model of Heavy Ion Collisions

Michael Cheng

E-mail: cheng24@llnl.gov
Lawrence Livermore Laboratory
7000 East Avenue, L-415
Livermore, CA 94550

Abstract. We report here on a recent lattice study of the QCD transition region at finite temperature and zero chemical potential using domain wall fermions (DWF). We also present a parameterization of the QCD equation of state obtained from lattice QCD that is suitable for use in hydrodynamics studies of heavy ion collisions. Finally, we show preliminary results from a multi-stage hydrodynamics/hadron cascade model of a heavy ion collision, in an attempt to understand how well the experimental data (*e.g.* particle spectra, elliptic flow, and HBT radii) can constrain the inputs (*e.g.* initial temperature, freezeout temperature, shear viscosity, equation of state) of the theoretical model.

1. Introduction

The aim of the various high energy heavy ion collision (HIC) programs at experimental facilities such as RHIC, SPS, LHC, and FAIR is to study the properties of nuclear matter under the extreme conditions of high energy and high density. In particular, at sufficiently high energy density, it is predicted that normal hadronic matter will undergo a transition into a wholly new state of matter, the quark-gluon plasma (QGP) [1; 2], where the constituent quarks and gluons will no longer be confined within hadronic states and chiral symmetry is restored.

In principle, the properties of the QGP can be calculated directly from the underlying theory, Quantum Chromodynamics (QCD). In practice, *ab initio* calculations via lattice QCD require very large-scale computing resources. It is only recently that high-precision lattice QCD results[3–8] have become available that have the potential to quantatively constrain models of heavy ion collisions. In Sec. 2 we discuss a lattice calculation[9] that uses the domain wall fermion method[10] to calculate the crossover temperature of QCD at zero chemical potential. In section 3 we present a parameterization of a high-precision lattice calculation of the QCD Equation of State (EoS) that is useful for modeling heavy ion collisions.

In addition to theoretical calculations of the QGP, a robust understanding of the dynamics of a heavy ion collision is needed in order to translate experimental results into constraints on

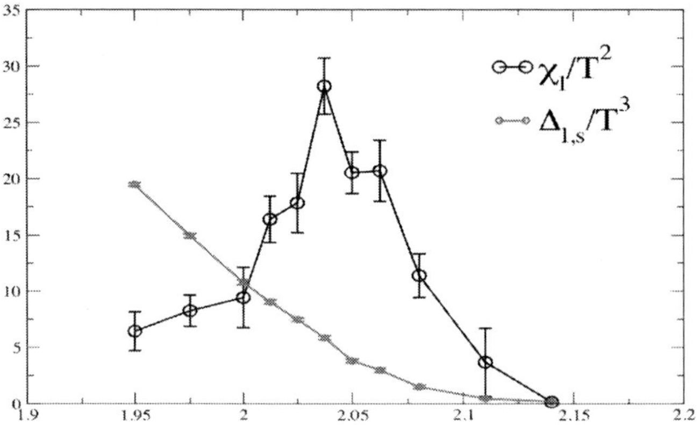

Figure 1. The disconnected chiral susceptibility, χ_l/T^2 and the subtracted chiral condensate, $\Delta_{l,s}/T^3$. as a function of the bare lattice coupling $\beta = 6/g_0^2$. The peak in the disconnected chiral susceptibility corresponds to $T \approx 170$ MeV.

the physical properties of hot QCD matter. Several approaches, such as Boltzman transport, hydrodynamics, and hadronic cascade [11–14] have been used to model heavy ion collisions. However, the relevant physics changes so drastically over the lifetime of a HIC that it is difficult to capture all the aspects of a collision with a single model. In Sec. 4 we present preliminary results on a multi-stage model of a HIC that incorporates Glauber initial conditions with pre-equilibrium flow, 2-d viscous hydrodynamics, Cooper-Frye freezeout, and a hadronic cascade.

2. QCD Transition using domain wall fermions

The location of the QCD crossover has been a subject of much debate in the past several years [4–6]. Because of computational expediency, many of these high-precision studies have been done using the staggered fermion formulation. Although computationally inexpensive, staggered fermions have the known flaw that they only preserve a $U(1)$ subgroup of the full $SU(N_f)$ chiral symmetry at finite lattice spacing. This manifests itself in large lattice artifacts in certain quantities, *e.g.* non-degeneracy in the pion spectrum. Domain Wall Fermions (DWF) preserve the full $SU(N_f)$ chiral symmetry at finite lattice spacing [10]. Thus, a DWF study at finite temperature is useful to test the robustness of the recent staggered studies of T_c and helps us to better understand the role of chiral symmetry in the QCD crossover.

In our DWF calculation [9], we perform simulations at 11 different temperatures in the range 120 MeV $< T <$ 220 MeV at zero chemical potential. We use a temporal lattice size of $N_t = 8$, which is related to the temperature via $T = 1/(N_t a)$. These temperatures correspond to lattice spacings in the range 0.11 fm $< a <$ 0.21 fm.

Various quantities are used to locate the chiral-symmetry restoring and deconfinement transitions. For the chiral transition, we use the subtracted chiral condensate,

$$\frac{\Delta_{l,s}(T)}{T^3} = \frac{\langle\bar{\psi}_l\psi_l\rangle - \frac{m_l}{m_s}\langle\bar{\psi}_s\psi_s\rangle}{T^3}, \tag{1}$$

where the subscripts l, s denote the light and strange quarks, respectively. In the phase where chiral symmetry is restored ($T > T_c$), we expect $\Delta_{l,s}(T)$ to vanish, while for $T < T_c$, it is non-zero. In addition to the chiral condensate, we also calculate the disconnected chiral

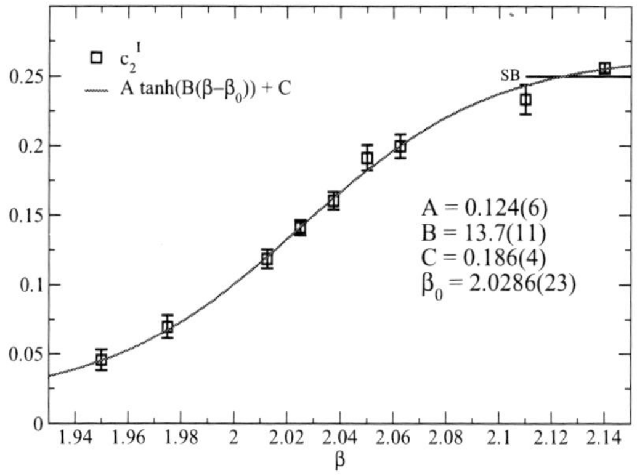

Figure 2. The isospin susceptibility, c_2^I versus bare coupling $\beta = 6/g_0^2$.

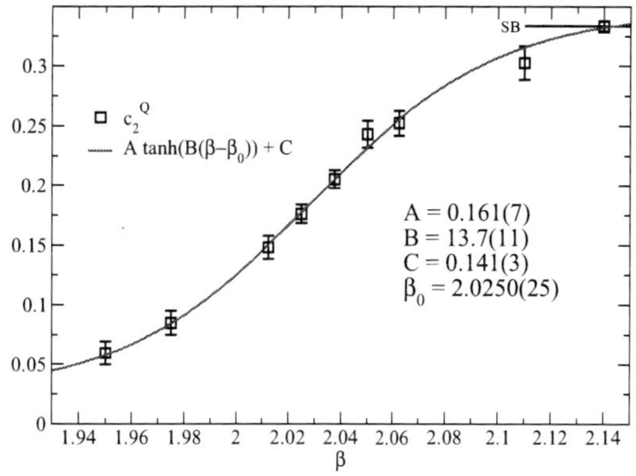

Figure 3. The electric charge susceptibility, c_2^Q versus bare coupling $\beta = 6/g_0^2$.

susceptibility,

$$\frac{\chi_l}{T^2} = VT^3\left(\left\langle(\bar{\psi}_l\psi_l)^2\right\rangle - \langle\bar{\psi}_l\psi_l\rangle^2\right). \tag{2}$$

This quantity should exhibit a peak at $T = T_c$. Fig. 1 shows both $\frac{\Delta_{l,s}}{T^3}$ and $\frac{\chi_l}{T^2}$. As can be seen, the disconnected chiral susceptibility has a clear peak near $\beta = 6/g_0^2 = 2.03$, where g_0 is the bare coupling on the lattice. This corresponds to $T \approx 170$ MeV.

For the deconfinement transition, we calculate the electric charge susceptibility (c_2^Q) and the isospin susceptibility (c_2^I):

$$\frac{c_2^X}{T^2} = \frac{1}{2}\frac{1}{VT^3}\frac{\partial^2 \ln Z}{\partial(\mu_X/T)^2}|_{\mu_X=0};\ \mathrm{X} = \mathrm{Q},\ \mathrm{I}, \tag{3}$$

where μ_Q and μ_I are the electric charge and isospin chemical potentials and $\ln Z$ is the QCD partition function. As can be seen in Figs. 2 and 3, these deconfinement observables show a rapid rise in the same general region as those observables related to chiral symmetry.

Although the chiral susceptibility has a clear peak near $T \approx 170$ MeV and the electric charge and isospin susceptibilities imply T_c in the same region, our calculation still suffers from significant uncertainties. The spatial volume used is quite small, from $V = (1.7\text{ fm})^3$ up to $V = (3.1\text{ fm})^3$, so our result may be polluted with uncontrolled finite-volume effects. In addition, the light quark masses are not physical, but correspond to $m_\pi = 300$ MeV near the susceptibility peak. Furthermore, the simulations are not done on a line of constant quark mass, so that m_π is effectively much heavier at low temperatures than at high temperatures. These effects introduce significant uncertainty into our estimate of T_c, so we are only able to quote a wide range of possible values: 150 MeV $< T <$ 190 MeV. Unfortunately, this does not do much to discriminate between the results of recent lattice calculations, but we hope to reduce these uncertainties in future calculations.

3. Parameterization of the lattice QCD equation of state

There have been several attempts to parameterize the lattice results for the QCD equation of state (EoS) in terms of a continuous function that can be conveniently input into hydrodynamic models. Our parameterization reproduces the expected asymptotic behavior of

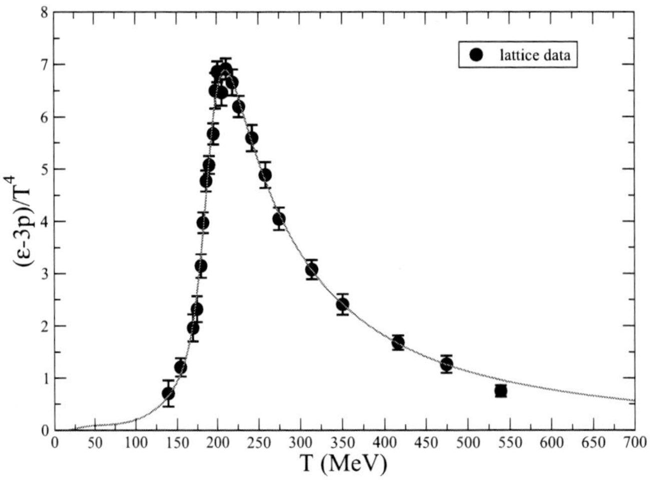

Figure 4. Lattice QCD results of the interaction measure $(\epsilon - 3p)/T^4$ with corresponding fit to the parameterization $I(T) = (1/I_{low}(T) + 1/I_{high}(T))^{-1}$.

the QCD equation of state in the limits $T \to 0$ and $T \to \infty$, smoothly joining them together in the intermediate regime near the QCD crossover.

In lattice calculations, the interaction measure (I(T) $= \epsilon - 3p$) is the thermodynamic quantity that is naturally calculated, from which all other thermodynamics quantities such as the pressure p, energy density ϵ, entropy density s, and speed of sound c_s can be derived. Thus, it is sufficient to parameterize the interaction measure.

In the high temperature limit, the interaction measure is expected to have the form [15; 16]:

$$\frac{I_{high}(T)}{T^4} = \frac{\alpha_1}{T^4} + \frac{\alpha_2}{T^4}. \tag{4}$$

The lattice QCD data agrees well with this expected behavior. At low temperatures, QCD can be described as a gas of mesons and baryons. A widely-used approximation is to assume that this gas is non-interacting. This is called the Hadron Resonance Gas (HRG) model [17].

The existing lattice results give an interaction measure that tends to be lower than that given by the HRG model at low temperatures. Thus, in our parameterization, deviations from the HRG model at low temperature are included in order to accurately fit the lattice data:

$$\frac{I_{low}(T)}{T^4} = \frac{I_{HRG}(T)}{T^4}\left(\alpha_3 + \left(\frac{T}{\alpha_4}\right)^{\alpha_5}\right). \tag{5}$$

The low and high temperature parameterizations are combined via:

$$I(T) = \left(\frac{1}{I_{low}(T)} + \frac{1}{I_{high}(T)}\right)^{-1} \tag{6}$$

and the parameters $(\alpha_1 \cdots \alpha_5)$ can be extracted through a fit to the lattice data. Note that fixing $\alpha_3 = 1$ is also an option, if one wants to force the parameterization onto the HRG result as $T \to 0$.

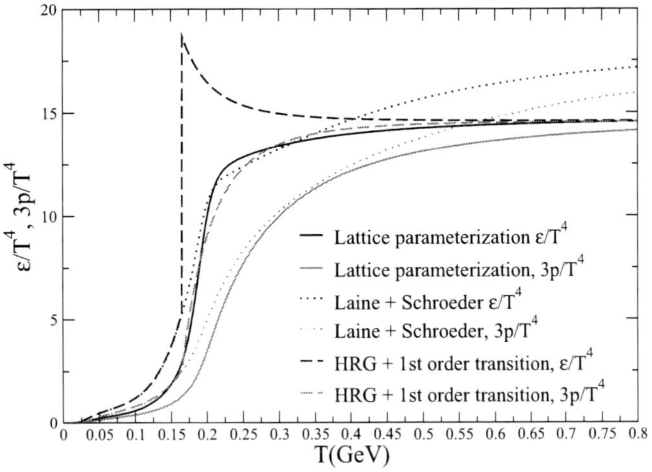

Figure 5. Comparison of the energy density, ϵ and three times the pressure, 3p, for our lattice parameterization, the EoS from Laine and Schroeder [18], and the Hadron Resonance Gas Model with a 1st order transition.

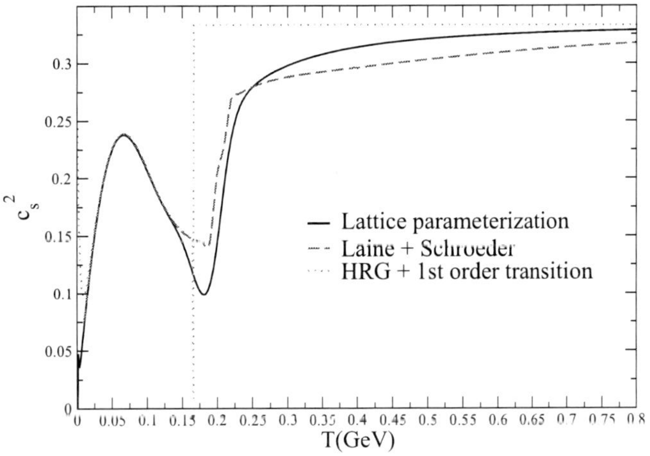

Figure 6. Comparison of the speed of sound squared, c_s^2, for our lattice parameterization, the EoS from Laine and Schroeder [18], and the Hadron Resonance Gas Model with a 1st order transition.

Table 1 tabulates the fit parameters ($\alpha_1 \cdots \alpha_5$) from a fit to the lattice EoS for the p4 action at $N_t = 8$ [8]. Fig. 4 shows the result of this parameterization plotted along with the actual lattice data. Fig. 5 shows a comparison of this lattice parameterization with two other forms for the equation of state often used in hydrodynamic modeling - a form proposed by Laine and Schroeder [18], and one where the HRG EoS at low temperature is joined to an ideal gas at high temperature with a first-order phase transition at $T_c = 165$ MeV. Figure 6 shows the same comparison for the speed of sound squared, c_s^2.

α_1	α_2	α_3	α_4	α_5
0.26867	0.00345	0.50120	0.18529	15.17516

Table 1. Fit parameters for $I(T)$.

4. Hydro+Cascade Model

We introduce a multi-stage model where a heavy ion collision is modeled from the first moment of collision until the final state hadrons are effectively free-streaming to the particle detectors. In constructing this model, we wish to systematically test the effect of varying various model inputs and to see which set of parameters give the best agreement with experimental results. Among the things that we wish to test are: the nature of the initial conditions, the inclusion of pre-equilibirum flow, the initial temperature T_i, the shear viscosity η, the freezeout temperature T_f, and the equation of state.

4.1. Initial Conditions

For initial conditions, we use an optical Glauber model [19; 20], where the initial energy density is proportional to the number of participant nuclei, N_{part} or the number of binary collisions, N_{coll}. The magnitude of the energy density is chosen so that the maximum energy density ϵ_{max} corresponds to some initial temperature, T_i. The impact parameter, b can be directly varied to obtain different centralities.

A commonly used assumption is that the initial flow velocities vanishes, neglecting the fact that flow velocities may develop in the first $\tau \approx 1$ fm/c, before hydrodynamic evolution is applicable [21]. Recently, it has been found that neglecting initial flow velocities might have significant effects on collective observables, particularly those related to the source size [22; 23]. Therefore, we allow the modification of our initial conditions to account for possible pre-thermalization flow.

4.2. Viscous Hydrodynamics

Because the relativistic generalizations of the Navier-Stokes equation are acausal, it is not entirely clear how to formulate relativistic dissipative hydrodynamics. One attempt is the Israel-Stewart formalism [26], where a relaxation time is introduced for every transport coefficient. However, there is still ambiguity as to which higher order terms to include. In the case of vanishing bulk viscosity, however, it has been shown [27] that the most general form includes five additional terms, $\tau_\Pi, \kappa, \lambda_1, \lambda_2, \lambda_3$, where $\kappa = 0$ in flat space-time. This formulation was implemented by Romatshke [28], and is the one that we utilize.

Romatschke's code [28] is two-dimensional, where the bulk viscosity (ζ) implicitly vanishes, but the shear viscosity (η) can be non-zero. For the relaxation times, we set $\tau_\Pi = 6\eta/sT$, $\lambda_1 = 0$, $\lambda_2 = -2\eta/\tau_\Pi$, $\lambda_3 = 0$, which are valid choices for a weakly-coupled plasma. The start time for the hydrodynamics evolution is taken to be $\tau_0 = 1fm/c$, with the initial conditions discussed above.

For the equation of state, we plan to test four different variations: 1) A HRG EoS for $T < T_c$ with a first order transition, 2) the "Laine-Schroder" EoS [18] used in [28], 3) a parameterization of the lattice QCD EoS discussed in Sec. 3, and 4) the parmeterization

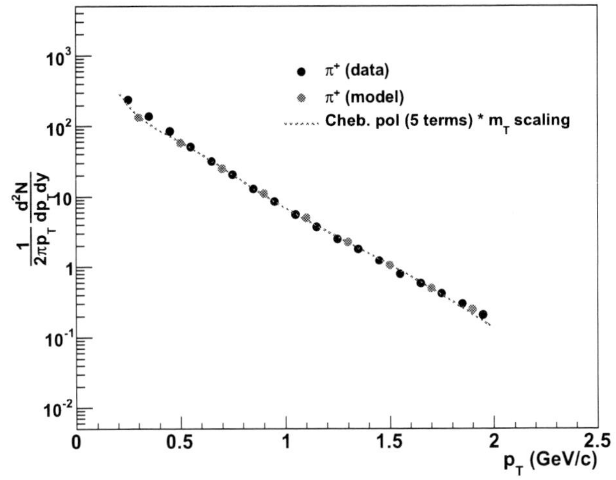

Figure 7. π^+ invariant yield. Experimental data from PHENIX results in the $10-15\%$ centrality bin [24]. Model results obtained with impact parameter $b = 4.4$fm, $T_i = 300$MeV, $T_f = 150$MeV, without initial flow and using the "lattice-inspired" EoS.

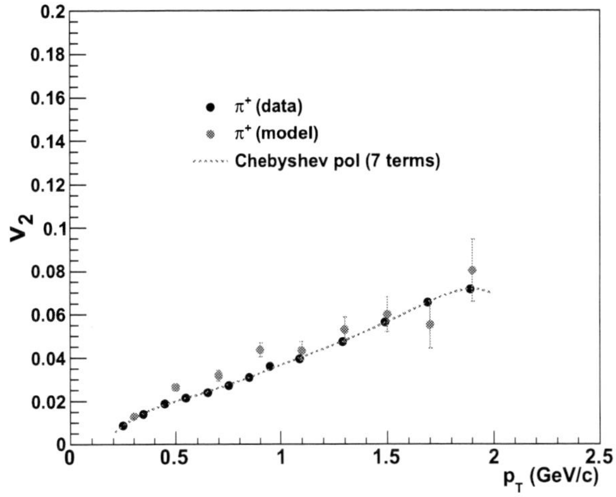

Figure 8. π^+ elliptic flow (v_2). Experimental data from PHENIX results in the $0-10\%$ centrality bin [25]. Model results obtained with impact parameter $b = 4.4$fm, $T_i = 300$MeV, $T_f = 150$MeV, without initial flow and using the "lattice-inspired" EoS.

of the lattice QCD EoS with $\alpha_3 = 1$ so that it is smoothly joined to the HRG EoS at low temperatures. A comparison of the first three of these can be seen in Figs. 5 and 6.

4.3. Cooper-Frye Freezeout

In modeling the freezeout of hadrons from the QGP, we use sudden freezeout and the Cooper-Frye prescription. In this method, the hadrons are frozen out on a hypersurface in (x, y, z, t) of constant temperature $T = T_f$, or equivalently constant energy density. This freezeout temperature can be varied to study the effects of early or late freezeout. The single particle spectrum using the Cooper-Frye method is given by:

$$E\frac{dN}{d^3\mathbf{p}} = \frac{d}{(2\pi)^3}\int p_\mu d\Sigma^\mu f(x^\mu, p^\mu), \tag{7}$$

where d is the degeneracy factor, $d\Sigma^\mu$ represents the normal to the freezeout hypersurface, and $f(x^\mu, p^\mu)$ is the phase space density with non-equilibrium corrections [30].

Because the hydrodynamic evolution occurs only in two spatial dimensions, we assume boost- invariance along the longitudinal direction to produce the full freezeout spectrum for the particles.

4.4. Hadron Cascade

After freezeout, one must take into account final state interactions and feed-down decays in order to extract the final particle spectra. In order to do this, we take the particles produced at freezeout and evolve them through a hadronic cascade. The code that we choose

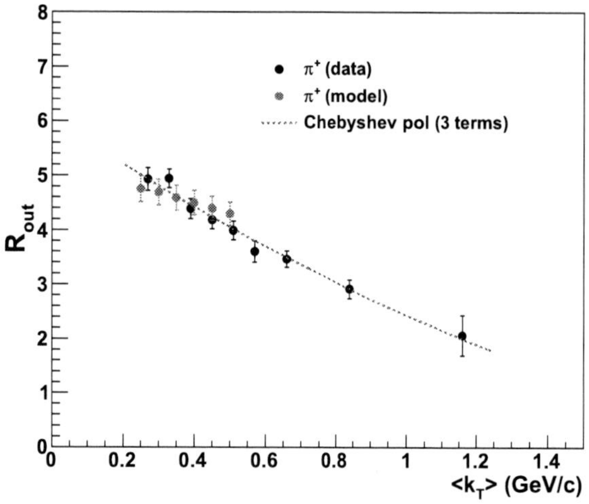

Figure 9. π^+ R_{out}. Experimental data from PHENIX results in the 0 – 30% centrality bin [29]. Model results obtained with impact parameter b = 4.4fm, T_i = 300MeV, T_f = 150MeV, without initial flow and using the EoS from Laine and Schroeder.

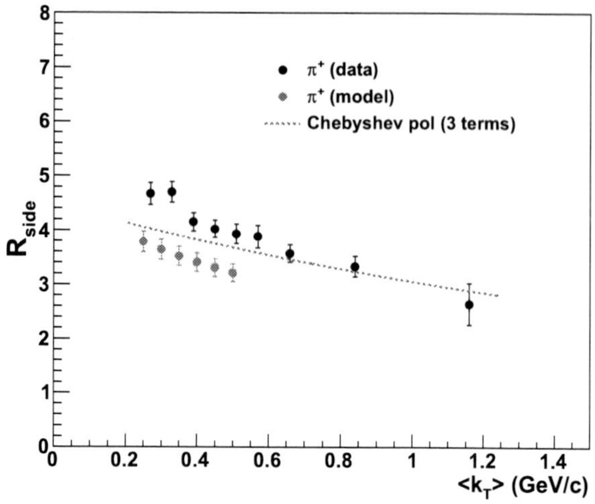

Figure 10. π^+ R_{side}. Experimental data from PHENIX results in the 0 – 30% centrality bin [29]. Model results obtained with impact parameter b = 4.4fm, T_i = 300MeV, T_f = 150MeV, without initial flow and using the EoS from Laine and Schroeder.

is the Ultrarelativistic Quantum Molecular Dynamics (UrQMD) code [13; 14], a microscopic transport code that explicitly takes into account particle decays and hadron-hadron collisions.

4.5. Preliminary Results

We have preliminary results for several model runs corresponding to Au+Au collisions at $\sqrt{s} = 200$ GeV per nucleon pair. The parameters that we have used are summarized in Tab. 2. However, we have not yet explored all possible variations of the model inputs. The impact parameter of $b = 4.4$ fm. was chosen to match as closely as possible to the mid-centraility bins for PHENIX results.

To match to experimental data, we concentrate on three classes of observables: 1) the invariant yield $(1/2\pi p_T)(dN/dp_T dy)$ at mid-rapidity, 2) the elliptic flow v_2, and 3) the HBT radii, R_{out}, R_{side}, and R_{long}. To determine the degree of agreement with experimental, we perform a joint fit of both the model and experimental results to a common smoothing function. For v_2 and the HBT radii, the smoothing function we use is a set of orthonormal polynomials, the Chebyshev polynomials. For the invariant yield, we found that the product of Chebyshev polynomials with $\exp(-m_T/T)$, where $m_T = m^2 + p_T^2$, produced the best fits. Fig. 7 shows the fit of the invariant yield for π^+ with one set of model parameters. Fig. 8 shows the fit for v_2 and Figs. 9 and 10 show HBT radii.

For each set of model parameters, the joint fit with the experimental data produces a value of χ^2, which is a loose measure of the difference between the model and experiment. Figs. 11 and 12 show the χ^2 distributions for π^+ for the invariant yield and v_2 as a function of T_i and η/s.

As this is a work in progress, we plan to perform a systematic exploration of the model

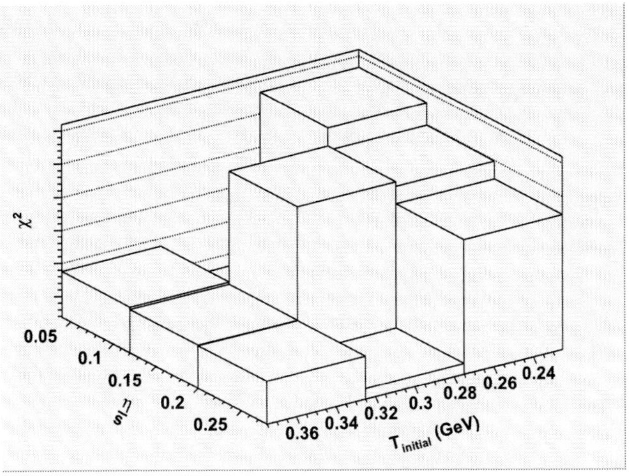

Figure 11. χ^2 distribution for model/data fits in the π^+ invariant yield as a function of T_i and η/s.

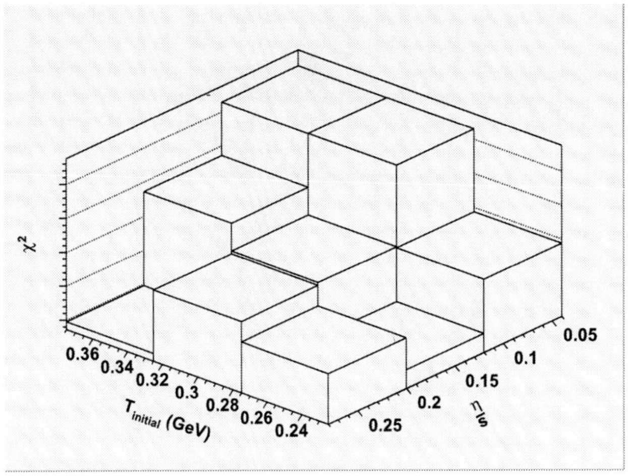

Figure 12. χ^2 distribution for model/data fits in π^+ v_2 as a function of T_i and η/s.

parameter space, and perform these joint fits with a larger set of experimental results. We intend to understand how sensitive the final results are to the various model inputs, and which model inputs give the best match to experimental results for various systems, energies, and centralities.

b (fm)	N_{part}	T_i (MeV)	η/s	T_f (MeV)	Initial Flow	EoS
4.4	270	250, 300, 350	0.08, 0.16, 0.24	150	yes, no	Laine & Schroder

Table 2. A summation of the model parameters that have been explored so far for Au+Au collisons at $\sqrt{s} = 200$ GeV per nucleon pair.

Acknowledgements
The work in Sec. 2 was done in collaboration with the RBC-Bielefeld Collaboration. We thank K. Rajagopal for suggesting the parameterization we use in Sec. 3, and R. Soltz for producing the fits. The work in Sec. 4 is done in collaboration with D. Brown, I. Garishvili, A. Glenn, J. Newby, S. Pratt, and R. Soltz. We thank P. Romatshke for providing his hydrodynamics code, and the UrQMD collaboration for providing the code for the hadron cascade. We also thank P. Huovinen and D. Molnar for useful discussions. Finally, we would in particular like to thank the organizers of the Winter Workshop on Nuclear Dynamics 2010, R. Lacey and S. Pratt, for allowing us to present this work. This work performed under the auspices of the U.S. Department of Energy by Lawrence Livermore National Laboratory under Contract DE-AC52-07NA27344.

References

[1] Collins J C and Perry M J 1975 *Phys. Rev. Lett.* **34** 1353

[2] Shuryak E V 1980 *Phys. Rept.* **61** 71–158

[3] Aoki Y, Fodor Z, Katz S D and Szabo K K 2006 *JHEP* **01** 089 (*Preprint* hep-lat/0510084)

[4] Aoki Y, Z, Katz S D and Szabo K K 2006 *Phys. Lett.* **B643** 46–54 (*Preprint* hep-lat/0609068)

[5] Aoki Y *et al.* 2009 *JHEP* **06** 088 (*Preprint* 0903.4155)

[6] Cheng M *et al.* 2006 *Phys. Rev.* **D74** 054507 (*Preprint* hep-lat/0608013)

[7] Cheng M *et al.* 2008 *Phys. Rev.* **D77** 014511 (*Preprint* 0710.0354)

[8] Bazavov A *et al.* 2009 *Phys. Rev.* **D80** 014504 (*Preprint* 0903.4379)

[9] Cheng M *et al.* 2010 *Phys. Rev.* **D81** 054510 (*Preprint* 0911.3450)

[10] Kaplan D B 1992 *Phys. Lett.* **B288** 342–347 (*Preprint* hep-lat/9206013)

[11] Molnar D and Gyulassy M 2002 *Nucl. Phys.* **A697** 495–520 (*Preprint* nucl-th/0104073)

[12] Teaney D, Lauret J and Shuryak E V 2001 *Phys. Rev. Lett.* **86** 4783–4786 (*Preprint* nucl-th/0011058)

[13] Bleicher M *et al.* 1999 *J. Phys.* **G25** 1859–1896 (*Preprint* hep-ph/9909407)

[14] Bass S A *et al.* 1998 *Prog. Part. Nucl. Phys.* **41** 255–369 (*Preprint* nucl-th/9803035)

[15] Pisarski R D 2007 *Prog. Theor. Phys. Suppl.* **168** 276–284 (*Preprint* hep-ph/0612191)

[16] Megias E, Ruiz Arriola E and Salcedo L L 2007 *Phys. Rev.* **D75** 105019 (*Preprint* hep-ph/0702055)

[17] Hagedorn R 1965 *Nuovo Cim. Suppl.* **3** 147–186

[18] Laine M and Schroder Y 2006 *Phys. Rev.* **D73** 085009 (*Preprint* hep-ph/0603048)

[19] Glauber R J 1955 *Phys. Rev.* **100** 242–248

[20] Miller M L, Reygers K, Sanders S J and Steinberg P 2007 *Ann. Rev. Nucl. Part. Sci.* **57** 205–243 (*Preprint* nucl-ex/0701025)

[21] Huovinen P and Ruuskanen P V 2006 *Ann. Rev. Nucl. Part. Sci.* **56** 163–206 (*Preprint* nucl-th/0605008)

[22] Pratt S 2009 *Phys. Rev. Lett.* **102** 232301 (*Preprint* 0811.3363)

[23] Vredevoogd J and Pratt S 2009 *Phys. Rev.* **C79** 044915 (*Preprint* 0810.4325)

[24] Adler S S *et al.* (PHENIX) 2004 *Phys. Rev.* **C69** 034909 (*Preprint* nucl-ex/0307022)

[25] Afanasiev S *et al.* (PHENIX) 2009 *Phys. Rev.* **C80** 024909 (*Preprint* 0905.1070)

[26] Israel W and Stewart J M 1979 *Ann. Phys.* **118** 341–372

[27] Baier R, Romatschke P, Son D T, Starinets A O and Stephanov M A 2008 *JHEP* **04** 100 (*Preprint* 0712.2451)

[28] Luzum M and Romatschke P 2008 *Phys. Rev.* **C78** 034915 (*Preprint* 0804.4015)

[29] Adler S S *et al.* (PHENIX) 2004 *Phys. Rev. Lett.* **93** 152302 (*Preprint* nucl-ex/0401003)

[30] Pratt S and Torrieri G 2010 (*Preprint* 1003.0413)

Phase Fluctuations near the Chiral Critical Point

Joseph Kapusta
School of Physics and Astronomy, University of Minnesota, Minneapolis, MN 55455, USA
E-mail: kapusta@physics.umn.edu

Abstract. The Helmholtz free energy density is parametrized as a function of temperature and baryon density near the chiral critical point of QCD. The parametrization incorporates the expected critical exponents and amplitudes. An expansion away from equilibrium states is achieved with Landau theory. This is used to calculate the probability that the system is found at a density other than the equilibrium one. Such fluctuations are predicted to be very large in heavy ion collisions.

1. Introduction

If the up and down quark masses are zero and the strange quark mass is not, the transition between quark-gluon plasma and hadronic matter may be first or second order at zero baryon chemical potential. If the up and down quark masses are small enough, there may be a phase transition at sufficiently large baryon chemical potential. This phase transition is predicted to be in the same universality class as liquid-gas phase transitions and the 3D Ising model. Diverse studies suggest that there is a curve of first-order phase transition in the μ-T plane that terminates in a second-order phase transition at some critical point (μ_c, T_c). The location of this chiral critical point has been estimated using various effective field theory models, such as the Namu Jona Lasinio model [1]-[3], a composite operator model [4], a random matrix model [5], a linear σ model [3], an effective potential model [6], and a hadronic bootstrap model [7], as well as various implementations of lattice QCD [8]-[11]. Reviews have been written by Stephanov [12] and Mohanty [13].

This subject is of experimental interest because collisions between heavy nuclei at medium to high energy, such as at the future Facility for Antiproton and Ion Research (FAIR), or in future low energy runs at the Relativistic Heavy Ion Collider (RHIC), may provide experimental information on the phase diagram in the vicinity of a critical point. One characteristic signature would be large fluctuations on an event by event basis [14]-[17].

The goal of this work [18] is to understand the equation of state of QCD near the chiral critical point and some of its implications for high energy heavy ion collisions. Among the requirements are the incorporation of critical exponents and amplitudes and to have sensible limits as temperature $T \to 0$ and baryon chemical potential $\mu \to 0$. This is accomplished by parametrizing the Helmholtz free energy density as a function of T and baryon density n in an appropriate fashion. Perhaps the closest work that addressed some of these issues blended a parameterization of the 3D Ising model equation of state into quark and hadron equations of state [19]. These two parameterizations can perhaps be viewed as alternatives which provide some idea as to the range of uncertainty in how to describe matter near the chiral critical point.

2. Critical curve

Numerous studies suggest that the curve in the T vs. μ plane separating the quark-gluon and hadron phases is approximately quadratic.

$$\left(\frac{T}{T_0}\right)^2 + \left(\frac{\mu}{\mu_0}\right)^2 = 1 \quad (1)$$

See figure 1. The parameters $T_0 = 180$ MeV and $\mu_0 = 1230$ MeV were chosen to match theoretical estimates and experimental data for the transition between the two phases at $\mu = 0$ and $T = 0$, respectively. In this example the critical point is chosen to have the temperature $T_c = 140$ MeV. Along the solid portion of the curve the transition is first order. It terminates at a second order transition at the critical point. The dashed portion of the curve indicates that there is no true thermodynamic phase transition separating the phases, only a rapid crossover.

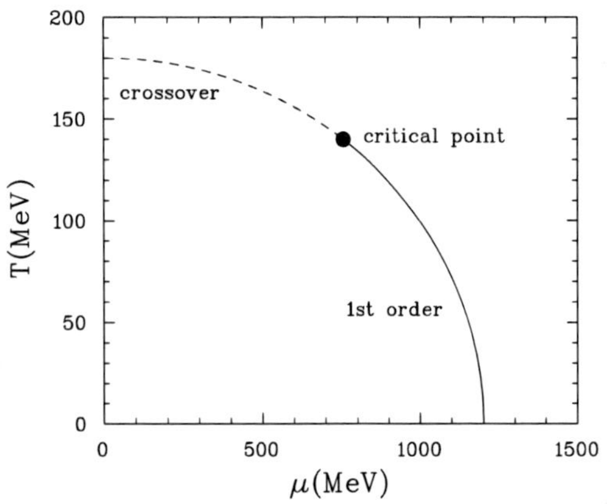

Figure 1. Temperature versus baryon chemical potential from the parameterization given in the text. The critical point lies somewhere along this curve.

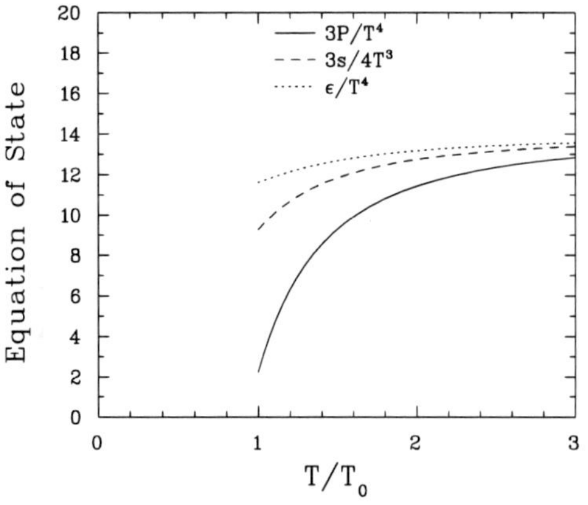

Figure 2. The pressure, entropy density, and energy density, normalized so that they all have the same asymptotic value, at $\mu = 0$. The parameterization is given in the text.

We will need the values of the pressure P_c, energy density ϵ_c, baryon density n_c, and entropy density s_c at the critical point. The parametrization of the equation of state near the critical point, to be given below, is normalized to these values. Let us assume that the high energy density equation of state can be parameterized as

$$P = A_4T^4 + A_2\mu^2T^2 + A_0\mu^4 - B_2T_0^2T^2 - B_4T_0^4. \quad (2)$$

The dimensionless coefficients A_n are adjusted to match a free gas of gluons and 2.5 flavors of massless quarks (the 1/2 to approximately take into account the mass of the strange quark). The dimensionless coefficients B_n are adjusted to reproduce lattice results near the crossover when $\mu = 0$ [20]-[22], and to make the pressure a constant along the critical curve when (1) is inserted. The resulting equation of state for $\mu = 0$ and $T > T_0$ is shown in figure 2.

3. Equation of state near the critical point

Given the Helmhotz free energy density $f(n, T)$ the other thermodynamic quantites follow uniquely from it, as summarized below.

$$P = n^2\frac{\partial}{\partial n}\left(\frac{f(n,T)}{n}\right) \quad (3)$$

$$\mu = \frac{\partial f(n,T)}{\partial n} \tag{4}$$
$$s = -\frac{\partial f(n,T)}{\partial T} \tag{5}$$
$$\epsilon = f(n,T) + Ts(n,T) \tag{6}$$

These satisfy the thermodynamic identity $\epsilon = -P + Ts + \mu n$. It is useful to define the dimensionless variables $t = (T - T_c)/T_c$ and $\eta = (n - n_c)/n_c$ which are a measure of distance from the critical point in temperature and density. Near the critical point the heat capacity c_V has the divergent behavior

$$c_V = T\frac{\partial s(n,T)}{\partial T} \rightarrow \begin{cases} c_-(-t)^{-\alpha} & \text{when } t \rightarrow 0^- \\ c_+ t^{-\alpha} & \text{when } t \rightarrow 0^+ \end{cases}, \tag{7}$$

while the isothermal compressibility κ_T and baryon susceptibility χ_B have the divergent behavior

$$\frac{\chi_B}{n^2} = \kappa_T = \left[n\frac{\partial P(n,T)}{\partial n}\right]^{-1} \rightarrow \begin{cases} \kappa_-(-t)^{-\gamma} & \text{when } t \rightarrow 0^- \\ \kappa_+ t^{-\gamma} & \text{when } t \rightarrow 0^+ \end{cases}. \tag{8}$$

Along the coexistence curve the density difference between the higher density liquid phase (quarks and gluons) and the lower density gas phase (hadrons) goes to zero as

$$n_l - n_g \sim (-t)^\beta. \tag{9}$$

Along the critical isotherm the pressure behaves as

$$P - P_c \sim |\eta|^\delta \operatorname{sign}(\eta). \tag{10}$$

Here α, β, γ and δ are critical exponents. These exponents are related by $\alpha + 2\beta + \gamma = 2$ and $\gamma = \beta(\delta - 1)$. Mean field theories normally give $\alpha = 0$, $\beta = 1/2$, $\gamma = 1$, and $\delta = 3$. Typical fluids are measured to have $\alpha \ll 1$, $\beta \approx 1/3$, $1.2 < \gamma < 1.3$, and $4 < \delta < 5$ [23]. The 3D Ising model has $\alpha = 0.11$, $\beta = 0.325$, $\gamma = 1.24$, and $\delta = 4.815$ [24], which are the values used here. Furthermore, the critical amplitudes are related by $\kappa_+/\kappa_- \approx 5$ and $c_+/c_- \approx 0.5$, which is universal to all theories within the same class. See [25].

The challenge now is to parametrize $f(\eta, t)$ so that it yields the correct critical behavior near $\eta = t = 0$ and still gives sensible results away from the critical point. Mean field theories normally result in a Taylor series expansion in integral powers of η with coefficients that depend on t. The simplest way to incorporate the correct critical behavior is with a modified expansion of the form

$$f = f_0(t) + f_1(t)\eta + f_2(t)\eta^2 + f_\sigma(t)|\eta|^\sigma. \tag{11}$$

The power $\sigma = \delta + 1 = 5.815$. Higher powers of η could be included but they would not affect the critical behavior and so are discarded here for simplicity. The coefficient functions must have the following behavior.

$$f_0(t) = \begin{cases} \bar{f}_0(t) - a_-(-t)^{2-\alpha} & \text{if } t < 0 \\ \bar{f}_0(t) - a_+ t^{2-\alpha} & \text{if } t > 0 \end{cases}. \tag{12}$$

$$f_1(t) = n_c\mu_0\sqrt{1 - \frac{T_c^2}{T_0^2}(1+t)^2}. \tag{13}$$

$$f_2(t) = \begin{cases} \bar{f}_2(t) - b_-(-t)^\gamma & \text{if } t < 0 \\ \bar{f}_2(t) + b_+ t^\gamma & \text{if } t > 0 \end{cases} \tag{14}$$

Here $\bar{f}_0(t)$ and $\bar{f}_2(t)$ are smooth functions of t with $\bar{f}_2(0) = 0$. They are parametrized so that $f(n,T)$ has sensible limits at $T = 0$ and at $n = 0$. The coefficient f_σ is assumed to be constant and proportional to P_c; for definiteness we shall take $f_\sigma = 5P_c \approx 512$ MeV/fm^3. Results for the critical curves, latent heat, thermal conductivity, and heat capacity are shown in figures 3 to 6.

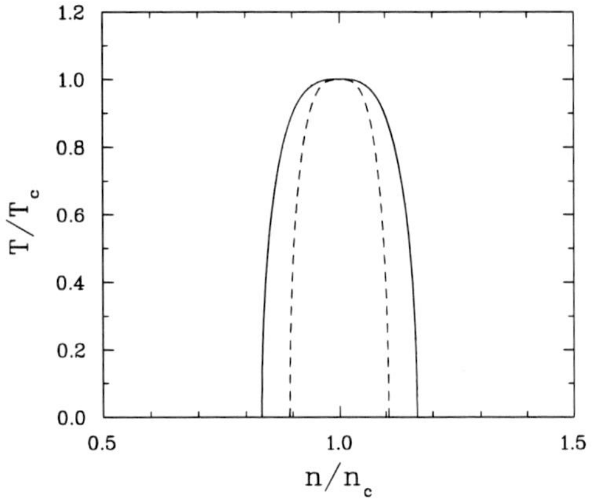

Figure 3. The solid curve denotes coexistence between high and low density phases and the dashed curve denotes the limits of metastability. The scaled curves are independent of the choice of critical temperature and density.

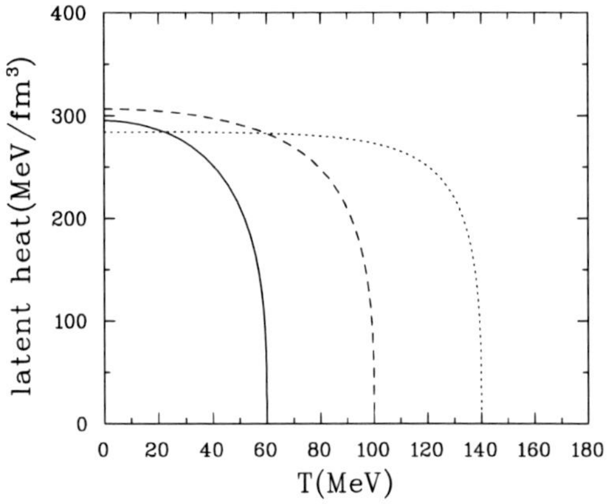

Figure 4. The latent heat per unit volume versus temperature for three choices of critical temperature.

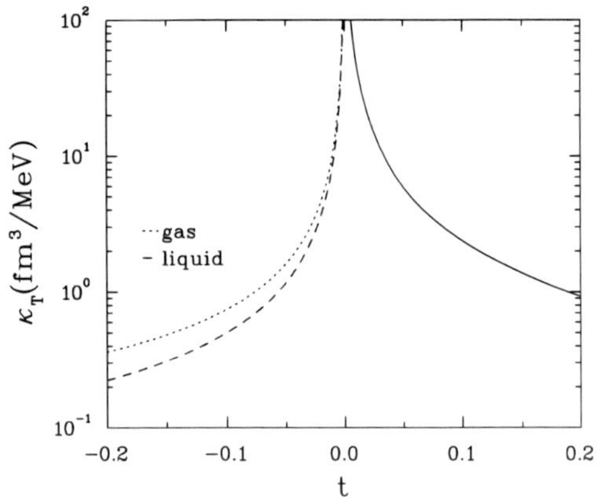

Figure 5. The isothermal compressibility. For $t < 0$ they are evaluated along the coexistence curve while for $t > 0$ it is evaluated at the critical density. The curves for different critical temperature fall on top of one another.

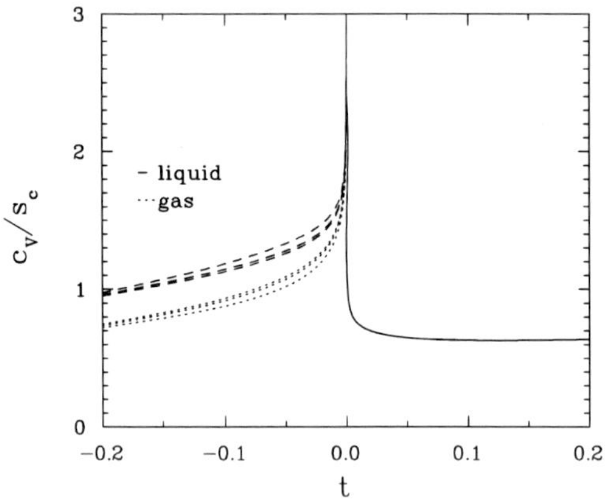

Figure 6. The heat capacity per unit volume. For $t < 0$ they are evaluated along the coexistence curve while for $t > 0$ it is evaluated at the critical density. When divided by the entropy density at the critical point the results are nearly independent of the choice of critical temperature.

4. Fluctuations

In any system of finite size there will be thermal fluctuations. These fluctuations become large as the critical point is approached. Fluctuations can be studied with Landau theory [26, 27].

The question is: What is the probability to find a system of volume V with a baryon density different from the equilibrium one? The expansion away from equilibrium states is determined by the thermodynamic potential

$$\Omega(\mu, T; \eta) - \Omega_0(\mu, T) = \left[(f_1 - n_c\mu)\eta + f_2\eta^2 + f_\sigma|\eta|^\sigma\right] V. \tag{15}$$

Here $\Omega_0(\mu, T) = (f_0 - n_c\mu)V$. Along the coexistence curve $f_1 = n_c\mu$. The relative probability to be at a density other than the equilibrium one, along the coexistence curve, is

$$\mathcal{P}(\eta)/\mathcal{P}(\eta_l) = \exp\left(-\Delta\Omega/T\right) \tag{16}$$

where

$$\Delta\Omega = \left[f_2\left(\eta^2 - \eta_l^2\right) + f_\sigma\left(|\eta|^\sigma - |\eta_l|^\sigma\right)\right] V. \tag{17}$$

For purposes of illustration the volume is taken to be 400 fm^3. The value of 400 fm^3 is about as large as one can imagine for high energy nuclear collisions. Considering that the critical density is estimated to be about $5n_0 \approx 0.75$ baryons/fm^3, this would mean that about 300 baryons participate in the fluctuation. That is a substantial fraction of the total in Au+AU, Pb+Pb, or in U+U collisions. Smaller volumes would have even larger fluctuations.

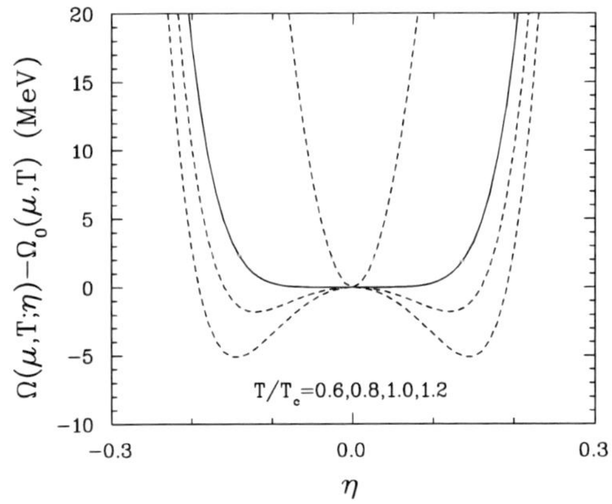

Figure 7. Thermodynamic potential as a function of η near the critical point when the volume is 400 fm^3. The stable phases are located at the minima of the potential. Four different temperatures are shown, with the solid curve representing the critical temperature.

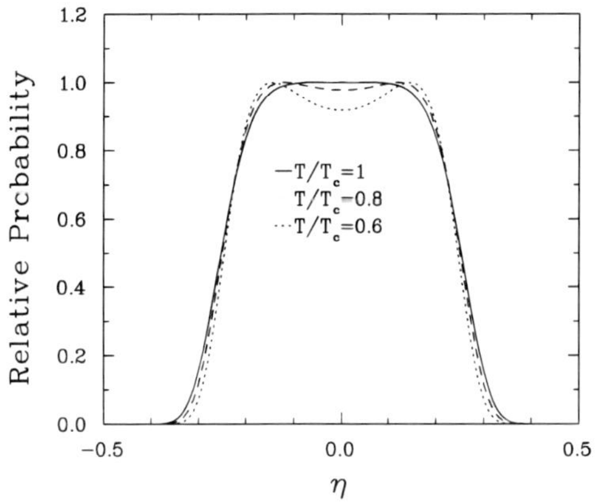

Figure 8. The probability to find the system at a particular density relative to the equilibrium densities at phase coexistence.

The thermodynamic potential is shown in figure 7 for several different temperatures. Phase coexistence occurs below T_c; the two phases are located at the minima of the potential. They are not very deep, indicating that fluctuations do not cost much free energy. The potential is so flat because $f_2(T_c) = 0$ and because the power $\sigma \approx 6$ is so large. The shallowness of the potential is reflected in the probabilities, as shown in figure 8. The probability to find the system with a density anywhere between $\pm 25\%$ of the critical density is greater than 50%. Fluctuations are large! They would be even larger for smaller volume systems.

5. Conclusions

An equation of state valid in the vicinity of the chiral critical point has been constructed. It incorporates correct values of the critical exponents and amplitudes. Since only certain properties of the equation of state are universal, there is some freedom to vary the noncritical functional dependence on temperature and density. Work on extending the equation of state to a wider range of T and μ is underway.

The Landau theory of fluctuations away from equilibrium states was used to determine the magnitude of the fluctuations one might expect in heavy ion collisions. These fluctuations are quite large, partly due to finite volume effects but mostly because the critical exponent δ is much larger than in mean field theories. This flattens the Landau free energy as a function of density away from the equilibrium densities and so decreases the cost to fluctuate away from them.

It will require careful thought as to how to incorporate fluctuations near the critical point into dynamical simulations of heavy ion collisions. What is the appropriate way to describe the transition in heavy ion collisions? Is it nucleation [28], spinodal decomposition [29], or something else? What are the best experimental observables and can they be measured at RHIC and/or FAIR? The future of this topic is exciting!

Acknowledgements

This work was supported by the US Department of Energy (DOE) under Grant No. DE-FG02-87ER40328.

References

[1] M. Asakawa and K. Yazaki, Nucl. Phys. **A504**, 668 (1989).
[2] J. Berges and K. Rajagopal, Nucl. Phys. **B538**, 215 (1999).
[3] O. Scavenius, A. Mòcsy, I. N. Mishustin, and D. H. Rischke, Phys. Rev. C **64**, 045202 (2001).
[4] A. Barducci, R. Casalbuoni, S. De Curtis, R. Gatto, and G. Pettini, Phys. Lett. **B231**, 463 (1989); Phys. Rev. D **41**, 1610 (1990); A. Barducci, R. Casalbuoni, G. Pettini, and R. Gatto, *ibid.* **49**, 426 (1994).
[5] M. A. Halasz, A. D. Jackson, R. E. Shrock, M. A. Stephanov, and J. J. M. Verbaarschot, Phys. Rev. D **58**, 096007 (1998).
[6] Y. Hatta and T. Ikeda, Phys. Rev. D **67**, 014028 (2003).
[7] N. G. Antoniou and A. S. Kapoyannis, Phys. Lett. **B563**, 165 (2003).
[8] Z. Fodor and S. D. Katz, J. High Energy Phys. 03 (2002) 014; *ibid.* 04 (2004) 050.
[9] S. Ejiri, C. R. Allton, S. J. Hands, O. Kaczmarek, F. Karsch, E. Laermann, and C. Schmidt, Prog. Theor. Phys. Suppl. **153**, 118 (2004).
[10] Ph. de Forcrand and O. Philipsen, Nucl. Phys. **B642**, 290 (2002); **B673**, 170 (2003); Nucl. Phys. Proc. Suppl. **129**, 521 (2004); J. High Energy Phys. 11 (2008) 012.
[11] R. V. Gavai and S. Gupta, Phys. Rev. D **71**, 114014 (2005).
[12] M. Stephanov, Prog. Theor. Phys. Suppl. **153**, 139 (2004); Int. J. Mod. Phys. A **20**, 4387 (2005); PoS(LAT2006)024.
[13] B. Mohanty, Nucl. Phys. **A830**, 899c (2009).
[14] M. A. Stephanov, K. Rajagopal, and E. Shuryak, Phys. Rev. Lett. **81**, 4816 (1998); Phys. Rev. D **60**, 114028 (1999).
[15] Y. Hatta and M. A. Stephanov, Phys. Rev. Lett. **91**, 102003 (2003); **91**, 129901(E) (2003).
[16] B. Berdnikov and K. Rajagopal, Phys. Rev. D **61**, 105017 (2000).
[17] M. A. Stephanov, Phys. Rev. Lett. **102**, 032301 (2009).
[18] J. I. Kapusta, preprint arXiv:1001.3650 and Phys. Rev. C (in press).
[19] C. Nonaka and M. Asakawa, Phys. Rev. C **71**, 044904 (2005).
[20] G. Boyd, J. Engels, F. Karsch, E. Laermann, C. Legeland, M. Lutgemeier, and B. Peterson, Nucl. Phys. **B469**, 419 (1996).
[21] R. D. Pisarski, Prog. Theor. Phys. Suppl. **168**, 276 (2007).
[22] A. Bazavov, *et al.*, Phys. Rev. D **80**, 014504 (2009).
[23] A. C. Flewelling, R. J. Defonseka, N. Khaleeli, J. Partee and D. T. Jacobs, J. Chem. Phys. **104**, 8048 (1996); C. A. Ramos, A. R. King and V. Jaccarino, Phys. Rev. B **40**, 7124 (1989).
[24] R. Guida and J. Zinn-Justin, Nucl. Phys. **B489** [FS], 626 (1997).
[25] J. Zinn-Justin, *Quantum Field Theory and Critical Phenomena*, Clarendon Press, Oxford, 3rd edition, 1996.

[26] E. M. Lifshitz and L. P. Pitaevskii, *Statistical Physics*, 3rd edition (Pergamon, New York, 1980), Part 1.
[27] A. L. Goodman, J. I. Kapusta, and A. Z. Mekjian, Phys. Rev. C **30**, 851 (1984).
[28] L. P. Csernai and J. I. Kapusta, Phys. Rev. D **46**, 1379 (1992); Phys. Rev. Lett. **69**, 737 (1992).
[29] J. Randrup, Phys. Rev. C **79**, 054911 (2009).

PHENIX and the Reaction Plane: Recent Results

David L. Winter, for the PHENIX Collaboration [1]
Columbia University, NY, NY 10027
E-mail: winter@nevis.columbia.edu

Abstract. During the past several years, experiments at RHIC have established that a dense partonic medium is produced in Au+Au collisions at $\sqrt{s} = 200$ GeV. Subsequently, a primary goal of analysis has been to understand and characterize the dynamics underlying this new form of matter. Among the many probes available, the measurements with respect to the reaction plane have proven to be crucial to our understanding of a wide range of topics, from the hydrodynamics of the initial expansion of the collision region to high-p_T jet quenching phenomena. Few tools have the ability to shed light on such a wide variety of observables as the reacion plane. In this talk, we discuss recent PHENIX measurements with respect to the reaction plane, and the implications for understanding the underlying physics of RHIC collisions.

1. Introduction

In relativistic heavy-ion collisions, the reaction plane is roughly defined by the plane that contains both the momenta of the nuclei and the impact parameter vector between them. In this view, the reaction plane is parallel to the short axis of the almond-shaped collision region and bisects the long axis of this region. In non-central collisions, the spatial anisotropy leads to a momentum anisotropy due to pressure gradients in the collision region tending to boost particles to higher p_T. We typically analyze the azimuthal distribution of particles with respect to the reaction plane via a Fourier series,

$$\frac{dN}{d\Delta\phi} \approx (1 + 2v_1 \cos\Delta\phi + 2v_2 \cos 2\phi + ...) , \tag{1}$$

where $\Delta\phi$ is the particle's azimuthal angle with respect to reaction plane, and the coefficients v_i are the Fourier coefficients. The second coefficient v_2 is sometimes referred to as the *elliptic flow* coefficient. The momentum anisotropy reflects the characteristics of the hot and dense medium created in these collisions; for example: small mean free path, early thermalization, and pressure gradients due to hydrodynamics. As a result, observables such as v_2 have long been considered powerful probes for studies of the quark gluon plasma.

An overview of the PHENIX experiment can be found in [2]. PHENIX uses four detectors to separately measure the reaction plane. These detectors are shown in Figure 1. The Beam-Beam Counters (BBCs) were installed earliest[3], and have provided a reaction plane measurement for all runs. The BBCs are 64 quartz Cherenkov radiators arranged in a hexagonal pattern around the beampipe, and the cover the range $3.0 < |\eta| < 4.0$. Starting in 2006 (Run 6), the first of the Muon Piston Calorimeters (MPCs) was installed, with the second in the following year. The MPCs consist of $PbW0_4$ PHOS crystal towers arranged in tiles located at $3.1 < |\eta| < 3.7$. With Run 7 (2007), the Reaction Plane Detector (RXPN) was installed. This detector was

Journal of Physics: Conference Series **230** (2010) 012017 doi:10.1088/1742-6596/230/1/012017

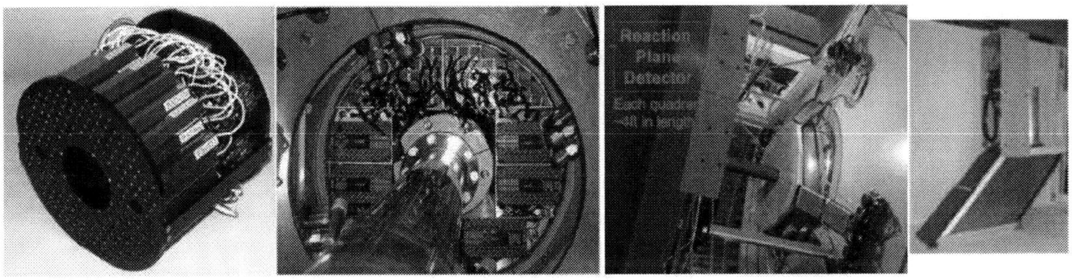

Figure 1. The PHENIX reaction plane detectors. From left to right: Beam-Beam Counters, Muon Piston Calorimeter, Reaction Plane Detector, and Zero-Degree Calorimeter.

specifically designed for measurement of the reaction plane, and consists of 12 plastic scintillators segmented in two ranges of η: $1.0 < |\eta| < 1.5$ (RXPNout) and $1.5 < |\eta| < 2.8$ (RXPNin). The η segmentation is to allow measurements that are unbiased by the potential presence of jets in the midrapidity region. Finally, there is a fourth detector system that can provide a reaction plane measurement: the Zero-Degree Calorimeter/ShowerMax Detector (ZDC-SMD), located at $|\eta| \sim 6.5$. This detector is typically used to provide systematic checks against the measurements in the other detectors.

All the reaction plane detectors have north and south installations, each capable of providing an independent measurement of the reaction plane orientation on an event-by-event basis. The actual measurement PHENIX uses from a detector system is the combined result, using the difference between the north and south subsystems to (statistically) calculate the resolution of the measurement. The procedure to correct the v_2 measurement resolution is described in [1]. The advantage of the newer reaction plane measurements over the BBCs is better resolution; for example the RXPNin has about 50% better resolution than the BBC.

2. Low-p_{T} results

Elliptic flow has been an observable studied for some time, and one of the earliest exciting results at RHIC was the observation of a large positive v_2 in Au+Au at $\sqrt{s}$ = 200 GeV [4]. One of the great theoretical successes has been the use of hydrodynamic models to predict the values of v_2, up to $p_{\mathrm{T}} \approx 1.5$ GeV/c. This excellent agreement implies the collision undergoes early thermalization (~ 0.6 fm/c) and that flow occurs at the quark level. Since its early observation, the phenomenon of large v_2 has been extensively measured, though the mass ordering of the values as well as the differences between mesons and baryons has been puzzling. An exciting breakthrough occurred when it was shown that different species' v_2 scaled with the quantity $\mathrm{KE_T}/n_q$, the transverse kinetic energy divided by the number of constituent quarks [5], as seen in Figure 2. At RHIC energies, this scaling appears to be independent of species, system size, and collision energy. Conversely, this same scaling does not appear to hold at SPS energies [6]. These results are further confirmation that at RHIC flow develops at the quark level.

It is also interesting to examine how the observed v_2 depends on the energy of the collision as well as the size of the collision system. PHENIX has measured identified hadron v_2 as a function of p_{T} in Au+Au collisions at both $\sqrt{s}$ = 200 and 62.4 GeV. An example of the comparison for between collision energies for Kaons, pions, and protons can be seen in Figure 3. The centrality dependence of the integrated v_2 for the two collision energies is also shown in Figure 4. We find that at a given centrality, there is no significant difference between the collision energies. The PHENIX measurements are placed in the context of world's data[7, 8, 9, 10, 11, 12, 13] in Figures 5 and 6. It appears from the data shown in Figure 6 that v_2 saturates above 62.4 GeV/c. This observation provides evidence that the matter created at RHIC reaches thermal equilibrium.

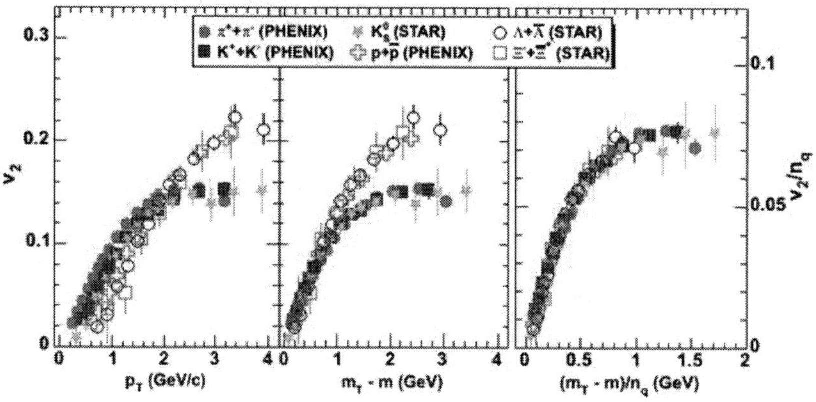

Figure 2. v_2 in Au+Au $\sqrt{s}$ = 200 GeV collisions for several identified species. From left to right the data are plotted as function of $p_{\rm T}$, $m_{\rm T} - m$ (KE_T), and $(m_{\rm T} - m)/n_q$ respectively. In addition, the right panel plots v_2/n_q.

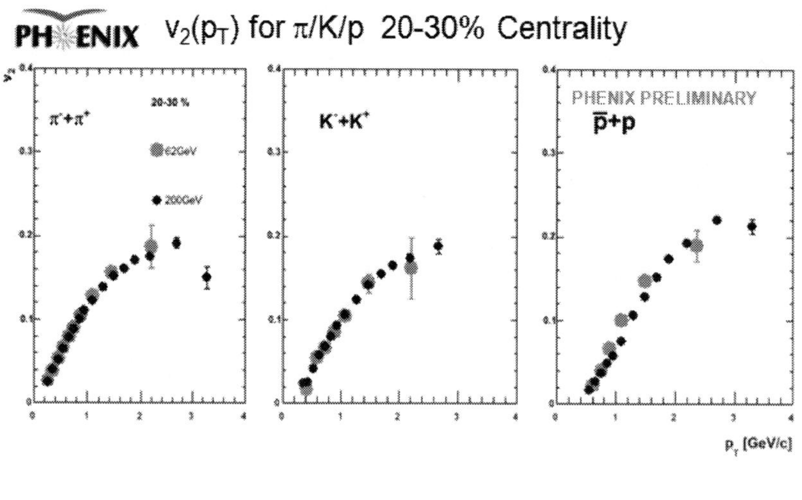

Figure 3. Energy dependence of $v_2(p_{\rm T})$ for Au+Au collisions at $\sqrt{s}$ = 200 and 62.4 GeV.

Further evidence can be seen when comparing the eccentrity scaling of v_2 between systems of different sizes: Au+Au and Cu+Cu. Plotted in Figure 2 is v_2 (v_2/ϵ) versus the number of participants for both Au+Au and Cu+Cu at $\sqrt{s}$ =200 GeV and 62.4 GeV. We see that the eccentricity scaling is independent of system size or energy.

3. High-$p_{\rm T}$ results

At low $p_{\rm T}$, v_2 is a probe of the early thermalization of matter created in the collision and the hydrodynamic evolution. High $p_{\rm T}$ is the realm of hard-scattering partons, which is also a critical probe of the early stages of the collision. Strong suppression of particle production in central collisions was observed early in the RHIC program, and PHENIX recently confirmed the suppression extends to very high $p_{\rm T}$ [14]. While measurements with respect to the reaction

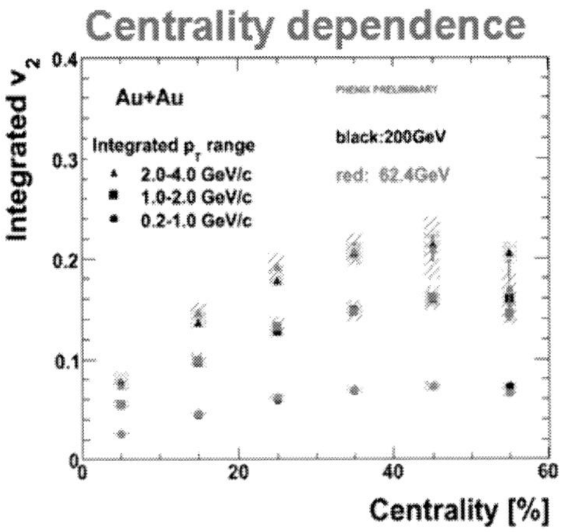

Figure 4. Centrality dependence of v_2 integrated over several p_T ranges for Au+Au collisions at $\sqrt{s} = 200$ and 62.4 GeV.

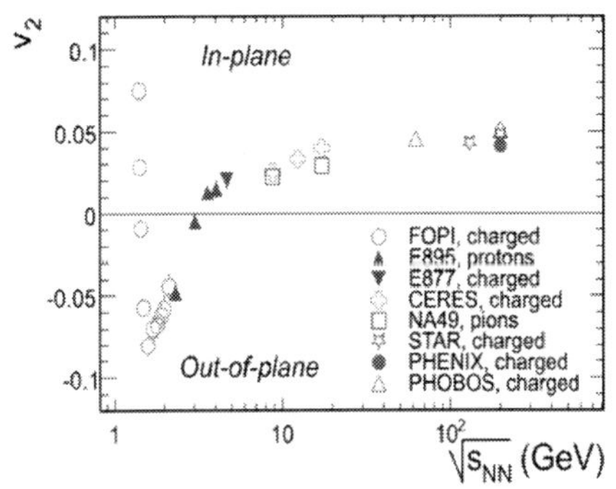

Figure 5. $\sqrt{s}$ dependence of p_T-integrated v_2 for Au+Au collisions.

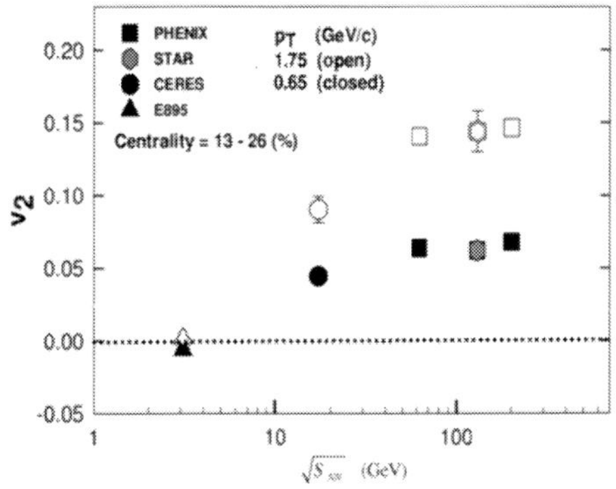

Figure 6. $\sqrt{s}$ dependence of v_2 in Au+Au collisions at $13 - 26\%$ centrality. The open markers are for $p_T \sim 1.75$ GeV/c and the closed markers data with $p_T \sim 0.65$ GeV/c.

plane have been traditionally done in conjuction with bulk observables, recent work has begun to focus on differential measurements involving jet and jet-related observables. In particular, measurements with respect to the reaction plane provide a handle on the path length through the medium traveled by the partons, as the geometry of the overlap region leads to the path length varying with angle of emission.

PHENIX has measured π^0 v_2 to high-p_T [1]. Most recently we have extended our measurement to $p_T = 14$ GeV/s in Run 7 [15], as shown in Figure 8, which has provided stronger evidence for non-zero v_2 at high p_T seen first in the Run 4 data. Application of the reaction plane orientation to the R_{AA} measurement provides a path to understand the effects of the geometry and path length in the suppression pattern of high-p_T particles [1]. An example is shown in Figure 9. This figure shows the high-p_T R_{AA} for 20-30% centrality, for π^0s emitted along the reaction

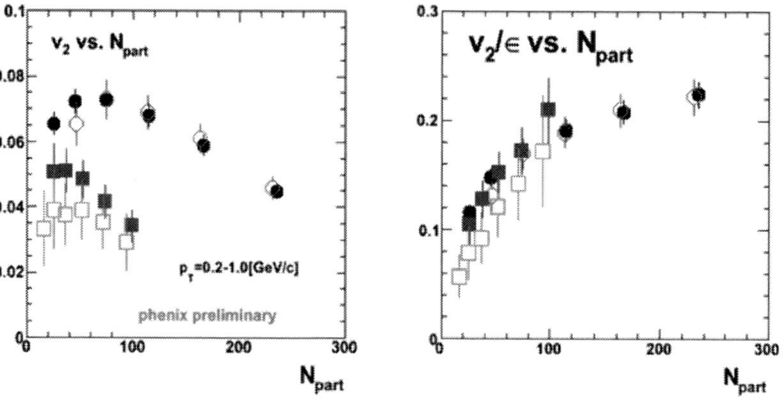

Figure 7. Charged hadron v_2 as a function of N_{part}, comparing system size and energies. The circles are Au+Au collisions and the squares are Cu+Cu collsions. The closed and open markers for each correspond to $\sqrt{s}$ =200 GeV and 62.4 GeV, respectively. The left panel is $v_2(N_{\mathrm{part}})$ and the right panel is $v_2/\epsilon(N_{\mathrm{part}})$.

plane and perpendicular to the reaction plane. This data is compared to several energy-loss models [16]. One limitation of inclusive R_{AA} measurements is that there are a number of models that can successfully reproduce R_{AA}, but differ in their assumptions for the initial conditions. The $\Delta\phi$ dependence of the R_{AA} provides greater discrimination between models. In Figure 9, the Armesto-Salgado-Wiedemann (ASW), Higher-Twist (HT), and Arnold-Moore-Yaffe (AMY) models are compared to PHENIX data. The results vary from model to model: the in-plane R_{AA} is flat with p_{T}, favoring the ASW and HT models; conversely, the out-of-plane R_{AA} has smaller energy loss with increasing p_{T}, favoring ASW and AMY.

4. Azimuthal Correlations

Because full reconstruction of jets is very difficult in the high-multiplicity of heavy-ion collisions, an important observable for accessing jet properties is azimuthal correlations. Much like the inclusive R_{AA} measurements, azimuthal correlations have benefited from the differential measurement with respect to the reaction plane. Fixing the angle of emission of the trigger particles places a tighter constraint on the average path length through the medium than offered by simply fixing just the centrality of the collision. For hadron triggers with $1 < p_{\mathrm{T}} < 2$, PHENIX has shown that the the away-side distributions in the "head" region ($\phi_{trig} \approx \pi$) show a clear effect of energy loss on the path length (the "shoulder" region, or $\phi_{trig} \approx 1.8$, is also very interesting, though more complicated to understand) [17]. Examining correlations with a higher p_{T} reach ($4 < p_{\mathrm{T}} < 7$), we can use away-side dependence on emission with respect to the reaction plane to discriminate between two production scenerios: one in which the observed jets are penetrating the medium, or one in which they are emitted purely along the surface of the collision region. In the former, we would expect the away-side yield to decrease from in-plane triggers to out-of-plane triggers, while the latter should exihibit the opposite trend. Figure 10 shows the per-trigger-yield for the away-side region in such a measurement, plotted with several theoretical predictions. In fact, the trends seen in the yields support a penetrating picture for jet production.

As an additional probe of the geometry of the collision we can study the distributions of associated particles to the "left" and "right" of the trigger particle, when it is held at a fixed

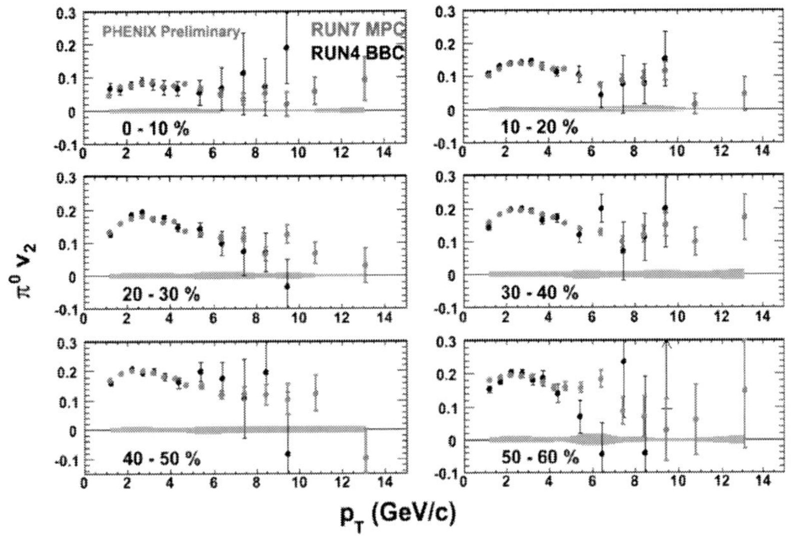

Figure 8. π^0 $v_2(p_{\rm T})$ for Au+Au collisions at $\sqrt{s}$ = 200 GeV. The red markers are Run 7 PHENIX preliminary results, and the black markers are Run 4 results. The error bars represent the statistical errors while the grey bands around $v_2 = 0$ represent the systematic uncertainties.

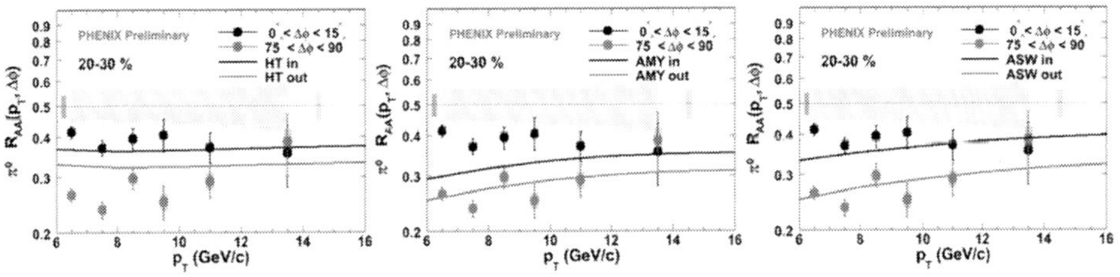

Figure 9. π^0 $R_{\rm AA}(p_{\rm T}, \Delta\phi)$ for Au+Au collisions at $\sqrt{s}$ = 200 GeV in 20-30% centrality. The black markers are in-plane $R_{\rm AA}$ and the red markers are out-of-plane $R_{\rm AA}$. The three panels each show the $R_{\rm AA}$ data compared with a single model: ASW, HT, and AMY (from left to right).

emission angle with respect to the reaction plane. In semi-central collisions, the associated partices will "see" different thicknesses of the region's almond-shaped collision zone. We expect different path lengths to result in variations in the away-side yields; this is indeed what we see, as shown in Figure 11.

5. Forward Rapidities and Heavy Flavor

In addition to detectors at mid-rapidity, PHENIX has two arms covering the rapidity region $1.2 < |\eta| < 2.4$. The forward arms are designed to access forward muon physics, but by analyzing tracks that only partially make it through the Muon Identifier absorbers, it is possible to study distributions of hadrons (as an example, see [18]). In fact, the rapidity acceptance fills in a region not covered by other forward measurements at RHIC. One recent example that uses the reaction plane is the measurement of forward hadron $v_2(p_{\rm T})$, as shown in Figure 12. Plotted with the

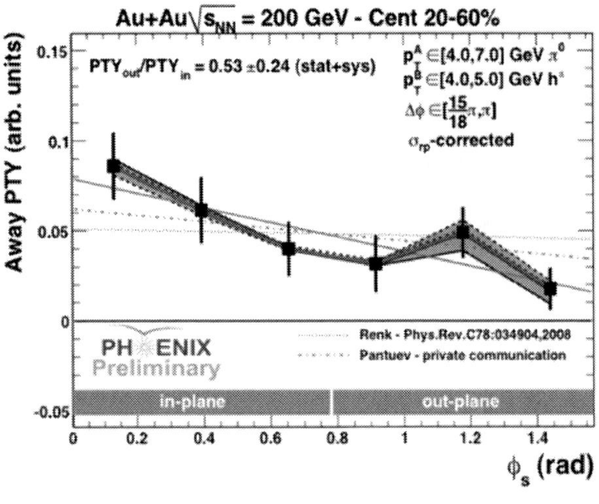

Figure 10. The per-trigger-yeild as a function of orientation with respect to the reaction plane, for hadron azimuthal correlations in 20-60% centrality $\sqrt{s}$ =200 GeV Au+Au collisions.

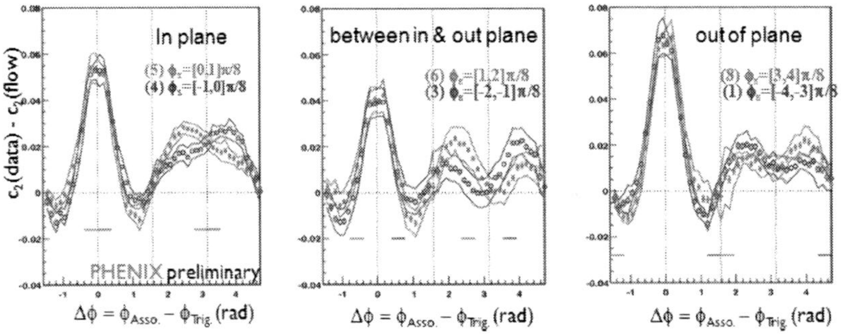

Figure 11. Hadron azimuthal correlation left-right asymmetries in $\sqrt{s}$ =200 GeV Au+Au collisions at 20-50% centrality. From left to right, the panels show the correlatations when the trigger is in-plane, oriented approximately at $\pi/4$, and out-of-plane, respectively. The red points are when the associated particle's angle with respect to the reaction plane is on one side of the trigger and blue when on the other side. The trigger p_T is 2-4 GeV/c and the associated p_T is 1-2 GeV/c.

forward results are mid-rapidity measurements of v_2 in similar centrality bins. The forward v_2 measurements are made using the RXNP detector, but in this case only the reaction plane from the opposite side is used, in order to avoid auto-correlations in the azimuthal distributions. In these mid-central collisions, we see the data exhibit a lower v_2 at forward rapidities.

The physics of the heavy quarks has emerged as one of the most puzzling and sought-after to understand. It was expected early on that because of the large mass of the charm and bottom quarks that hydrodynamics–and therefore elliptic flow–would not apply to them. One of the most exciting results from Quark Matter 2005 was the presentation of the v_2 of non-photonic electrons [19]. The PHENIX measurements from both Run 4 and Run 7 favor models that allow for the flow of charm quarks. In addition to single electrons, PHENIX has measured the v_2

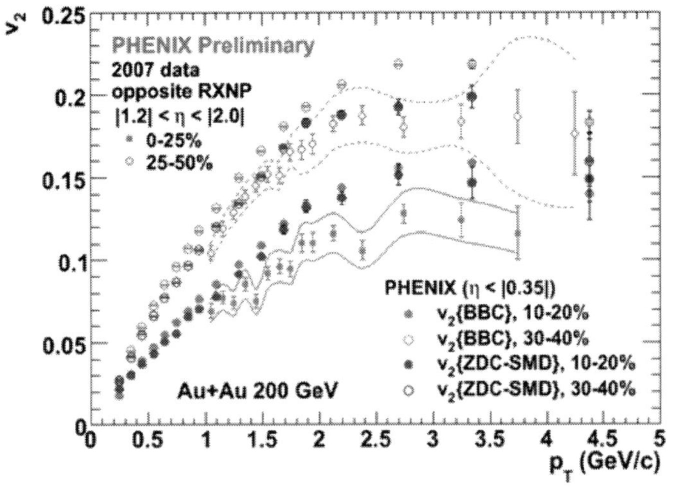

Figure 12. Hadron v_2 as a function of p_T for mid-central Au+Au $\sqrt{s}$ =200 GeV collisions. The red closed and open circles represent 0-25% and 25-50% centrality bins, respectively. The systematic error bands include reaction plane resolution, background estimation, and choice of reaction plane angle.

of the J/Ψ, shown in Figure 13. The J/Ψ is measured in both the mid-rapidity and forward regions. This measurement is fascinating, but unfortunately suffers from the lack of statistics. Thus it is hard to draw any firm conclusions on the flow of the J/Ψ.

6. Summary

The reaction plane in heavy-ion collisions is an extremely powerful tool. Analyses studying the anisotropy of emission with respect to the reaction plane's orientation provide access to a wide variety of probes and measurements of the hot, dense matter created in RHIC collisions. PHENIX has multiple overlapping and complementary systems for measuring the reaction plane on an event-by-event basis, and we have applied them to a variety of analyses. In this article we have presented recent results on the elliptic flow parameter v_2 at both low and high p_T, and its dependence on centrality, system size, and energy of the collision. In addition to flow measurements at mid-rapidity, we have measured flow for forward hadrons and heavy-flavor. We have also shown how the reaction plane can be used to probe parton energy loss as a function of the geometry, through azimuthal correlations and R_{AA}.

References

[1] Afanasiev S *et al.* (PHENIX) 2009 *Phys. Rev.* **C80** 054907 (*Preprint* 0903.4886)
[2] Morrison D P *et al.* (PHENIX) 1998 *Nucl. Phys.* **A638** 565–570 (*Preprint* hep-ex/9804004)
[3] Ikematsu K *et al.* 1998 *Nucl. Instrum. Meth.* **A411** 238–248 (*Preprint* physics/9802024)
[4] Adler S S *et al.* (PHENIX) 2003 *Phys. Rev. Lett.* **91** 182301 (*Preprint* nucl-ex/0305013)
[5] Adare A *et al.* (PHENIX) 2007 *Phys. Rev. Lett.* **98** 162301 (*Preprint* nucl-ex/0608033)
[6] Blume C and the NA49 Collaboration 2008 *Journal of Physics G: Nuclear and Particle Physics* **35** 044004 URL http://stacks.iop.org/0954-3899/35/i=4/a=044004
[7] Pinkenburg C *et al.* (E895) 1999 *Phys. Rev. Lett.* **83** 1295–1298 (*Preprint* nucl-ex/9903010)
[8] Ray R L (STAR) 2003 *Nucl. Phys.* **A715** 45–54 (*Preprint* nucl-ex/0211030)
[9] Adamova D *et al.* (CERES) 2002 *Nucl. Phys.* **A698** 253–260

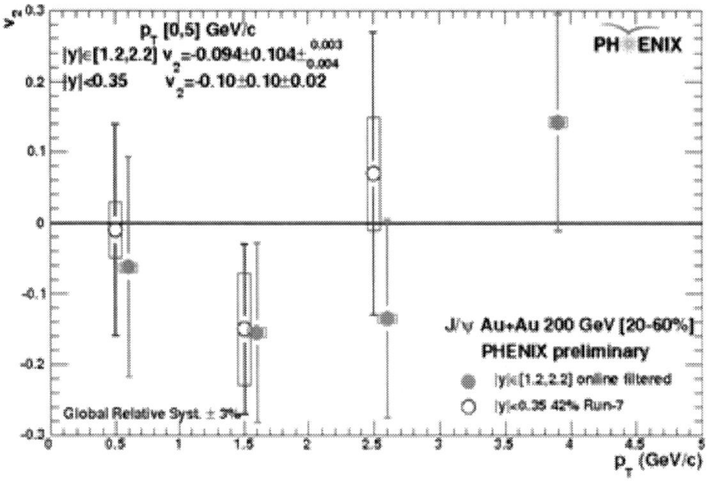

Figure 13. The elliptic flow parameter for the J/Ψ. The closed magenta circles represent measurement at forward rapidities while the open blue circles represent the mid-rapidity measurement. The error bars represent statistical uncertainties and the boxes systematic uncertainties.

[10] Alt C *et al.* (NA49) 2003 *Phys. Rev.* **C68** 034903 (*Preprint* nucl-ex/0303001)
[11] Adler S S *et al.* (PHENIX) 2005 *Phys. Rev. Lett.* **94** 232302 (*Preprint* nucl-ex/0411040)
[12] Andronic A *et al.* (FOPI) 2005 *Phys. Lett.* **B612** 173–180 (*Preprint* nucl-ex/0411024)
[13] Alver B *et al.* (PHOBOS) 2007 *Phys. Rev. Lett.* **98** 242302 (*Preprint* nucl-ex/0610037)
[14] Adare A *et al.* (PHENIX) 2008 *Phys. Rev. Lett.* **101** 232301 (*Preprint* 0801.4020)
[15] Wei R (PHENIX) 2009 *Nucl. Phys.* **A830** 175c–178c (*Preprint* 0907.0024)
[16] Bass S A *et al.* 2009 *Phys. Rev.* **C79** 024901 (*Preprint* 0808.0908)
[17] Holzmann W G (PHENIX) 2009 *Nucl. Phys.* **A830** 781c–784c (*Preprint* 0907.4833)
[18] Adler S S *et al.* (PHENIX) 2005 *Phys. Rev. Lett.* **94** 082302 (*Preprint* nucl-ex/0411054)
[19] Butsyk S A 2006 *Nucl. Phys.* **A774** 669–672 (*Preprint* nucl-ex/0510010)

Journal of Physics: Conference Series **230** (2010) 012018 doi:10.1088/1742-6596/230/1/012018

Directed Flow of Identified Particles in Au + Au Collisions at $\sqrt{s_{\rm NN}} = 200$ GeV

Aihong Tang (for the STAR Collaboration)
Physics Department, Brookhaven National Laboratory, PO Box 5000, Upton, NY 11973
E-mail: aihong@bnl.gov

Abstract. STAR's measurements of directed flow (v_1) for pions, kaons (charged and K_s^0), protons and antiprotons for Au + Au collisions at $\sqrt{s_{\rm NN}} = 200$ GeV are presented. Negative $v_1(y)$ slope is observed for pions, antiprotons, protons and kaons. The proton $v_1(y)$ slope at midrapidity is found extremely small. Sizable difference is seen between v_1 of protons and anti-protons in 5-30% central collisions. Anti-flow can explain the negative slope, however, it has difficulties explaining the centrality dependence of the difference between the $v_1(y)$ slopes of protons and anti-protons. The v_1 excitation function is presented. Comparisons to model calculations (RQMD, uRQMD, AMPT and QGSM) are made, and it is found that none of the four models can successfully describe the data.

1. Introduction

In relativistic heavy ion collisions, anisotropic flow describes the collective motion of particles with respect to the reaction plane, and it is conventionally characterized [1] by Fourier coefficients

$$v_n = \langle \cos n(\phi - \psi) \rangle \qquad (1)$$

where ϕ denotes the azimuthal angle of an outgoing particle, ψ is the orientation of the reaction plane, and n denotes the harmonic. So far three such coefficients are found finite at RHIC, namely, directed flow v_1, elliptic flow v_2 and 4th order harmonic flow v_4. This paper will focus on the directed flow, the first Fourier coefficient. Directed flow describes the sideward motion of fragments in ultra-relativistic nuclear collisions, and it carries early information from the collision [2]. The shape of directed flow at midrapidity is of special interest because it might reveal a signature of a possible phase transition from normal nuclear matter to a Quark-Gluon Plasma (QGP) [3, 4, 5]. It is argued that directed flow, as a function of rapidity, may exhibit a flatness at midrapidity due to a strong, tilted expansion of the source. Such expansion gives rise to anti-flow [3] or a 3rd flow [4] component. The anti-flow (3rd flow component) is perpendicular to the source surface, and is in the direction that is opposite to the direction of the bouncing-off motion of nucleons. If the tilted expansion is strong enough, it can even overcome the bouncing-off motion and results in a negative $v_1(y)$ slope at midrapidity, producing a wiggle-like structure. In [5], it is emphasized that proton v_1 can serve as a probe for the first order phase transition, as calculation shows that proton $v_1(y)$ does not show negative slope for a hadronic Equation of State (EoS) without a QGP phase transition. Note that although the calculation is done with a first order phase transition at SPS energies, the direct cause of the negative slope is the strong, tilted expansion, which is also relevant at top RHIC energies. A wiggle structure is also seen in RQMD [6], and it is attributed to the baryon stopping and positive space-momentum correlation. In this picture, no phase transition is needed, and pions and nucleons flow in opposite directions. To distinguish between baryon stopping and anti-flow associated with a phase transition, it is desirable to measure the $v_1(y)$ for identified particles and compare the signs of their slope at midrapidity. The centrality dependence of proton $v_1(y)$ may have implications for a possible first order phase transition. It is expected that in very peripheral collisions, protons flow in the same direction as spectators. In mid-central collisions, if there is a phase transition, the proton $v_1(y)$ slope at midrapidity may change sign and become negative. Eventually the slope diminishes in central collisions due to the symmetry of collisions.

At low energies, it is shown by the E895 collaboration [7] that K_s^0 has a negative $v_1(y)$ slope around midapidity, while lambda and protons have positive slope [8]. It is explained by a repulsive kaon-nucleon potential and attractive lambda-nucleon potential. The NA49 collaboration [9] has measured the v_1 for pions and protons, and negative $v_1(y)$ slope is observed by the standard event plane method. The three particle correlation method $v_1\{3\}$, which is believed to be less sensitive to non-flow effects, gives negative slope too, but with larger statistical error. At top RHIC energies, v_1 has been studied for charged particles by both the STAR and the PHOBOS collaborations [10, 11, 12, 13]. It is found that v_1 in the forward region follows the limiting fragmentation hypothesis well, and $v_1(\eta)$ depends only on the incident energy, but not on the size of the colliding system at a given centrality. The system size independence of v_1 can be explained by Hydrodynamic calculation with an titled initial condition [14]. The systematic study of v_1 for identified particles at RHIC did not begin until recently because it is more challenging for two reasons, 1) v_1 for some identified particles (for example, protons) is much smaller than that of charged particles, thus is more difficult to measure , 2) it demands more statistics to measure v_1 for identified particles other than poins.

2. Data Set

In this analysis we used 62 million events for Au + Au collisions at $\sqrt{s_{\rm NN}} = 200$ GeV, all taken by STAR's minimum bias trigger during RHIC run 2007. Charged particles are identified by their energy loss inside STAR's Time Projection Chamber (TPC) [15], and $K_s^0(\rightarrow \pi^+\pi^-)$ are reconstructed by their charged daughter tracks inside the TPC. Track quality cuts are the same as used in [16]. The centrality definition of an event was based on the number of charged global tracks in the TPC with track quality cuts: $|\eta|$ <0.5, a Distance of Closest Approach (DCA) to the global vertex less than 3 cm, and 10 or more fit points. Additional weight is assigned to each event in the analysis accounting for the bias in the vertex Z direction, due to the inclusion of hits of Silicon Vertex Tracker in tracking for year 2007 data production. In addition, the transverse momentum p_T for protons are required to be larger than 400 MeV/c, and DCA are required to be less than 1 cm, in order to avoid the beam-pipe background. The same cut is applied to anti-protons as well to enssure a fair comparison with protons. The high end of the p_T cut is 1 GeV/c where protons and pions have the same energy loss in the TPC and thus become indistinguishable. The event plane angle is determined from the sideward deflection of spectator neutrons measured by STAR's Shower Maximum Detector at Zero Degree Calorimeters (ZDCSMD), which are located close to beam rapidity and have the minimum contribution from non-flow. The description of measuring v_1 using the ZDCSMD event plane can be found in [12].

3. Result

Fig. 1 shows the pion v_1 as a function of rapidity. The result is compared to four model calculations, namely, RQMD [6], UrQMD [17], AMPT [18] and QGSM [19]. Errors in this plot (and other data figures in this paper) are statistical only. Systematical uncertainties are studied by varying track quality cuts, and are found to be negligible. Calculations from RQMD and

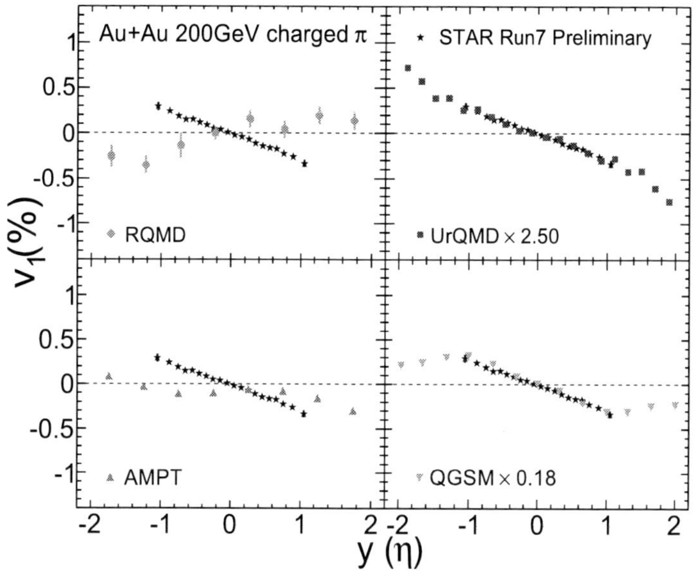

Figure 1. Pion v_1 as a function of rapidity (or η, for the QGSM model only) for 10-70% Au + Au collisions at $\sqrt{s_{NN}} = 200$ GeV. Also plotted are corresponding model calculations. The calculation from UrQMD and QGSM are scaled by a factor of 2.5 and 0.18, respectively. Errors are statistical only.

QGSM are scaled by a factor of 2.5 and 0.18, respectively. Following convention, the sign of spectator v_1 in the forward region is chosen to be positive, to which the measured sign of v_1 for particles of interest is only relative. The pion $v_1(y)$ result follows closely to the published charged particle result [12] and it shows a negative slope at midrapidity. Most models either predict the wrong sign of pion v_1 or the wrong magnitude by a signifiant factor [1], which indicates that it is a challenging job to understand the dynamics happening at early stages. In this sense the measurement presented in this paper will offer strong constraints on models.

To address the shadowing effect and the anti-flow, in Fig.2, proton and anti-proton v_1 are plotted together with pion v_1. Pions are the dominant particles produced and they carry the flow of the bulk. It is not a surprise to see that anti-protons, also as produced particles, follow the flow direction of pions. They both have negative slope at midrapidity, which is consistent with the anti-flow picture. $v_1(y)$ of protons, however, exhibits a flatter shape than others. The mass difference between protons and pions cannot explain the flatness of proton $v_1(y)$ because the anti-protons have the same mass as that of protons, yet they flow close to pions. Indeed, the observed v_1 for protons is a convolution of directed flow of produced protons with that of transported protons, and the flatness of inclusive proton $v_1(y)$ around midrapidity can be explained by the negative flow of produced protons being compensated by the positive flow of protons transported from spectator rapidity, as a feature expected in the anti-flow picture. Fig. 3 shows v_1 of kaon and K_s^0 as a function of rapidity, for centrality 10-70%. Pion v_1 is also plotted for comparison. Kaon v_1 is a unique measurement because the kaon/proton cross section is smaller than the pion/proton cross section, thus its directed flow is less likely to be "induced" by the shadowing effect. The plot shows that both kaon and K_s^0 have negative v_1

[1] However, in a recent paper [14] posted after the 26th Winter Workshop on Nuclear Dynamics, it is shown that the directed flow of charged particles can be explained by Hydrodynamic calculation with a tilted initial condition.

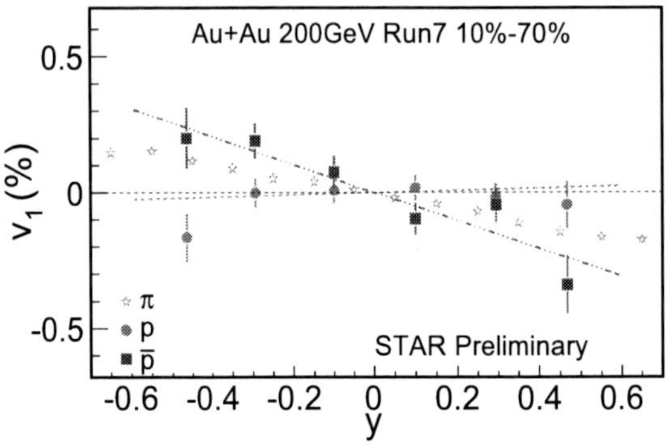

Figure 2. Proton (solid circles) and anti-proton (solid squares) v_1 as a function of rapidity for 10-70% Au + Au collisions at $\sqrt{s_{\rm NN}}$ = 200 GeV, pion $v_1(y)$ (empty stars) is shown as a reference.

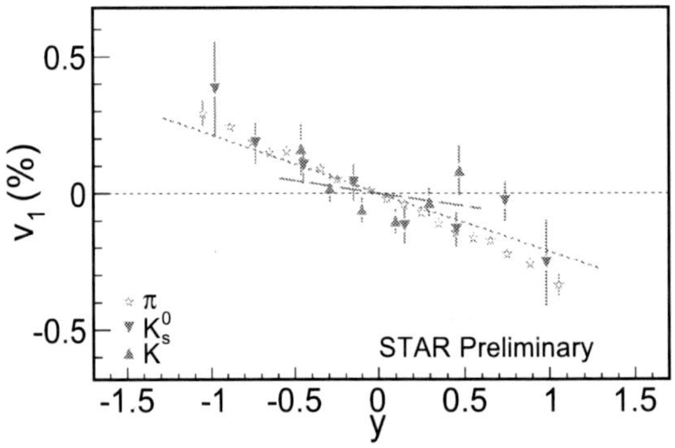

Figure 3. Charged kaon (up triangles) and K^0_s (down triangles) v_1 as a function of rapidity for 10-70% Au + Au collisions at $\sqrt{s_{\rm NN}}$ = 200 GeV, pion $v_1(y)$ (empty stars) is shown as a reference.

slope, which means that the observed v_1 is not likely to be caused by the shadowing effect.

On the other hand, anti-flow has difficulties in explaining the centrality dependence of v_1, as shown by Fig. 4 , in which v_1 slope at midrapidity is plotted as a function of centrality for protons, anti-protons and charged particles. If there is an anti-flow due to the strong, tilted expansion, one expects such an effect is larger in mid-central collisions than that in peripheral collisions. As a consequence, proton v_1 slope, which is expected to be positive in very peripheral collisions, will change its sign to negative in mid-central collisions and approach zero in central collisions. Due to the large statistical error, we cannot exclude the possibility that proton v1 can be positive in 70-80% central collisions. However, in 30-80% central collisions, proton v_1 slope is found mostly negative and the magnitude decreases with decreasing centrality. In more central (5-30%) collisions, proton v_1 slope becomes extremely small, while anti-proton v_1 slope remains negative and continues to follow that of charged particles (mostly pions). Anti-flow may cause a difference between v_1 of protons and anti-protons, and such difference is expected to be accompanied by strongly negative v_1. In data, the large difference between proton and anti-proton v_1 is seen in 5-30% centrality, while strongly negative v_1 is found for protons, anti-protons

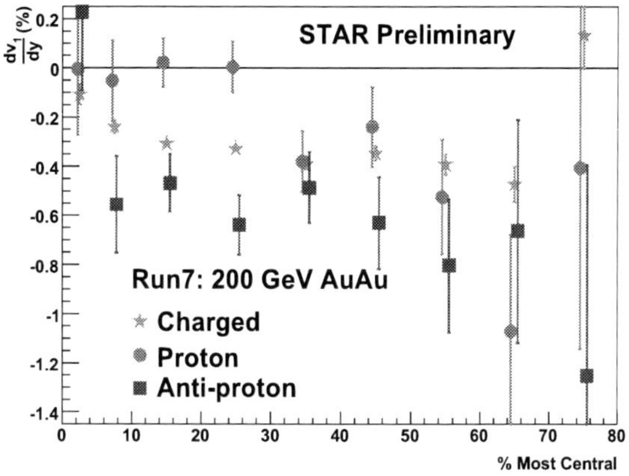

Figure 4. Charged (solid stars), proton (solid circles) and anti-proton (solid squares) $v_1(y)$ slope (dv_1/dy) at midrapidity as a function of centrality for Au + Au collisions at $\sqrt{s_{NN}} = 200$ GeV.

and charged particles in a different centrality (30-70%).

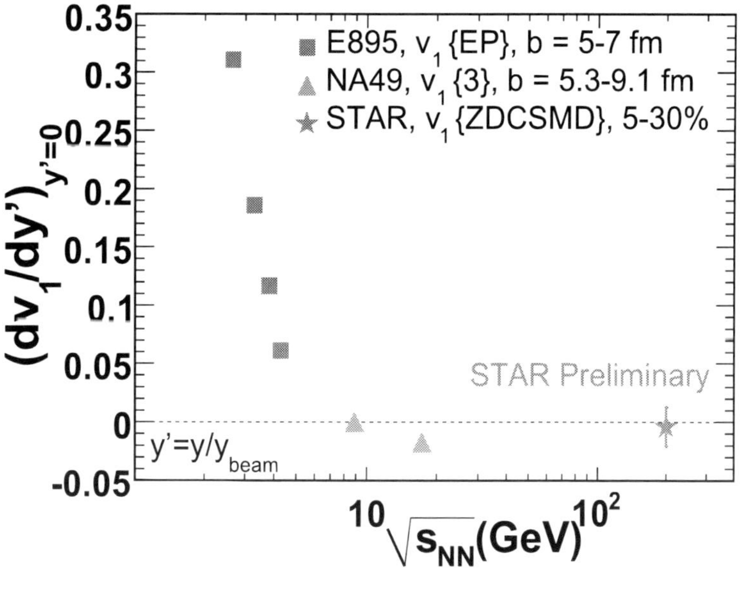

Figure 5. Proton $v_1(y^{'})$ slope $(dv_1/dy^{'})$ at midrapidity as a function of center of mass collision energy, where $y^{'} = y/y_{beam}$.

In Fig. 5, proton $v_1(y/y_{beam})$ slope at midrapidity is plotted as a function of collision energy. We see that proton v_1 slope decreases rapidly with increasing energy, reaching zero around $\sqrt{s_{NN}} = 9$ GeV. It changes its sign to negative as shown by the data point at $\sqrt{s_{NN}} = 17$ GeV, measured by NA49 experiment. With previous measurements, which include only one point above $\sqrt{s_{NN}} = 9$ GeV, one cannot conclude if proton v_1 slope continues to decrease or stays close to zero when energy increases, the addition of the data point from RHIC indicates that proton v_1 slope remains close to zero at $\sqrt{s_{\rm NN}} = 200$ GeV. Judging over the broad energy range, the transition of the trend happens around $\sqrt{s_{\rm NN}} = 9$ GeV, making it a discontinuous point, and interestingly, it coincides with the energy vicinity where $\langle k^+\rangle/\langle\pi^+\rangle$ exhibits a kink [20]. Hopefully the on-going Beam Energy Scan program will map out the region of interest with details.

4. Summary

For 10-70% central Au + Au collisions at $\sqrt{s_{\rm NN}} = 200$ GeV, directed flow for pions, kaons (charged and K_s^0), and anti-protons are found to have negative slope at mid-rapidity. Protons, however, exhibit a flat shape around mid-rapidity. Sizable difference is seen between v_1 of protons and anti-protons in 5-30% central collisions. Anti-flow can explain the negative slope, however, it has difficulties explaining the centrality dependence of the difference between the $v_1(y)$ slopes of protons and anti-protons.

References

[1] A. M. Poskanzer and S. A. Voloshin, *Phys. Rev. C* **58**, 1671 (1998).
[2] H. Sorge, *Phys. Rev. Lett.* **78**, 2309 (1997).
[3] J. Brachmann, S. Soff, A. Dumitru, H. Stöcker, J. A. Mahruhn, W. Greiner, L. V. Bravina, and D. H. Rischke, *Phys. Rev. C* **61**, 024909 (2000).
[4] L. P. Csernai and D. Röhrich, *Phys. Lett. B* , **458**, 454 (1999).
[5] H. Stöcker, Nucl. Phys. A **750**, 121 (2005).
[6] R. J. Snellings, H. Sorge, S. A. Voloshin, F. Q. Wang and N. Xu, *Phys. Rev. Lett* **84**, 2803 (2000).
[7] P. Chung *et al.* [e895 Collaboration], *Phys. Rev. Lett.* **85**, 940 (2000).
[8] P. Chung *et al.* [e895 Collaboration], *Phys. Rev. Lett.* **86**, 2533 (2001).
[9] C. Alt *et al.* [na49 Collaboration], *Phys. Rev. Lett.* **68**, 034903 (2003).
[10] J. Adams *et al.* [STAR Collaboration], *Phys. Rev. Lett.* **92**, 062301 (2004).
[11] B. B. Back *et al.* [phobos Collaboration], *Phys. Rev. Lett* **97**, 012301 (2006).
[12] J. Adams *et al.* [STAR Collaboration], *Phys. Rev. C* **73**, 034903 (2006).
[13] B. I. Abelev *et al.* [STAR Collaboration], *Phys. Rev. Lett.* **101** 252301 (2008).
[14] Piotr Bożek and Iwona Wyskiel, arXiv:1002.4999 [nucl-th] (2010).
[15] M. Anderson *et al.*, *Nucl. Instrum. Method. A* **499** 659 (2003).
[16] J. Adams *et al.* [STAR Collaboration], *Phys. Rev. C* **72** 014904 (2005).
[17] M. Bleicher and H. Stöcker, *Phys. Rev. B* **526** 309 (2002).
[18] J. Y. Chen, J. X. Zuo, X. Z. Cai, F. Liu, Y. G. Ma and A. H. Tang, *Phys. Rev. C* **81**, 014904 (2010).
[19] G. Burau, J. Bleibel, C. Fuchs, A. Faessler, L. V. Bravina, and E. E. Zabrodin, *Phys. Rev. C* **71** 054905 (2005).
[20] S. V. Afanasiev *et al.* [NA49 Collaboration], *Phys. Rev. C* **66**, 054902 (2002); C. Alt *et al.* [NA49 Collaboration], *Phys. Rev. C* **77**, 024903 (2008); *Phys. Rev. C* **73**, 044910 (2006); T. Anticic *et al.* [NA49 Collaboration], *Phys. Rev. C* **69**, 024902 (2004).

Journal of Physics: Conference Series **230** (2010) 012019 doi:10.1088/1742-6596/230/1/012019

Investigation of jet quenching and elliptic flow within a pQCD-based partonic transport model

Oliver Fochler[1], Zhe Xu[2,1], and Carsten Greiner[1]

[1] Institut für Theoretische Physik, Johann Wolfgang Goethe-Universität Frankfurt, Max-von-Laue-Str. 1, D-60438 Frankfurt am Main, Germany

[2] Frankfurt Institute for Advanced Studies, Ruth-Moufang-Str. 1, D-60438 Frankfurt am Main, Germany

E-mail: `fochler@th.physik.uni-frankfurt.de`

Abstract. The partonic transport model BAMPS (a Boltzmann approach to multiparton scatterings) is employed to investigate different aspects of heavy ion collisions within a common framework based on perturbative QCD. This report focuses on the joint investigation of the collective behavior of the created medium and the energy loss of high–p_T gluons traversing this medium. To this end the elliptic flow and the nuclear modification factor of gluons in heavy ion collisions at 200 AGeV are simulated with BAMPS.

The mechanism for the energy loss of high energy gluons within BAMPS is studied in detail. For this, purely elastic interactions are compared to radiative processes, $gg \rightarrow ggg$, that are implemented based on the matrix element by Gunion and Bertsch. The latter are found to be the dominant source of energy loss within the framework employed in this work.

1. Introduction

Experiments at the Relativistic Heavy Ion Collider (RHIC) have established that jets with high transverse momenta are suppressed in $Au+Au$ collisions with respect to a scaled $p+p$ reference [1, 2]. This quenching of jets is commonly attributed to energy loss on the partonic level as the hard partons produced in initial interactions are bound to traverse the hot medium, the quark–gluon plasma (QGP), created in the early stages of the heavy ion collision. Another major discovery has been the strong collective flow of the created medium [3, 4], that is usually quantified in terms of the Fourier parameter v_2 and in this context often referred to as elliptic flow.

Due to the large momentum scales involved the energy loss of partonic jets can be treated in terms of perturbative QCD (pQCD) and most theoretical schemes attribute the main contribution to partonic energy loss to radiative processes [5, 6, 7, 8, 9, 10, 11]. The bulk properties of the medium, such as the elliptic flow, are on the other hand usually investigated within hydrodynamical models. The comparison of hydrodynamic calculations to data indicates that the viscosity of the QGP is quite small [12], possibly close to the conjectured lower bound $\frac{\eta}{s} = \frac{1}{4\pi}$ from a correspondence between conformal field theory and string theory in an Anti-de-Sitter space [13].

It is a major challenge to combine these two aspects, jet physics on the one and bulk evolution on the other hand, within a common framework. Recently the efforts to combine pQCD-based jet physics with hydrodynamic modeling of the medium have been intensified, for instance results from hydrodynamical simulations are used as an input for the medium evolution in jet–quenching calculations (see [14] for an overview) and as ingredients in Monte Carlo event generators [15]. However, these approaches still treat medium physics and jet physics in the QGP on very different grounds. Moreover, so far no schemes are available that cover the full dynamics of the interplay between jets and the medium.

Partonic transport models might provide means to investigate bulk properties of the QGP and high–energy parton jets within a common physical framework automatically including the full dynamics of the system evolution. In previous publications [16, 17] we have explored the capabilities of the transport model BAMPS (a Boltzmann approach to multiparton scatterings) with this goal in mind.

2. The transport model BAMPS

BAMPS [18, 19] is a microscopic transport model aimed at simulating the QGP stage of heavy ion collisions via pQCD interactions consistently including parton creation and annihilation processes. Partons within BAMPS are treated as semi-classical and massless Boltzmann particles and at this stage the model is limited to gluonic degrees of freedom. Thus $N_f = 0$ is understood throughout this work. The strong coupling is fixed to $\alpha_s = 0.3$. Please also note that the term "jet" as used throughout this paper refers to a single gluon with high energy that traverses the medium and does thus not fully coincide with the experimental notion.

The interactions between partons are based on leading order pQCD matrix elements from which transition probabilities are computed. These are used to sample the interactions of particles in a stochastic manner [18]. The test particle method is introduced to reduce statistical fluctuations. For elastic interactions of gluons, $gg \leftrightarrow gg$, we use the Debye screened cross section in small angle approximation $\frac{d\sigma_{gg\to gg}}{dq_\perp^2} = \frac{9\pi\alpha_s^2}{(\mathbf{q}_\perp^2+m_D^2)^2}$. The Debye screening mass is computed from the local particle distribution $f = f(p,x,t)$ via $m_D^2 = d_G \pi \alpha_s \int \frac{d^3p}{(2\pi)^3}\frac{1}{p} N_c f$, where $d_G = 16$ is the gluon degeneracy factor for $N_c = 3$.

Inelastic $gg \leftrightarrow ggg$ processes are treated via the Gunion-Bertsch matrix element [20]

$$\left|\mathcal{M}_{gg\to ggg}\right|^2 = \frac{72\pi^2\alpha_s^2 s^2}{(\mathbf{q}_\perp^2 + m_D^2)^2}\,\frac{48\pi\alpha_s \mathbf{q}_\perp^2}{\mathbf{k}_\perp^2[(\mathbf{k}_\perp - \mathbf{q}_\perp)^2 + m_D^2]}\Theta\left(\Lambda_g - \tau\right), \tag{1}$$

where $\mathbf{q}_\perp$ and $\mathbf{k}_\perp$ denote the perpendicular components of the momentum transfer and of the radiated gluon momentum in the center of momentum (CM) frame of the colliding particles, respectively. Detailed balance between gluon multiplication and annihilation processes is ensured by the relation $\left|\mathcal{M}_{gg\to ggg}\right|^2 = d_G \left|\mathcal{M}_{ggg\to gg}\right|^2$.

The Theta function in (1) is an effective implementation of the LPM (Landau, Pomeranchuk, Migdal) effect [21], that describes coherence effects in multiple bremsstrahlung processes. This interference effect cannot be incorporated directly into a semi–classical microscopic transport model such as BAMPS, so the cut–off $\Theta\left(\Lambda_g - \tau\right)$ ensures that successive $gg \to ggg$ processes are independent of each other. τ is the formation time of the gluon emitted with transverse momentum $k_\perp$ and Λ_g denotes the mean free path, i.e. the time between successive interactions, of the parent gluon. When comparing the formation time to the mean free path of the parent gluon special attention needs to be paid to the frames of reference. This ultimately renders the cut–off dependent on the boost $\vec{\beta}$ between the plasma rest frame and the center of momentum (CM) frame in which (1) is evaluated [17] and numerically further complicates the calculations.

3. Gluon jets in a static medium

In order to obtain baseline calculations and to gain a better understanding of the energy loss mechanisms in BAMPS, we study the evolution of high energy gluons in a simplified brick setup, i.e. the propagation within a static thermal medium with fixed temperature T. Also, for reasons of computing time, possible effects of the propagating jet on the medium are neglected in most computations presented in this section. For example, the mean energy loss per unit path length dE/dx is then calculated as follows ($c = 1$)

$$\frac{dE}{dx} = \frac{dE}{d(ct)} = \sum_i \langle \Delta E^i \rangle R^i \tag{2}$$

where i denotes the interaction type ($gg \to gg$, $gg \to ggg$ and $ggg \to gg$) and R^i is the interaction rate for process i and $\langle \Delta E^i \rangle$ is the mean energy loss in a single collision of type i.

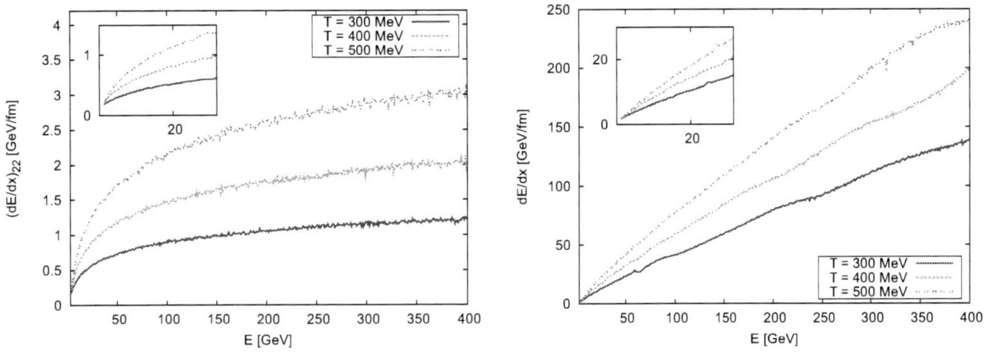

Figure 1. Differential energy loss of a gluon jet in a static and thermal medium of gluons at $T = 400\,\mathrm{MeV}$, $T = 500\,\mathrm{MeV}$ and $T = 600\,\mathrm{MeV}$.
Left panel: Elastic interactions only. Right panel: Including $gg \leftrightarrow ggg$ processes.

The differential energy loss from elastic interactions as shown in the left panel of Fig. 1 exhibits the expected (see [11] for an overview) logarithmic dependence on the jet energy E and the dominant quadratic dependence on the medium temperature T, $\left.\frac{dE}{dx}\right|_{2\to2} \propto C_R \pi \alpha_s^2 T^2 \ln\left(\frac{4ET}{m_D^2}\right)$, where C_R is the quadratic Casimir of the propagating jet, $C_R = C_A = N_c$ for gluons. For $T = 400\,\mathrm{MeV}$ and $E = 50\,\mathrm{GeV}$ we find an elastic energy loss of $\left.\frac{dE}{dx}\right|_{2\to2} \approx 1.2\,\mathrm{GeV/fm}$ that increases to $\left.\frac{dE}{dx}\right|_{2\to2} \approx 2\,\mathrm{GeV/fm}$ at $E = 400\,\mathrm{GeV}$.

The energy loss caused by radiative $gg \to ggg$ interactions in BAMPS is much stronger and by far dominates the total differential energy loss (right panel of Fig. 1) of a high energy gluon. The differential energy loss from $gg \to ggg$ is rising almost linearly with the energy, for example resulting in a total $dE/dx \approx 32.6\,\mathrm{GeV/fm}$ at $E = 50\,\mathrm{GeV}$ and $T = 400\,\mathrm{MeV}$.

The large differential energy loss in $gg \to ggg$ processes, however, is not governed by excessively strong cross sections. The individual cross sections increase only slowly with the jet energy as seen in the left panel of Fig. 2. For instance at $E = 50\,\mathrm{GeV}$ and $T = 400\,\mathrm{MeV}$ a gluon jet interacts with cross sections $\langle \sigma_{gg\to gg} \rangle \approx 1.3\,\mathrm{mb}$ and $\langle \sigma_{gg\to ggg} \rangle \approx 3.5\,\mathrm{mb}$. This emphasizes that BAMPS does indeed operate with reasonable partonic cross sections based on pQCD matrix elements.

Rather it is the mean energy loss per single radiative process, $\langle \Delta E_{23} \rangle$, shown in the right panel of Fig. 2, that determines the magnitude of the differential energy loss dE/dx. It is noteworthy that the strong and linear rise in the energy loss due to $gg \to ggg$ is only present when identifying the outgoing gluon with the highest energy as the jet gluon and thus using $\Delta E = E^{\mathrm{in}} - \max\left(E_1^{\mathrm{out}}, E_2^{\mathrm{out}}, E_3^{\mathrm{out}}\right)$. This is the most natural choice and is employed throughout all calculations in this work. The average energy ω of the radiated gluon, however, is rising

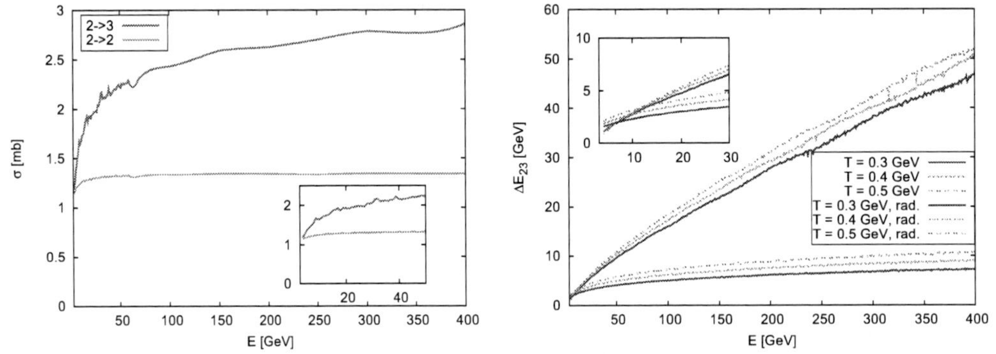

Figure 2. Left panel: Cross sections for $gg \to gg$ $(2 \to 2)$ and $gg \to ggg$ $(2 \to 3)$ processes. Right panel: Mean energy loss $\langle \Delta E_{23} \rangle$ in a single $gg \to ggg$ process. Upper group of lines: $\Delta E = E^{\text{in}} - \max\left(E_1^{\text{out}}, E_2^{\text{out}}, E_3^{\text{out}}\right)$. Lower group of lines (labeled "rad."): $\Delta E = \omega$, with ω being the energy of the radiated gluon.
All quantities given as a function of the energy of the gluon jet inside a thermal medium with $T = 400\,\text{MeV}$.

much slower with the jet energy. This is due to the fact that the energy is distributed among three outgoing particles, the gluon emitted with energy ω being only one of them. See [17] for an in–depth discussion on how the complex interplay between kinematics, Gunion–Bertsch matrix element and phase space restriction imposed by the effective LPM cut–off gives rise to a fat tail in the ΔE_{23} distribution, that ultimately leads to large mean values $\langle \Delta E_{23} \rangle$.

The full time evolution of the energy distribution of a jet particle propagating through the medium obviously contains more detailed information than the mean energy loss per unit path length. Fig. 3 shows $p(E)\,dE|_t$, the probability that a parton starting with $E(t = 0\,\text{fm/c}) = E_0 = 50\,\text{GeV}$ has an energy $E \leq E(t) < E + dE$ at a given time t.

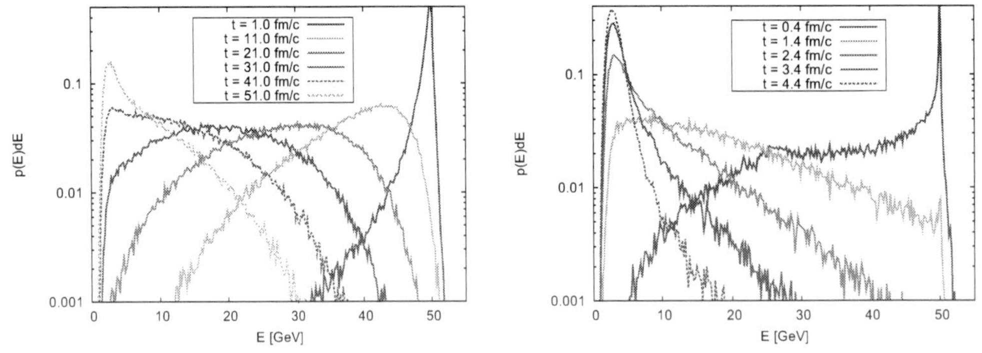

Figure 3. Time evolution of the energy distribution of a gluon jet that traverses a static and thermal medium of gluons ($T = 400\,\text{MeV}$). The initial ($t = 0\,\text{fm/c}$) energy of the gluon jet is $E_0 = 50\,\text{GeV}$.
Left panel: Elastic interactions only. Right panel: Including $gg \leftrightarrow ggg$ processes.

For both $gg \to gg$ and $gg \to ggg$ the distribution of the jet energy induced by collisions with the constituents of the medium becomes rather broad. The distributions significantly differ from Gaussian shapes and a simple shift of the mean energy accompanied with momentum diffusion could not account for the behavior depicted in Fig. 3. A distinct peak at lower energies only re-emerges at very late times. The mean energy loss as depicted in Fig. 1 is therefore a valuable observable but contains only limited information. It is noteworthy that there exists a finite probability for the jet to gain energy by collisions with the thermal gluons.

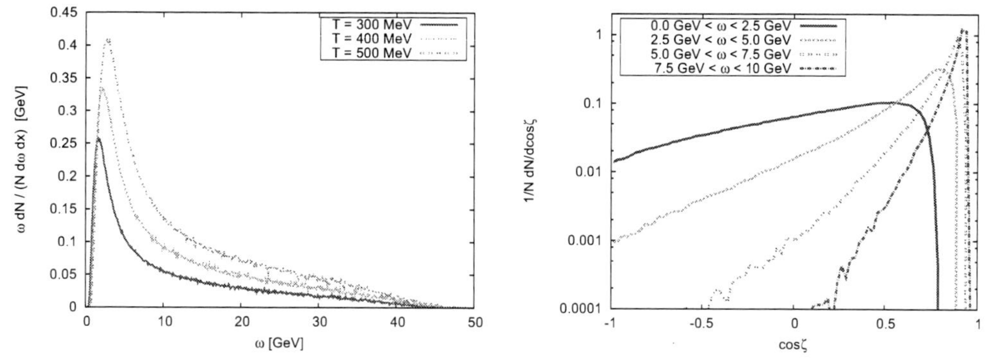

Figure 4. Left panel: Energy spectrum $\omega \frac{dN}{N\,d\omega\,dx}$ of radiated gluons per energy interval $d\omega$ and distance dx. ω is the energy (lab frame) associated with the radiated gluon according to the Gunion–Bertsch matrix element. The energy of the gluon jet is $E = 50\,\mathrm{GeV}$.
Right panel: Angular distribution of the radiated gluon in the lab frame with respect to the original jet direction for different energies ω of the radiated gluon. Jet energy $E = 50\,\mathrm{GeV}$, medium $T = 400\,\mathrm{MeV}$.

Accompanying the above discussions, Fig. 4 shows the energy spectrum of gluons radiated in $gg \to ggg$ processes and the angular distribution of the radiated gluons for different ranges of their energy ω. It is clearly visible that, due to the cut–off in transverse momentum (1), the gluons cannot be emitted at very forward angles, an effect that is more pronounced for low ω. The spectra are peaked at energies $\omega \ll E$, with a small tail reaching out to high energies. With increasing temperature the peak of the spectrum shifts towards higher energies in an apparently linear way, favoring the emission of gluons with higher energies.

4. Au+Au collisions at 200 AGeV

BAMPS has been applied to simulate elliptic flow and jet quenching at RHIC energies [16], for the first time using a consistent and fully pQCD–based microscopic transport model to approach both key observables on the partonic level within a common setup. The left panel of Fig. 5 shows that the medium simulated in the parton cascade BAMPS exhibits a sizable degree of elliptic flow in agreement with experimental findings at RHIC as established in [22, 23]. And η/s of the gluon matter in BAMPS has been shown to be small [24].

For simulations of heavy ion collisions the initial gluon distributions are sampled according to a mini–jet model with a lower momentum cut-off $p_0 = 1.4\,\mathrm{GeV}$ and a K–factor of 2. The test particle method [18] is employed to ensure sufficient statistics and to allow for the resolution of adequate spatial length scales. The underlying nucleon-nucleon collisions follow a Glauber-model with a Wood-Saxon density profile and the results by Glück, Reya and Vogt [25] are used as parton distribution functions. Quarks are discarded after sampling the initial parton distribution since currently a purely gluonic medium is considered. To model the freeze out of the simulated fireball, free streaming is applied to regions where the local energy density has dropped below a critical energy density ε_c ($\varepsilon_c = 1.0\,\mathrm{GeV/fm^3}$ unless otherwise noted). This setup has been successfully checked against experimental findings such as the distribution of transverse energy in rapidity and the flow parameter v_2 at various centralities in [19, 22, 23].

The right panel of Fig. 5 shows the gluonic R_{AA} simulated in BAMPS for central, $b = 0\,\text{fm}$, collisions. It is roughly constant at $R_{AA}^{\text{gluons}} \approx 0.053$ and in reasonable agreement with recent analytic results for the gluonic contribution to the nuclear modification factor R_{AA} [11], though the suppression of gluon jets in BAMPS appears to be slightly stronger. We expect improved agreement in future studies when employing a carefully averaged $\langle b \rangle$ that will be better suited for comparison to experimental data than the strict $b = 0\,\text{fm}$ case.

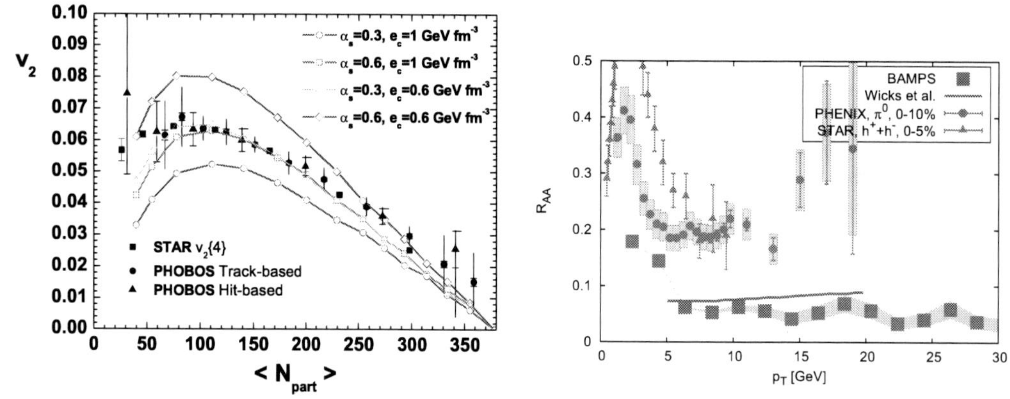

Figure 5. Left panel: Elliptic flow v_2 as a function of the number of participants for Au+Au at 200 AGeV for different combinations of the strong coupling α_s and the critical energy density ε_c. See [23] for more information.
Right panel: Gluonic R_{AA} at midrapidity ($y \in [-0.5, 0.5]$) as extracted from simulations for central Au+Au collisions at 200 AGeV. For comparison the result from Wicks et al. [11] for the gluonic contribution to R_{AA} and experimental results from PHENIX [26] for π^0 and STAR [27] for charged hadrons are shown.

As a first step towards making more extensive comparisons with experimental data and analytic models possible, we have computed the gluonic R_{AA} for non–central Au + Au collisions at the RHIC energy of $\sqrt{s} = 200\text{AGeV}$ with a fixed impact parameter $b = 7\,\text{fm}$, which roughly corresponds to 20% to 30% experimental centrality. The results is shown in the right panel of Fig. 6.

A comparison in terms of the magnitude of the jet suppression for $b = 7\,\text{fm}$ is difficult since there are no published results for the gluonic contribution to R_{AA} from analytic models available. Taking the ratio of the $b = 7\,\text{fm}$ to the $b = 0\,\text{fm}$ results as a rough guess indicates that the decrease in quenching is more pronounced in BAMPS compared to experimental data. The ratio of the nuclear modification factor between central (0% - 10%) and more peripheral (20% - 30%) collisions is $R_{AA}|_{0\%-10\%} \,/\, R_{AA}|_{20\%-30\%} \approx 0.6$ for the experimental data, while for the BAMPS results $R_{AA}|_{b=0\,\text{fm}} \,/\, R_{AA}|_{b=7\,\text{fm}} \approx 0.4$. However, the issue of detailed quantitative comparison needs to be re-investigated once light quarks and a fragmentation scheme are included into the simulations.

To complement the investigations of R_{AA} at a non–zero impact parameter $b = 7\,\text{GeV}$, we have computed the elliptic flow parameter v_2 for gluons at the same impact parameter and extended the range in transverse momentum up to roughly $p_T \approx 10\,\text{GeV}$, see left panel of Fig. 6. For this calculation we have used a critical energy density $\varepsilon_c = 0.6\,\text{GeV/fm}^3$ in order to be comparable to previous calculations.

The v_2 of high–p_T gluons is rising up to $p_T \approx 4\,\text{GeV}$. Afterwards, from about $p_T \approx 5\,\text{GeV}$ on, the elliptic flow slightly decreases again with p_T. This behavior is in good qualitative agreement

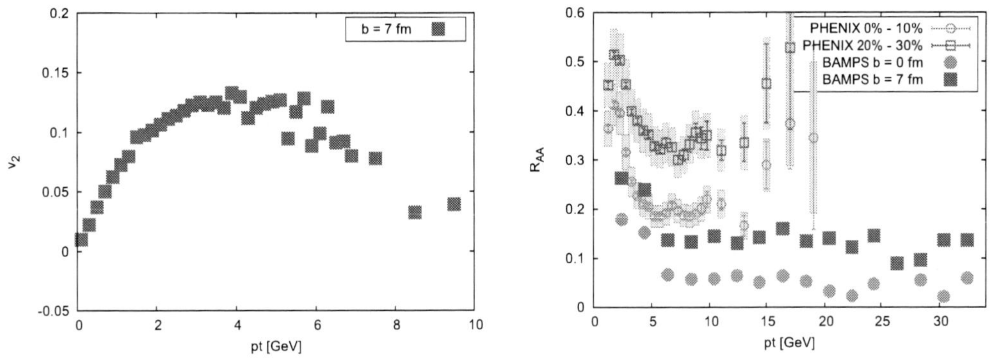

Figure 6. Left panel: Elliptic flow v_2 for gluons in simulated Au+Au collisions at 200 AGeV with $b = 7\,\mathrm{fm}$. $\varepsilon_c = 0.6\,\mathrm{GeV/fm^3}$.
Right panel: Gluonic R_{AA} as extracted from BAMPS simulations for $b = 0\,\mathrm{fm}$ and $b = 7\,\mathrm{fm}$, $\varepsilon_c = 1.0\,\mathrm{GeV/fm^3}$. For comparison experimental results from PHENIX [26] for π^0 are shown for central (0% - 10%) and off–central (20% - 30%) collisions.

with recent RHIC data [28] that for charged hadrons shows v_2 to be rising up to $v_2 \approx 0.15$ at $p_T \approx 3\,\mathrm{GeV}$ followed by a slight decrease.

5. Summary

We have computed the gluonic contribution to the nuclear modification factor R_{AA} and the elliptic flow v_2 employing the pQCD based transport model BAMPS. This model provides means to investigate various characteristics of the evolution of the partonic medium created in heavy ion collisions, ranging from bulk properties to high–p_T physics, consistently including the full dynamics of the system.

The gluonic R_{AA} is found to be flat over a wide range in p_T at $R_{AA} \approx 0.13$ in off–central events ($b = 7\,\mathrm{GeV}$) and at $R_{AA} \approx 0.053$ in central events ($b = 0\,\mathrm{GeV}$) for a critical energy density of $\varepsilon_c = 1.0\,\mathrm{GeV/fm^3}$. Since BAMPS allows for the simultaneous investigation of high–p_T observables and bulk properties of the medium, we have also studied the elliptic flow parameter for gluons up to roughly 10 GeV for Au+Au at $b = 7\,\mathrm{fm}$. v_2 peaks at a $p_T \approx 4 \div 5\,\mathrm{GeV}$ and slowly drops towards larger transverse momenta.

In order to systematically investigate the energy loss of gluons as implemented in BAMPS we have studied the evolution of high energy gluons within thermal and static media of gluons. Inelastic $gg \to ggg$ processes are found to be the dominant source of energy loss for high energy gluons in computations within the BAMPS model resulting in a strong differential energy loss that rises almost linearly with the jet energy. The strong mean energy loss in $gg \to ggg$ processes is due to a heavy tail in the ΔE distribution for single interactions, caused by the phase space configurations of outgoing particles dictated by the Gunion-Bertsch matrix element in combination with the effective LPM cutoff [17].

The characteristics of the strongly interacting, but still fully pQCD based, medium within the BAMPS description will be studied in further detail in upcoming works including light quark degrees of freedom. While a consistent modeling of low–p_T hadronization needs careful consideration, the application of fragmentation functions to the high–p_T sector will be straightforward once light quarks are included and will allow for more direct comparison to hadronic observables. Also the application of BAMPS to heavy quark elliptic flow and quenching will provide further valuable insight and is underway [29]. Additionally the medium response to high–p_T particles will be studied in more detail. It has already been demonstrated [30] that

BAMPS offers the ability to describe collective shock phenomena in a viscous hydrodynamic medium.

Acknowledgments

This work has been supported by the Helmholtz International Center for FAIR within the framework of the LOEWE program launched by the State of Hesse. The simulations have been performed at the Center for Scientific Computing (CSC) at the Goethe University Frankfurt.

References

[1] Adler C *et al.* (STAR) 2002 *Phys. Rev. Lett.* **89** 202301 (*Preprint* nucl-ex/0206011)
[2] Adcox K *et al.* (PHENIX) 2002 *Phys. Rev. Lett.* **88** 022301 (*Preprint* nucl-ex/0109003)
[3] Adler S S *et al.* (PHENIX) 2003 *Phys. Rev. Lett.* **91** 182301 (*Preprint* nucl-ex/0305013)
[4] Adams J *et al.* (STAR) 2004 *Phys. Rev. Lett.* **92** 052302 (*Preprint* nucl-ex/0306007)
[5] Zakharov B G 1996 *JETP Lett.* **63** 952–957 (*Preprint* hep-ph/9607440)
[6] Baier R, Dokshitzer Y L, Mueller A H, Peigne S and Schiff D 1997 *Nucl. Phys.* **B484** 265–282 (*Preprint* hep-ph/9608322)
[7] Baier R, Dokshitzer Y L, Mueller A H and Schiff D 1998 *Phys. Rev.* **C58** 1706–1713 (*Preprint* hep-ph/9803473)
[8] Gyulassy M, Levai P and Vitev I 2001 *Nucl. Phys.* **B594** 371–419 (*Preprint* nucl-th/0006010)
[9] Jeon S and Moore G D 2005 *Phys. Rev.* **C71** 034901 (*Preprint* hep-ph/0309332)
[10] Salgado C A and Wiedemann U A 2003 *Phys. Rev.* **D68** 014008 (*Preprint* hep-ph/0302184)
[11] Wicks S, Horowitz W, Djordjevic M and Gyulassy M 2007 *Nucl. Phys.* **A784** 426–442 (*Preprint* nucl-th/0512076)
[12] Romatschke P and Romatschke U 2007 *Phys. Rev. Lett.* **99** 172301 (*Preprint* 0706.1522)
[13] Kovtun P, Son D T and Starinets A O 2005 *Phys. Rev. Lett.* **94** 111601 (*Preprint* hep-th/0405231)
[14] Bass S A *et al.* 2009 *Phys. Rev.* **C79** 024901 (*Preprint* 0808.0908)
[15] Schenke B, Gale C and Jeon S 2009 *Phys. Rev.* **C80** 054913 (*Preprint* 0909.2037)
[16] Fochler O, Xu Z and Greiner C 2009 *Phys. Rev. Lett.* **102** 202301 (*Preprint* 0806.1169)
[17] Fochler O, Xu Z and Greiner C 2010 (*Preprint* 1003.4380)
[18] Xu Z and Greiner C 2005 *Phys. Rev.* **C71** 064901 (*Preprint* hep-ph/0406278)
[19] Xu Z and Greiner C 2007 *Phys. Rev.* **C76** 024911 (*Preprint* hep-ph/0703233)
[20] Gunion J F and Bertsch G 1982 *Phys. Rev.* **D25** 746
[21] Migdal A B 1956 *Phys. Rev.* **103** 1811–1820
[22] Xu Z, Greiner C and Stocker H 2008 *Phys. Rev. Lett.* **101** 082302 (*Preprint* 0711.0961)
[23] Xu Z and Greiner C 2009 *Phys. Rev.* **C79** 014904 (*Preprint* 0811.2940)
[24] Xu Z and Greiner C 2008 *Phys. Rev. Lett.* **100** 172301 (*Preprint* 0710.5719)
[25] Gluck M, Reya E and Vogt A 1995 *Z. Phys.* **C67** 433–448
[26] Adare A *et al.* (PHENIX) 2008 *Phys. Rev. Lett.* **101** 232301 (*Preprint* 0801.4020)
[27] Adams J *et al.* (STAR) 2003 *Phys. Rev. Lett.* **91** 172302 (*Preprint* nucl-ex/0305015)
[28] Abelev B I *et al.* (STAR) 2008 *Phys. Rev.* **C77** 054901 (*Preprint* 0801.3466)
[29] Uphoff J, Fochler O, Xu Z and Greiner C 2010 (*Preprint* 1003.4200)
[30] Bouras I *et al.* 2009 *Phys. Rev. Lett.* **103** 032301 (*Preprint* 0902.1927)

Search for local parity violation with STAR ZDC-SMD

Gang Wang (for the STAR Collaboration)
Department of Physics and Astronomy, University of California, Los Angeles, CA 90095 U.S.A.
E-mail: gwang@physics.ucla.edu

Abstract. Parity-odd domains are predicted to lead to charge separation of quarks along the orbital momentum of the system created in non-central relativistic heavy ion collisions. A signal consistent with several of the theoretical expectations has been detected by STAR, with a three particle azimuthal correlator which is a $\mathcal{P}$-even observable, but still sensitive to the charge separation effect. In this correlator, the first two particles are used to reveal the physics of interest, and the third particle serves as a reference for the reaction plane. In this work, to minimize the non-parity correlation between the third particle and the first two, we utilize the spectator neutrons detected by STAR ZDC-SMDs to define the event plane. The 1st-order event plane thus obtained can also be used to study the global strong parity violation effect, and to provide further systematic checks for this analysis. We report measurements of both $\mathcal{P}$-even and $\mathcal{P}$-odd observables using the STAR ZDC-SMDs in Au+Au at 200 GeV. The results are presented as a function of collision centrality, particle separation in rapidity and transverse momentum (p_T). Systematic checks on the impact from the directed flow will also be discussed.

1. Introduction

The concept of local parity ($\mathcal{P}$) violation in high-energy heavy ion collisions was brought up by Lee *et al.* [1, 2, 3] and elaborated by Kharzeev *et al.* [4]. Recently it was argued [5] that in non-central collisions such a $\mathcal{P}$-odd domain can manifest itself via preferential same charge particle emission for particles moving along the system's angular momentum, due to the strong magnetic field produced in the collision. Phenomenologically, the charge separation can be described by adding $\mathcal{P}$-odd sine terms to the Fourier decomposition of the particle azimuthal distribution with respect to the reaction plane angle, Ψ_{RP} , which is often used in the description of anisotropic flow:

$$\frac{dN_{\pm}}{d\phi} = 1 + 2v_1\cos(\Delta\phi) + 2v_2\cos(2\Delta\phi) + 2a_{1,\pm}\sin(\Delta\phi) + ..., \tag{1}$$

where $\Delta\phi = \phi - \Psi_{RP}$ is the particle azimuthal angle relative to the reaction plane, v_1 and v_2 are coefficients, assumed here to be the same for positive and negative particles, accounting for the so-called directed and elliptic flow, respectively. The a parameters, $a_- = -a_+$, are $\mathcal{P}$-odd quantities, and describe the $\mathcal{P}$-violating effect.

According to the theory, the sign of $a_\pm$ varies event to event following the fluctuations in the domain's topological charge, so $\langle a_\pm\rangle$ must be zero, and the observation of the effect is possible only via correlations, e.g. measuring $\langle a_\alpha a_\beta\rangle$, where α and β denote the particle type. To study this effect, a three particle mixed harmonics azimuthal correlator was proposed [6]:

$$\langle\cos(\phi_\alpha + \phi_\beta - 2\Psi_{RP})\rangle = [\langle v_{1,\alpha}v_{1,\beta}\rangle + B_{in}] - [\langle a_\alpha a_\beta\rangle + B_{out}], \tag{2}$$

which is a $\mathcal{P}$-even observable, but can be sensitive to the charge separation effect. The $\langle v_{1,\alpha}v_{1,\beta}\rangle$ term should vanish when measured over a rapidity region symmetric around zero, since v_1 is a anti-symmetric function of rapidity. B_{out} represents all other (than parity violation) contributions by correlations projected onto the direction perpendicular to the reaction plane ("out of plane"). The effects contributing to B_{out} may be large and are difficult to estimate reliably. B_{in} is the contribution of the in-plane correlations which are analogous to B_{out}. B_{in} and B_{out} contain correlations not related to the reaction plane orientation, and they largely cancel out. Only the parts of such correlations that depend on azimuthal orientation with respect to the reaction plane remain as backgrounds.

To minimize the reaction-plane independent backgrounds produced by direct three (or more) particle clusters, we utilize the spectator neutrons detected by the Shower Maximum Detectors (SMD) of the Zero Degree Calorimeters (ZDC) [7] to define the event plane in STAR. The advantage of the event plane from the ZDC over that from the TPC [8] lies in the large rapidity gap between the spectators and the particles of interest in the analysis. Furthermore, the spectator v_1 helps estimate the first-order event plane, instead of the second-order event plane from v_2 measured in TPC, so that the global $\mathcal{P}$-violation quantity $\langle a\rangle$ can be directly measured, and the systematics from the $\langle v_{1,\alpha}v_{1,\beta}\rangle$ term can be studied.

In the following sections, we will compare the correlator measured in STAR as a function of centrality with the event plane from ZDC, along with the STAR published results where the reaction plane is approximated with a third particle from TPC [9]. The vanishing of the $\langle a\rangle$ and $\langle v_{1,\alpha}v_{1,\beta}\rangle$ terms will also be demonstrated. If the two particles truly come from the same $\mathcal{P}$-odd domain, we expect them to be close to each other in rapidity space and momentum space. The correlator as functions of the pseudo-rapidity gap and of the transverse momentum gap will thus be discussed.

2. Data sets and analyses

This study is based on 36 million 200 GeV Au+Au events taken during RHIC run VII (2007). Charged particle tracks were reconstructed in STAR's main TPC, with pseudo-rapidity coverage $|\eta| < 1.3$. The centrality definition (in which zero represents the most central collisions) is the same as in Ref. [10], and the track quality cuts are the same as in Ref. [9]. The reconstruction and utilization of the first-order event plane from STAR ZDC-SMDs has previously been extensively discussed [11, 12, 13]. Due to the poor event plane resolution in 0 – 10% collisions [14], we only show results for 10% – 80% collisions in this work.

3. Results and discussions

In practice, we use an event plane (an estimated reaction plane) or a third particle to approximate the reaction plane Ψ_{RP} in Eq. (2), and then correct the measurement with the event plane resolution or the v_2 of the third particle. Figure 1 shows the correlator as a function of centrality in 200 GeV Au+Au collisions, with the thrid particle from STAR TPC [9] and the event plane estimated from ZDC. The consistency between the results with different reaction plane approximations (and very different correction factors for the event plane resolution or v_2) illustrates that the measured correlator is truly related to the reaction plane orientation, and that the non-flow/non-parity three-particle correlations largely cancel out between B_{in} and B_{out}. The like- and unlike-sign correlations clearly exhibit very different behaviors. According to Eq. (2), a negative value of the correlator for the like-sign particle pairs means a positive correlation between them with respect to the reaction plane, or in other words that they tend to be emitted to the same side of the reaction plane. On the other hand, a positive value of the correlator for the unlike-sign pairs means that they go to the opposite sides. The difference between the like-sign and unlike-sign pairs is consistent with the charge separation with respect to the reaction plane. The fact that the correlator for the unlike-sign pairs is close to zero for

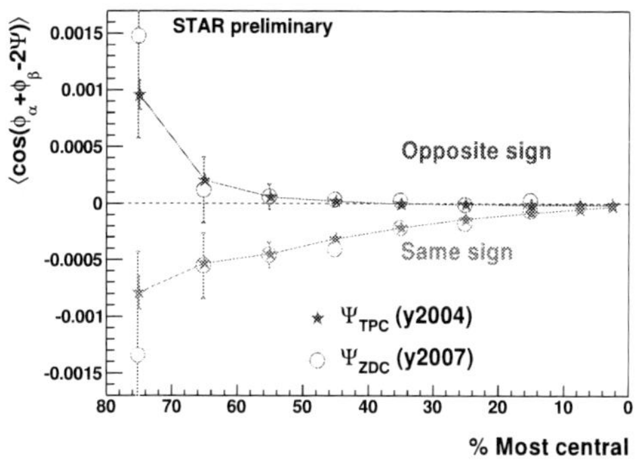

Figure 1. $\langle\cos(\phi_\alpha + \phi_\beta - 2\Psi_{RP})\rangle$ in Au+Au collisions at 200 GeV, with the thrid particle from STAR TPC [9] and the event plane estimated from ZDC.

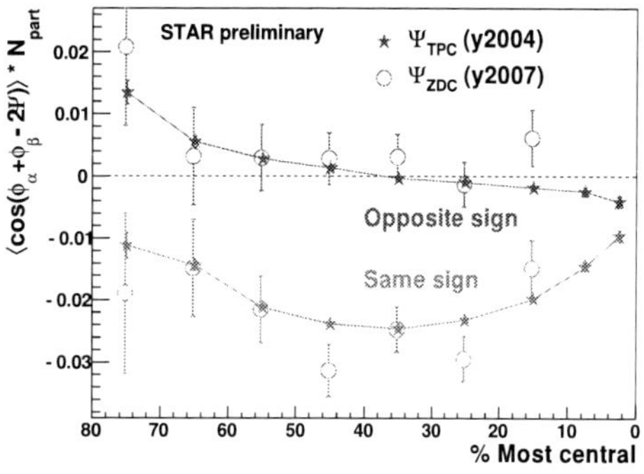

Figure 2. $\langle\cos(\phi_\alpha + \phi_\beta - 2\Psi_{RP})\rangle$ scaled with the number of participants in Au+Au collisions at 200 GeV with the thrid particle from STAR TPC [9] and the event plane estimated from ZDC.

most centrality bins, instead of a similar magnitude as that for the like-sign pairs, could be explained by the suppression of back-to-back correlations due to the medium [15].

The correlations are weaker in more central collisions compared with more peripheral collisions, which can be attributed partially to the dilution of correlations in the case of particle production from multiple sources. If more than one $\mathcal{P}$-odd domain is created in the medium, and the sign of $a_\pm$ in each domain fluctuates, the superposition of the correlations from different domains acts like a random walk. In a random walk of N steps, the final position R follows a Gaussian distribution with the expectation to be 0 and the RMS to be $\sqrt{N}$ at large N. Our measurement of the $\mathcal{P}$-even quantity is in analogy to R^2 which has an expectation of N according to the Gaussian distribution. Compared with going in a fixed direction, where $R^2 = N^2$, the random-walk observable is diluted by a factor of N. As to our measurement, N here is the number of $\mathcal{P}$-odd domains in one event. In practice, we assume N is proportional to the number of participants (N_{part}) in the following compensation for this effect. To present a more complete picture of the centrality dependence, we show in Fig. 2 results multiplied by N_{part} [9]. The decrease of the correlations in the most central collisions is expected as the magnetic field weakens. The like- and unlike-sign correlations in the peripheral Au+Au collisions where the medium is thin, are found to be at similar magnitudes. This supports the picture of the suppression of back-to-back correlations due to the dense medium leading to the difference in the magnitudes of like- and unlike-sign correlations in central collisions [15]. The negative values of the unlike-sign correlator in central collisions may be caused by some residual non-$\mathcal{P}$ effects, such as radial flow.

The global $\mathcal{P}$-violation effect has also been investigated with the first-order event plane, and the $\mathcal{P}$-odd quantities $\langle a_\pm\rangle$ are plotted in Fig. 3 as a function of centrality. Within errors, the measured $\langle a_+\rangle$ is consistent with $\langle a_-\rangle$, supporting the picture of the random fluctuations in the $\mathcal{P}$-odd domain's topological charge. In most centrality bins, the values of $\langle a_\pm\rangle$ are on the order of 10^{-4}, leading to a negligible contribution ($\sim 10^{-8}$) to the correlator.

If $v_1(\eta)$ is not anti-symmetric around $\eta = 0$, then the $\langle v_{1,\alpha}v_{1,\beta}\rangle$ term in Eq. (2) will not vanish, and it will have a finite contribution to the correlator. We have measured $v_1(\eta)$ for positively and negatively charged particles in the same events where the correlator is calculated, and fit $v_1(\eta)$ with a straight line function for each centrality. If $v_{1,\alpha}$ and $v_{1,\beta}$ are independent of each other, we can factorize the $\langle v_{1,\alpha}v_{1,\beta}\rangle$ term as $\langle v_{1,\alpha}\rangle\langle v_{1,\beta}\rangle$. The slope of the fit function has no contributions

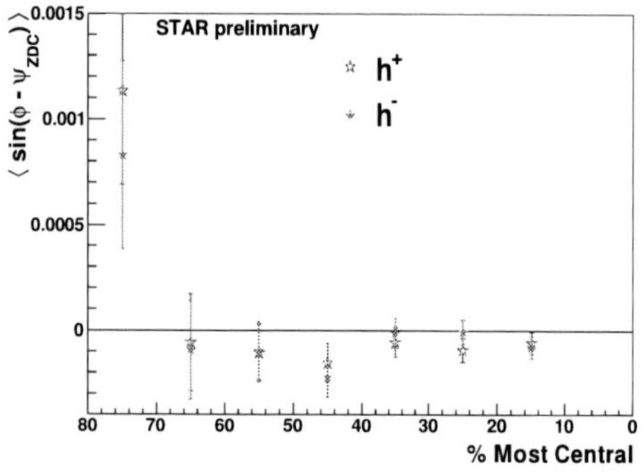

Figure 3. $\langle \sin(\phi - \Psi_{\rm ZDC}) \rangle$ in Au+Au collisions at 200 GeV, for positively and negatively charged particles.

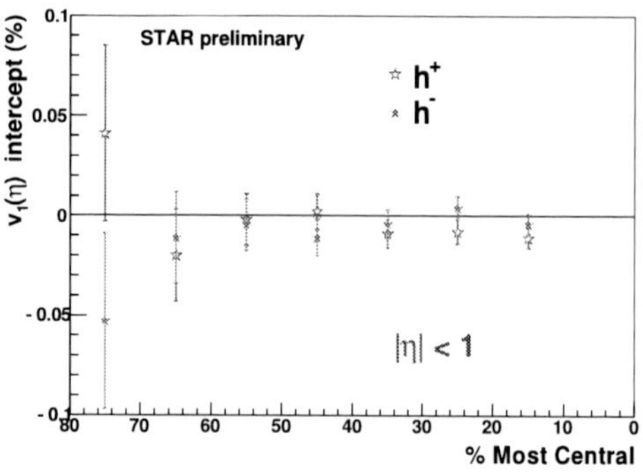

Figure 4. $v_1(\eta)$ intercept in Au+Au collisions at 200 GeV, for positively and negatively charged particles.

to the integration over a symmetric η range. The intercept of the fit quantifies the deviation of directed flow from anti-symmetry due to non-flow effects such as momentum conservation and Coulomb effects. The $v_1(\eta)$ intercept is shown in percentage in Fig. 4 as a function of centrality. The intercept is consistent with zero in all centrality bins, and has a similar magnitude with $\langle a_\pm \rangle$, and thus a similarly negligible contribution to the correlator. If $v_{1,\alpha}$ and $v_{1,\beta}$ have correlations between them other than flow-wise, such as non-flow effects produced by direct two particle clusters, then the $\langle v_{1,\alpha} v_{1,\beta} \rangle$ term contains an extra reaction-plane independent background. From the above experience of the direct three (or more) particle clusters, we expect a similar background in the $\langle a_\alpha a_\beta \rangle$ term which largely cancels out its counterpart in the $\langle v_{1,\alpha} v_{1,\beta} \rangle$ term.

In the case where the difference between the like- and unlike-sign pair correlators truly results from the charge separation due to a $\mathcal{P}$-odd domain in the medium, we expect that the two particles from the same domain are close to each other in the rapidity space. In other words, there should be a dependence of the signal on the separation of the two particles. In Fig. 5, we show the correlator as a function of pseudo-rapidity separation for 20% – 70% collisions. While the unlike-sign correlator is always close to zero, the like-sign correlator shows a clear trend: the correlator has the largest magnitude when the two particles are close to each other, and approaches zero when they are further apart. This dependence on the pseudo-rapidity separation supports the picture of the local parity violation.

When the two particles have a small p_T gap (e.g. below 50 MeV/c), the correlator may be dominated by quantum interference (HBT) or Coulomb effects. To study this type of systematics, we show in Fig. 6 the correlator as a function of transverse momentum separation for 20% – 70% collisions. At small p_T gap, both like- and unlike-sign pair correlators show a structure possibly induced by HBT or Coulomb effects. Above a p_T gap $\sim$ 200 MeV/c, the unlike-sign correlator is consistent with zero, and the like-sign correlator is sizably negative, excluding HBT or Coulomb effects as possible explanations for the signal. Furthermore, the smaller difference between the like- and unlike-sign correlator at a small p_T gap compared with the sizable difference at a higher p_T gap means that HBT or Coulomb effects would reduce the integrated signal instead of increasing it. There are no specific theoretical predictions on this dependence for the parity violation effect, though naively one would expect that the signal should not extend to large values of p_T gap. Our measurement at high p_T gap seems to indicate that the like-sign correlator is approaching zero, but the statistical errors are too large to be conclusive.

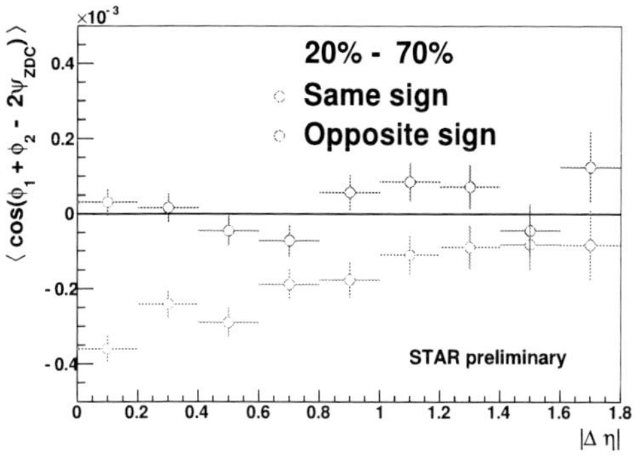

Figure 5. $\langle\cos(\phi_\alpha + \phi_\beta - 2\Psi_{\mathrm{ZDC}})\rangle$ in Au+Au collisions at 200 GeV, as a function of the pseudo-rapidity gap between the two particles.

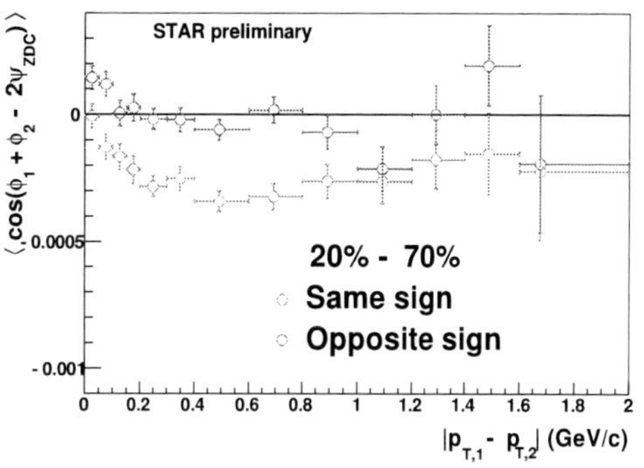

Figure 6. $\langle\cos(\phi_\alpha+\phi_\beta-2\Psi_{\mathrm{ZDC}})\rangle$ in Au+Au collisions at 200 GeV, as a function of the transverse momentum gap between the two particles.

4. Conclusion

To study the possible local $\mathcal{P}$-violation effects in heavy-ion collisions, we have employed the first-order event plane from STAR ZDC-SMDs to calculate the three-particle correlator which is sensitive to the predicted charge separation with respect to the reaction plane. The correlator thus measured is consistent with that estimated with a third particle from TPC, indicating that the reaction-plane independent non-flow and non-parity correlations largely cancel out. Qualitatively the results agree with the gross features of the theoretical predictions for $\mathcal{P}$-violation in heavy-ion collisions with charge separation across the reaction plane suppressed by the medium opacity. We have demonstrated that the global $\mathcal{P}$-violation term and the directed flow have negligible contributions to the correlator. The dependence of the correlator on the pseudo-rapidity separation of the two particles supports the picture of local parity violation. We have checked the p_T-gap dependence of the correlator, and found that HBT or Coulomb effects reduce the signal instead of increasing it.

Acknowledgments

I thank the STAR collaborators for the support.

References

[1] Lee T D 1973 *Phys. Rev.* D **8** 1226
[2] Lee T D and Wick G C 1974 *Phys. Rev.* D **9** 2291
[3] Morley P D and Schmidt I A 1985 *Z. Phys.* C **26** 627
[4] Kharzeev D, Pisarski R D and Tytgat M H G 1998 *Phys. Rev. Lett.* **81** 512; Kharzeev D and Pisarski R D 2000 *Phys. Rev.* D **61** 111901
[5] Kharzeev D 2006 *Phys. Lett.* B **633** 260; Kharzeev D and Zhitnitsky A 2007 *Nucl. Phys.* A **797** 67
[6] Voloshin S A 2004 *Phys. Rev.* C **70** 057901
[7] Adler C *et al.* (STAR Collaboration) 2001 *Nucl. Instr. Meth.* A **470** 488; Allgower C E *et al.* (STAR Collaboration) 2003 *Nucl. Instr. Meth.* A **499** 740; STAR Collaboration 2003 STAR ZDC-SMD proposal *STAR Note* **SN-0448**
[8] Anderson M *et al.* (STAR Collaboration) 2003 *Nucl. Instr. Meth.* A **499** 659
[9] Abelev B I *et al.* (STAR Collaboration) 2009 *Phys. Rev. Lett.* **103** 251601; Abelev B I *et al.* (STAR Collaboration) 2009 Preprint arxiv:0909.1717 [nucl-ex]
[10] Tang A H *et al.* (STAR Collaboration) 2010 these proceedings
[11] Adams J *et al.* (STAR Collaboration) 2006 *Phys. Rev.* C **73** 034903
[12] Adams J *et al.* (STAR Collaboration) 2008 *Phys. Rev. Lett.* **101** 252301

[13] Wang G *et al.* 2007 *Phys. Rev.* C **76** 024907
[14] Wang G *et al.* (STAR Collaboration) 2007 *J. Phys.* G **34** S1093
[15] Kharzeev D E, McLerran L D and Warringa H J 2008 *Nucl. Phys.* A **803** 227

Journal of Physics: Conference Series **230** (2010) 012021 doi:10.1088/1742-6596/230/1/012021

Local strong parity violation and new possibilities in experimental study of non-perturbative QCD

Sergei A. Voloshin

Department of Physics and Astronomy, Wayne State University, Michigan 48201, USA

E-mail: voloshin@wayne.edu

Abstract. Quark interaction with topologically non-trivial gluonic fields, instantons and sphalerons, violates $\mathcal{P}$ and $\mathcal{CP}$ symmetry. In the strong magnetic field of a non-central nuclear collision such interactions lead to the charge separation along the magnetic field, the so called chiral magnetic effect, which manifests local parity violations. An experimental observation of the chiral magnetic effect would be a direct proof for the existence of such physics. Recent STAR results on charge and the reaction plane dependent correlations are consistent with theoretical expectations for the chiral magnetic effect. In this paper I discuss other approaches to experimental study of the local parity violation, and propose future measurements which can clarify the picture. In particular I propose to use central body-body U+U collisions to disentangle correlations due to chiral magnetic effect from possible background correlations due to elliptic flow. Further more quantitative studies can be performed with collision of isobaric beams.

1. Introduction. Local parity violation.

It is widely accepted that Quantum Chromodynamics (QCD) is the theory of strong interactions. Perturbative QCD is firmly established and thoroughly tested experimentally. In the non-perturbative sector, QCD links chiral symmetry breaking and the origin of hadron masses to the existence of topologically non-trivial classical classical gluonic fields describing the transitions between the vacuum states with different Chern-Simons numbers. Quark interactions with such fields change the quark helicity and are $\mathcal{P}$ and $\mathcal{CP}$ odd. For a review, see [1, 2]. It was first suggested in [3] to look for such metastable $\mathcal{P}$ and $\mathcal{CP}$ odd domains in ultra-relativistic heavy ion collisions. The possibilities for an experimental detection of this *local strong parity violation* was discussed in [3, 4, 5].

Originally [3], it was proposed to look for the effect by detecting the non-statistical fluctuations in the variable

$$J = \sum_{\pi^+,\pi^-} \frac{(\vec{p}_{\pi^+} \times \vec{p}_{\pi^-})_z}{p_{\pi^+} p_{\pi^-}}. \tag{1}$$

In [4] it was shown that J is directly related to the difference in the event planes reconstructed from positive and negative particles. This difference is similar to the effect of a (electro) magnetic field oriented along the beam direction (parallel or anti parallel). The existence of such a field in a symmetric collision would constitute parity violation. Experimental measurements of the

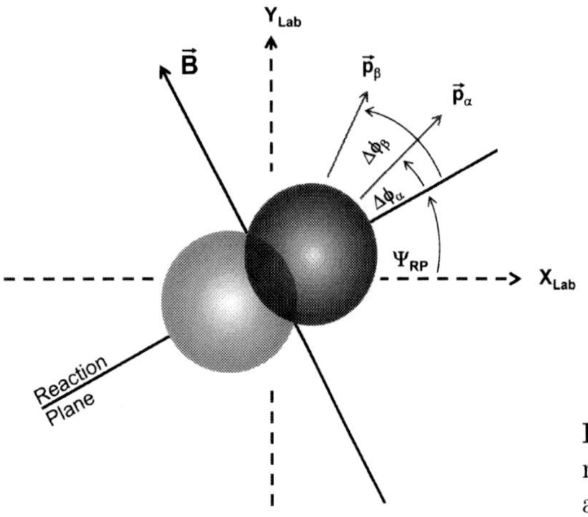

Figure 1. Schematic view of non-central nuclear collision showing the definition of angles in Eqs. 2, 4.

effect by NA49 Collaboration revealed a signal consistent with zero [6]. Unfortunately, this observable have not been measured at RHIC yet.

More recently, it was noticed [7, 8] that in non-central nuclear collisions such domains can demonstrate themselves via the asymmetry in the emission of positively or negatively charged particle perpendicular to the reaction plane. Such charge separation is a consequence of the difference in the number of particles with positive and negative helicities positioned in the strong magnetic field ($\sim 10^{15}$ T) of a non-central nuclear collision, the so-called *chiral magnetic effect* [7, 9]. The same phenomenon can also be described in terms of the induced electric field that is parallel to the static external magnetic field, which occurs in the presence of topologically non-trivial vacuum solutions [10]. The direction of the separation varies event by event in accord with the sign of the topological charge of the domain, and the observation of the effect is possible only by correlation techniques. An observable directly sensitive to the charge separation effect, has been proposed in [13]. It is discussed in more detail below.

According to Refs. [7, 8, 9] the charge separation could lead to asymmetry in particle production $\sim Q/N_{\pi^+}$, where $Q = 0, \pm 1, \pm 2, ...$ is the topological charge and N_{π^+} is the positive pion multiplicity in one unit of rapidity – the typical scale of such correlations. The charge separation effect is expected to depend strongly on deconfinement and chiral symmetry restoration [9], and the signal might be greatly suppressed or completely absent at an energy below that at which a quark-gluon plasma can be formed.

2. Chiral magnetic effect and charge dependent azimuthal correlations

In non-central nuclear collisions particle distribution in azimuthal angle is not uniform. The deviation from a flat distribution is called anisotropic flow and often is described by the Fourier decomposition [11] (for a review, see [12]):

$$\frac{dN_\alpha}{d\phi} \propto 1 + 2v_{1,\alpha}\cos(\Delta\phi) + 2\,v_{2,\alpha}\cos(2\Delta\phi) + ... \\ + 2a_{1,\alpha}\sin(\Delta\phi) + 2\,a_{2,\alpha}\sin(2\Delta\phi) + ..., \tag{2}$$

where $\Delta\phi = (\phi - \Psi_{RP})$ is the particle azimuth relative to the reaction plane (see Fig. 1), v_1 and v_2 account for directed and elliptic flow. Subscript α is used to denote the particle type. Due to the "up-down" symmetry of the collisions a_n coefficients are usually omitted. Chiral magnetic effect violates such a symmetry. Although the "direction" of the violation fluctuates event to event and on average is zero, in events with a particular sign of the topological charge, the average is not zero. As a result, it leads to a non-zero contribution to correlations, e.g. $\langle a_{n,\alpha} a_{n,\beta} \rangle$, where α and β denote the particle type. One expects that the first harmonic would account for the most of the effect. Below, if not explicitly mentioned, it is assumed that $n = 1$. Note that only particles originated from (interacted with) the same domain are correlated. The size of the domain is expected to be less than 1 fm, which means that the correlated particles are likely to be within about one unit of rapidity from each other. To measure $\langle a_\alpha a_\beta \rangle$, it was proposed [13] to use the correlator:

$$\langle \cos(\phi_\alpha + \phi_\beta - 2\Psi_{RP}) \rangle = \langle \cos \Delta\phi_\alpha \cos \Delta\phi_\beta \rangle - \langle \sin \Delta\phi_\alpha \sin \Delta\phi_\beta \rangle \tag{3}$$

$$= [\langle v_{1,\alpha} v_{1,\beta} \rangle + B_{in}] - [\langle a_\alpha a_\beta \rangle + B_{out}] \approx - \langle a_\alpha a_\beta \rangle + [B_{in} - B_{out}]. \tag{4}$$

This correlator represents the difference between correlations "projected" onto the reaction plane and the correlations projected onto an axis perpendicular to the reaction plane. The key advantage of using such a difference is that it removes all the correlations among particles α and β that are not related to the reaction plane orientation.The contribution given by the term $\langle v_{1,\alpha} v_{1,\beta} \rangle$ can be neglected because directed flow averages to zero in a rapidity region symmetric with respect to mid-rapidity.

RP dependent, "physics", background. The remaining background in Eq. 4, $B_{in} - B_{out}$, are due to processes in which particles α and β are products of a cluster (e.g. resonance, jet, di-jets) decay, and the cluster itself exhibits elliptic flow or decays (fragments) differently when emitted in-plane compared to out-of-plane. The corresponding contribution to the correlator can be estimated as:

$$\langle \cos(\phi_\alpha + \phi_\beta - 2\Psi_{RP}) \rangle = \frac{N_{\frac{clust}{event}} N_{\frac{pairs}{clust}}}{N_{\frac{pairs}{event}}} \langle \cos(\phi_\alpha + \phi_\beta - 2\phi_{clust}) \rangle_{clust} \, v_{2,clust}, \tag{5}$$

where $\langle ... \rangle_{clust}$ indicates that the average is performed only over pairs consisting of two daughters from the same cluster. This kind of background can not be easily suppressed. To address its contribution one has to rely on model calculations or perform experiments where the relative contribution of chiral magnetic effect and background can be varied, see Section 4.

RP independent background. The reaction plane is not known experimentally, and has to be estimated event-by event. Note that the second order reaction plane [14], reconstructed from strong elliptic flow [15], is sufficient for this measurement. Usually, the correlator Eq. 4 is evaluated with the help of 3-particle correlations:

$$\langle \cos(\phi_\alpha + \phi_\beta - 2\phi_c) \rangle \approx \langle \cos(\phi_\alpha + \phi_\beta - 2\Psi_{RP}) \rangle v_{2,c}. \tag{6}$$

This factorization can be broken, e.g. by correlations from a cluster decay to all three, α, β, and c particles. Such contribution can be greatly suppressed by appropriate choice of particle c, see [16].

3. STAR results.

Recently, the charge and reaction plane dependent correlations for Au+Au and Cu+Cu collisions at $\sqrt{s_{NN}}$=200 GeV and $\sqrt{s_{NN}}$=62 GeV have been published by the STAR Collaboration [17, 18]. The correlations are reported for charged particle in the region $|\eta| < 1.0$ with $p_t > 0.15$ GeV/c. Figure 2 shows STAR results for Au+Au collisions at $\sqrt{s_{NN}}$=200 GeV compared to predictions from different event generators. Note that the latter are not zero, and is due to discussed above reaction plane dependent background. To illustrate this better, the results from UrQMD event generator are connected by dashed lines. Reaction plane independent background, for

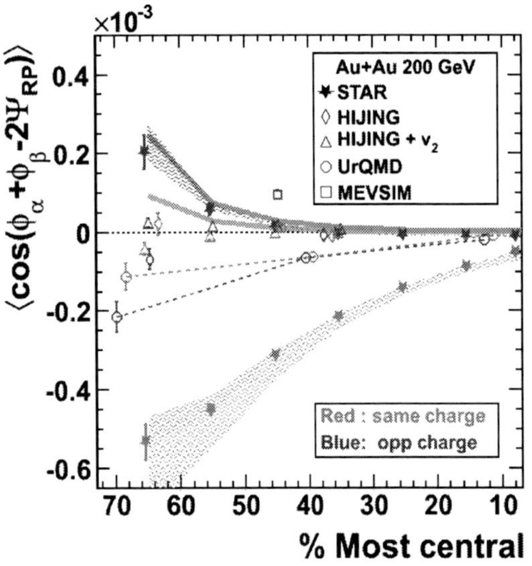

Figure 2. (Taken from [17]) STAR results compared to simulations for 200 GeV Au+Au. Blue symbols mark opposite-charge correlations, and red are same-charge. The shaded bands show the systematic error due to uncertainty in v_2 measurements. In simulations the true reaction plane from the generated event was used. Thick solid lighter colored lines represent non reaction-plane dependent contribution as estimated by HIJING. Corresponding estimates from UrQMD are about factor of two smaller.

a particular method employed in this analysis, is shown by thick lines. Opposite-charge correlations are smaller than same-charge correlations. This observation led to the proposal [9] that back-to-back correlations may be suppressed due to the opacity of the medium. The correlations are weaker in more central collisions compared to more peripheral collisions, which partially can be attributed to dilution of correlations which occurs in the case of particle production from multiple sources. To compensate for the dilution effect, in particular when comparing Au+Au results to Cu+Cu, STAR also presented results multiplied by the number of participants (for a plot, see [17]). There, the same and opposite sign correlations clearly exhibit very different behavior. The opposite sign correlations in Au+Au and Cu+Cu collisions are found to be very close at similar values of N_{part} in rough qualitative agreement with the picture in which their values are mostly determined by the suppression of back-to-back correlations. The difference in magnitude between same and opposite sign correlations is considerably smaller in Cu+Cu than in Au+Au, qualitatively in agreement with the scenario of stronger suppression of the back-to-back correlations in Au+Au collisions.

Somewhat unexpected result was the dependence of the signal on the transverse momentum of the two particles, see Fig. 3. It was found that the signal is not concentrated in the low p_t region as naively might be expected for $\mathcal{P}$-violation (non-perturbative) effects, and that the correlation depends very weakly on $|p_{t,\alpha} - p_{t,\beta}|$. An interesting possible explanation for such a dependence was found in [19]. It was found that such a dependence can be naturally understood if one assumes that the p_t distribution of correlated pairs has somewhat harder spectrum compared to that of all (random) pairs. The "trick" is that even if the distribution of pairs from clusters is only slightly harder, the relative weight of correlated pairs increases with p_t. Figure 4 shows the pair distributions in transverse momentum that was obtained in [19] using STAR data as an input.

4. Future program.

The STAR observation of the charge dependent azimuthal correlations consistent with the theoretical expectations for the chiral magnetic effect can be a beginning of an exciting program. Of course, one would need first to confirm that the observed correlations indeed are related to the local parity violation. If confirmed, it will open the door for *direct* experimental measurements

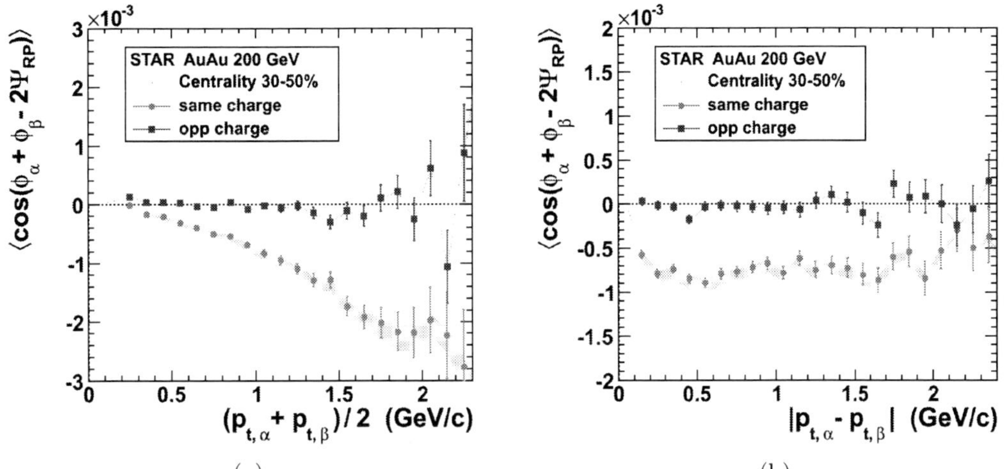

Figure 3. (Taken from [17, 18]). Au+Au at 200 GeV [17]. The correlations dependence (a) on sum and (b) the difference of the magnitude of transverse momenta.

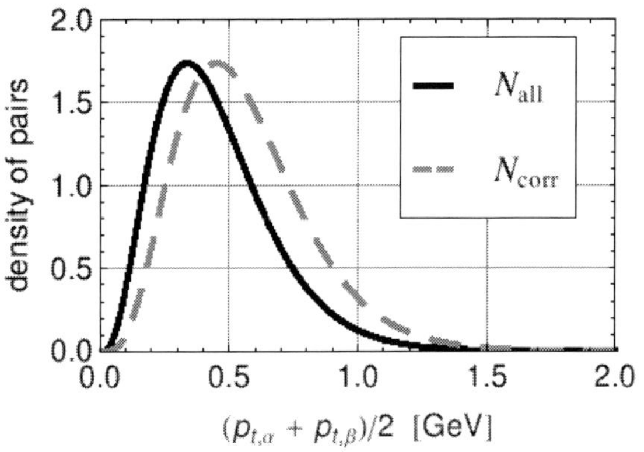

Figure 4. Taken from [19]. Transverse momentum distribution of pairs from clusters, shown in red, compared to distribution of all pairs.

of the effects from non-perturbative sector of QCD, so far not possible.

4.1. **Experiment.**

Below I discuss several directions for a future experimental program. In particular I propose to use central body-body U+U collisions to test if the observed correlations are indeed related to the strong magnetic field, as they must be in chiral magnetic effect. More detailed study of the dependence of the effect on the magnetic field can be achieved with collisions of isobaric nuclei. Measurements aimed on understanding the properties of the clusters in multiparticle production would be another important part of the program.

- *Central body-body U+U collisions.* Ambiguity in the interpretation of the STAR results lies in the difficulty to eliminate/suppress the RP dependent background, $B_{in} - B_{out}$, which by itself originates in the anisotropy of particle production relative to the reaction plane – elliptic flow. To eliminate/suppress elliptic flow, and at the same time preserve strong magnetic field needed for the chiral magnetic effect does not look realistic. But the opposite

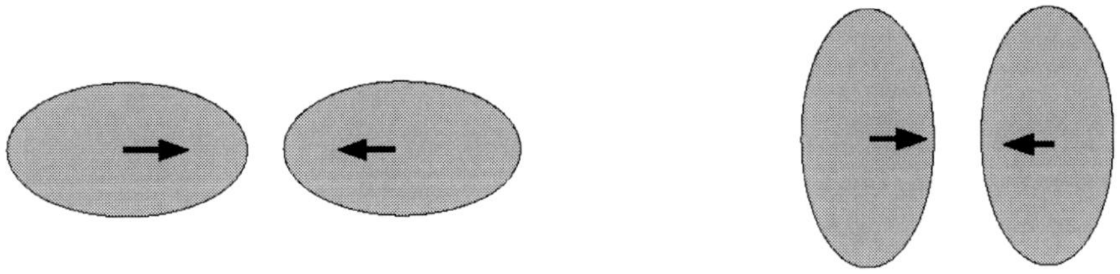

(a) (b)

Figure 5. Schematic view of central U+U collisions: (a) tip-tip and (b) body-body.

is possible, one can have collisions with strong elliptic flow and no (or almost no) magnetic field. This can be achieved in central body-body U+U collisions. Uranium nuclei are not spherical and have roughly ellipsoidal shape. Central collision, when most of the nucleons interact, can have different geometry, ranging from the so called tip-tip collisions to body-body collisions [20], see Fig. 5. Unlike tip-tip collisions, body-body ones would exhibit strong elliptic flow. Neither would lead to a strong magnetic field; consequently, a very weak signal due to chiral magnetic effect is expected. At RHIC one can select central collisions by requiring a low signal in the zero degree calorimeters that detect spectator neutrons. Then one can analyze the dependence of the signal on the elliptic flow present in the events. If the signal is due to elliptic flow one should find a direct relations between the two. Calculations of the relative strength of the effect in different scenarios are under way [21].

- *Collision of isobaric nuclei.* The charge separation dependence on the strength of the magnetic field can be further studied with collision of isobaric nuclei, such as $^{96}_{44}Ru$ and $^{96}_{40}Zr$. These nuclei have the same mass number, but differ by the charge. The multi-particle production in the midrapidity region would be affected very little in collision of such nuclei, and in particular one would expect very similar elliptic flow. At the same time the magnetic field would be proportional to the nuclei charge and can vary by more than 10%, which can results in 20% variation in the signal. Such variations should be readily measurable. The collisions of $^{96}_{44}Ru$ and $^{96}_{40}Zr$ isotopes have been successfully used at GSI [22, 23] in a study of baryon stopping. Collisions of isobaric nuclei at RHIC will be also extremely valuable for understanding the initial conditions, and in particular the initial velocity fields, the origin of directed flow, etc.
- *Beam energy scan.* The charge separation effect might depend strongly on the formation of a quark-gluon plasma and chiral symmetry restoration [9], and the signal can be greatly suppressed or completely absent at an energy below that at which a quark-gluon plasma is formed. Taking also into account that the life-time of the strong magnetic field is larger at smaller collision energies, it could lead to an almost threshold effect: with lowing the energy the signal might slowly increase with an abrupt drop thereafter. This questions can (and will) be addressed, for example, during the RHIC beam energy scan. But, unfortunately, the exact energy dependence of the chiral magnetic effect is not calculated yet.
- *Identified particle studies.* Large statistics recorded by RHIC experiments during the last runs(s) will allow identified and multiparticle correlations studies. There are several questions to be addressed in such analyses. With multiparticle correlations one can try to estimate the size of the cluster (the average multiplicity). The correlations using neutral particles should be mostly determined by background effects, and thus provide an estimate of those. The most interesting from my point of view, but also difficult, would be a study

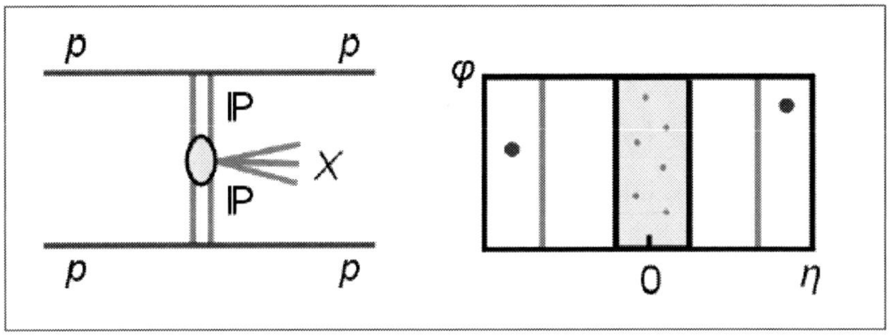

Figure 6. Double Pomeron exchange reaction and particle distributing in $\eta - \phi$ plane.

of the "quark content" of the cluster. The topological cluster decays in equal number of $q\bar{q}$-pairs of all flavors, the so-called 't Hooft interaction (e.g. used in instanton model to explain the difference in $\bar{d}/\bar{u}$ in the nucleon "sea").

- *Clusters, pp2pp experiment and double Pomeron collisions.* From early FNAL and ISR measurements it has been known that cluster formation plays an important role in multiparticle production at high energies, for a review, see [24]. These clusters, with a size of about 2–3 charged particles per cluster, may account for production of a significant fraction of all particles. It is interesting and important to establish how these clusters are related to the topological clusters as suggested in [25] ("turning points" – QCD sphalerons). Recent progress in describing of the soft Pomeron as a multi-instanton ladder [26] also suggests that the topologically nontrivial gluonic configurations play important role played in multiparticle production. The properties of such clusters in principle can be studied in any multiparticle production processes, but one possibility stands out of the list, namely the double Pomeron exchange [27], as illustrated in Fig. 6. A specific kinematics of double Pomeron exchange process can be addressed by pp2pp experiment [28]. This reaction allows to investigate the properties of the clusters in a very clean environment, with one cluster per event and well controlled kinematics. The goal is to compare the measurements to expectations for sphaleron decays – invariant mass, angular distribution of the decay products, quark composition consistent with 't Hooft interaction, etc. Two pion Bose-Einstein correlation analysis will be of a particular interest. If the particle production indeed is due to decay of classical fields, the so-called chaoticity parameter λ (intercept of the correlation function at zero relative momentum) could show a significant decrease.
- *CP forbidden decays.* Not going into theoretical details, I only mention here a very exciting possibility of an observation of $\mathcal{CP}$-forbidden decays, e.g. $\eta \to \pi\pi$, which become possible in the presence of $\mathcal{CP}$-odd domains [30, 31].

4.2. Need for a "better" theory.

For new theoretical developments, including recent lattice results, I refer to the talk of D. Kharzeev [32]. Here I only emphasize, that detailed interpretation of the experimental data is not possible without realistic theoretical calculations. Many of needed calculations, such as dependence on centrality and system system size, look fully doable though require significant computing and man power (e.g. 3d hydrodynamics is needed for the calculation of the magnetic field). Detailed predictions on the transverse momentum and particle type dependence of the effect is also essential in differentiating the signal from possible background contributions. A good theoretical understanding of the background correlations themselves is also required, as at present all event generators lack a good description of correlation results.

5. Summary.

Experimental observation of the chiral magnetic effect may provide a unique opportunity for a direct observation of the topological structure of QCD. The theoretical predictions are well within reach of the experiment. STAR has reported results that agree with the magnitude and gross features of the theoretical predictions for local $\mathcal{P}$-violation in heavy-ion collisions, but better theoretical calculations of the expected signal and potential physics backgrounds are essential for further experimental study of this phenomenon.

A very exciting future program dedicated to detail study of the effect is emerging. It includes U+U collisions, which could serve as a test of the chiral magnetic effect relation to the correlations observed by STAR.

Acknowledgments. Discussions with J. Dunlop, D. Kharzeev, S. Pratt, and E. Shuryak are gratefully acknowledged.

[1] E. V. Shuryak and T. Schafer, Ann. Rev. Nucl. Part. Sci. **47**, 359 (1997).
[2] D. Diakonov, Prog. Part. Nucl. Phys. **51**, 173 (2003).
[3] D. Kharzeev, R. D. Pisarski and M. H. G. Tytgat, Phys. Rev. Lett. **81**, 512 (1998) [arXiv:hep-ph/9804221]; D. Kharzeev and R.D. Pisarski, Phys. Rev. D **61**, 111901 (2000).
[4] S. A. Voloshin, Phys. Rev. C **62**, 044901 (2000).
[5] L. E. Finch, A. Chikanian, R. S. Longacre, J. Sandweiss and J. H. Thomas, Phys. Rev. C **65**, 014908 (2001).
[6] S. Voloshin and the NA49 Collaboration, *Search for parity violation in minimum bias Pb-Pb collisions at SPS*, LBNL 1998 annual report, http://ie.lbl.gov/nsd1999/rnc/RNC.htm, report R10.
[7] D. Kharzeev, Phys. Lett. B **633**, 260 (2006).
[8] D. Kharzeev and A. Zhitnitsky, Nucl. Phys. A **797**, 67 (2007).
[9] D. E. Kharzeev, L. D. McLerran and H. J. Warringa, Nucl. Phys. A **803**, 227 (2008).
[10] K. Fukushima, D. E. Kharzeev and H. J. Warringa, Phys. Rev. D **78**, 074033 (2008).
[11] S. Voloshin and Y. Zhang, Z. Phys. C **70**, 665 (1996) [arXiv:hep-ph/9407282].
[12] S. A. Voloshin, A. M. Poskanzer and R. Snellings, arXiv:0809.2949 [nucl-ex].
[13] S. A. Voloshin, Phys. Rev. C **70**, 057901 (2004).
[14] A. M. Poskanzer and S. A. Voloshin, Phys. Rev. C **58**, 1671 (1998) [arXiv:nucl-ex/9805001].
[15] K. H. Ackermann *et al.* [STAR Collaboration], Phys. Rev. Lett. **86**, 402 (2001) [arXiv:nucl-ex/0009011].
[16] Gang Wang (STAR Collaboration) This proceedings.
[17] B. I. Abelev *et al.* [STAR Collaboration], arXiv:0909.1717 [nucl-ex].
[18] B. I. Abelev *et al.* [STAR Collaboration], Phys. Rev. Lett. **103**, 251601 (2009) [arXiv:0909.1739 [nucl-ex]].
[19] A. Bzdak, V. Koch and J. Liao, arXiv:0912.5050 [nucl-th].
[20] C. Nepali, G. I. Fai and D. Keane, Phys. Rev. C **76**, 051902 (2007) [Erratum-ibid. C **76**, 069903 (2007)] [arXiv:0709.1497 [hep-ph]].
[21] S. A. Voloshin, in preparation.
[22] B. Hong [FOPI Collaboration], Phys. Rev. C **66**, 034901 (2002);
[23] B. Hong [FOPI Collaboration], Nucl. Phys. A **721**, 317 (2003).
[24] L. Foa, Phys. Rept. **22**, 1 (1975).
[25] D. M. Ostrovsky, G. W. Carter and E. V. Shuryak, Phys. Rev. D **66**, 036004 (2002) [arXiv:hep-ph/0204224].
[26] D. E. Kharzeev, Y. V. Kovchegov and E. Levin, Nucl. Phys. A **690**, 621 (2001) [arXiv:hep-ph/0007182].
[27] E. Shuryak and I. Zahed, Phys. Rev. D **68**, 034001 (2003) [arXiv:hep-ph/0302231].
[28] W. Guryn, Acta Phys. Polon. B **40**, 1897 (2009).
[29] G. Bonvichini, private communication, 2008;
[30] R. Millo and E. Shuryak, arXiv:0912.4894 [hep-ph].
[31] G. Bonvichini, private communication, 2008;
[32] D. Kharzeev, these proceedings;

Journal of Physics: Conference Series **230** (2010) 012022 doi:10.1088/1742-6596/230/1/012022

Measurement of ϕ and ω at RHIC-PHENIX

Misaki Ouchida for the PHENIX Collaboration
Hiroshima University, 1-3-1 Kagamiyama, Higashi-Hiroshima, Hiroshima 739-0402, Japan
E-mail: ouchida@hepl.hiroshima-u.ac.jp

Abstract. The PHENIX experiment at RHIC measured the production of ω and ϕ mesons via both leptonic and hadronic decay modes in p+p, d+Au, Cu+Cu and Au+Au at C.M.S. collision energy per nucleon pair of 200 GeV. The recent systematic measurements of these mesons are shown and discussed.

1. Introduction

The measurement of hadrons under extreme conditions created by relativistic heavy-ion collisions is an intriguing study being carried out as a part of the quest to observe the QCD phase transition to Quark Gluon Plasma (QGP) [1].

RHIC provided systematic measurement of hadrons in p+p, d+Au, Cu+Cu and Au+Au. Solid baseline results in p+p, d+Au and comparison study of the invariant yield measurement in heavy-ion collisions suggested that the particle production is effected by jet quenching which is considered to be an effect of the matter created by heavy-ion collisions [2]. Low mass vector mesons, ϕ and ω are one of the interesting probes for the systematic study of hadron suppression to understand particle production mechanism in heavy-ion collisions.

Another possible study of low mass vector mesons is to see partial chiral symmetry restoration via mass and/or branching ratios modification. Especially in low transverse momentum region, a significant number of mesons decay inside the medium since ω and ϕ mesons have short lifetimes, 23 and 46 fm/c respectively. Several theoretical calculations predicted [3][4][5] properties (mass and/or width) are modified in either a hot or dense nuclear medium.

By using both leptonic and hadronic decay modes for the ϕ and ω measurement, PHENIX can cover wide kinematic range (low p_T by the leptonic decay mode and high p_T by hadronic decay mode). thus able to address a number of interesting physics topics.

In this article, we present recent results of the systematic measurement of ϕ by using e^+e^- for leptonic and K^+K^- for hadronic decay mode, and ω by using e^+e^- for leptonic and $\pi^0\gamma$, $\pi^+\pi^-\pi^0$ for hadronic decay mode in p+p, d+Au, Cu+Cu and Au+Au collected in 2003-2008.

2. Experimental Setup

RHIC provided various collisions of a broad range of nuclear species from proton, deuteron to Cu and Au ions at energies up to $\sqrt{s_{NN}}$=200 GeV. The PHENIX is designed specifically to measure electromagnetic probes of the collisions such as electrons, muons and photons using multi-purpose detectors [6]. The setup of the PHENIX experiment can be grouped into three categories; global detectors close to the beam pipe, 2 muon arms which have pseudo-rapidity coverage of $\pm$(1.2-2.4) and 2 central arms which has pseudo-rapidity coverage of $\pm$0.35, each subtending 90 in azimuthal angle.

Global detectors, beam-beam counters (BBC) and zero degrees calorimeters (ZDC) measure the vertex and centrality of the interactions. Central arms are capable of measuring a variety of particles by Ring Imaging Cerenkov detectors (RICH) for the separation of electrons, Drift Chambers (DC) and Pad Chambers (PC) for the tracking of charged particles and the measurement of their momentum, by Electro Magnetic Calorimeter(EMCal) for the measurement of spatial positions and energies of photons and electrons. EMCal comprises six sectors of Lead Scintillator Calorimeter(PbSc) and two sectors of Lead Glass Calorimeter (PbGl), which have an energy resolution of $\sigma_{PbSc}(E)/E \approx 2.1\% \oplus 8.1\%/\sqrt{E(GeV)}$ and $\sigma_{PbGl}(E)/E \approx 0.8\% \oplus 5.9\%/\sqrt{E(GeV)}$, respectively.

3. Analysis

3.1. Leptonic analysis

Electrons and positrons are identified mainly by the Čerenkov photons emitted in the RHIC. An energy-momentum match between the momentum measured by the tracking system and the energy measured in EMC is required to be within 2 times the resolution of the EMC. Finally the time-of-flight as measured by the EMC was required to be consistent with the flight time of an electron (10 ns $\sim$ 25 ns). The uncorrelated combinatorial background can be explicitly calculated by use of a mixed event technique, by taking candidate tracks from different events [7].

Figure 1 and 2 show an invariant mass spectrum of e^+e^- pairs in p+p, d+Au and Au+Au collisions at $\sqrt{s_{NN}}$ = 200 GeV after subtracting the combinatorial background. The solid line shows the fit of the ϕ, ω, ρ with widths fixed to the PDG values. Underneath the ω in p+p, we fit a ρ component with mass 770 MeV/c^2 and width of Γ=150MeV/c^2 (cyan line) where we assume ω and ρ have the same yield and have the same branching ratios as in vacuum.

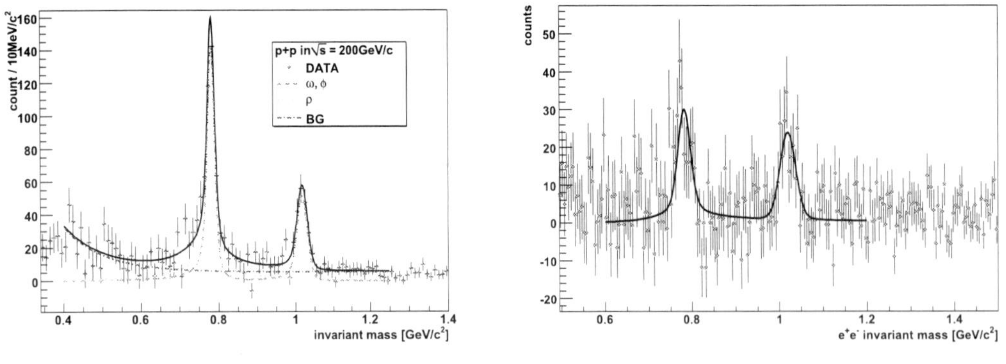

Figure 1. Invariant mass of e^+e^- pairs detected in p+p (left) [10] and d+Au collisions (right) [9]. Breit-Wigner fit to the ϕ and ω resonance is made with the resonance natural width set to the PDG values. The contribution from the residual background (blue line) and the estimated ρ contribution (assuming the same yield as the ω)(light blue) are also included in the fit.

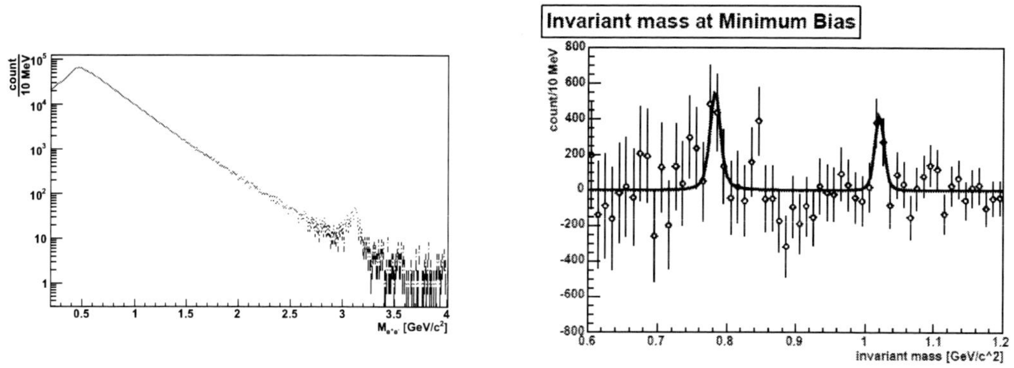

Figure 2. Invariant mass of e^+e^- pairs detected in Au+Au collisions before (left) and after (right) subtracting combinatorial background [9]. Breit-Wigner fit to the ϕ and ω resonance is made with the mass and width set to the PDG values.

In the baseline measurements in p+p and d+Au collisions, clear peaks of ϕ and ω are observed thanks to good momentum resolutions ($\approx$ 10 MeV/c), however, the statistical significance of the peaks in Au+Au collisions is only marginal to discuss mass modification.

3.2. Hadronic analysis

In $\phi \rightarrow K^+K^-$ channel, charged kaons are identified using a combination of the momentum measured by the tracking system and the timing from the TOF. Three techniques were used to reconstruct the invariant mass; explicitly requiring particle ID for both kaon, for only one kaon and without particle ID for both kaons [7] [8] . For the Au+Au data, a mixed event technique was used to subtract combinatorial background.

In $\omega \rightarrow \pi^+\pi^-\pi^0$ and $\omega \rightarrow \pi^0\gamma$ channel, first analysis step is to reconstruct π^0 candidates by combining photons pairs and applying a p_T-dependent cut around the mass of the π^0. Candidates (which include combinatorial background) are combined with a third photon for $\omega \rightarrow \pi^0\gamma$ or with two unidentified charged tracks (assumed to be π mesons) for $\omega \rightarrow \pi^+\pi^-\pi^0$. Due to high multiplicity and low S/B associated with it, cut optimization was performed in case of Au+Au analysis. Also, background subtraction is executed mixed by estimating three background sources; correlated background (for example, $\pi^0\gamma$ pairs from one of the photons from true π^0 or η making a fake π^0 candidate) , uncorrelated background which comes from the combination of independent 3γs, and K_s^0 contribution ($K_s^0 \rightarrow \pi^0\pi^0 \rightarrow 4\gamma s$), then each amount of background is determined by fitting.

In p+p and d+Au analysis, raw yields are extracted by a Breit-Wigner convoluted with a Gaussian fitting to the p_T slices of the invariant mass distribution shown in the Figure 3 (widths are about 5 MeV/c^2 for ϕ and about 9 MeV/c^2 for ω). The Figure 4 shows the invariant mass distribution in Au+Au analysis with(a) and without(b) combinatorial background. Combinatorial backgrounds are estimated by a mixed event technique which is explained previously.

For both leptonic and hadronic analyses, the extracted raw spectra are corrected for geometrical and reconstruction efficiency to produce the final yield. The efficiency is determined using a full GEANT simulation of the PHENIX detector. For electron-RHIC triggered (ERT) data sample, additional bias for this trigger was estimated by simulated Minimum Bias event sample with applying the same trigger requirements as the ERT.

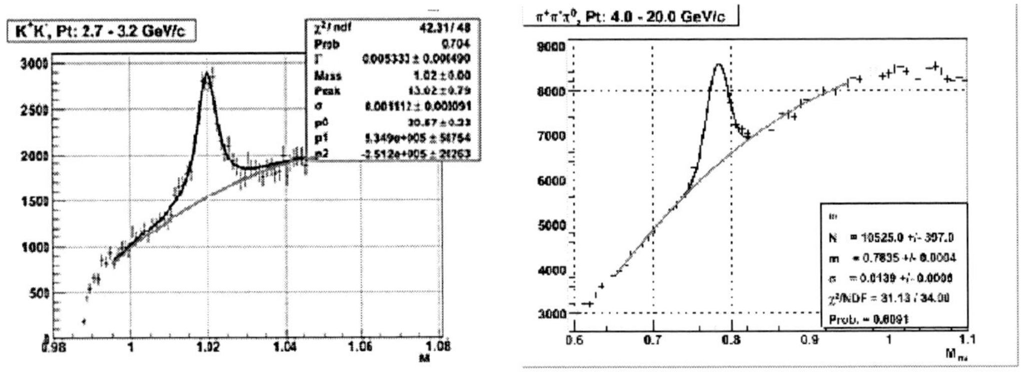

Figure 3. Example of invariant mass distribution in the $\phi \to K^+K^-$ in d+Au (left) and $\omega \to \pi^+\pi^0\pi^-$ in p+p (right).A Breit-Wigner convoluted with a Gaussian fit was performed.

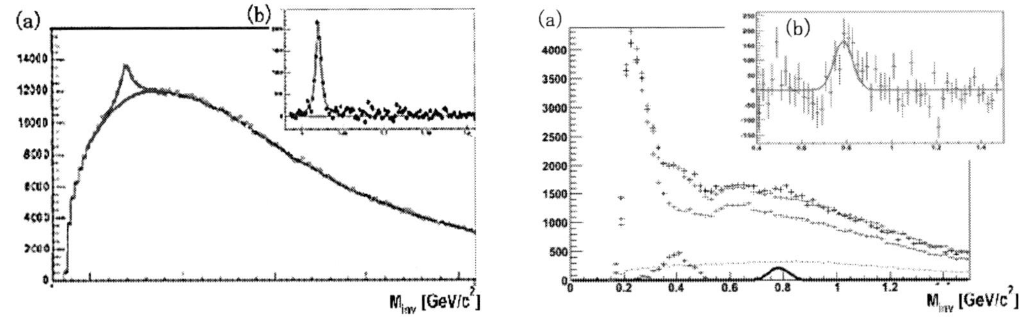

Figure 4. Invariant mass distribution in Au+Au for $\phi \to K^+K^-$ (left) and $\omega \to \pi^0\gamma$ (right) before(a) and after(b) subtracting combinatorial background. Combinatorial background is estimated by event-mixed technique (see text).

4. Results

Figure 5 shows invariant cross section for ϕ and ω production in p+p and minimum bias d+Au at $\sqrt{s_{NN}}$=200 GeV, as a function of p_T. Figure 6 shows invariant yield in Au+Au at $\sqrt{s_{NN}}$=200 GeV for ϕ and ω in different centrality; 0-20%, 20-40%, 40-92% and MB for ϕ spectra and 0-20%, 20-60%, 60-92% and MB for ω spectra. The spectra in p+p are fitted by a Levy function[15] (solid lines in the plots) and scaled by the expected N_{coll} for each collision (dashed lines in the plots). Although some points of $\phi \to e^+e^-$ at low p_T in p+p and d+Au may have a questionable divergence compared to the fitted line in ϕ spectra, overall spectra of ϕ and ω in p+p and d+Au are consistent within the errors. In Au+Au collisions, no signature of mass modification is observed beyond the errors, as shown in the Figure 6. The difference, if any, between the two decay modes is to be quantified via careful examination of centrality dependences of integrated yields, transverse momentum slopes and the spectra themselves.

Figure 7 shows nuclear modification factor, R_{dAu} for ϕ and ω as a function of p_T in minimum bias d+Au. Data for ϕ shown with both $\phi \to e^+e^-$ and $\phi \to K^+K^-$(left) and ω with both $\omega \to e^+e^-$ and $\omega \to \pi^0\gamma$(right). All R_{dAu} are consistent within errors and there are no significant signature of suppression or enhancement.

Figure 8 shows nuclear modification factor, R_{AA} of ϕ and ω with other hadrons; proton[12], kaon[12] π^0 [13] and η [14]. Protons are enhanced at $p_T > 1.5$ GeV/c and kaon is less suppressed than π^0 and η. π^0 and η mesons follow the same suppression pattern while ϕ mesons appear to be less suppressed in the range of $2.45 < p_T < 4.5$ GeV/c and become similar to π^0 and η at

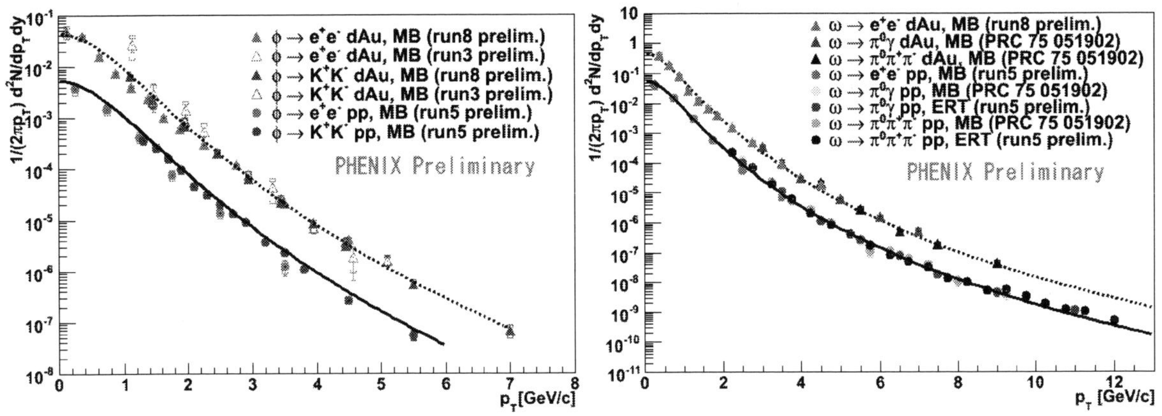

Figure 5. Invariant transverse momentum spectra of ϕ(left) and ω(right) in p+p and d+Au. The solid line represents a Levy fit[15] to both hadron and electron channels in p+p collisions. The dashed line is the same fit scaled by the expected N_{coll} for the case of d+Au.

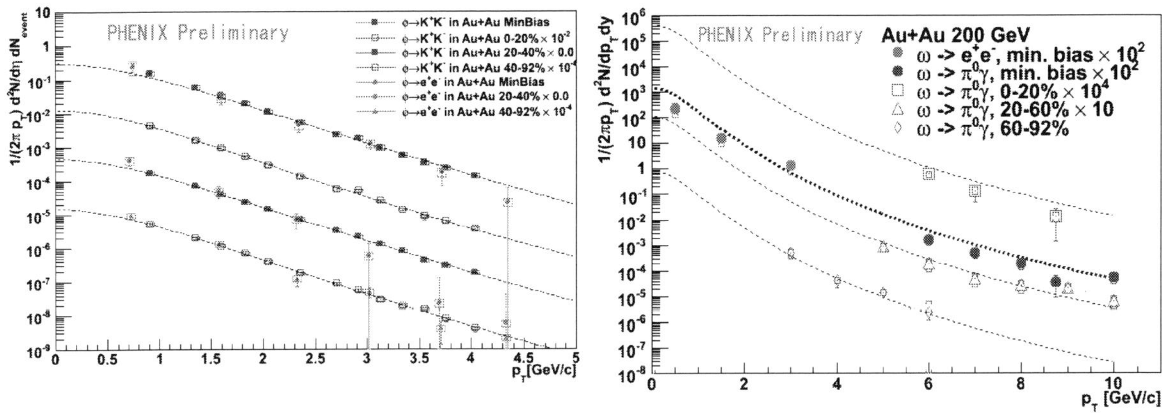

Figure 6. Invariant transverse momentum spectra of ϕ(left) and ω(right) in Au+Au. The dashed lines are a Levy fit to p+p results scaled by the expected N_{coll} for corresponding centralities in Au+Au.

high p_T, so do ω mesons within errors. The different behavior of kaon and ϕ suppression pattern ($p_T < 4.5$ GeV/c) is under discussion. Since there is no overlap p_T region between them, the results cannot attribute to an onset of flavor.

ω analysis is going to include more statistics in Au+Au to reduce statistical errors and also add Cu+Cu analysis for the systematic study to affirm that all hadron production in high p_T follows the picture of the parton fragmentation mechanism. On the other hand, comparison study of Au+Au and Cu+Cu data for ϕ was performed and it shows no significant difference (See the Figure 9).

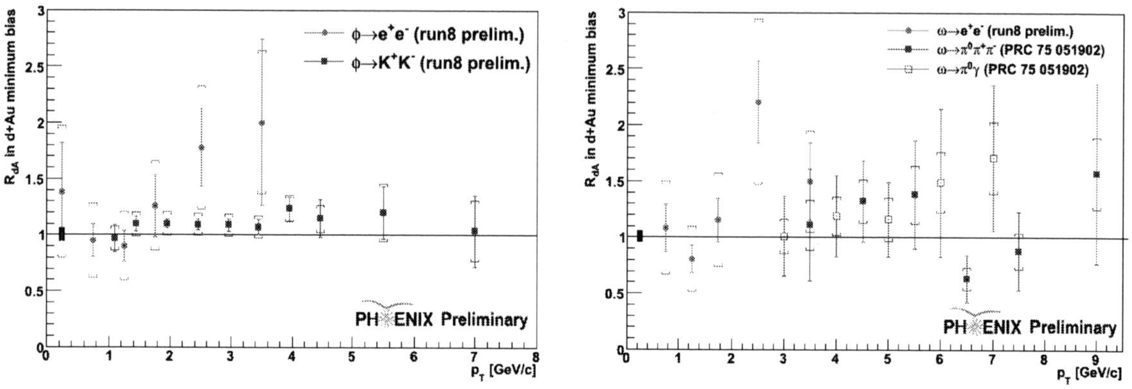

Figure 7. Nuclear modification factor, R_{dAu}, versus transverse momentum in minimum bias d+Au. Data for ϕ shown with both $\phi \to e^+e^-$ and $\phi \to K^+K^-$(left) and ω with both $\omega \to e^+e^-$ and $\omega \to \pi^0\gamma$ [11].

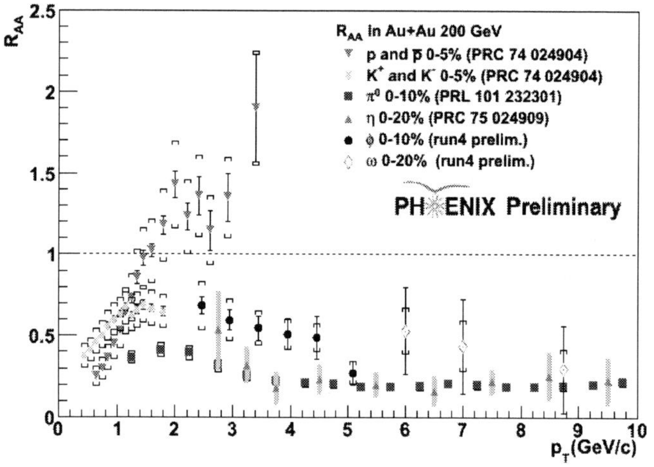

Figure 8. Nuclear modification factor R_{AA} as a function of p_T for ϕ and ω with π^0[13], η[14], $(K^+ + K^-)/2$ and $(p + \bar{p})/2$[12] (explanation of each suppression pattern is in text).

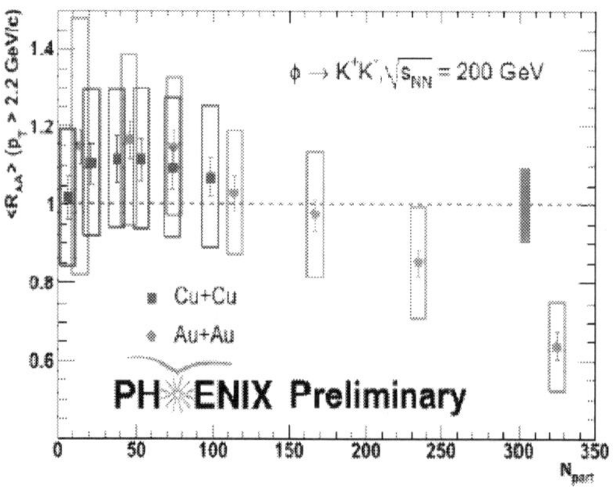

Figure 9. ϕ mesons's integrated R_{AA} for Au+Au and Cu+Cu collisions at $\sqrt{s_{NN}} = 200$ GeV for similar number of participating nucleons $< N_{part} >$ [8].

5. Summary

The PHENIX has measured the production of ϕ and ω mesons via both leptonic and hadronic decay modes in p+p, d+Au, Cu+Cu and Au+Au at C.M.S. collision energy per nucleon pair of 200 GeV. The results show a good agreement for invariant yield spectra in different decay modes in p+p and in d+Au respectively. The modification factor, R_{AA} suggests suppression of ϕ/ω at high p_T. More statistics will be available in Au+Au data which is taken in 2007-2008 and 2009-2010 with a HBD detector which has a capability to reject the π^0 dalitz background, thus able to reduce the uncertainties.

References

[1] Mikhail A.Stephanov, Int.J.Mod.Phys.A20:4387-4392(2005)
[2] K.Adcox *et al.*, Nucl.Phys.A757,184-282(2005)
[3] R.Rapp and J.Wambach, hep-ph/9909229v1(1999)
[4] Y.Nambu and G.Jona-Lasinio, Phys.Rev.122,345-358(1961)
[5] T.Hatsuda and S.H.Lee, Phy.Rev.C46(1992)R34
[6] K.Adcox *et al.*, Nuclear Instruments and Methods A499 469-479(2003)
[7] S.S.Adler *et al.*, Phys.Rev.C72,014903(2005)
[8] M.Naglis for the PHENIX Collaboration, Nucl.Phys.A830,430c(2009)
[9] Y.Tsuchimoto for the PHENIX Collaboration, Nucl.Phys.A830,487c(2009)
[10] D.Sharma for the PHENIX Collaboration, arXive:0901.3360v1
[11] S.S.Adler *et al.*, Phys.Rev.C75,051902(2007)
[12] S.S.Adler *et al.*, Phys.Rev.C74,024904(2006)
[13] S.S.Adler *et al.*, Phys.Rev.Letter 101,232301(2008)
[14] S.S.Adler *et al.*, Phys.Rev.C 75,024909(2007)
[15] J.Adams *et al.*, Phys.Lett.B637,161(2006)

Journal of Physics: Conference Series **230** (2010) 012023 doi:10.1088/1742-6596/230/1/012023

The ALICE TPC: Status and Perspectives

Hans Rudolf Schmidt, for the ALICE Experiment

GSI Helmholtz Centre for Heavy Ion Research GmbH, Darmstadt

h.r.schmidt@gsi.de

Abstract. The ALICE TPC is the largest Time Projection Chamber ever built. Given the performance requirements it is also, and justifiably so, dubbed the most challenging one. In this paper we browse through the often contradictory optimization strategies and outline the solutions taken to meet the specifications. Mainly on the basis of the commissioning data taken in 2008 and 2009 we will examine to which extend the ALICE TPC came up to its performance expectations, which were outlined elsewhere [1, 14]. First results on the performance of the TPC with proton-proton collisions in December 2009 are presented.

1. Introduction

The physics of hot and dense matter has been investigated over the last two decades at the AGS [2], the SPS [3] and at RHIC [4, 5, 6, 7]. The results obtained in recent years at RHIC have confirmed the previous SPS findings and, beyond that, given exciting new insights into the physics of strongly interacting matter. ALICE (A Large Ion Collider Experiment) is designed to study the physics of hot and dense matter and the transition to a quark-gluon plasma at the CERN Large Hadron Collider (LHC). The ALICE detector is optimized to cope with the highest particle multiplicities anticipated for PbPb collisions (dN_{ch}=dy up to 5000). In addition to heavy systems, the ALICE Collaboration will study collisions of lower-mass ions, which are a means of varying the energy density, and protons (both pp and pA), which primarily provide reference data for the nucleus-nucleus collisions. The pp data will in addition allow for a number of genuine pp physics studies.

2. The ALICE Experiment

The detector (cf. Figure 1) consists of a central part, which measures event-by-event hadrons, electrons and photons, and of a forward spectrometer to measure muons. The central part, which covers polar angles from 45° to 135° ($|\eta| < 0.9$) over the full azimuth, is embedded in the large L3 solenoidal magnet. The central barrel consists of: an Inner Tracking System (ITS) of high-resolution silicon detectors; a cylindrical Time-Projection Chamber (TPC), a Transition Radiation Detector (TRD) and a Time-Of-Flight (TOF) detector. A large lead-scintillator Electromagnetic Calorimeter covering $\Delta\phi = 110°$ and $|\eta| < 0.9$ complements the central barrel of ALICE. The single-arm detectors PHOS, a lead-tungsten crystal electromagnetic calorimeter and HMPID, a ring imaging Cherenkov hodoscope add to the central detectors.

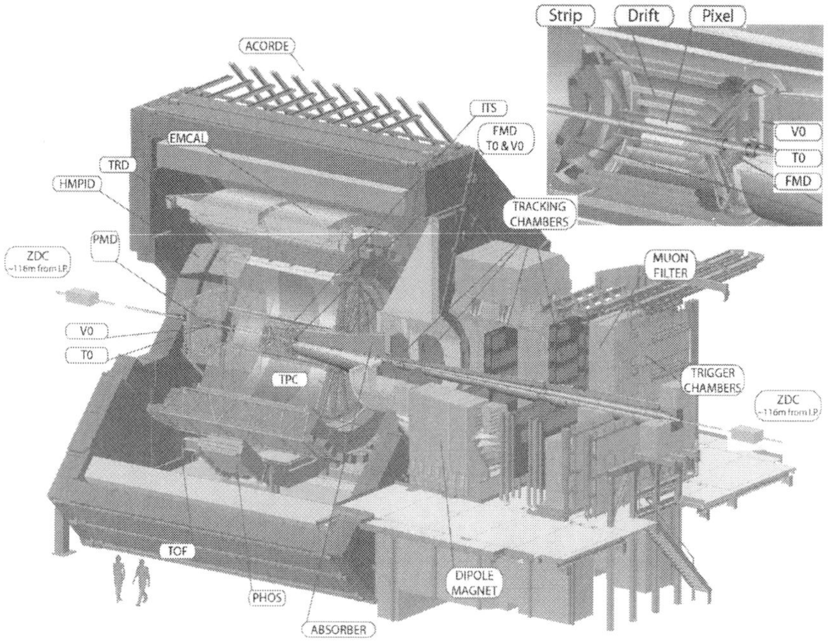

Figure 1. Sketch of the ALICE detector. The insert shows an expanded view of the inner tracking system (ITS).

The forward muon arm (covering polar angles ϑ from 171° to 178°) consists of a complex arrangement of absorbers, a large dipole magnet, and 14 planes of tracking and triggering chambers. Several smaller detectors (PMD, ZDC, FMD, T0, V0) for global event characterization and triggering are located at forward angles. An array of scintillators (ACORDE) on top of the L3 magnet will be used to trigger on cosmic rays. The central detectors ITS, TPC and TOF are completely installed and commissioned. The TRD has 7 out of 18 super-modules installed and will be completed in 2011. Furthermore, the forward muon spectrometer, the central HMPID single-arm spectrometer and the Photon Multiplicity Detector PMD are installed and ready. Both the scintillator-lead calorimeter and the high resolution PHOS photon spectrometer are at present partially installed and are schedule to be completed in 2010/11. The ZDC (Zero Degree Calorimeter) as well as the other trigger detectors are ready as well. A High Level Trigger System (HLT) is fully operational for pp-collisions. Thus, at the restart of the LHC in fall 2009 the ALICE detector had full hadron and muon and partial electron and photon capability.

3. The ALICE Time Projection Chamber

3.1. Design

A comprehensive overview of the TPC design and some results on the performance with cosmic rays during commissioning are contained in [8]. The TPC is cylindrical in shape with an active radial range from about 85 to 250 cm, and an overall length along the beam direction of 500 cm, divided by the central HV electrode into two drift regions of 250 cm length (cf. Figure 2). Multi-wire proportional chambers are mounted into 18 trapezoidal sectors in each end-plate. The field cage is based on a design with a central high-voltage electrode and two opposite axial resistive potential dividers which create a highly uniform electrostatic field in the common gas volume. The central electrode is an aluminized, stretched Mylar foil. The potential of the drift region is defined by Mylar strips wound around 18 inner and outer support rods which contain the resistor chains, gas in- and outlet and a

Journal of Physics: Conference Series **230** (2010) 012023 doi:10.1088/1742-6596/230/1/012023

system for distribution of laser rays. Because of the drift gas used in the TPC, the field cage has to be operated at a rather high gradient of 400 V/cm. The high at the central electrode is voltage 100 kV which results in a maximum drift time of about 100 μs. The field cage is surrounded by double-shelled containment vessels with CO_2 as insulator. Composite materials were chosen for high mechanical stability at a low material budget (only 3.5% of a radiation length for tracks with normal incidence). The readout chambers instrument the two end plates of the TPC cylinder with an overall active area of 32.5 m^2 with 18 sectors on either side. The chambers are multi-wire proportional chambers with cathode pad readout. The sectors are segmented radially into two chambers with varying pad sizes, optimized for the radial dependence of the track density. There are 557568 pads, radially pointing, with 3 sizes: 4 × 7.5 for the inner chambers, 6 × 10 and 6 × 15 mm^2 ($r \times \varphi$) for the outer chambers.

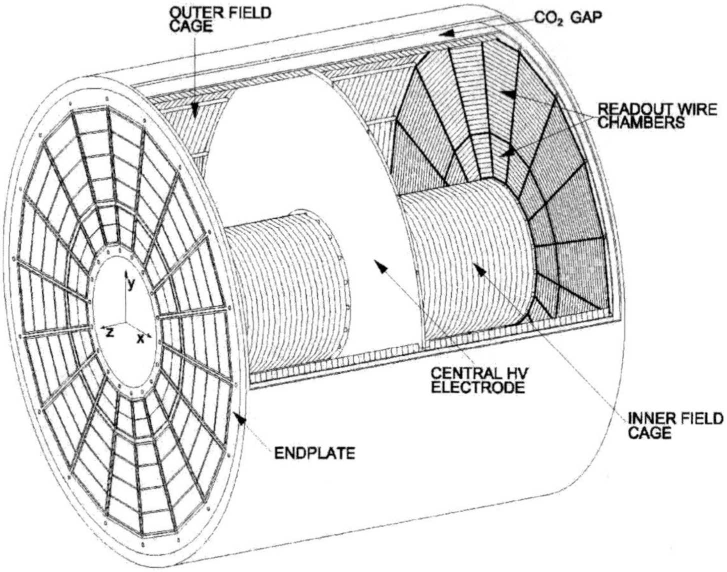

Figure 2. 3D view of the TPC field cage. The high voltage electrode is located at the center of the drift volume. The endplates with 18 sectors and 36 readout chambers on each end are shown.

A schematic layout of the front-end electronics readout chain is shown in Figure 3. The pad signals are treated in a charge-sensitive amplifier and shaper in 0.35 μm CMOS technology with 16 channels per chip. One 10-bit ADC per channel samples the analog signal at a rate of 5 or 10 MHz (500 or 1000 samples). The following pipelined digital processor performs pedestal subtraction, tail cancellation, zero-suppression, baseline restoration, formatting and buffering. The ADC's and digital circuits of 16 channels are implemented in one chip (ALTRO) in a 0.25 μm CMOS process. The digital signal processing is essential for the performance at high occupancy and rate. An important feature is that processing parameters and algorithms can be reconfigured. The overall system noise value is about 600e rms for the typical pad capacitance of 12 pF. The front-end electronics chain consists of a 128 channel front-end card (FEC) with only 40 mW/channel, a frontend bus for rows of up to 25 FEC's linked to one readout control unit (RCU), with 6 RCUs for each of the 36 readout sectors. Connection to the data acquisition system is via optical links (one per RCU).

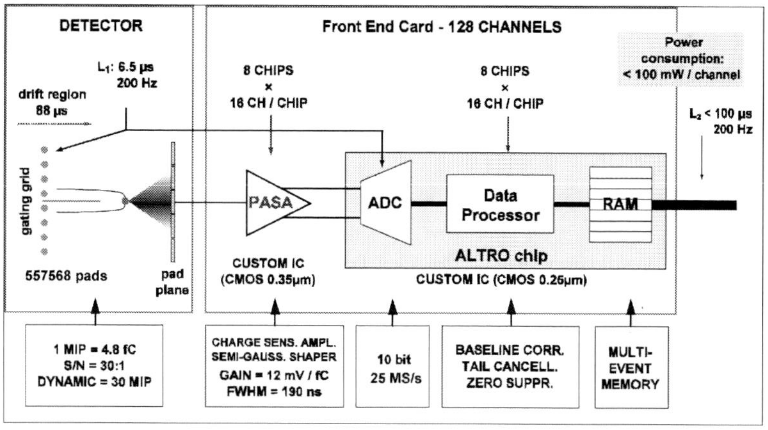

Figure 3. An overview of the ALICE TPC front-end electronics.

The design of the TPC is thus ‘conventional’ in overall structure but innovative in many details. This is necessitated by the extreme environment expected for Pb-Pb collisions at center-of-mass energy of 1144 TeV. In the following section we will outline in detail the specific challenges imposed by this environment and the attempts of technical solutions.

3.2. Challenges

The ALICE Time-Projection Chamber is designed for operation in the extreme environment of heavy ion collisions up to Pb–Pb at LHC. The TPC has been designed for a maximum multiplicity $dN_{ch}/dy=8000$, resulting in 20000 charged primary and secondary tracks in the acceptance, an unprecedented track density for a TPC. Although this multiplicity is by now likely be higher by a factor of 4 than the extrapolation from RHIC data, this upper limit has been kept for the design and simulations. At the Pb–Pb design luminosity, a minimum bias interaction rate of 8 kHz is expected of which about 10% are to be considered as central collisions. Extreme multiplicities and high rates set new demands on the design of a TPC.

The obvious answer to high multiplicity events is to adapt the granularity of the readout chambers, i.e., the pad sizes and time bins, to the expected spatial and temporal track density. This approach has, however, the following caveats:

- due to diffusion of the electron clouds while drifting towards the anode grid, the granularity is limited by the transverse diffusion. The width of the charge cloud after 250 cm of drift is of the order of mm depending on the choice of the drift gas and voltage, i.e., any reduction of the pad beyond a certain size results in an oversampling of the track without any gain of information;
- small pad sizes imply a small (primary) signal, hence the pad size a limited by achievable signal-to-noise ratio. Moreover, a sufficient number of pads per charge cluster in terms of position resolution is 2-3. Thus an increase in the number of pads is sensible only if the induced charge from the (point-like) avalanche spreads over no more the 2-3 pads. This can be achieved by reducing the distance between anode wire and pad plane. However, at a certain point the distance HV-GND gets critical;
- similarly, one could think of increasing the frequency of the time sampling. However, as diffusion occurs also in longitudinal direction this would result as well in an oversampling of the pulse. The choice of a shorter shaping time is limited by the fact that below 150-200 ns shaping time the signal/noise ratio becomes critical.

Another serious difficulty imposed by high multiplicity events (at high rate) is the accumulated space charge in the drift volume. Owing to the much smaller mobility of the ions as compared to the electrons a quasi-stationary positive charge will distort the drift field significantly.

3.3. Solutions

From the above constraints it is obvious that the gas choice is a critical issue. The requirements to the gas are low diffusion (a "cold" gas) and low space charge. A straightforward choice, also taken previously by the NA49 experiment [9] for their vertex TPC, is $NeCO_2$ as drift gas, which fulfils the above two requirements. However, this selection is not without severe drawbacks:

- as a low Z gas the number a primary electrons is scarce (about $20e^-$/cm track length). This aggravates the above mentioned already small signal/pad due to the pad size itself. A possible solution is an increase in amplification, i.e., gas gain. This, however, is by itself again not with out problems since CO_2 is not a very good quencher. CH_4 would, in this respect be much better, but its diffusion properties and the increased ageing due to its organic nature are not acceptable;
- the drift velocity in $NeCO_2$ is, differently from the $ArCH_4$ mixture used in STAR, not saturated, which means that it depends sensitively on the gas temperature. The change in drift velocity is about 0.35%/K at 400 V/cm. This means that a temperature difference larger than 0.1 K over the full drift length of 250 cm results in longitudinal position variations Δz of the order of 1 mm at v_{Drift}= 2.64 ms/cm. This exceeds the internal resolution of the readout chambers. Therefore the detector cooling and temperature stabilization systems are involved [10];
- the drift velocity is comparatively small. This limits the internal rate capability of the TPC. While STAR achieved v_{Drift}= 5.5 cm/μs at 150 V/m drift field, the field inside the ALICE TPC has to be raised to 400 V/cm to reach a drift velocity v_{Drift} of 2.64 cm/μs. This, of course, requires sophisticated insulation of the field cage vessel, e.g., the double walled field cage vessels are flushed with CO_2 as an insulation gas.

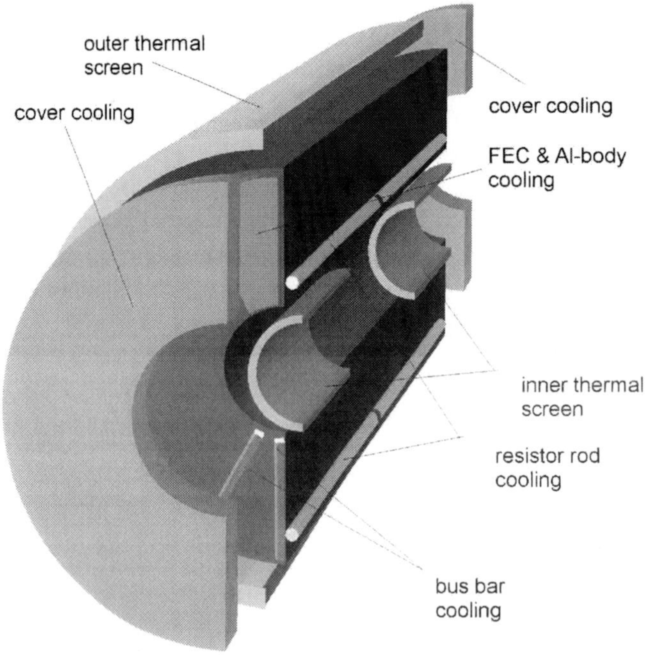

Figure 4. Schematic view of the various TPC cooling elements.

4. Commissioning and Calibration of the Time Projection Chamber

First particles from the LHC were seen on June 15th, when the ALICE Silicon Pixel Detector received a significant load of muons after the injection of particles from the SPS into the LHC. During LHC tune-up until October 2008, and thereafter from July 2009 until the end of November 2009 the detectors were mostly operated under "realistic" conditions, i.e., continuous 24 hours data taking in global mode (all detectors participating in the data stream). The data taken were comics, laser, source or pulser data and were used to calibrated and align the detector. Altogether, more then 5×10^8 event have been recorded.

4.1. Temperature Stabilization

The strong dependence of the drift velocity on the temperature for the gas mixture used in the TPC requires to stabilize the temperature to better than 0.1 K over the full drift volume of 88 m^3. This is achieved by very efficient heat removal from the front-end electronics and by well-defined iso-thermal surfaces around the TPC barrel via thermal screens. In addition, the Al-bodies of the readout chambers and the resistive voltage divider chain of the field cage ("resistor rod"), which face or are inside of the gas volume, respectively, are set to precisely defined temperatures. The cooling system (cf. Figure 4) consists of 60 individual sub-atmospheric cooling circuits, which can be regulated in flow and temperature. The histogram of the skirt temperature sensor distribution is shown in Fig. 5. The so-called skirt sensors are arrays of Pt1000 sensors, which are installed within the gas volume of the TPC close to the outer circumference of the readout-plane. The temperatures were sampled over a period of 24 hrs with stable environmental conditions. The RMS value of the histogram is below 50 mK showing that the desired temperature homogenization of the TPC gas volume below 0.1 K is within reach.

4.2. Noise Performance

In Fig. 6 the histogrammed noise distributions obtained during the final stage of commissioning are shown for all pads and for the different pad sizes separately. Only 1.7 (0.14)% of the channels show noise values above 1(1.5) ADC channels. The largest pads with 5.7 (0.4)% contribute most to this value. The mean ENC in the TPC is about 730 e. A systematic variation in the noise level is observed increasing from the center of each readout partition towards its edges. This variation is directly related to the variation in length of the traces on the pad-plane PCB board, which connects the pads with the connectors on its back side. With the trace length the capacitance at the input of the charge-sensitive preamplifier/shaping chips (PASAs) rises and hence the noise level rises. A detailed description of the noise performance is contained in [11].

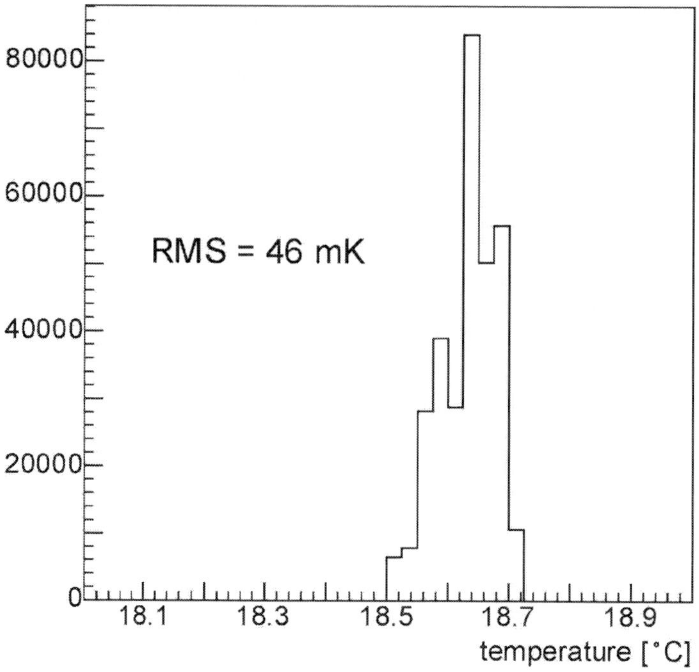

Figure 5. Temperatures distribution measured with the skirt PT1000 sensors.

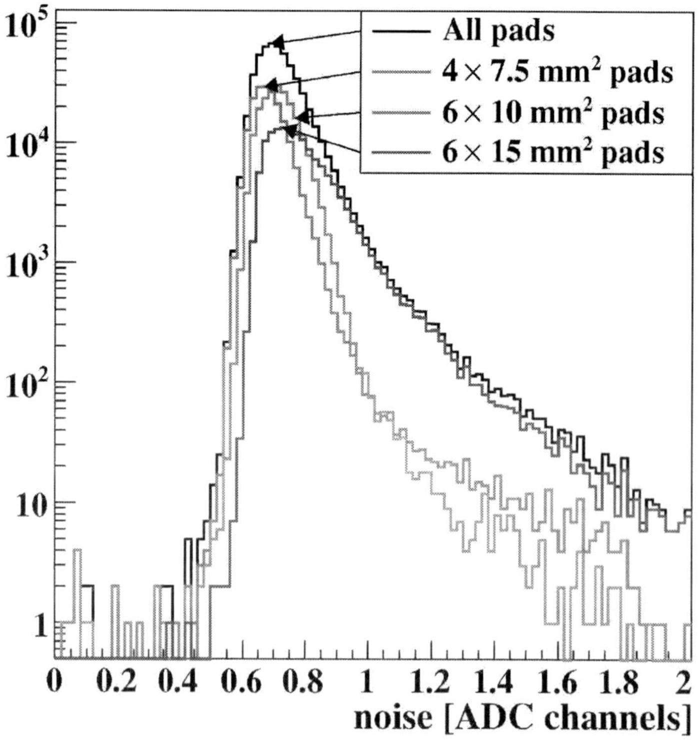

Figure 6. Noise distribution; for all chambers and separately for the different pad sizes.

Journal of Physics: Conference Series **230** (2010) 012023 doi:10.1088/1742-6596/230/1/012023

4.3. Gain Calibration and dE/dx calibration

The gain of 557568 individual readout pads located at the two TPC end-planes was calibrated injecting radioactive ^{83}Rb into the TPC gas. ^{83}Rb ($\tau_{1/2}$=86.2 d) decays to ^{83m}Kr via electron capture. The decay of the isomeric Krypton state ($\tau_{1/2}$=1.83 hrs) produces via IC with a large probability an electron cluster of 41 keV, which is used to equalize the gains. This calibration procedure reveals an average pad-to-pad gain variation of 15%, which is well within the range expected from the production tolerances.

Having the gain equalized, the response of the pads to the differential energy loss dE/dx of charged tracks in the gas is calibrated as a function of γ to match the Bethe-Bloch curves. The resolution in dE/dx is close to the design value of 5.5%. This allows to separate particles (protons, kaons, pions and electrons) on a statistical basis in the region of the relativistic rise, i.e., at momenta above 1-2 GeV/c.

4.4. Space Point and Momentum Resolution

Both the space point and the momentum resolution of individual tracks are determined from cosmic rays showers measured in the TPC. The space point resolution is determined from cosmic rays. It is, both in rϕ- and z –direction, in agreement with simulations. It is limited by diffusion. The resolution relevant for high p_t tracks (large drift and diffusion) is of the order 300-800 μm.

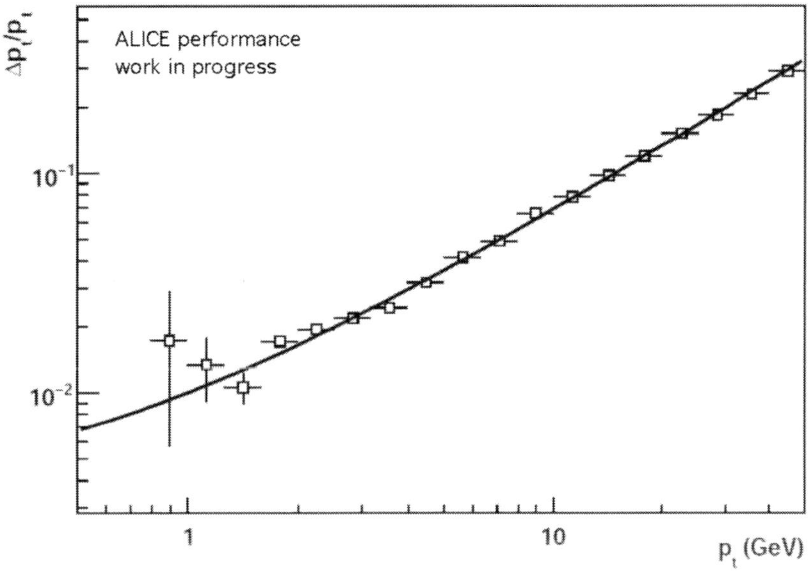

Figure 7. Transverse momentum resolution obtained from cosmics tracks.

The momentum resolution has been estimated from the correlation of semi-tracks, i.e., cosmic tracks crossing the TPC through the center have been split into to half tracks. Those are reconstructed in two parts due to the optimization of the reconstruction code for the collider geometry. The extracted momentum resolution is shown in Figure 7 and is 6.5% for 10 GeV/c tracks and 1 % for 1 GeV/c tracks. This is, with the present set of corrections, somewhat worse than the design resolution (4.5% at 10 GeV/c).

5. Proton-Proton Collisions

First proton-proton collisions were recorded in the ALICE experiment on Nov. 23rd 2009. For the initial data taking period the TPC had been switched off as a precaution. The data published in [12] refer to those first 285 events taken.

On Dec. 6th 2009 the TPC saw its first tracks from collisions at $\sqrt{s}$=900 GeV/c. The TPC trigger rate from the collisions of 4 on 4 counter-rotating proton bunches was up to 10 Hz. Each of the bunches containing initially up to 10^{11} protons. The trigger included both empty events and beam-gas interaction. The events size was in average of the order of 170 kB owing to the excellent low noise behaviour of the TPC readout chain. The magnetic field was set to ±0.5 T and, for alignment purposes, to B=0 T. An online monitor plot of tracks of a central collision is shown in Fig. 8.

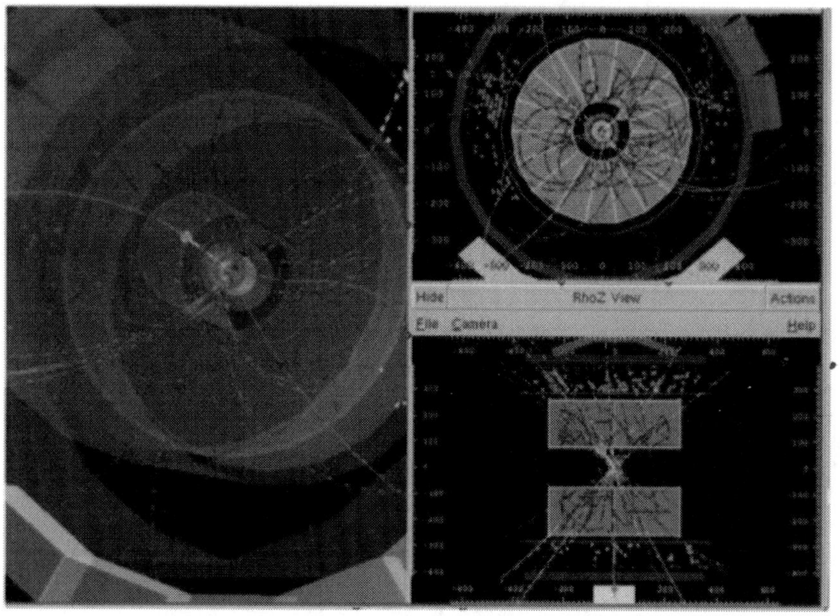

Figure 8. Online monitor picture of a central collision.

To ensure immediate feedback on the data quality and to optimize the calibration of the detector, the raw data of all events was immediately transferred to GSI. The events were directly reconstructed and analyzed using the entire GSI Tier 2 batch farm and High Performance Cluster where 2000 cores and 1 PB of storage on Lustre were available. Results on the detector performance were available within a few hours after the data had been recorded.

In parallel, calibration parameters were extracted. Several reconstruction passes with iteratively improved values were performed yielding continuously better results. All steps were automated as much as possible and all results were constantly communicated to the CERN detector control room. Thanks to the outstanding performance of the GSI computing infrastructure, the whole chain was running uninterrupted during the complete 2009 data taking period.

Fig. 9 shows a dE/dx vs. momentum spectrum from pp collision at 900 GeV/c employing the calibration data from the cosmic and krypton runs. As can be seen the excellent particle separation anticipated from the cosmic runs is also manifest in the collision data, e.g., by the π/μ-separation visible at low momentum.

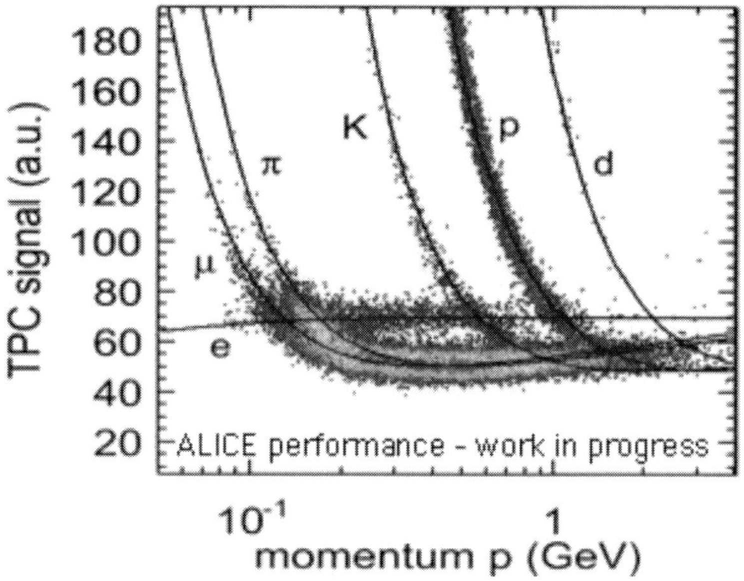

Figure 9. Particle identification the TPC.

This, together with precise tracking, readily allows reconstructing invariant mass spectra. As a further example of the TPC performance with data we show in Fig. 10 the K_S^0 -invaiant mass plot from the $\pi^+\pi^-$ V0 decays vertices. The narrow width and the perfect agreement with the PDG mass should be taken as certification of the readiness of the TPC for the upcoming pp and PbPb runs in 2010.

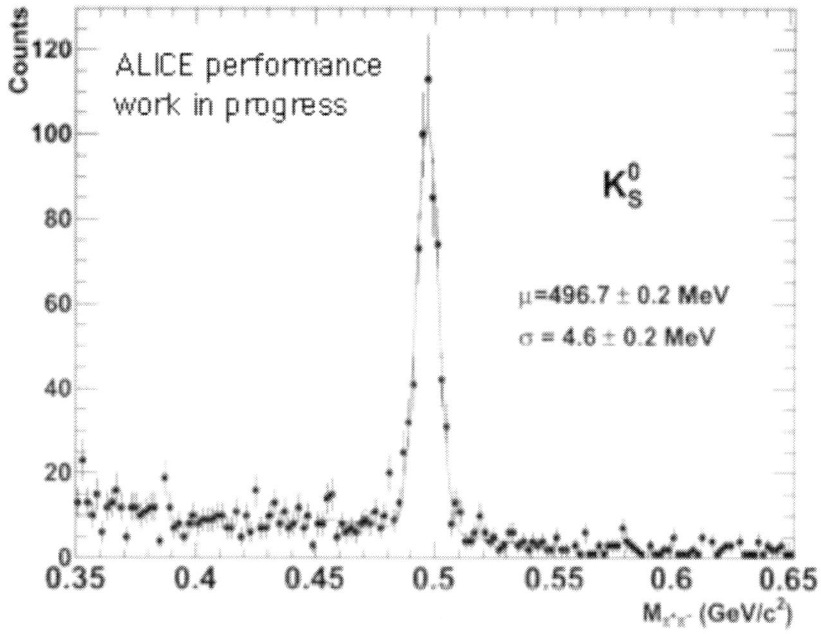

Figure 10. Reconstructed K_S^0 from $\pi^+\pi^-$-V0 decays vertices.

6. Summary

From the very first ideas on an ALICE Time Projection Chamber as outlined in the Letter-of-Intent [13] in 1993 it took more then 15 years to develop, build, test and commission the detector. The very intense decade of R&D and testing of an exceptionally ambitious detector is now rewarded by the very successful operation both during the commissioning with cosmic rays as well as with the first pp collisions in Dec. 2009. The TPC fulfills the promised technical specifications and performance figures as outlined in the Technical Proposal [14] and the Physics Performance Reports [1].

References

[1] ALICE Collaboration et al, 2004 J. Phys. G: Nucl. Part. Phys. **30** 1517-1763
ALICE Collaboration et al, 2006 J. Phys. G: Nucl. Part. Phys. **32** 1295-2040
[2] CA Ogilvie et al 2002 Nucl. Phys. A 698 3
[3] C Lourenco et al 2002 Nucl. Phys. A 698 13
[4] BB Back et al, (PHOBOS Collaboration) 2005 Nucl. Phys. A 757 28
[5] I Arsene et al, (BRAHMS Collaboration) 2005 Nucl. Phys. A 757 1
[6] J Adams et al, (STAR collaboration) 2005 Nucl. Phys. A 757 102183
[7] K Adcox et al, (PHENIX collaboration) 2005 Nucl. Phys. A 757 184283
[8] J Alme et al, "The ALICE TPC, a large 3-dimensional tracking device with fast readout for ultra-high multiplicity events", 2010, e-Print: arXiv:1001.1950, submitted for publication to Nucl. Instrum. Meth. A.
[9] S Afanasev et al, "The NA49 large acceptance hadron detector", Nucl. Instrum. Methods Phys. Res. A430, 210 (1999)
[10] U Frankenfeld et al, Experimental Evaluation of the ALICE TPC Front-End Electronics Cooling Strategy, ALICE-INT-2005-001, 2005.
[11] J. Wiechula, Commissioning and Calibration of the ALICE-TPC, PhD thesis, Goethe-Universität Frankfurt am Main, 2009
[12] ALICE Collaboration, K Aamodt et al, Eur. Phys. J. C (2010) 65: 111–125
[13] ALICE Letter of Intent, CERN/LHCC/93-16 LHCC/I4
[14] TPC Technical Design Report, CERN/LHCC 2000-001

Journal of Physics: Conference Series **230** (2010) 012024 doi:10.1088/1742-6596/230/1/012024

A Gas of Bags: Weirdness and Criticality[1]

L. G. Moretto, J. B. Elliott[2], P. T. Lake, L. Phair

Nuclear Science Division,
Lawrence Berkeley National Laboratory
Berkeley, CA, USA 94720

E-mail: `lgmoretto@lbl.gov`

Abstract. A gas of MIT bags has peculiar properties, one of which is that of not existing. We shall discuss its instability against coalescence. The absence of surface energy makes the bags already critical at their unique temperature T_H. The addition of a surface energy does not seem to cure the problem.

1. Introduction

It was shown elsewhere that both Hagedorn's model [1, 2] and the MIT bag model [3] have an exponential spectrum of the form

$$g(m) = \exp\left(+\frac{m}{T_H}\right). \tag{1}$$

This is a spectrum of a thermostat with temperature T_H. It will impart its unique, immutable temperature to any other "normal" system in contact with it.

It was also shown that the evaporation of the bag into "normal" particles produces a *saturated* vapor with a concentration [4, 5]

$$\frac{N(m)}{V} = g(m)\left(\frac{mT_H}{2\pi}\right)^{3/2} \exp\left(-\frac{m}{T_H}\right). \tag{2}$$

2. A Gas of Bags

The question arises as to whether particles with the same exponential spectrum can give rise to a stable gas akin to a physical gas of resonances. It is clear that, apart from translational degrees of freedom, bags are indifferent towards coalescence or fragmentation as shown in figure 1. Now let us include the translational modes and consider no interactions other than coalescence or fragmentation.

[1] Dedicated to W. Swiatecki in fond memory.

[2] Current Address:
Lawrence Livermore National Laboratory,
94551 Livermore, CA, USA

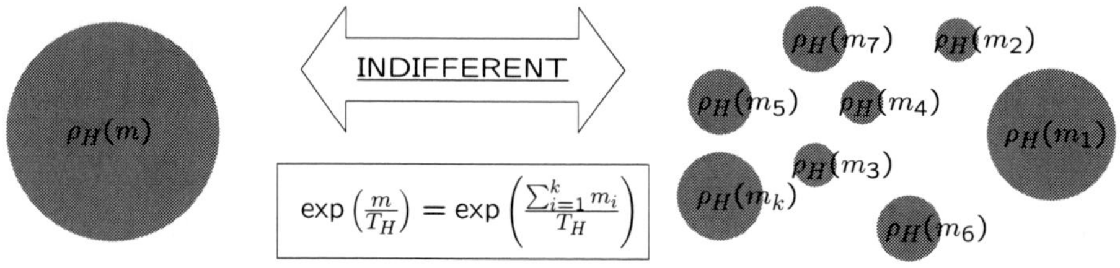

Figure 1. An MIT bag is indifferent to coalescence or fragmentation.

Let us consider the equilibrium between a bag A and the n equal bags B resulting from its fragmentation. The equilibrium constant is

$$K = \frac{c_B^n}{c_A} = \frac{q_B^n}{q_A} = \frac{q_A^{n-1}}{n^{\frac{3}{2}n}} \tag{3}$$

$$\text{where: } q_B = \left(\frac{2\pi m_A T}{h^2 n}\right)^{3/2} = \frac{1}{n^{3/2}} q_A,$$

c_A, c_B are the concentrations of the two species and q_A, q_B are the corresponding reduced partition functions.

It follows that

$$\lim_{n\to\infty} c_b = \lim_{n\to\infty} \frac{c_A^{1/n} q_A^{\frac{n-1}{n}}}{n^{3/2}} = 0 \tag{4}$$

and that the percent dissociation, α, is

$$\alpha = \frac{1}{n^{5/2}} \frac{c_A^{1/n}}{c_0} q_A^{\frac{n-1}{n}} \tag{5}$$

which also goes to 0 as $n \to \infty$.

Thus the translational degrees of freedom strongly bias the system towards coalescence.

From equation 2 we obtain the mass distribution

$$\frac{N(m)}{V} = \left(\frac{m T_H}{2\pi}\right)^{3/2}, \tag{6}$$

again strongly biased towards the largest possible bag or, grand canonically, towards an infinite bag.

3. Surface Energy and Criticality

Normal van der Waals fluids exhibit a first order coexistence line between liquid and vapor which terminates at the critical temperature. Above the critical temperature the two phases are not distinguishable any longer but a scar remains in the phase diagram which is visualized with, for instance, density fluctuations or heat capacities.

The steam tables give a good example [6]. Figure 2 shows the isobaric heat capacity overlaid on the phase diagram of water. At and below the critical temperature, there is a singularity in the heat capacity in going from one phase to another. Above the critical point, the heat capacity stays finite, but there is a maximum value given a constant pressure. This figure should be kept in mind when thinking about criticality in QCD.

But what about MIT bags? Let us recall how, in Fisher's cluster picture, criticality comes about.

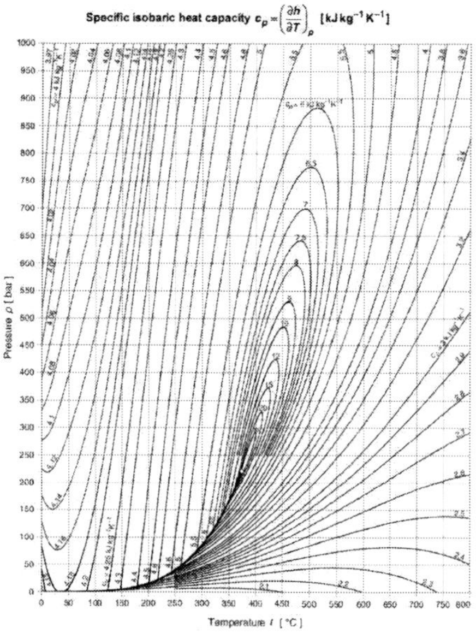

Figure 2. The isobaric heat capacity of water shows how even though the heat capacity stays finite everywhere above the critical temperature, a scar remains in that there is a maximum value for any given pressure.[6]

4. Fisher cluster model

To describe cluster formation in a simple fluid, Fisher proposed the degeneracy of clusters of a given size A to be of the form [7, 8]

$$g(A) = q_0 A^{-\tau} \exp\left[\frac{c_0}{T_c} A^{\sigma}\right]. \tag{7}$$

This is in contrast to the MIT bag picture, unless $\sigma = 1$. In the case Fisher considered, $c_0 A^{\sigma}$ is the surface energy, and $\sigma = 2/3$. What gives rise to this discrepancy?

A simple model to study clusters is the Ising model. Being confined to a lattice, a cluster is just a collection of connected lattice points with the same spin. The Ising model also exhibits both a first order and a second order phase transition, making it an invaluable tool in studying phase transitions in terms of clusters.

The study of the degeneracies of clusters on a grid, also known as lattice animals, is an active field of research. There is no closed form to enumerate all the lattice animals of a given size and surface. Much work has gone in finding these values, both exactly and approximately.

4.1. Lattice animal degeneracy

One method of approximating lattice animal degeneracies is by studying pure percolation on a grid [9]. A close relative to the Ising model, the percolation model is the case when each lattice site is occupied with a certain probability, p, uncorrelated to any other site. Such a lattice is completely described by its percolation probability and its volume, V. The Ising model at the limit of infinite temperature is isomorphic with such a percolating grid.

Let us consider a specific lattice animal, X, at a specific location on the grid, (x_1, y_1), as in figure 3. The animal is defined not only by the its area, A, but also by the surrounding unoccupied spaces, s_s, marked by circles in the figure. The probability of finding said animal is

$$P(X; x_1, y_1) = p^A (1-p)^{s_s}. \tag{8}$$

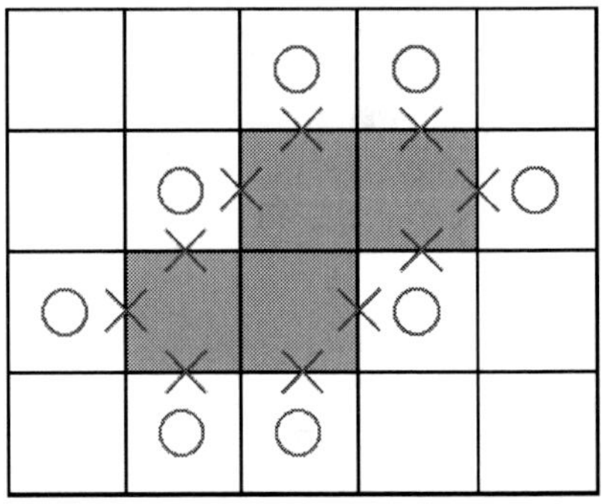

Figure 3. This lattice animal has $A = 4$, $s_s = 8$, and $s_b = 10$. To show the difference between the site and bond surfaces, the site surface is marked with circles whereas the bond surface is marked with crosses.

If periodic boundary conditions are used, there are V unique locations where animal X can exist. Furthermore, since $P(X; x_1, y_1)$ is only a function of A and s_s, any other animal with the same A and s_s will have the same probability. Putting this together, the probability of finding any animal of size A and surface s_s anywhere on the grid is

$$P(A, s_s) = V\,p^A(1-p)^{s_s}\,g(A, s_s) \tag{9}$$

Notice that the degeneracy of the clusters, $g(A, s_s)$, enters the previous equation.

With high statistics of a percolating grid, we can measure $P(A, s_s)$ and solve for $g(A, s_s)$:

$$g(A, s_s) = \frac{P(A, s_s)}{V\,p^A(1-p)^{s_s}}. \tag{10}$$

One further note needs to be made before these values can be applied to the Ising model. The parameter s_s is seen to be a site surface, which is different from the surface used in the Ising model. The Ising model considers the bond surface, s_b, of the cluster. Instead of the number of sites surrounding the animal, the number of bonds between the cluster and the outside determine the energy. Figure 3 gives an example of an animal with different values for its site and bond surfaces.

Further classification of the animals as a function of A, s_s, and s_b yields

$$g(A, s_s, s_b) = \frac{P(A, s_s, s_b)}{V\,p^A(1-p)^{s_s}}. \tag{11}$$

To find the physically significant degeneracy for the Ising model, $g(A, s_b)$, the degeneracies are summed over the different s_s:

$$g(A, s_b) = \sum_{s_s} g(A, s_s, s_b). \tag{12}$$

4.2. Lattice animals in the Ising model

Taking the degeneracies of these lattice animals, first consider the total degeneracy of animals as a function of size, as seen in figure 4. It can be seen that the data fits to the case $\sigma = 1$, a volume effect.

This method does not model the Ising system properly. In the case of the Ising model, the appearance of a lattice animal is affected by the amount of surface the animal possesses. The

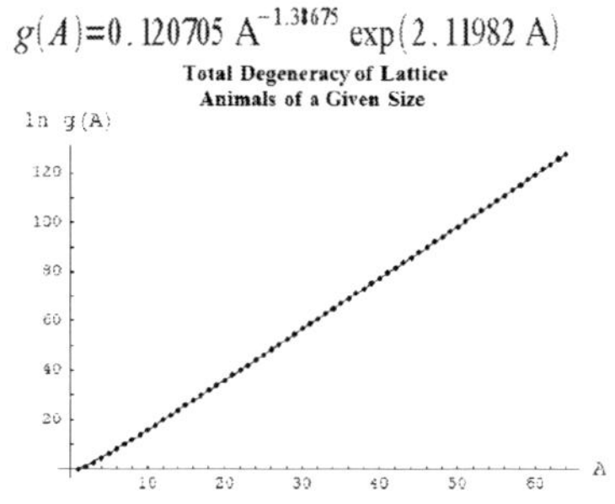

Figure 4. The total degeneracy of lattice animals follows the same trend as the Hagedorn spectrum.

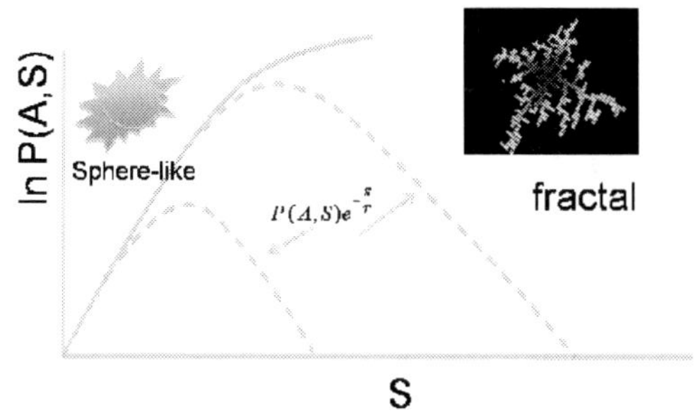

Figure 5. The shape of the clusters changes as a function of temperature, controlled by the Boltzmann factor for the surface energy.

concentration of a cluster as a function of size and (bond) surface, $n(A, s)$, is thus weighted by a Boltzmann factor

$$\begin{aligned} n(A, s) &= g(A, s) \exp\left[\frac{-E(A, s)}{T}\right] \\ &= g(A, s) \exp\left[\frac{-2s}{T}\right] \end{aligned} \tag{13}$$

As an example, figure 5 shows the degeneracy of clusters as a function of surface and the effects of applying a Boltzmann factor at various temperatures.

A result of this procedure is that the average surface of the clusters changes as a function of temperature. In fact, for a given temperature, plotting the average surface as a function of cluster size yields the relation

$$\langle s \rangle \propto A^{\sigma(T)} \tag{14}$$

where the power σ becomes a function of the temperature.

Figure 6 shows the values of σ over a large temperature range. At low temperatures $\sigma = 2/3$, as originally stated by Fisher. There is then a transition as the temperature increases where $\sigma \to 1$, which corresponds to the second order transition in the Ising model. The presence of the surface energy causes the clusters to be compact at low temperatures. Then as the temperature increases, the degeneracy of the clusters favors the dendritic, fractal form. This is analogous to the second order phase transition in that there is a limiting temperature where the surface free energy goes to zero.

Herein lies the contradiction. Since there is no surface energy present in the MIT bag model, the natural shape of the bags should be fractal. This implies that the natural temperature of

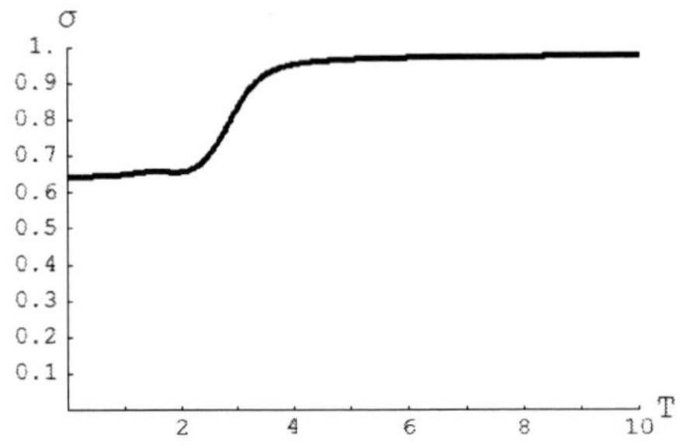

Figure 6. Surface dimensionality ($S \propto A^\sigma$) of 3-dimensional lattice clusters with surface as a function of temperature.

the system, T_H, is above the critical temperature. Yet, the formation of a bag is described well as a first order phase transition, which do not exist above the critical temperature.

5. Effects of surface energy

How does the picture change if surface energy is added to the MIT bags? First of all, it would make the situation of a gas of bags even more precarious, since there would be an ever greater tendency to maximize the drop size and to minimize the surface. Even more interesting is the effect of the surface on the bag temperature. The surface translates into an additional pressure on the bag. The enthalpy of the bag becomes:

$$E\,V = H = [f(T) + B]\,V + c_s\,V^{2/3}. \tag{15}$$

The pressure is:

$$p = \frac{1}{3}f(T) - \left(B + \frac{2}{3}c_s\,V^{-1/3}\right) = 0 \text{ at equilibrium,} \tag{16}$$

for which:

$$T = f^{-1}\left[3\left(B + \frac{2}{3}c_s\,V^{-1/3}\right)\right]. \tag{17}$$

In figure 7 we show the dependence of the bag size on the temperature, energy, and heat capacity. Notice that the temperature of a bag increases and tends to infinity with decreasing bag size.

The introduction of a temperature dependent surface energy is also easily implemented. If:

$$c_s = c_s^0\left(1 - \frac{T}{T_c}\right), \tag{18}$$

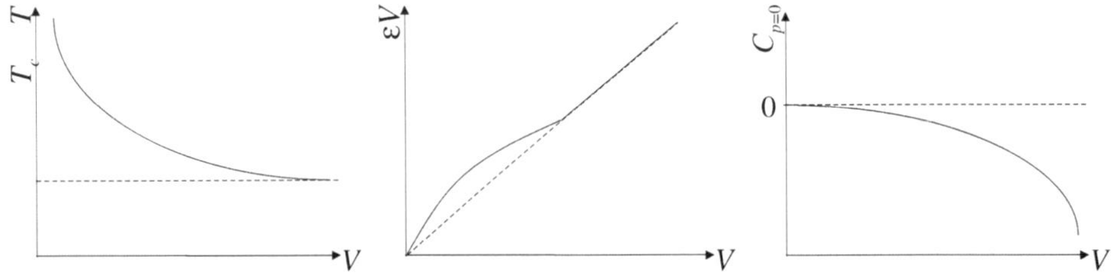

(a) Temperature (b) Energy (c) Heat Capacity

Figure 7. Effects of bags having a surface energy on various thermodynamic properties.

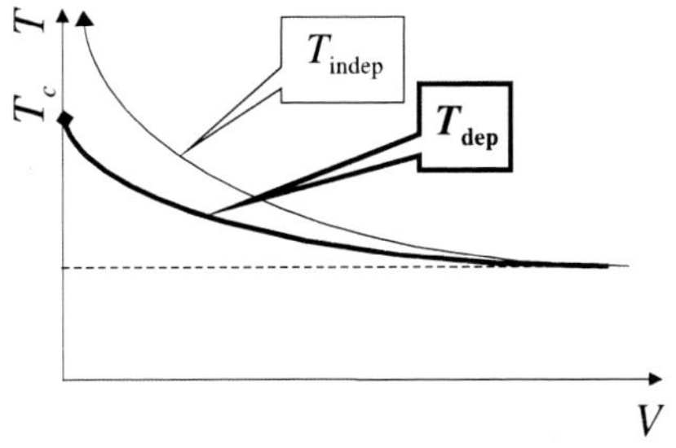

Figure 8. Effects of the surface energy varying with temperature. Notice how the temperature stays finite with this change when previously it diverges.

(a) Stability of a gas of bags (b) The decay of a bag with surface

Figure 9. Schematic picture of the effects of a bag having surface energy.

we have for the stability condition:

$$\sigma T^4 = 3\left[B + \frac{2}{3}c_s^0\left(1 - \frac{T}{T_c}\right)V^{-2/3}\right]. \tag{19}$$

Compared to the previous situation, the bag temperature goes to T_c rather than infinity as the bag becomes small, as seen in figure 8. It also preserves the trend that the large bag sizes go to a temperature of T_H.

This temperature dependence of the bag leads to two interesting conclusions:

1) A gas of bags of different sizes, and thus *of different temperatures*, is out of equilibrium. The system will tend to make one single bag of maximum size and minimum surface.

2) If a bag decays, its temperature, as manifested by the decay products such as pions, will progressively increase as the bag evaporates.

These conclusions are shown in figure 9.

Acknowledgments

This work was supported by the US Department of Energy under Contract No. DE-AC02-05CH11231.

References

[1] Hagedorn R 1965 *Suppl. Nuovo Cimento* **3** 147
[2] Hagedorn R and Ranft J 1968 *Suppl. Nuovo Cimento* **6** 169
[3] Chodos A, Jaffe R L, Johnson K, Thorn C B and Weisskopf V F 1974 *Phys. Rev. D* **9** 3471
[4] Moretto L G, Bugaev K A, Elliot J B and Phair L 2006 *Europhys. Let.* **76** 402
[5] Moretto L G, Bugaev K A, Elliot J B and Phair L 2006 *Proc. Sci.* **CPOD2006** 37

Journal of Physics: Conference Series **230** (2010) 012024 doi:10.1088/1742-6596/230/1/012024

[6] Wagner W and Kretzscmar H 2008 *International Steam Tables - Properties of Water and Steam based on the Industrial Formulation IAPWS-IF97* (Berlin, Heidelberg: Springer-Verlag Berlin Heidelberg) p 360
[7] Fisher M E 1967 *Rep. Prog. Phys.* **30** 615
[8] Fisher M E 1967 *Physics (N. Y.)* **3** 255
[9] Stauffer D 1985 *Introduction to Percolation Theory* (London, Philadelphia: Taylor & Francis)

26th Winter Workshop on Nuclear Dynamics IOP Publishing
Journal of Physics: Conference Series **230** (2010) 012025 doi:10.1088/1742-6596/230/1/012025

Recent Results of Fluctuation and Correlation Studies from the STAR Experiment

Terence J. Tarnowsky

National Superconducting Cyclotron Laboratory, Michigan State University, East Lansing, MI 48824, USA

E-mail: tarnowsk@nscl.msu.edu

Abstract. Enhanced fluctuations and correlations have been observed in the phase transitions of many systems. Their appearance at the predicted QCD phase transition (especially near the expected critical point) may provide insight into the nature of the phase transition. Recent results from the STAR experiment will be presented with a focus on particle ratio (K/π and p/π) fluctuations, forward-backward multiplicity correlations, and balance functions. Also discussed will be possibilities for measuring these correlations and fluctuations in the recently initiated beam energy scan at RHIC.

1. Introduction

Fluctuations and correlations are well known signatures of phase transitions. Specifically, the quark/gluon to hadronic phase transition may lead to significant fluctuations [1]. As part of an ongoing beam energy scan at the Relativistic Heavy Ion Collider (RHIC), the search for the QCD critical point will make use of the study of correlations and fluctuations, particularly those that could be enhanced during a phase transition that passes close to the critical point. The STAR experiment has studied a number of correlation and fluctuation observables as a function of energy, system-size, and centrality. Results from the STAR experiment on particle ratio fluctuations (K/π and p/π), forward-backward multiplicity correlations, balance functions, and net charge fluctuations are discussed.

2. Measurements

2.1. Particle Ratio Fluctuations

Dynamic particle ratio fluctuations, specifically fluctuations in the K/π and p/π ratio, can provide information on the quark-gluon to hadron phase transition [2, 3, 4]. These can be measured using the variable ν_{dyn}, originally introduced to study net charge fluctuations [5, 6]. ν_{dyn} quantifies deviations in the particle ratios from those expected for an ideal statistical Poissonian distribution. The definition of $\nu_{dyn,K/\pi}$ (describing fluctuations in the K/π ratio) is,

$$\nu_{dyn,K/\pi} = \frac{< N_K(N_K-1) >}{< N_K >^2} + \frac{< N_\pi(N_\pi-1) >}{< N_\pi >^2} - 2\frac{< N_K N_\pi >}{< N_K >< N_\pi >} \tag{1}$$

where N_K and N_π are the number of kaons and pions in a particular event, respectively. In this proceeding, N_K and N_π are the total charged multiplicity for each particle species. A

Journal of Physics: Conference Series **230** (2010) 012025 doi:10.1088/1742-6596/230/1/012025

formula similar to (1) can be constructed for p/π fluctuations. $\nu_{dyn} = 0$ for the case of a Poisson distribution of kaons and pions and is largely independent of detector acceptance and efficiency in the region of phase space being considered [6]. An in-depth study of K/π fluctuations in Au+Au collisions at $\sqrt{s_{NN}} = 200$ and 62.4 GeV was carried out by the STAR experiment [7].

Previous measurements of particle ratio fluctuations utilized the variable σ_{dyn} [8],

$$\sigma_{dyn} = sgn(\sigma^2_{data} - \sigma^2_{mixed})\sqrt{|\sigma^2_{data} - \sigma^2_{mixed}|} \tag{2}$$

where σ is the reduced width of the K/π or p/π distribution in either real data or mixed events. It has been shown that ν_{dyn} and σ_{dyn} are related as,

$$\sigma^2_{dyn} \approx \nu_{dyn} \tag{3}$$

The excitation function for K/π fluctuations expressed as σ_{dyn} is shown in Figure 1(a). To convert ν_{dyn} to σ_{dyn}, the relationship in (3) was used. Errors were also propagated from ν_{dyn} to σ_{dyn}. There is a strong decrease with increasing incident energy for the NA49, Pb+Pb 0-3.5% results (solid blue squares). The STAR results (solid red stars) for 0-5% Au+Au at $\sqrt{s_{NN}} =$ 19.6, 62.4, 130, and 200 GeV, and 0-10% Cu+Cu at $\sqrt{s_{NN}} = 22.4$ GeV all show the same value of σ_{dyn} within errors. σ_{dyn} is approximately flat above $\sqrt{s_{NN}} = 10$ GeV for all systems. Also shown are model predictions for UrQMD processed for the NA49 (open blue squares) and STAR (open red stars) experimental acceptances, respectively. A third model prediction, Hadron String Dynamics (HSD) [9] processed using the STAR experimental acceptance (open black triangles) is also plotted.

Figure 1(b) shows the excitation function for p/π fluctuations expressed using the variable σ_{dyn}. The general trend demonstrated by the data is smaller negative fluctuations with increasing incident energy for both NA49, Pb+Pb 0-3.5% results (solid blue squares) and STAR results (solid red stars) for 0-5% Au+Au collisions. The outlier data point is from 0-10% Cu+Cu collisions at $\sqrt{s_{NN}} = 22.4$ GeV, which is much more strongly negative. Both UrQMD with the STAR experimental acceptance (open red stars), and HSD with the NA49 and STAR experimental acceptances (open diamonds and triangles, respectively, reproduce most of the low energy data and over predict the high energy data points.

2.2. Forward-Backward Multiplicity Correlations

Forward-backward multiplicity correlations have been studied in elementary particle collisions [10, 11, 12, 13, 14, 15, 16], and more recently, from the STAR experiment in heavy ion collisions at RHIC [17]. FB correlations are measured as a function of pseudorapidity (η). Because they probe the longitudinal characteristics of the system produced in heavy ion collisions, they provide unique insight into the space-time dynamics and the earliest stages of particle production. A relationship between the multiplicity in the forward and backward hemisphere was found in hadron-hadron collisions [18],

$$< N_b(N_f) >= a + bN_f \tag{4}$$

where the coefficient b is referred to as the correlation strength. Eq. 4 can be expressed in terms of the expectation values,

$$b = \frac{< N_f N_b > - < N_f >< N_b >}{< N_f^2 > - < N_f >^2} = \frac{D^2_{bf}}{D^2_{ff}} \tag{5}$$

where D^2_{bf} and D^2_{ff} are the backward-forward and forward-forward dispersions, respectively. All quantities in Eq. (5) are measurable experimentally. The presence of long-range FB correlations has been predicted in some particle production models, specifically the Dual Parton

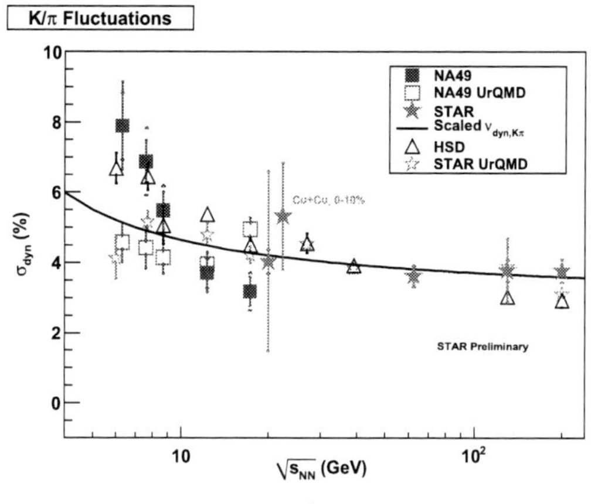

(a)

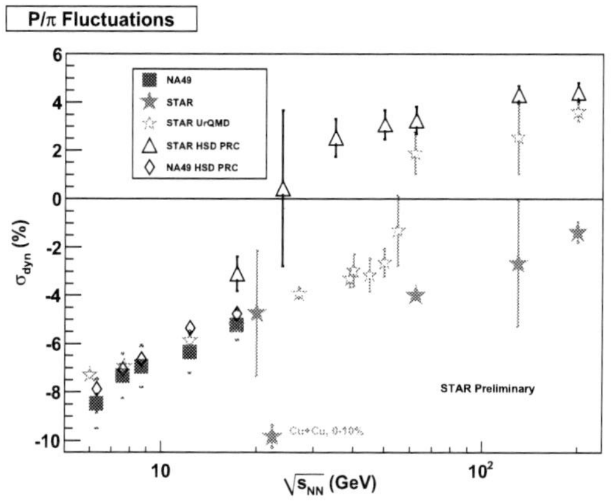

(b)

Figure 1. K/π (left) and p/π (right) fluctuations expressed as σ_{dyn}, as a function of incident energy. Data is shown from both the NA49 (solid blue squares) and STAR experiments (solid red stars) from central collisions, Pb+Pb, 0-3.5% for NA49 and Au+Au, 0-5% for STAR, except for the Cu+Cu point at $\sqrt{s_{NN}}$ = 22.4 GeV (0-10%). Also shown are model calculations from UrQMD using the NA49 (open blue squares) and STAR (open red stars) experimental acceptances and HSD (open black triangles) with the STAR and NA49 (open black diamonds) acceptance. Models labeled "STAR" are Au+Au collisions, while models labeled "NA49" are Pb+Pb. The solid black line in Figure 1(a) shows the relationship between ν_{dyn} at high energies converted to σ_{dyn} and extrapolated to lower energies. Errors include both statistical and systematic effects, except for the STAR point in Cu+Cu at $\sqrt{s_{NN}}$ = 22.4 GeV, which has only statistical errors.

Model (DPM) and the Color Glass Condensate/Glasma (CGC) model [19, 20]. Both models contain a form of longitudinal color flux tubes: color strings in the DPM and Glasma flux tubes in the CGC [21]. These longitudinal structures are predicted to be the origin of the long-range correlation.

Figure 2 shows the forward-backward (FB) correlation strength (b) as a function of centrality from $\sqrt{s_{NN}}$ = 200 GeV Au+Au collisions. From most central (0-10%) to peripheral (50-80%) collisions there is a general decrease in the FB correlation strength. The shape of the FB correlation strength for the 50-80% bin as a function of $\Delta\eta$ indicates the presence of mostly

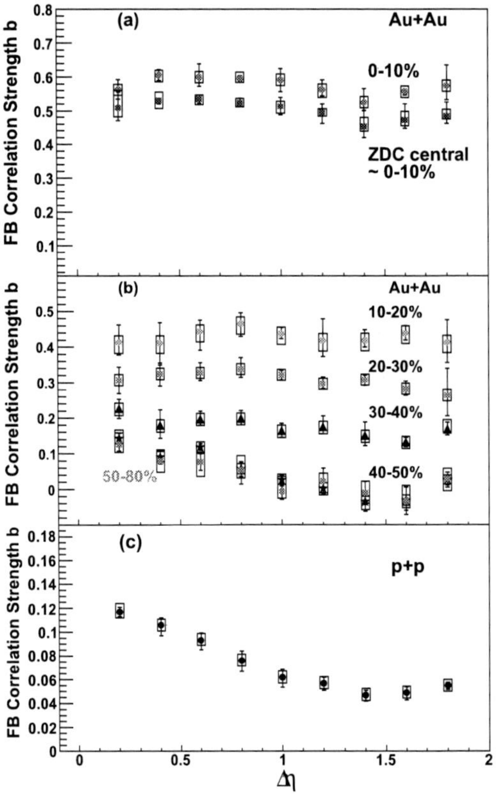

Figure 2. The centrality dependence of the FB correlation strength for Au+Au at $\sqrt{s_{NN}} =$ 200 GeV for central 0-10%, 10-20%, 20-30%, 30-40%, 40-50%, and peripheral 50-80%. The long-range correlations are seen in the most central Au+Au data, but behave similar to that of only short-range correlations beginning at 30-40% most central data. The correlation strength in the most peripheral Au+Au collisions resembles that of *pp*.

short-range correlations, which are expected to decrease exponentially with increasing η gap. If only short-range correlations contributed to the correlation, all centralities would resemble the 50-80% result. The presence of a strong correlation that grows with centrality and is flat beyond $\Delta\eta > 1.0$ is indicative of a long-range correlation produced by multiparton interactions in central $\sqrt{s_{NN}} = 200$ GeV Au+Au collisions.

2.3. Balance Function

The charge balance function has been predicted to be sensitive to the creation of a deconfined system of quarks and gluons by indicating a delay in the time to hadronization [22]. If hadrons are produced locally in pairs of oppositely charged particles, those produced early in the collision will experience more interactions (such as scattering and/or diffusion). Those pairs produced later will experience less of these effects. The width of the balance function is controlled by this, as well as the relative momentum between the produced pair, which is dictated by the thermal properties of the system. This leads to a prediction of a wide balance function for early production of hadrons and a narrow balance function for delayed hadronization. Delayed

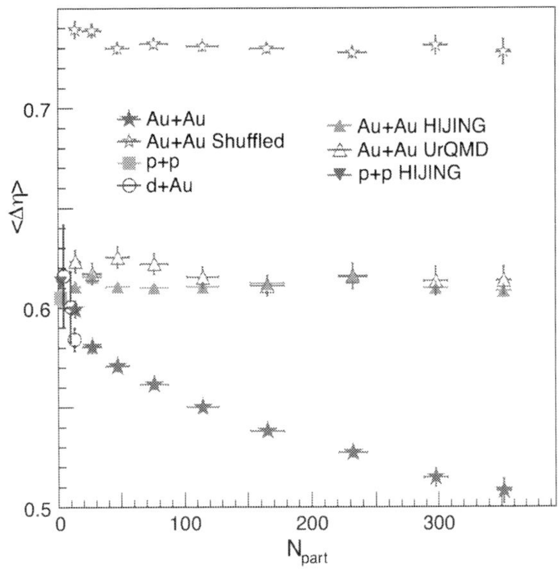

Figure 3. The centrality dependence of the width of the inclusive charged particle balance function ($< \Delta\eta >$) in $\sqrt{s_{NN}} = 200$ GeV Au+Au collisions, compared to model predictions and to pp and d+Au collisions at the same energy. The narrowing of the balance function width in central Au+Au collisions may be an indication of delayed hadronization.

hadronization can result from a deconfined system of quarks and gluons that prevents the production of hadrons for some amount of time. The balance function can be expressesd as,

$$B(\Delta\eta) = \frac{1}{2}\left\{\frac{N_{+-}(\Delta\eta) - N_{++}(\Delta\eta)}{N_{+}} + \frac{N_{-+}(\Delta\eta) - N_{--}(\Delta\eta)}{N_{-}}\right\} \tag{6}$$

where N_{xy} (x,y = + or -) is the number of charged particle pairs in the pseudorapidity range, $\Delta\eta = \eta_2 - \eta 1$. The terms in Equation 6 are calculated from pairs on an event-by-event basis and the results are summed over all events. The width is quantified using the weighted average, $< \Delta\eta >$. Figure 3 shows the width of the balance function in $\Delta\eta$ for inclusive charged particles, for pp, d+Au, and Au+Au collisions at $\sqrt{s_{NN}} = 200$ GeV. For Au+Au and d+Au, it is shown as a function of centrality (N_{part}) and for pp only minimum bias data is shown. The Au+Au data is compared to two models: HIJING and URQMD. Also shown are the results from shuffled Au+Au events. Shuffled events remove the charge correlation among particles in that event. The result from event shuffling demonstrates the widest width of the balance function that is possible inside the finite detector acceptance. Figure 3 shows that the width of the balance function peripheral Au+Au collisions at $\sqrt{s_{NN}} = 200$ GeV is in good agreement with the width in pp and d+Au collisions at the same energy. The width of the balance function decreases as a function of increasing centrality in Au+Au collisions, and is narrowest in the most central (top 0-5%) collisions. This effect is not reproduced by the models, which demonstrate a constant balance function width as a function of centrality. . This is consistent with the idea of delayed hadronization in central $\sqrt{s_{NN}} = 200$ GeV Au+Au collisions due to the formation of a partonic system.

3. Outlook

Beginning in 2010 RHIC embarked on a Beam Energy Scan (BES) program, motivated primarily by the possibility of finding evidence of a QCD critical point, as shown by lattice QCD predictions

[23, 24, 25]. The current program will be split over two years and cover a range in energies from $\sqrt{s_{NN}}$ = 200 GeV down to as low as $\sqrt{s_{NN}}$ = 5 GeV. The STAR detector is currently in a configuration with a minimum amount of material between the beam pipe and the main detector, the Time Projection Chamber (TPC). With large acceptance in η and ϕ and a fully installed Time-of-Flight detector, STAR is fully prepared to study fluctuations and correlations at all energies, while searching for signatures of the QCD critical point.

4. Summary

Several different fluctuation and correlation measurements from the STAR experiment have been discussed. Large changes in event-by-event fluctuation and correlation observables are expected in the hadron to quark-gluon phase transition, and also near the critical point. STAR continues to search for rapid changes of these observables as a function of incident collision energy and centrality. In the next two years, STAR will have large statistics datasets, acquired data across a range in energies from $\sqrt{s_{NN}}$ = 200 to $\sqrt{s_{NN}}$ = 5 GeV. This spans a broad area of the QCD phase diagram, including that of a predicted QCD critical point, though experimental limitations limit study to the one particular value of μ_B for each chosen collision energy. The study of fluctuations and correlations (including those not discussed here), are expected to be primary signatures of the proposed QCD critical point accessible at RHIC.

References

[1] V. Koch. *arXiv:00810.2520 [nucl-th]*, (2008).
[2] M.I. Gorenstein, M. Ga?dzicki, and O.S. Zozulya. *Phys. Lett.*, B585:237, (2004).
[3] V. Koch, A. Majumder, and J. Randrup. *Phys. Rev.*, C72:064903, (2005).
[4] V. Koch, A. Majumder, and J. Randrup. *Phys. Rev. Lett.*, 95:182301, (2005).
[5] C. Pruneau, S. Gavin, and S. Voloshin. *Phys. Rev.*, C66:044904, (2002).
[6] J. Adams et al. (STAR Collaboration). *Phys. Rev.*, C68:044905, (2003).
[7] B.I. Abelev et al. (STAR Collaboration). *Phys. Rev. Lett.*, 103:092301, (2009).
[8] C Alt et al. (NA49 Collaboration). *arXiv:0808.1237 [nucl-ex]*, (2008).
[9] M.I. Gorenstein, M. Hauer, V.P. Konchakovski, and E.L. Bratkovskaya. *arXiv:0811.3089 [nucl-th]*, (2008).
[10] W. Braunschweig et al. (TASSO Collaboration). *Z. Phys.*, C45:193, (1989).
[11] P. Abreu et al. (DELPHI Collaboration). *Z. Phys.*, C50:185, (1991).
[12] R. Akers. *Phys. Lett.*, B320:417, (1994).
[13] C.M. Bromberg et al. *Phys. Rev.*, D9:1864, (1974).
[14] S. Uhlig et al. *Nucl. Phys.*, B132:15, (1978).
[15] R.E. Ansorge et al. (UA5 Collaboration). *Z. Phys.*, C37:191, (1988).
[16] T. Alexopoulos et al. *Phys. Lett.*, B353:155, (1995).
[17] B.I. Abelev et al. (STAR Collaboration). *Phys. Rev. Lett.*, 103:172301, (2009).
[18] G.J. Alner et al. *Phys. Rept.*, 154:247, (1987).
[19] A. Capella, U. Sukhatme, C.-I. Tan, and J. Tran Thanh van. *Phys. Report*, 236:225, (1994).
[20] Lappi T. and L. McLerran. *Nucl. Phys.*, A772:200, (2006).
[21] N. Armesto, L. McLerran, and Pajares C. *Nucl. Phys.*, A781:201, (2007).
[22] Steffen A. Bass, Pawel Danielewicz, and Scott Pratt. *Phys. Rev. Lett.*, 85:2689, (2000).
[23] F. Karsch et al. *Nucl. Phys. Proc. Suppl.*, 129:614, (2004).
[24] Z. Fodor and S.D. Katz. *JHEP 04*, page 050, (2004).
[25] R.V. Gavai and S. Gupta. *Phys. Rev.*, D71:114014, (2005).

Journal of Physics: Conference Series **230** (2010) 012026 doi:10.1088/1742-6596/230/1/012026

Collective Effects in Proton Proton and Heavy Ion Scattering, and the "Ridge" at RHIC

K. Werner[(a)], **Iu. Karpenko**[(a,b)], **T. Pierog**[(c)]

[(a)] SUBATECH, University of Nantes – IN2P3/CNRS– EMN, Nantes, France
[(b)] Bogolyubov Institute for Theoretical Physics, Kiev 143, 03680, Ukraine
[(c)] Forschungszentrum Karlsruhe, Institut fuer Kernphysik, Karlsruhe, Germany

Abstract. We discuss why the hydrodynamic description should not only be applicable to heavy ion collisions at RHIC, but also for proton proton scattering at the LHC. A new realistic treatment of the hydrodynamic evolution is tested by comparing to AuAu data at RHIC, in particular concerning the "ridge" phenomenon. We apply our new approach to obtain first results for pp at LHC energies.

1. Introduction

Let us consider the energy density at an early time in a Au-Au scattering at RHIC, as obtained from an EPOS simulation [1]. In fig. 1, we plot the energy density at different values of space-time rapidity η_s, as a function of the transverse coordinates x and y . We observe a very bumpy structure concerning the $x-y-$dependence, whereas the variation with η_s is small. There are in particular peaks in the $x-y-$plane, which show up at the same position at different values of η_s So we have sub-flux-tubes which exhibit a long range structure in the longitudinal variable η_s.

In fig. 1 , we clearly identify several sub-flux-tubes, with a typical width of the order of a fermi. This is exactly the width we obtain if we compute the initial energy density in proton scattering at the LHC. This means, if a hydrodynamic treatment is justified for Au-Au collisions at RHIC, it is equally justified for pp scattering at the LHC, provided the energy densities are high enough. This latter condition can easily be satisfied, since in proton-proton scattering on has the possibility to trigger on high multiplicity events, with ten or twenty times the multiplicity compared to an average event.

We are going to employ a new tool for treating the hydrodynamic evolution, based on the following features (see [1]):

- initial conditions obtained from a flux tube approach (EPOS), compatible with the string model used since many years for elementary collisions (electron-positron, proton proton), and the color glass condensate picture;
- consideration of the possibility to have a (moderate) initial collective transverse flow;
- event-by-event procedure, taking into the account the highly irregular space structure of single events, being experimentally visible via so-called ridge structures in two-particle correlations;
- core-corona separation, considering the fact that only a part of the matter thermalizes;

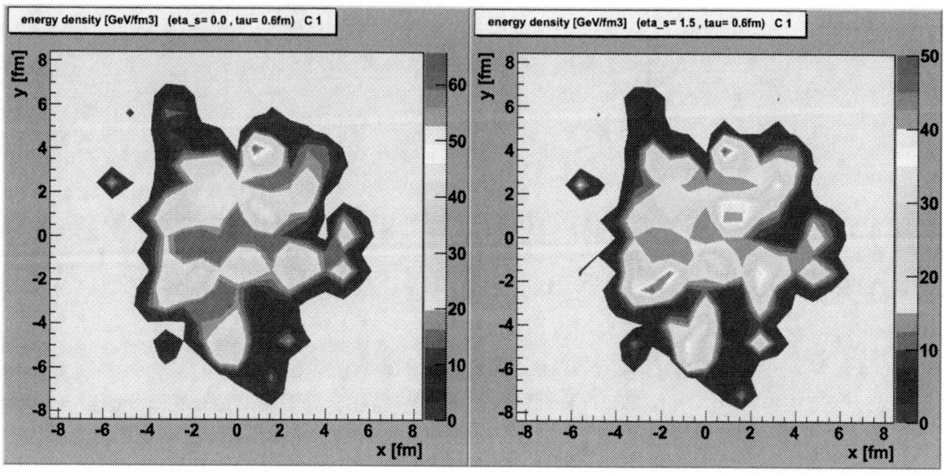

Figure 1. Energy density in central Au-Au scattering at 200 GeV.

- use of an efficient code for solving the hydrodynamic equations in 3+1 dimensions, including the conservation of baryon number, strangeness, and electric charge;
- employment of a realistic equation-of-state, compatible with lattice gauge results – with a cross-over transition from the hadronic to the plasma phase;
- use of a complete hadron resonance table, making our calculations compatible with the results from statistical models;
- hadronic cascade procedure after hadronization from the thermal system at an early stage.

In [1], we test the approach by investigating all soft observables of heavy ion physics, in case of AuAu scattering at 200 GeV. Here, we are going to discuss some selected (and interesting) topics.

2. Test case: the "ridge" at RHIC

A remarkable feature of an event-by-event treatment of the hydrodynamical evolution based on random flux tube initial conditions is the appearance of a so-called ridge-structure, found in Spherio calculations based on Nexus initial conditions [2, 3]. We expect to observe a similar structure doing an event-by-event hydrodynamical evolution based on flux-tube initial conditions from EPOS. The result is shown in fig. 2, where we plot the dihadron correlation $dN/d\Delta\eta\, d\Delta\phi$, with $\Delta\eta$ and $\Delta\phi$ being respectively the difference in pseudorapidity and azimuthal angle of a pair of particles. Here, we consider trigger particles with transverse momenta between 3 and 4 GeV/c, and associated particles with transverse momenta between 2 GeV/c and the p_t of the trigger, in central Au-Au collisions at 200 GeV. Our ridge is very similar to the structure observed by the STAR collaboration [4].

In the following we will discuss a particular event, which can, however, be considered as a typical example, with similar observations being true for randomly chosen events. Important for understanding the strong $\Delta\eta - \Delta\phi$ correlation is the observation, that the initial energy density has a very bumpy structure as a function of the transverse coordinates x and y. However, this irregular structure is the same at different longitudinal positions. This can be clearly seen in fig. 1, where we show for a given event the energy density distributions in the transverse planes at different space-time rapidities, namely $\eta_s = 0$ and $\eta_s = 1.5$: we observe almost the same structure. For different events, the details of the bumpy structures change, but we always find an approximate "translation invariance": the distributions of energy density in the

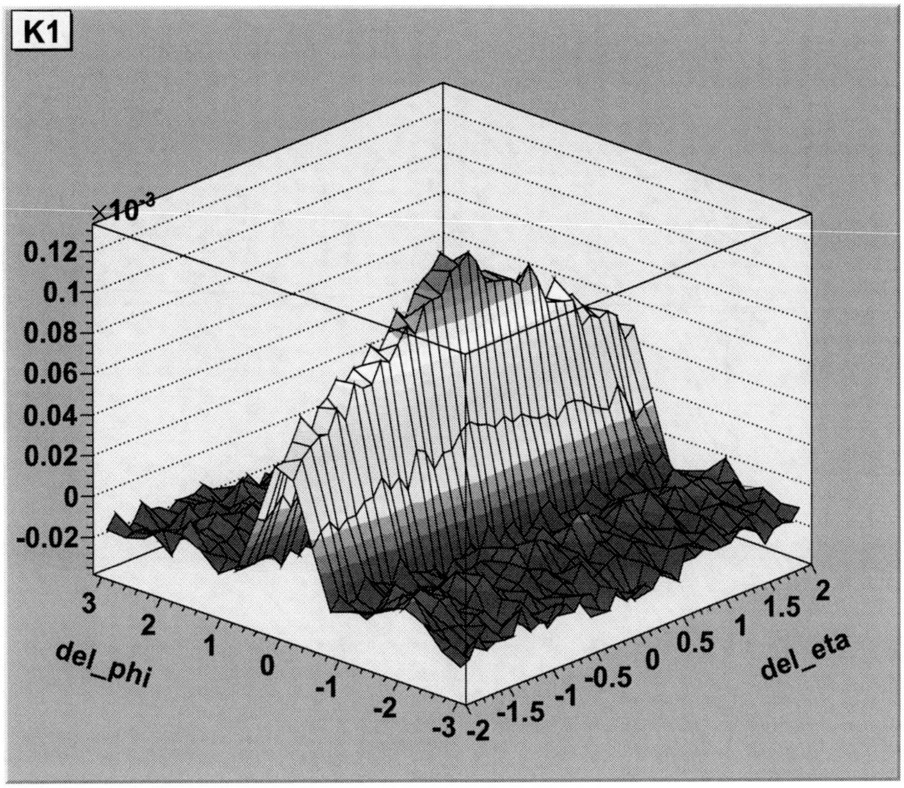

Figure 2. Dihadron $\Delta\eta - \Delta\phi$ correlation in a central Au-Au collision at 200 GeV, as obtained from an event-by-event treatment of the hydrodynamical evolution based on random flux tube initial conditions. Trigger particles have transverse momenta between 3 and 4 GeV/c, and associated particles have transverse momenta between 2 GeV/c and the p_t of the trigger.

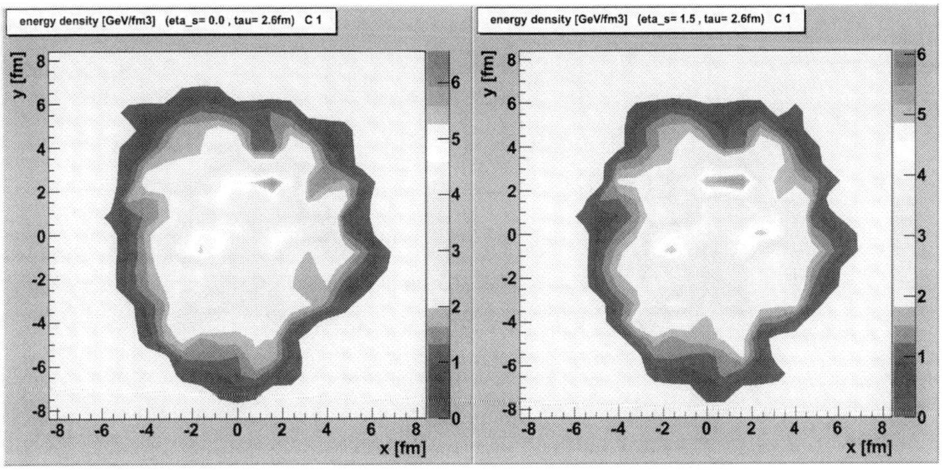

Figure 3. Energy density at a proper time $\tau = 2.6\,\mathrm{fm/c}$, at a space-time rapidities $\eta_s = 0$ and $\eta_s = 1.5$.

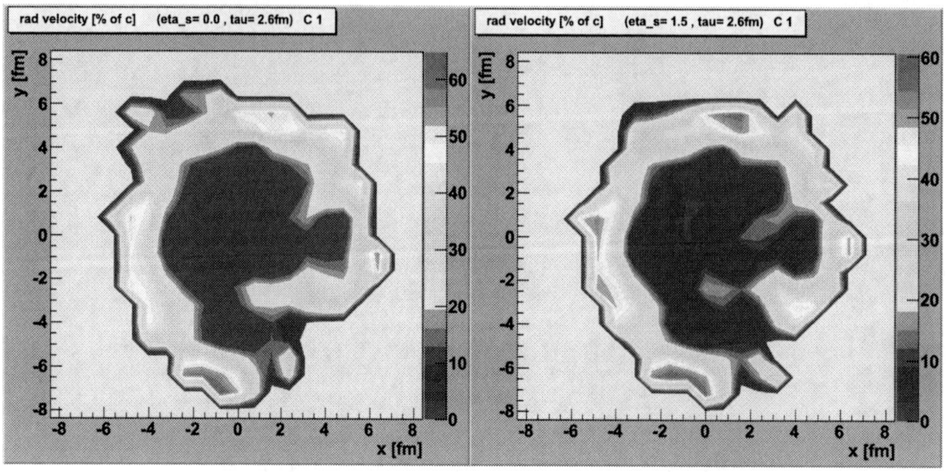

Figure 4. Radial flow velocity at a proper time $\tau = 2.6\,\mathrm{fm/c}$, at a space-time rapidities $\eta_s = 0$ and $\eta_s = 1.5$.

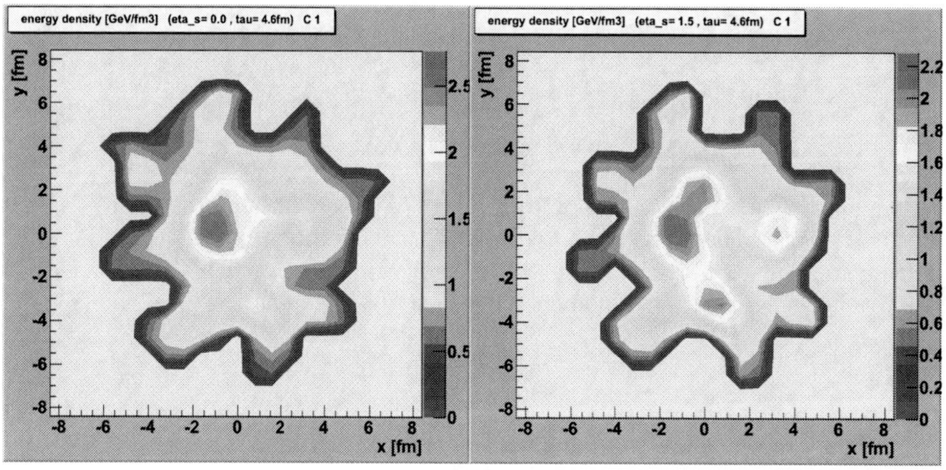

Figure 5. Energy density at a proper time $\tau = 4.6\,\mathrm{fm/c}$, at a space-time rapidities $\eta_s = 0$ and $\eta_s = 1.5$.

transverse planes vary only little with the longitudinal variable η_s. It should be noted that the colored areas represent only the interior of the hadronization surface, the outside regions are white. Hadronization is meant to be an intermediate step, before the hadronic cascade. An approximate translational invariance is also observed when we go to larger values of η_s, so for example when we compare the energy density at $\eta_s = 1.5$ with the one at $\eta_s = 3.0$: the form of the energy distributions is similar, however, the magnitude at large η_s is smaller.

Considering later times, we see in figs. 3 and 4, that the approximate translational invariance is conserved, for both energy densities and radial flow velocities. It is remarkable (and again true in general, for arbitrary events) that the energy distribution in the transverse plane is much smoother than initially, the distribution looks more homogeneous. Very important for the following discussion is the flow pattern, seen in fig. 4, for $\eta_s = 0$ and $\eta_s = 1.5$: the radial flow is as expected largest in the outer regions. Closer inspection of the outside ring of large radial flows reveals an irregular atoll-like structure: there are well pronounced peaks of large flow over the

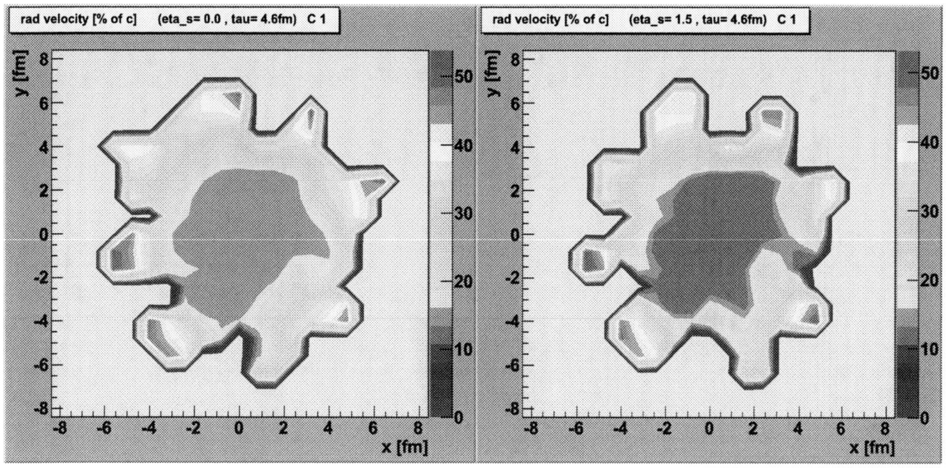

Figure 6. Radial flow velocity at a proper time $\tau = 4.6\,\mathrm{fm/c}$, at a space-time rapidities $\eta_s = 0$ and $\eta_s = 1.5$.

background ring. At even later times, as seen in figs. 5 and 6, the outer surfaces get irregular, due to the irregular flows discussed above, again with well identified peaks of large radial flows.

The well isolated peaks of the radial flow velocities have two important properties: they sit close to the hadronization surface, and they sit at the same azimuthal angle, when comparing different longitudinal positions η_s. As a consequence, particles emitted from different longitudinal positions get the same transverse boost , when their emission points correspond to the azimuthal angle of a common flow peak position. And since longitudinal coordinate and (pseudo)rapidity are correlated, one obtains finally a strong $\Delta\eta - \Delta\phi$ correlation.

To summarize the above discussion: the flux tube initial conditions provide a bumpy structure of the energy density in the transverse plane, which shows, however, an approximate translational invariance (similar behavior at different longitudinal coordinates). Solving the hydrodynamic equations preserves this invariance, leading in the further evolution to an invariance of the transverse flow velocities. These identical flow patterns at different longitudinal positions lead to the fact that particles produced at different values of η_s profit from the same collective push, when they are emitted at an azimuthal angle corresponding to a flow maximum.

Finally we have to address the question, why we have a irregular transverse structure with an approximate translational invariance. The basic structure of EPOS is such that each individual nucleon-nucleon collision results in a projectile and target remnant, and two or more elementary flux tubes (strings). The higher the energy the bigger the number of strings. Most of the energy of the reaction is carried by the remnants, the flux tubes cover only a limited range in rapidity, but their "lengths" (in rapidity) vary enormously. Nevertheless we obtain a very smooth variation of the energy density with the longitudinal coordinate η_s. This is due to the fact that the transverse positions of a string is given by the position of the nucleon pair, who's interaction gave rise the the formation of the flux tube. These "pair positions" fluctuate considerably, event-by-event, and one obtains typically a situation where there are areas with a high density of interaction points, and areas which are less populated. These transverse positions of interacting pairs define also the corresponding positions of the flux tubes associated to the pairs. The flux tubes have variable longitudinal lengths, they do not cover the full possible length between projectile and target, but only a portion. But even then, the transverse structure (minima and maxima of the energy density) is to a large extent determined by the density of nucleon-nucleon pairs.

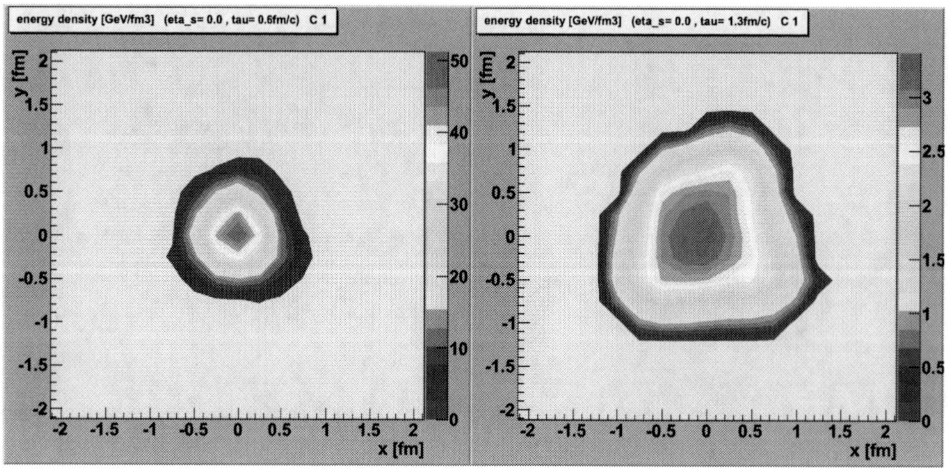

Figure 7. Energy density at a proper times $\tau = 0.6\,\mathrm{fm/c}$ and $\tau = 1.3\,\mathrm{fm/c}$, at a space-time rapidity $\eta_s = 0$, in a "central" pp scattering.

We just discussed one interesting observable, namely $\Delta\eta$–$\Delta\phi$ correlations. Many more exist, like azimuthal asymmetries, particle spectra, and yields. All these have been carefully studied to thoroughly test our approach.

3. Hydrodynamic evolution in proton proton scattering

Having tested our tool against heavy ion data at RHIC, we now proceed to proton-proton scattering at the LHC. This is work in progress, so for the moment we only show some very preliminary results concerning the hydrodynamic evolution in "central" pp scattering at 7 TeV. The word "central" has here a somewhat different meaning than in heavy ion scattering. The impact parameter plays also a role in proton proton scattering, but collisions with large b lead to low multiplicity events. All interesting cases correspond to small impact parameters, because only then we have a considerable probability of a large number of elementary scatterings, corresponding to a high multiplicity of produced particles. Only these events exhibit large energy densities, and allow us to observe similar collective effects as in heavy ion collisions at RHIC.

In the following, we investigate proton proton collisions at 7 TeV, considering a case with eight elementary scatterings. We show a single event, however, a typical one for this "centrality" class. In fig. 7, we plot the energy density at $\tau = 0.6\,\mathrm{fm/c}$ and at $\tau = 1.3\,\mathrm{fm/c}$, at a space-time rapidity $\eta_s = 0$. Indeed the initial energy density is high – up to 50 $\mathrm{GeV/fm^3}$, comparable to what one reaches in central AuAu collisions at RHIC. The expansion is fast, at $\tau = 1.3\,\mathrm{fm/c}$ the maximum energy density has already dropped to about 3 $\mathrm{GeV/fm^3}$. Remarkable is the very large collective flow which develops very quickly. At $\tau = 1.3\,\mathrm{fm/c}$ we observe already a maximum of 80% of the velocity of light, at (relatively) large distances from the center, see fig. 8. Having these large radial flows close to the hadronization surface, we expect big effect concerning particle spectra.

As already said, this is work in progress, full calculations with comparison to LHC data will come soon.

Acknowledgments
This research has been carried out within the scope of the ERG (GDRE) "Heavy ions at ultra-relativistic energies", a European Research Group comprising IN2P3/CNRS, Ecole des Mines de

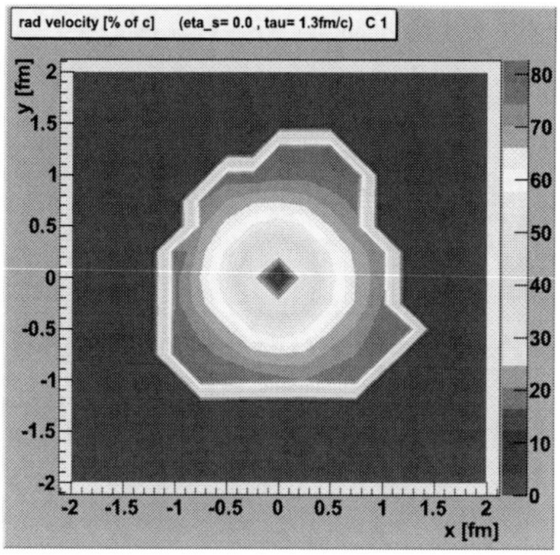

Figure 8. Radial flow velocity at a proper time $\tau = 1.3\,\mathrm{fm/c}$, at a space-time rapidity $\eta_s = 0$, in a "central" pp scattering.

Nantes, Universite de Nantes, Warsaw University of Technology, JINR Dubna, ITEP Moscow, and Bogolyubov Institute for Theoretical Physics NAS of Ukraine. Iu. K. acknowledges partial support by the MESU of Ukraine, and Fundamental Research State Fund of Ukraine, agreement No F33/461-2009. Iu.K. and K.W. acknowledge partial support by the Ukrainian-French grant "DNIPRO", an agreement with MESU of Ukraine No M/4-2009. T.P. and K.W. acknowledge partial support by a PICS (CNRS) with KIT (Karlsruhe).

[1] K. Werner, Iu. Karpenko, T. Pierog, M. Bleicher, K. Mikhailov, arXiv:1004.0805, submitted to Phys. Rev. C
[2] J. Takahashi et al. Phys. Rev.Lett. 103 (2009) 242301
[3] R.P.G. Andrade, F. Grassi, Y. Hama, W.L. Qian, arXiv:0912.0703
[4] STAR Collaboration, B. Abelev et al., Phys. Rev. C 80 (2009) 64192

Kinetic Theory Based Modeling of Type II Core Collapse Supernovae

Terrance Strother
Department of Physics and Astronomy and National Superconducting Cyclotron Laboratory, Michigan State University East Lansing, MI 48842, USA
E-mail: strothe6@pa.msu.edu

Wolfgang Bauer
Department of Physics and Astronomy and National Superconducting Cyclotron Laboratory, Michigan State University East Lansing, MI 48842, USA
E-mail: bauer@pa.msu.edu

Abstract. Motivated by the success of kinetic theory in the description of observables in intermediate and high energy heavy ion collisions, we use kinetic theory to model the dynamics of core collapse supernovae. The specific way that we employ kinetic theory to solve the relevant transport equations allows us to explicitly model the propagation of neutrinos and a full ensemble of nuclei and treat neutrino-matter interactions in a very general way. With these abilities, our preliminary calculations have observed dynamics that may prove to be an entirely new neutrino capture induced supernova explosion mechanism.

1. Introduction

Hydrodynamic models have been the tool of choice for supernova modeling for many decades. Most insights the field of supernova science have been made by hydrodynamic based calculations, however a great deal more must be done before these simulations can be regarded as complete. A truly complete hydrodynamic simulation would in principle have to model the dynamics of multiple fluids with strongly time dependent viscosities to model the propagation of a full ensemble of nuclei and neutrinos in the full six-dimensional phase space in a completely general way. No hydrodynamic simulation is currently capable of this. Since state-of-the-art hydrodynamic calculations already strain the capabilities of high performance supercomputers, there may be a long wait before more complex models that simulate the dynamics of hundreds of different species of nuclei and neutrinos in a general way in the full six-dimensional phase space. This motivates us to move away from the traditional hydrodynamic approach that we are familiar with and draw from other disciplines of physics in an attempt to circumvent this technological roadblock. It turns out that the field nuclear collision modeling is an ideal candidate for this purpose.

Intermediate and high energy nuclear collisions have been very accurately modeled by simulations that made use of transport theories based on a semiclassical implementation of kinetic theory [1, 2, 3]. Given the similarities between the requirements that must be satisfied by simulations of nuclear collisions and supernovae [4, 5, 6], such as the ability to model particle production, shock wave formation, collective deflection, as well as the interplay between regular and chaotic collective dynamics, it is tempting to implement these types of kinetic theory based approaches to model the physics and astrophysics of supernova explosions. This is the aim of our work.

As discussed in our previous work [7], the full potential of our code will be realized when it is run large multiprocessor installations. Once it is fully parallelized, it will be capable of efficiently calculating all desired statistical distributions in the full three-dimensional coordinate space while propagating test particles in the full six-dimensional phase space. However, for testing purposes, we also want to provide ways to implement our ideas on a single processor. The preliminary calculations discussed here were performed exactly for that purpose and were simulations of the collapse and early stages of the explosion of a non-rotating spherically symmetric core [8]. While test particles were still propagated in the full six-dimensional phase space, due to statistical limitations imposed by working on a single processor [7], we assumed that all statistical distributions were spherically symmetric. The algorithms employed to calculate the spherically symmetric statistical distributions are discussed in detail in our previous work [7]. Furthermore, our weak nuclear reaction network was only partially activated during these preliminary simulations, and no strong nuclear reactions (i.e. fission and fusion channels) were modeled. Since our approach to modeling these reactions is entirely new, we have elected to implement and test them in stages. The remaineder of the weak and strong nuclear reaction networks will be activated and tested after the ongoing parallelization process is complete.

The output generated by this series of single processor test calculations is well understood and consistent with the physics included in the model thus far, however not all of the results were expected. The emergence of what may prove to be a new supernova explosion mechanism was observed in all test calculations. This potential new explosion mechanism is discussed in sections 6 and 7.

2. Equations of Motion

The one-body transport equation for the baryon phase space density $f_b(xp)$ is given by [9]

$$\begin{aligned}\frac{\partial f_b(xp)}{\partial t} + \frac{\Pi^i}{E_b^*(p)}\nabla_i^x f_b(xp) &- \frac{\Pi^\mu}{E_b^*(p)}\nabla_i^x U_\mu(x)\nabla_p^i f_b(xp) + \frac{M_b^*}{E_b^*(p)}\nabla_i^x U_s \nabla_p^i f_b(xp) \\ &= I_{bb}^b(xp) + I_{b\nu}^b(xp) \qquad (1)\end{aligned}$$

for the particular state b of the baryon. The left hand side of (1) describes the temporal changes in the baryon phase space density due to interactions due to the interactions of the nucleons with the mean field vector and scalar potentials U_μ and U_s. The two source terms on the right hand side of (1) are collision integrals that that represent the effects that correlations due to the two-body baryon-baryon and baryon-neutrino collisions have the baryon phase space density respectively. For any neutrino species, the transport equation simplifies to an equation of motion that contains only the streaming and baryon-neutrino collision terms since there are no mean field contributions and the possible effects of neutrino-neutrino collisions are neglected,

$$\frac{\partial f_\nu(xk)}{\partial t} + \frac{k \cdot \nabla^x}{E_\nu(k)} f_\nu(xk) = I_{b\nu}^\nu(xk) \qquad (2)$$

We want to point our, however, that neutrino oscillations (vacuum and/or matter-induced) between different lepton flavors can be explicitly included. It is through the baryon-neutrino collisions that the baryon and neutrino phase space densities can affect one another and the transport equations are coupled by the source terms that represent these effects.

3. Test Particle Method

To numerically solve the coupled baryon and neutrino transport equations (1) and (2), the so-called test particle method is used [10]. Instead of fully discritizing the relevant six-dimensional phase space and calculating the phase space densities in each grid cell in every time step, the test particle method only follows the initially occupied phase space cells in time and represent them by a finite number of imaginary test particles. These imaginary test particles are propagated in a way that models the physical evolution of the phase space. They interact with one another via mean field one-body potentials and scatter with realistic cross sections.

3.1. Test Particle Method Applicability

The test particle approach was originally used to model individual intermediate and high energy nuclear collisions [10]. In that microscopic system, there were many more test particles than there were physical particles. If we wish to use the test particle approach to model a macroscopic system containing a very large number of physical particles, it is clearly impossible to have the number of test particles, N_{tp}, exceed the number of physical particles, N_{phys}. Computational limitations require in such cases that $N_{phys}/N_{tp} \gg 1$.

The test particle approach can still be applicable in cases where $N_{phys}/N_{tp} \gg 1$, so long as N_{tp} is sufficiently large to capture the gross dynamics of the macroscopic system's phase space. The ratio N_{phys}/N_{tp} effectively determines a scale cutoff of sorts below which details cannot be resolved. When N_{phys}/N_{tp} becomes sufficiently large, some truly microscopic phenomena become impossible to directly simulate with test particles. Therefore it must be established that these unresolvable details do not impact the gross phase space dynamics and/or can be taken into account indirectly. This can be accomplished with convergence tests. In cases where convergence tests fail, we will have to implement multi-scale methods. This can be the case, for example, in the presence of thermodynamic instabilities or strongly chaotic dynamics. However, we have not reached this point as of now.

3.2. Test Particle Method Formalism

The test particle method formally approximates the phase space density with a sum over delta functions

$$f(\vec{r},\vec{p},t) = \sum_{i=0}^{N_{tp}} \delta^3(\vec{r}-\vec{r}_i(t))\delta^3(\vec{p}-\vec{p}_i(t)) \qquad (3)$$

Insertion of this approximation of the phase space density into the transport equations (1) and (2) yields simple semi-classical first-order linear differential equations of motion for the centroid coordinates of each test particle. The fact that the baryon and neutrino transport equations are solved in an identical fashion means that the treatment of neutrino dynamics is on equal footing with that of baryons. This is a significant advantage that the test particle approach has over traditional hydrodynamic models.

4. Matter Test Particles

Matter test particles are used to propagate a full ensemble of nuclei and free baryons. It is through the propagation of an entire ensemble of nuclei and free baryons that our calculations are able to retain full knowledge of nuclear composition everywhere in the core at all times. This is superior to the standard tracking of the abundances of free protons, neutrons, alpha particles, and a "representative heavy nucleus" employed by most hydrodynamic calculations. Nuclear structure effects significantly impact electron and neutrino capture rates [11], and many of these effects can be missed if one speaks of a "representative heavy nucleus" instead of an ensemble of nuclei.

4.1. Matter Test Particle Properties
An exhaustive description of matter test particle properties can be found in our previous works [12, 13]. Here it suffices to say the following. Matter test particle explicitly represents a fixed number of nuclei of a given species and a variable numbers of free protons and neutrons. They also implicitly represent the number of electron required to render them charge neutral as well as electron-positron pairs. Each matter test particle has its own set of nuclear properties and temperature. Initially the nuclear properties and temperature of each matter test particle are determined by the initial conditions of the progenitor [8]. At later times, a matter test particle temperature can be changed by thermal mixing and both its nuclear properties and temperatures can be changed by weak and strong nuclear reactions.

4.2. Matter Test Particle Equations
After the insertion of the delta function approximation of the baryon phase space density into equation (1), it is found that the centroid of each matter test particle is subject to three mean field forces: a gravitational force, a nucleonic force, and a force generated by the pressure exerted by the local matter and radiation gas. Matter test particles can also scatter with one another. The equations of motion for the centroid coordinates of the matter test particles are given by the following first order differential equations

$$\begin{aligned} \frac{d}{dt}\vec{p}_j &= \vec{F}_{G,j} + \vec{F}_{nuc}(\vec{r}_j) + \vec{F}_{gas}(\vec{r}_j) + \vec{\mathcal{C}}(\vec{p}_j) \\ \frac{d}{dt}\vec{r}_j &= \frac{\vec{p}_j}{\sqrt{m^2 + p_j^2/c^2}} \qquad (4) \\ j &= 1, \ldots, N \end{aligned}$$

where $\vec{F}_{G,j}$ is the gravitational force acting on the jth matter test particle, $\vec{\mathcal{C}}(\vec{p}_j)$ symbolizes the effects that two-body collisions with other matter test particles have on the jth matter test particle's momentum, and N is the number of matter test particles used to model the core and is constant. The evaluations of the spherically symmetric forces $\vec{F}_{nuc}(\vec{r}_j)$, $\vec{F}_{gas}(\vec{r}_j)$, and $\vec{F}_{G,j}$ are discussed in out previous works [12, 14, 15] respectively. The way we model test particle scattering is explained at length in our previous works [13, 16]. Here we simply state that test particle scatterings are modeled relativistically and semi-classically in a way similar to those used in the simulation of heavy ion collisions [17].

5. Neutrino Test Particles
Unlike the net baryon number, the number of neutrino test particles is not constant. Neutrino test particles can be created and destroyed. The latter process can be induced by a weak interaction or by a neutrino test particle escaping the core. The number of neutrino test particles therefore varies greatly at different times during the simulation. The only neutrino production mechanism included in our preliminary test calculations is electron capture by nuclei and free protons. It is clear how to include other neutrino production mechanisms into our model as well as how to model the presence of all flavors of neutrinos and anti-neutrinos, but we have elected to proceed incrementally when constructing the weak reaction network for this completely new model. Subroutines that model other neutrino production process such as weak decay and pair production have been written, but they will not be integrated into the code until the ongoing parallelization process is complete.

All present model of neutrino flavor oscillations indicate that these are strongly suppressed in the core [13, 18], only the presence of electron neutrinos is simulated by our test calculations.

(However, there are indications [19] that baryon densities are sufficiently low ($\approx$1 kg/cm^3) just outside the iron core and may have a significant impact on the relative isotope abundance of heavy nuclei produced in the final stages of the supernova explosion.) Neutrino test particles are assumed to be massless, move at the speed of light, and subject to no mean-field-type forces. The only way they can interact with other test particles is through scattering with or being captured by matter test particles. Thus the propagation of neutrino test particles between weak reaction sites is quite simple. Merely multiplying the intermediate unit momentum vector of a neutrino test particle by the speed of light and propagation time determines its new location. No complicated numerical method of approximating the solutions to differential equations is required. This light speed propagation does put limits in on the time step size. To realistically model the propagation of neutrino test particles within the constraints of our coordinate space cells, the time step size should be no larger than 10^{-5} s.

5.1. Neutrino Test Particle Creation

Since we do not explicitly simulate the presence of electrons, we rely upon electron capture rate tables to model the production of test particles representing neutrinos produced via electron capture. Because our simulation models the propagation of a full ensemble of nuclei, electron capture rates are needed for many rare isotopes far from the valley of beta stability. Currently no table exists that can satisfy our input needs in this regard. Until more comprehensive electron capture rate tables become available, we must extrapolate existing tabulated rates to the drip lines. For preliminary test calculations, the widely available Fuller-Fowler-Newman (FFN) table [20] is used as the source for electron capture rates. We extrapolate the rates from the FFN table to the nuclei not included in their table that we include in our simulation, and then reduce all entries by an order of magnitude. Our motivation for doing this is as follows. More recent calculations of weak reaction rates using new shell models of the distribution of Gamow-Teller strength have resulted in an improved and often reduced estimate of its strength compared to those the FFN calculations yielded using extrapolations of the known experimental rates and a simple single-state representation of this resonance [21]. The difference is often an order of magnitude or more, so this table serves as a reasonable estimate of the rates.

5.2. Neutrino Test Particle Propagation

Beam attenuation arguments are ideally suited for calculating neutrino-matter interaction probabilities for neutrinos represented by a neutrino test particle. The probability that the neutrinos represented by a neutrino test particle that moves from location $\vec{x}_1$ to $\vec{x}_2$ is taken to be given by

$$P_{int} = 1 - \exp\left[-\int_{\vec{x}_1}^{\vec{x}_2} \sum_i \sigma_i(\vec{x}) n_i(\vec{x}) d\vec{x}\right] \tag{5}$$

where the sum over i runs over all of the interaction channels available to the neutrino test particle between $\vec{x}_1$ and $\vec{x}_2$, n_i is the local number density of the ith species of particle, electrons, free baryons, or nuclei, and σ_i is the average effective interaction cross section corresponding to the ith interaction channel. The n_i's are readily calculable and the σ_i's are interpolated from tables [22]. Once the interaction probability (5) is calculated, a simple Monte-Carlo algorithm determines if the neutrino test particle interacts with matter. Interaction channels are selected by constructing relative probabilities out of weighted average effective cross sections and using another simple Monte-Carlo algorithm to choose a channel. The significant appeal that this very general way of modeling neutrino-matter interactions is that it applicable everywhere in the core. This is a major advantage this kinematic model has over its hydrodynamic counterparts. Typical hydrodynamic treatments of neutrinos are sufficient only in the extremely short and long mean free path limits and are quite problematic at intermediate values [23, 24].

6. New Dynamics

All of our test simulations have observed new dynamics that may prove to be an entirely new neutrino capture induced supernova explosion mechanism. The early stages of the collapse calculated by our code unfold identically to the accepted picture of supernova collapse. Electron capture rates initially slowly increased as the collapse progressed and at later times rapidly increased and eventually led to rapid deleptonization of the inner core. Not long after the inner region of the core begins to rapidly deleptonize, an unexpected phenomenon is observed in each test calculation. It is always found that neutrino captures at intermediate radii inside the inner region of the core deposit a large amount of electrons in a narrow spherical shell centered about a radius of approximately 45-55 km containing neutron-rich matter with a density on the order of 0.1 nuclear matter density. This radially localized accumulation of electrons alters the electron number density gradient in such a way that substantially increases the outward pressure exerted by the electron gas in the region just outside the radius at which the most electrons are produced. The resultant electron gas pressure profile generates an outward explosion of matter at radii of approximately 55 km. Matter inside this region always collapses inward and formed a proto-remnant.

This neutrino capture induces explosion mechanism substantially differs from the accepted bounce mechanism in many ways. The density in the region where the outward explosion of matter forms was always found to be on the order of 10^{-3} nuclear matter density while the central densities at that time were consistently found to be on the order of 0.1 nuclear matter density. This is a significant deviation from the accepted picture of bounce in which the initial outward explosion of matter is generated by the pressure exerted by matter at supernuclear densities accumulated in a small volume near the origin. Thus the role of the nuclear equation of state if fundamentally different in this scenario. Rather then directly dictating the how the outward explosion of matter is formed by determining the maximum density the central region can assume, it indirectly influences the neutrino capture induced explosion by governing the rate at which electron capture occur that power the explosion.

To gain further insight into the dynamics of the new explosion mechanism our calculations are consistently yielding thus far, we study two plots of the electron fraction, average radial beta, and density of matter in the spherical shells the core is divided into for the purpose of calculating statistical distributions. The plot displayed in figure 1 corresponds to the time time the outward explosion of matter begins shortly after the spike develops and the matter between the spike's peak and the outer edge of the rapidly deleptonized inner region of the core where the electron fraction is depressed has clearly decelerated. It is important to point out that nowhere in the core have we achieved supernuclear densities. The highest density at this point in our reference calculation only slightly more than 0.36 nuclear matter density and more importantly at the radius at which the outward explosion of matter forms the density is on the order of 10^{-3} nuclear matter density. All of or calculations yield results similar to this. The maximum density in the core when the explosion begins is always well below nuclear matter density and the density at the radius where it is always on the order of 10^{-3} nuclear matter density.

The plot displayed in figure 2 corresponds to a time when the outward explosion of matter is well underway. The matter at radii near the peak in the spike of the electron fraction distribution in fully releptonized by this time and a large outward moving density wave is seen beginning its separation from the dense proto-remnant. As the proto-remnant "gently" assumes the the configuration of minimum energy, supernuclear densities are achieved in its center. Typically maximum densities are on the order of 1.5 nuclear matter density.

7. Analysis of New Dynamics

For the spike in the electron fraction to be created, it must be that in the narrow shell in which the spike forms, neutrino capture strongly dominates electron capture. Recall that in all

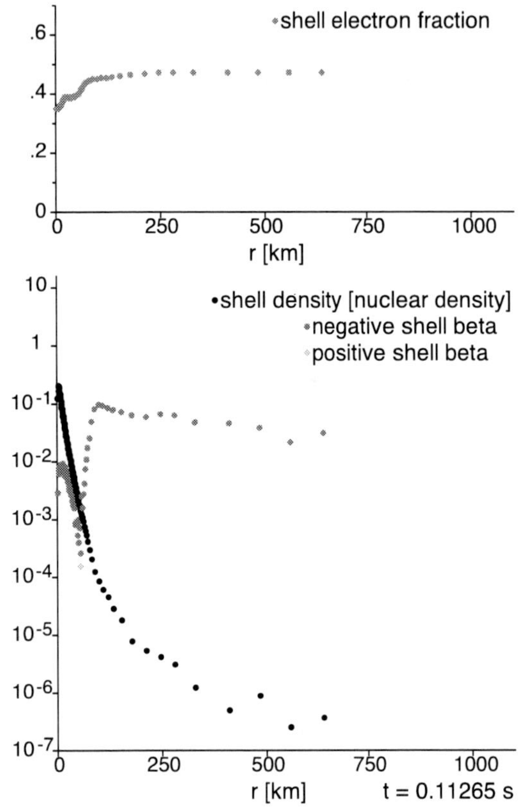

Figure 1. Plot of the electron fraction, average radial beta, and density of matter in the spherical shells at the time the outward explosion of matter begins shortly after the spike develops and the matter between the spike's peak and the outer edge of the rapidly deleptonized inner region of the core where the electron fraction is depressed has clearly decelerated.

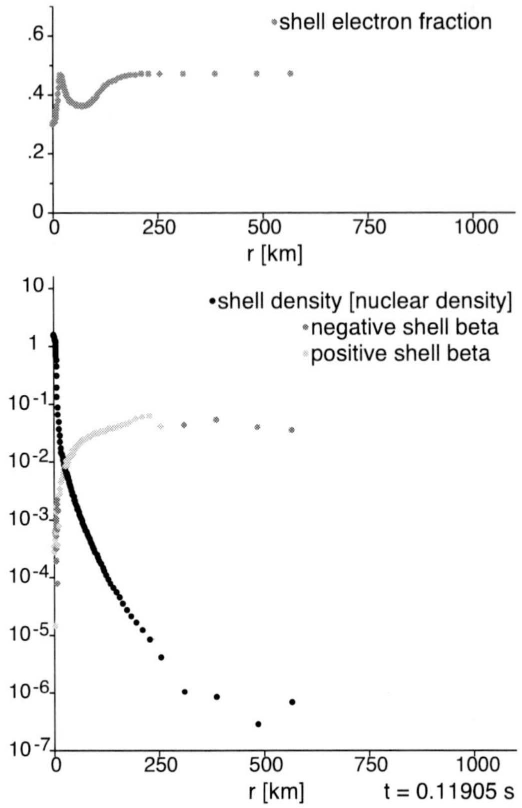

Figure 2. Plot of the electron fraction, average radial beta, and density of matter in the spherical shells showing fully releptonized matter in the spike having an electron fraction approximately equal to matter in the outermost shells where the effects of electron capture are negligible and a large outward moving density wave is seen beginning its separation from the dense proto-remnant.

calculations the spike always formed in a region containing neutron-rich matter with a density on the order of 10^{-3} nuclear matter density. The importance of this is fourfold. Obviously the farther away a nucleus is from the valley of beta stability and the closer it is to the neutron drip line, the lower its electron capture rate is. Furthermore, since it is energetically favorable for very neutron-rich nuclei to capture neutrinos and move towards the valley of stability, the neutrino capture Q-value becomes positive and large. This increases the bare capture cross section [22] and therefore the neutrino capture probability increases in the region. Additionally at densities on the order of 10^{-3} nuclear matter density none of the electron capture rates are insurmountably large and neutrino captures are not likely to be forbidden by the Pauli Exclusion Principle. This sets the stage for neutrino capture dominance.

Recall that the region of the core comprised of neutron-rich matter with a density approximately equal to 10^{-3} nuclear matter density where the spike in the electron fraction distribution forms is centered about a radius roughly 50 km from the origin. To its interior is a region of hotter denser matter in which the degeneracy of the electron gas suppresses neutrino capture rates and electron capture rates are strongly dominant. The energy of the neutrinos escaping this hotter denser region increases as it continues to contract as does the rate at which it emanates neutrinos. Thus the region where the electron fraction spike forms is bathed in an increasingly intense flux of increasingly energetic neutrinos. For the reasons stated in the previous paragraph, neutrino capture rates in this region quickly outpace electron capture rates and the local electron fraction rises. Not only can neutrino captures push neutron-rich nuclei back along isobars towards the valley of stability, as the energies of the neutrinos propagating through the region increase, they can push some nuclei past the valley of stability into proton-rich territory. In addition to this leptonization mechanism, it turns out free neutrons are highly abundant in this region and they capture a large amount of neutrinos as well. The heavy presence of free neutrons is the result of numerous electron captures by neutron drip line nuclei in the region. Through these two mechanisms, the region can become hyper-leptonized. As it does, it absorbs nearly all of the high energy neutrinos from the flux bathing it. This deprives outlying regions of the opportunity to become hyper-leptonized and is the reason that only the base of the outer edge of the electron fraction spike thickens after it is formed. The tallest part of the electron fraction spike remains narrow and radically alters the radial derivative of the electron number density distribution. This in turn divides the core into a proto-remnant and ejecta in the fashion described in section 6.

8. Summary

The preliminary results generated by our kinetic theory based model and presented here are encouraging. The neutrino capture driven mechanism that it is uniquely poised to observe is robust enough to launch explosions in all test simulations conducted so far using three different nucleon potentials, four electron capture rate tables generated with the FFN table using different extrapolation techniques, and three different numbers of spherical shells used to calculate statistical distributions. We do not yet regard this work as complete, but we are confident that we are working in the right direction with our approach. The ability of the of the test particle method to treat the dynamics of baryons and neutrinos on equal footing, explicitly model the propagation of neutrinos in a general way that is applicable in all regions of the core at all times, and explicitly model the propagation of a full ensemble of nuclei are all significant steps forward. Without all of these assets that only our code possesses, it would not be capable of generating a neutrino capture driven explosion. They each play a critical role in its realization.

Before we can contend that this picture is complete, in addition to completing the parallelization process and activating the three-dimensional subroutines, there is still more physics that needs to be added to our model. In particular fusion and several additional weak reactions must be built in. It is clear how to accomplish all of this with the test particle approach.

We are confident that this new explosion mechanism will survive the inclusion the aforementioned physics. The additional degrees of freedom introduced when three-dimensional distributions are calculated may destabilize the region in which electron are deposited by neutrino capture as it may be unstable to convection. An entropy analysis can confirm if this is a possibility. However even if convection does destabilize this region, the fact that convection begins at only $\sim$ 1100 ms after the collapse begins at a radius on the order of 50 km would be a significant discovery in and of itself. The future of this work looks extremely bright, and it promises to provide a new intellectual bridge between nuclear physics and astrophysics. In particular, it has the capability to serve as the nexus for efforts made at nuclear physics laboratories such as NSCL and FRIB and laboratories that study neutrino physics like DUSEL, the Deep Underground Science and Engineering Laboratory. As this model advances, we hope to see its kinetic theory based approach establish itself as the new standard for supernova calculations.

References

[1] Bertsch G F, Kruse H, and Das Gupta S 1984 *Phys. Rev.* C **29** 673
[2] Kruse H, Jacak B V, and Stöcker H 1985 *Phys. Rev. Lett.* **54** 289
[3] Bauer W, Bertsch G F , Gassing W, and Mosel U 1986 **34** 2127
[4] Wilson J R 1985 *Numerical astrophysics* (Boston, MA: Jones and Bartlett)
[5] Herant M, Benz W, Hix W R, Fryer C L, and Colgate S A 1994 *Astrophys.* J **435** 339
[6] Fryer C L and Heger A 2000 *Astrophys.* J **541** 1033
[7] Strother T and Bauer W 2009 *Prog. Part. Nucl. Phys.* **62** 468
[8] Woosley S E and Weaver T A 1986 *Ann. Rev. Astron. Astrophys.* **24** 205
[9] Bauer W 2004 *Acta Phys. Hung.* A **21** 371
[10] Wong C Y 1982 *Phys. Rev.* C **25** 1460
[11] Matrínez G, Liebendörfer M, and Frekers D, 2006 *Nucl. Phys. A* **777** 395
[12] Strother T and Bauer W *Neutrino capture induced supernova explosions*: submitted to *Int. J. Mod. Phys.* D
[13] Strother T 2009 *A kinetic theory based numerical study of core collapse supernova dynamics* (Michigan State University: Doctoral Dissertation)
[14] Strother T and Bauer W 2010 *A Kinetic Theory Model for Type II Core Collapse Supernovae*: submitted to *XLVIII Int. Winter Meeting on Nucl. Phys.*
[15] Bauer W and Strother T 2005 *Int. J. Mod. Phys.* E **14** 129
[16] Strother T and Bauer W 2008 *Nuclear physics and supernova explosions: Unified dynamics*: submitted to *Int. J. Mod. Phys.* D
[17] Kortemeyer G, Daffin F, and Bauer W, 1996 *Phys. Lett.* **B374** 25
[18] Wolfenstein L 1978 *Phys. Rev.* D **17** 2369
[19] Yoshida T, Kajino T, Yokomakura H, Kimura K-I, Takamura A, and Hartmann D 2006 *Astrophys. J.* **649** 319
[20] Fuller G, Fowler W, and Newman M 1985 *Astrophys. J.* **293** 1
[21] Heger A and Woosley S E 2001 *Astrophys. J.* **560** 307
[22] Burrows A and Thompson T *Neutrino-matter interaction rates in supernovae: The essential microphysics of core collapse* (to be published in Core Collapse of Massive Stars) ed. C. Fryer (Kluwer Academic Publishers)
[23] Herant M, Benz W, and Colgate S 1992 *Astrophys. J.* **395** 642
[24] Herant M, Benz W, Hix W, Fryer C, and Colgate S 1994 *Astrophys. J.* **435** 339

Journal of Physics: Conference Series **230** (2010) 012028 doi:10.1088/1742-6596/230/1/012028

Astrophysically important reactions and nucleosynthesis processes

Karlheinz Langanke

GSI Helmholtzzentrum für Schwerionenforschung, Darmstadt, Germany
Institut für Kernphysik, Technische Universität Darmstadt, Darmstadt, Germanny
Frankfurt Institute for Advanced Studies, Frankfurt, Germany

E-mail: k.langanke@gsi.de

Abstract.
Neutrino reactions play an important role at various stages of core-collapse supernova. During infall, neutrinos are produced by electron capture mainly on nuclei and contribute significantly to the cooling of the collapsing core. After core bounce the nascent neutron star cools by neutrino emission. It is a major goal to observe such neutrinos from a future supernova by earthbound detectors and to establish their spectra. Recently it has been shown that the spectrum of electron neutrinos from the early neutrino burst is significantly altered if inelastic neutrino-nucleus scattering is considered in supernova simulations. The neutrino wind from the nascent neutron star is the site of explosive nucleosynthesis, including the recently discovered νp process and likely the r-process. Finally spallation reactions induced by neutrinos when passing through the outer burning shells can produce certain nuclides in what is called neutrino nucleosynthesis.

1. Some introductory remarks

Massive stars end their lifes as type II supernovae, triggered by a collapse of their central iron core with a mass of more than $1M_\odot$ [1]. Despite of improved description of the microphysics entering the simulations and sophisticated neutrino transport treatment spherically symmetric simulations of core-collapse supernovae fail to explode [2, 3, 4]; i.e. the energy transported to the stalled shock, which triggers the explosion, by absorption of neutrinos on free neutrons and protons is not sufficient to revive the shock wave which has run out of energy by dissociating the matter it traverses into free nucleons. However, the revival is successful in two-dimensional simulations as convection supported by hydrodynamical instabilities increases the efficiency by which neutrinos carry energy to the shock region. Successful two-dimensional simulations are reported in [5, 6, 7].

2. Electron capture in core-collapse supernovae

During most of the collapse the equation of state is given by that of a degenerate relativistic electron gas [1]; i.e. the pressure against the gravitational contraction arises from the degeneracy of the electrons in the stellar core. However, the electron chemical potential at core densities in excess of 10^8 g/cm^3 is of order MeV and higher, thus making electron captures on nuclei energetically favorable. As these electron captures occur at rather small momentum transfer, the process is dominated by Gamow-Teller transitions; i.e. by the GT_+ transitions, in which a proton is changed into a neutron.

When the electron chemical potential μ_e (which grows with density like $\rho^{1/3}$) is of the same order as the nuclear Q-value, the electron capture rates are very sensitive to phase space and require a description of the detailed GT_+ distribution of the nuclei involved which is as accurate as possible. Furthermore, the finite temperature in the star requires the implicit consideration of capture on excited nuclear states, for which the GT_+ distribution can be different than for the ground state. It has been demonstrated [8, 9] that modern shell model calculations are capable to describe GT_+ distributions rather well [10] (an example is shown in Figure 1) and are therefore the appropriate tool to calculate the weak-interaction rates for those nuclei ($A \sim 50 - 65$) which are relevant at such densities [11]. At higher densities, when μ_e is sufficiently larger than the respective nuclear Q values, the capture rate becomes less sensitive to the detailed GT_+ distribution and depends practically only on the total GT strength. Thus, less sophisticated nuclear models might be sufficient. However, one is facing a nuclear structure problem which has been overcome only very recently. Once the matter has become sufficiently neutronrich, nuclei with proton numbers $Z < 40$ and neutron numbers $N > 40$ will be quite abundant in the core. For such nuclei, Gamow-Teller transitions would be Pauli forbidden [12] (GT_+ transitions change a proton into a neutron in the same harmonic oscillator shell) were it not for nuclear correlation and finite temperature effects which move nucleons from the pf shell into the gds shell. To describe such effects in an appropriately large model space (e.g. the complete $fpgds$ shell) is currently only possible by means of the Shell Model Monte Carlo approach (SMMC) [13, 14]. In [15] SMMC-based electron capture rates have been calculated for more than 100 nuclei. Together with the shell model results for fd shell nuclei [9] a compilation of electron rates is available which covers the composition during the early collapse phase well (until densities to a few 10^{10} g/cm^3). However, at even higher densities the continuous electron capture drives the matter composition to more neutron-rich and heavier nuclei than considered in [15]. The neglect of these nuclei could lead to a systematic overestimate of the capture rates as the neglected nuclei have larger Q values and enhanced Pauli blocking due to increased neutron excess than the nuclei considered. Unfortunately an SMMC evaluation of the several thousands of nuclei present in the matter composition during the late phase of collapse before neutrino trapping is numerically not feasible. Hence a simpler approach has to be adopted. This is based on the observation that the single particle occupation numbers obtained in the SMMC calculations can be well approximated by a parametrized Fermi-Dirac distribution. By adjusting the parameters of this distribution to the SMMC results for about 250 nuclei, occupation numbers were derived for nearly 3000 nuclei and the respective individual rates have been calculated within an RPA calculation based on these partial occupations. It is found that this simplified approach reproduces the SMMC results quite well at electron chemical potential $\mu_e > 15$ MeV [16]. This corresponds to the density regime where the previously neglected nuclei become abundant in the matter composition. At smaller electron chemical potentials (i.e. at lower densities) details of the GT distribution are important which are not recovered by the simple parametrized approach. However, at these conditions electron capture is dominated by nuclei for which individual shell model rates exist.

Fig. 2 compares the capture rate derived for the pool of more than 3000 nuclei (i.e. combining the rates from shell model diagonalization, SMMC, and from the parametrized approach) with those obtained purely on the basis of the shell model results. While the agreement is excellent at small electron chemical potentials (here the shell model rates dominate), the rates for the large pool are slightly smaller at higher μ_e values due the presence of neutron-rich heavy nuclei with smaller individual rates. Furthermore the new rates also include plasma screening effects which lead to an increase of the effective Q values and a reduction of the electron chemical potential, which both reduce the electron capture rates. However, the effect is rather mild and does not alter the conclusion that electron capture on nuclei dominates over capture on protons during the collapse.

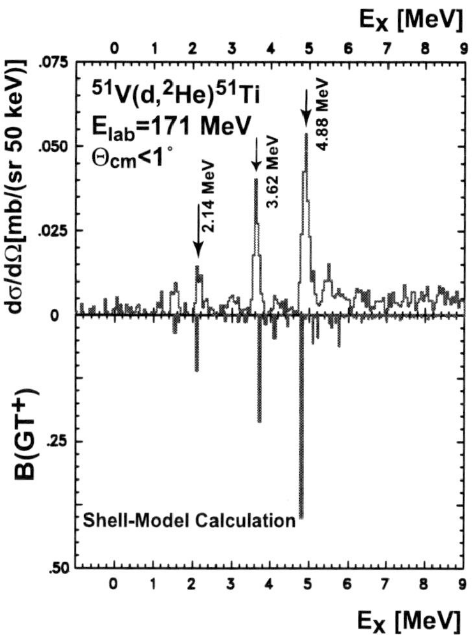

Figure 1. Comparison of the measured ^{51}V(d,^{2}He)^{51}Ti cross section at forward angles (which is proportional to the GT_+ strength) with the shell model GT distribution in ^{51}V (from [21]).

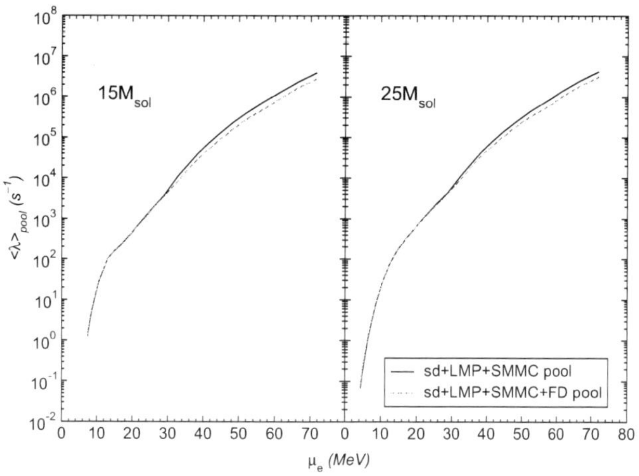

Figure 2. Comparison of NSE-averaged electron capture rates calculated for about 3000 individual nuclei (solid, see text) with those obtained for the restricted set of nuclei (dashed) considered in [15] (from [16]).

The shell model capture rates have significant impact on collapse simulations. In the presupernova phase ($\rho < 10^{10}$ g/cm^3) the captures proceed slower than assumed before and for a short period during silicon burning β-decays can compete [17, 18]. As a consequence, the core is cooler, more massive and less neutronrich before the final collapse. However, until recently simulations of this final collapse assumed that electron captures on nuclei are prohibited by the Pauli blocking mechanism, mentioned above. However, based on the SMMC calculations it has been shown in [15] that capture on nuclei dominates over capture on free protons. The changes compared to the previous simulations are significant [15, 19, 20]. Importantly the shock is now created at a smaller radius with more infalling material to traverse, but also the density, temperature and entropy profiles are strongly modified [19].

The possibility to capture electrons on nuclei and free protons leads to a strong self-regulation during the collapse and it is observed that the collapse approaches the same core trajectory for stars between 11 and 25 solar masses before neutrino trapping and thermalization is reached at densities around 10^{12} g/cm^3 (Fig. 3, [20]). Thus it is expected that supernovae, independent of the stellar mass, have a quite similar neutrino burst spectrum. This spectrum, however, has also significantly changed after inelastic neutrino-nucleus reactions have been included in the simulations.

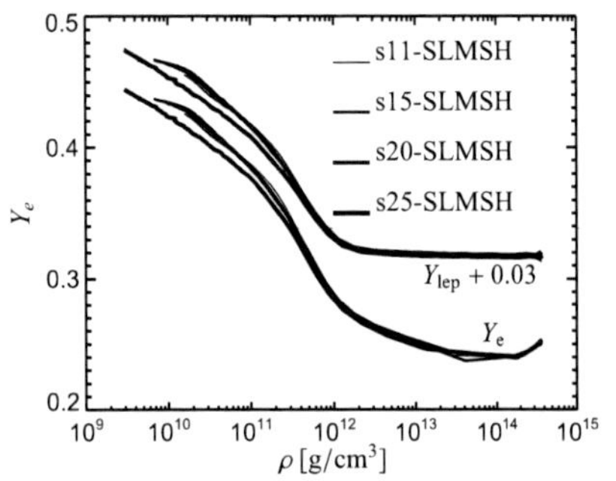

Figure 3. The evolution of the electron-to-baryon ratio Y_e and the lepton-to-baryon ratio Y_{lep} as function of central density for stars with masses between 11 and 25 $M_\odot$. (from [20]).

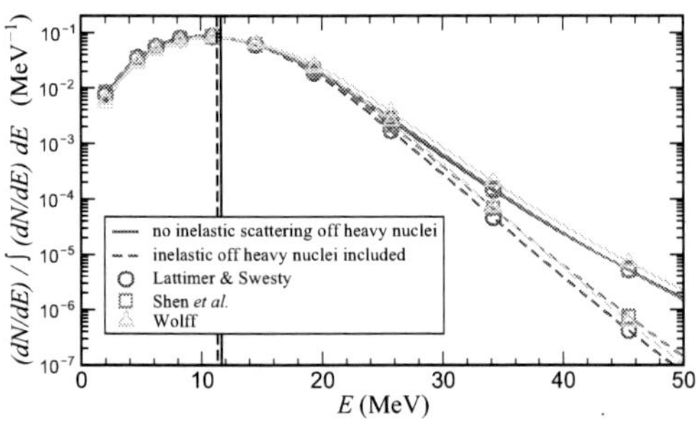

Figure 4. Comparison of the normalized neutrino spectra, arising from the ν_e burst shortly after bounce, without (solid) and with (dashed) consideration of inelastic neutrino-nucleus scattering in the supernova simulations. (from [25]).

3. Inelastic neutrino-nucleus reactions

When, with progressing collapse, matter gets denser, neutrino interactions with matter become increasingly important and require careful treatment which is usually done within Boltzmann transport theory [3, 4]. Here elastic neutrino-nucleus and inelastic neutrino-electron scattering have been identified as the sources for neutrino trapping (which occurs at densities around a few 10^{11} g/cm^3) and neutrino thermalization. Recently inelastic neutrino-nucleus scattering, as another mode of energy exchange between neutrinos and matter, has been included for the first time in supernova simulations. The relevant cross sections have been calculated based on large-scale shell model calculations for the allowed GT transitions and within the random phase approximation for forbidden transitions [22], taking special care of finite temperature effects [23]. At low and modest neutrino energies E_ν the cross sections are dominated by GT$_0$ contributions for which the shell model has been validated by detailed comparison to precision M1 data derived from electron scattering on spherical nuclei which are mainly due to the same isovector response [24].

Although inelastic neutrino-nucleus scattering contributes to the thermalization of neutrinos with the core matter, the inclusion of this process has no significant effect on the collapse trajectories. However, it increases noticeably the opacity for high-energy neutrinos in the ν_e neutrino burst just after the bounce. As these neutrinos excite the nuclei, they are down-scattered in energy, in this way significantly reducing the high-energy tail of the spectrum of emitted supernova neutrinos (see Fig. 4, [25]). This makes the detection of supernova neutrinos by earthbound detectors more difficult, as the neutrino detection cross section scales with E_ν^2.

4. Supernova nucleosynthesis

When in a successful explosion the shock passes through the outer shells, its high temperature induces an explosive nuclear burning on short time-scales. This explosive nucleosynthesis can alter the elemental abundance distributions in the inner (silicon, oxygen) shells. More interestingly the shock has sufficiently high temperatures that nuclear binding cannot withstay and matter is ejected as free protons and neutrons from the neutron star surface. As the outflow is adiabatic, the ejected matter reaches cooler regions at larger distances and nuclei can be assembled from free nucleons. The abundance outcome obviously depends strongly on the ratio of neutrons and protons which is set by the competition of neutrino and anti-neutrino absorption on free nucleons.

Recently explosive nucleosynthesis has been investigated consistently within supernova simulations, where a successful explosion has been enforced by slightly enlargening the neutrino absorption cross section on nucleons or the neutrino mean-free path, which both increase the efficiency of the energy transport to the stalled shock. The results presented in [26, 27] showed that in an early phase after the bounce the ejected matter is actually protonrich as the ν_e neutrinos have sufficiently large energies to drive the matter protonrich. In later stages, i.e. a few seconds after bounce, the neutrino opacities in the neutron-rich matter ensure that $\bar{\nu}_e$ have larger average energies than ν_e, and it is believed that $\bar{\nu}_e$ absorption on protons dominates driving the matter neutronrich (allowing for the r-process to occur).

4.1. The νp process

The studies presented in [26, 27, 28] show that matter with Y_e larger than 0.5 will always be present in core-collapse supernovae explosions with ejected matter irradiated by a strong neutrino flux, independently of the details of the explosion. As this proton-rich matter expands and cools, nuclei can form resulting in a composition dominated by $N = Z$ nuclei, mainly ^{56}Ni and ^{4}He, and protons. Without the further inclusion of neutrino and antineutrino reactions the composition of this matter will finally consist of protons, alpha-particles, and heavy (Fe-group) nuclei (in nucleosynthesis terms a proton- and alpha-rich freeze-out), with enhanced abundances of ^{45}Sc, ^{49}Ti, and ^{64}Zn [26, 27]. In these calculations the matter flow stops at ^{64}Ge with a small proton capture probability and a beta-decay half-life (64 s) that is much longer than the expansion time scale ($\sim$ 10 s) [26].

As noted by Martinez-Pinedo and explored in [29] the synthesis of nuclei with $A > 64$ can be obtained, if one explores the previously neglected effect of neutrino interactions on the nucleosynthesis of heavy nuclei. $N \sim Z$ nuclei are practically inert to neutrino capture (converting a neutron into a proton), because such reactions are endoergic for neutron-deficient nuclei located away from the valley of stability. The situation is different for antineutrinos that are captured in a typical time of a few seconds, both on protons and nuclei, at the distances at which nuclei form ($\sim$ 1000 km). This time scale is much shorter than the beta-decay half-life of the most abundant heavy nuclei reached without neutrino interactions (e.g. ^{56}Ni, ^{64}Ge). As protons are more abundant than heavy nuclei, antineutrino capture occurs predominantly on protons, causing a residual density of free neutrons of 10^{14}–10^{15} cm^{-3} for several seconds, when the temperatures are in the range 1–3 GK. The neutrons produced via antineutrino absorption on protons can easily be captured by neutron-deficient $N \sim Z$ nuclei (for example ^{64}Ge), which have large neutron capture cross sections. The amount of nuclei with $A > 64$ produced is

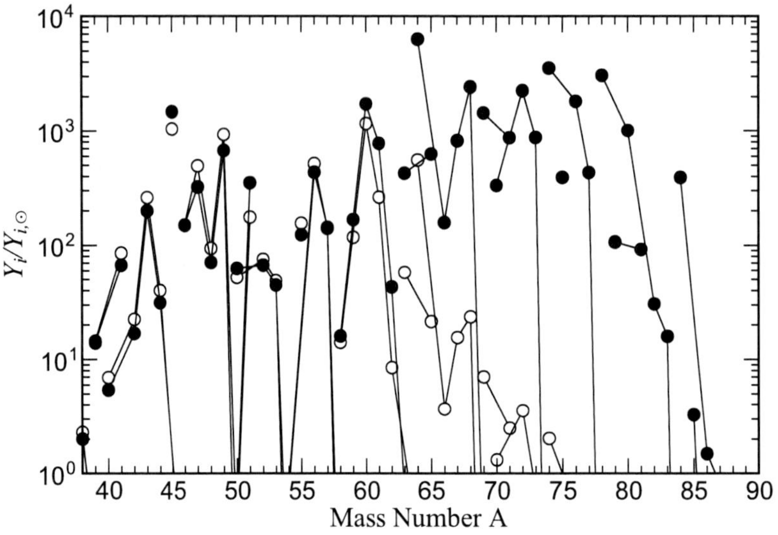

Figure 5. Elemental abundance yields (normalized to solar) for elements produced in the protonrich environment shortly after the supernova shock formation. The matter flow stops at nuclei like ^{56}Ni and ^{64}Ge (open circles), but can proceed to heavier elements if neutrino reactions are included during the network (full circles) (from [29].

then directly proportional to the number of antineutrinos captured. While proton capture on ^{64}Ge takes too long, the (n,p) reaction dominates (with a lifetime of 0.25 s at a temperature of 2 GK), permitting the matter flow to continue to nuclei heavier than ^{64}Ge via subsequent proton captures.

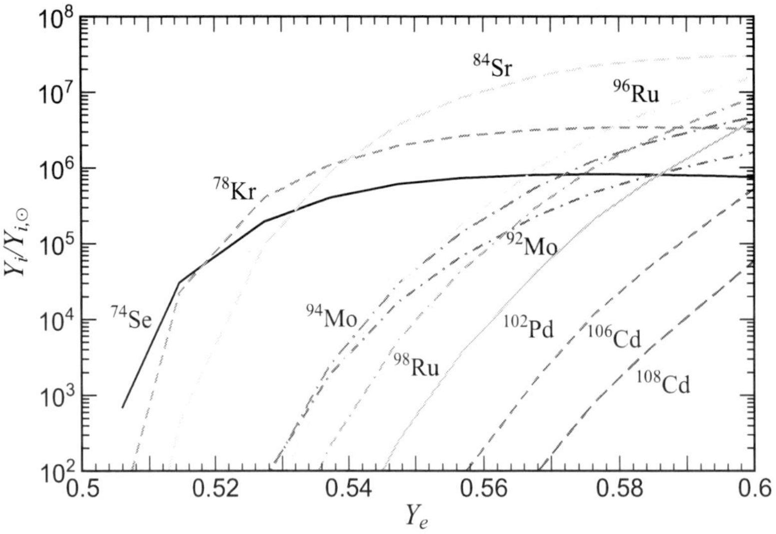

Figure 6. Light p-nuclei abundances in comparison to solar abundances as a function of Y_e. The Y_e-values given are the ones obtained at a temperature of 3 GK that corresponds to the moment when nuclei are just formed and the νp-process starts to act. (from [29].

Fröhlich *et al.* argue that all core-collapse supernovae will eject hot, explosively processed matter subject to neutrino irradiation and that this novel nucleosynthesis process (called νp process) will operate in the innermost ejected layers producing neutron-deficient nuclei above $A > 64$. However, how far the massflow within the νp process can proceed, strongly depends on the environment conditions, most noteably on the Y_e value of the matter. Obviously the larger Y_e, the larger the abundance of free protons which can be transformed into neutrons by antineutrino absorption. Fig. 6 shows the dependence of the νp-process abundances as a function of the Y_e value of the ejected matter. Nuclei heavier than $A = 64$ are only produced for $Y_e > 0.5$, showing a very strong dependence on Y_e in the range 0.5–0.6. A clear increase in the production of the light p-nuclei, ^{92,94}Mo and ^{96,98}Ru, is observed as Y_e gets larger. Thus the νp process offers the explanation for the production of these light p-nuclei, which was yet unknown. It might also explain the presence of strontium in the extremely metalpoor, and hence very old, star HE 1327-2326 [30]. The νp process results of Fröhlich *et al.* have been recently confirmed by Pruet *et al.* [31], in a study of the nucleosynthesis that occurs in the early proton-rich neutrino wind.

4.2. The r-process

About half of the elements heavier than mass number $A \sim 60$ are made within the $r-process$, a sequence of rapid neutron captures and β decays [32, 33]. The process is thought to occur in environments with extremely high neutron densities. Then neutron captures are much faster than the competing decays and the r-process path runs through very neutron-rich, unstable nuclei. Once the neutron source ceases, the process stops and the produced nuclides decay towards stability producing the neutron-rich heavier elements. Many parameter studies of the reaction network, aimed at reproducing the observed r-process abundances, have progressed our general understanding. So it is generally accepted that the r-process operates in an environment with temperatures of order 10^9 K and with neutron densities in excess of $10^{20}/cm^3$ [34]. Simulations of the r-process reaction network indicate that under such conditions for temperature and neutron densities the reaction path runs through extremely neutron-rich nuclei with separation energies, $S_n < 2-3$ MeV [34], for which most of the required properties (i.e mass, lifetime and neutron capture cross sections) are experimentally unknown. Thus, these quantities have to be modelled, based on experimental guidance. As we will discuss in the next subsection some progress has been achieved recently, but the real breakthrough is expected from future radioactive ion-beam facilities. Finally we mention that the observed peaks in the r-process abundance distribution are related to the magic neutron numbers, $N = 50, 82, 126$, where the r-process runs through several, rather long β decays at $N = 50, 82$ and 126 (the respective nuclei are called r-process waiting points) and a relatively large amount of matter is built up.

Despite many promising attempts the actual site of the r-process has not been identified yet. However, parameter studies have given clear evidence that the observed r-process abundances cannot be reproduced at one site with constant temperature and neutron density [35]. Thus the abundances require a superposition of several (at least three) r-process components. This likely implies a dynamical r-process in an environment in which the conditions change during the duration of the process. The currently favored r-process sites (type II supernovae [36] and neutron-star mergers [37]) offer such dynamical scenarios. However, recent meteoritic clues might even point to more than one distinct site for our solar r-process abundance [38]. The same conclusion can be derived from the observation of r-process abundances in low-metallicity stars (e.g. [39]), a milestone of r-process research.

4.2.1. Nuclear r-process input Arguably the most important nuclear ingredient in r-process simulations are the nuclear masses as they determine the flow-path. Unfortunately nearly all of them are experimentally unknown and have to be theoretically estimated. Traditionally this is done on the basis of parametrizations to the known masses. Although these empirical mass formulae achieve rather remarkable fits to the data (the standard deviation is of order 700 keV), extrapolation to unknown masses appears less certain and different mass formulae can predict quite different trends for the very neutron-rich nuclei of relevance to the r-process. The most commonly used parametrizations are based on the finite range droplet model (FRDM), developed by Möller and collaborators [40], and on the ETFSI (Extended Thomas-Fermi with Strutinski integral) model of Pearson [41]. A new era has been opened very recently, as for the first time, nuclear mass tables have been derived on the basis of nuclear many-body theory (Hartree-Fock-Bogoliobov model) [42, 43, 44] rather than by parameter fit to data.

The nuclear halflives strongly influence the relative r-process abundances. In a simple β-flow equilibrium picture the elemental abundance is proportional to the halflife, with some corrections for β-delayed neutron emission [45]. As r-process halflives are longest for the magic nuclei, these waiting point nuclei determine the minimal r-process duration time; i.e. the time needed to build up the r-process peak around $A \sim 200$. We note, however, that this time depends also crucially on the r-process path and can be as short as a few 100 milliseconds if the r-process path runs close to the neutron dripline.

There are a few milestone halflife measurements including the N=50 waiting point nuclei ^{80}Zn, ^{79}Cu, and ^{78}Ni, and the N=82 waiting point nuclei ^{130}Cd and ^{129}Ag [46, 47]. Although no halflives for $N = 126$ waiting points have yet been determined, there has been decisive progress towards this goal recently [48].

These data play crucial roles in constraining and testing nuclear models which are still necessary to predict the bulk of halflives required in r-process simulations. It is generally assumed that the halflives are dominated by allowed Gamow-Teller (GT) transitions, with forbidden transitions contributing noticeably for the heavier r-process nuclei [49]. The β decays only probe the weak low-energy tail of the GT distributions and provide quite a challenge to nuclear modeling as they are not constrained by sumrules. Traditionally the estimate of the halflives are based on the quasiparticle random phase approximation on top of the global FRDM or ETFSI models. Recently halflives for selected (spherical) nuclei have been presented using the QRPA approach based on the microscopic Hartree-Fock-Bogoliubov method [50] or a global density functional [51, 52]; in particular the later approach achieved quite good agreement with data for spherical nuclei in different ranges of the nuclear chart. Applications of the interacting shell model [53, 54, 55] have yet been restricted to waiting point nuclei with magic neutron numbers. Here, however, this model which accounts for correlations beyond the QRPA approach, obtains quite good results (for an example see Fig. 7). We remark that, besides a good description of the allowed (and forbidden) transition matrix elements, an accurate reproduction of the Q_n values is required from the models.

Further r-process simulations require rates for neutron capture, and for the various fission processes (neutron-induced, beta-delayed, spontaneous, perhaps neutrino induced) together with the corresponding yield distributions. If the r-process occurs in strong neutrino fluences, different neutrino-induced charged-current (e.g. (ν_e, e^-)) and neutral-current (e.g. (ν, ν')) reactions, which are often accompanied by the emission of one or several neutrons [56, 57, 58], have to be modelled and included as well. We mention that the occurence of low-lying dipole strength in neutron rich nuclei, as predicted by nuclear models [59], can have noticeable effects on neutron capture cross sections [60, 61].

4.3. R-process simulations and the role of fission

Recently r-process simulations, which for the first time included detailed fission yield distributions derived on the basis of the abration-ablation model code ABLA [62], have been performed [63, 64]. The studies have been performed assuming the neutrino-driven wind scenario

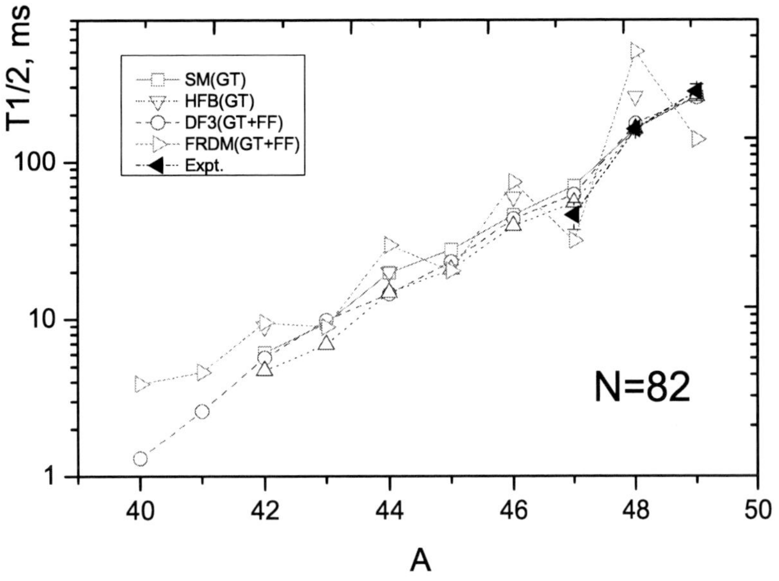

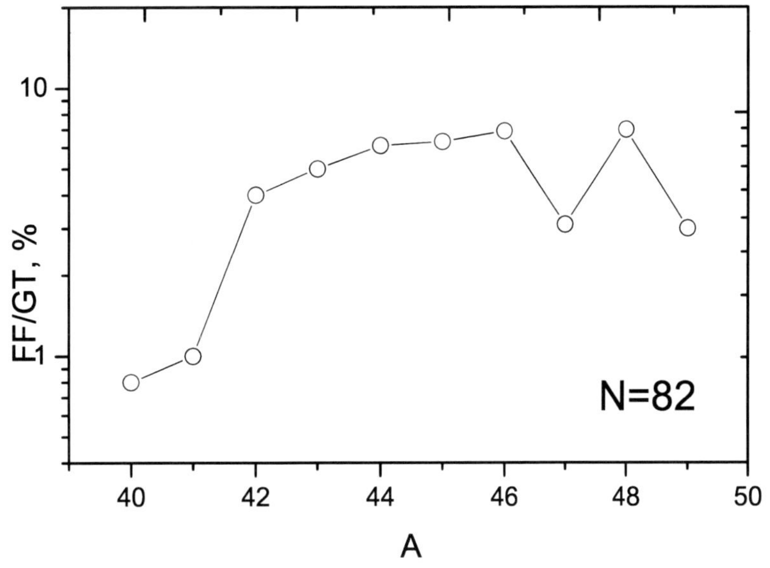

Figure 7. top: Comparison of various theoretical halflife predictions with data for the $N = 82$ r-process waiting points. Bottom: Ratio of forbidden and allowed Gamow-Teller contributions to the halflives calculated with the Fayans density functional (courtesy I.N. Borzov).

(i.e. matter ejected from the neutron star in a core-collapse supernova) as the r-process site. The entropy of the ejected matter, or equivalently the neutron-to-seed (n/s) ratio, has been treated as a free parameter. To explore the potential effect of fission on the r process, this ratio has to be relatively large, ($n/s > 250$). If this condition is fulfilled, then the r-process runs quite fast along a path close to the dripline reaching heavy nuclei with neutron numbers close to the magic number $N = 184$ and above. Once the neutron source is used, i.e. the ejected matter has reached radii where the neutron number density is too low to support fast neutron captures, the produced nuclides decay. For the heavy nuclei the dominant decay mode

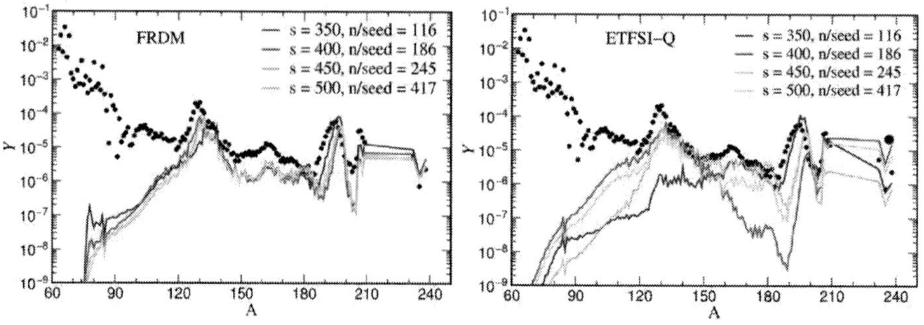

Figure 8. R-process abundances obtained in simulations which explore the potential impact of fission. The calculations are performed for different neutron-to-seed ratios and two different mass tabulations. The solid line is the solar r-process abundance distribution. (from [64])

is neutron-induced fission which via their fission decay products produce peaks around mass numbers $A \sim 130$ and 195, corresponding to the second and third peaks in the observed r-process abundance distributions (Fig. 8). These simulations clearly stress the importance of the nuclear masses. While the studies, which use the FRDM masses, give quite similar abundance distributions for different n/s ratios once a threshold value is overcome, the calculations based on the ETFSI masses yield abundance results which vary strongly with the assumed n/s value. This strikingly different behavior can be traced back to differences in the predicted masses for very neutron rich nuclei. The FRDM mass tabulation predicts that certain nuclei just above the magic neutron number $N = 82$ act as obstacles for the r-process matter flow holding back material and ensuring that more free neutrons are available when part of the matter reaches heavy nuclei with $N = 184$ to guarantee production of nuclides even beyond these waiting points. The ETFSI mass tabulation does not identify such obstacles on the r-process path and, as no matter is kept back at relatively small mass numbers, less neutrons are available once the matter flow reaches $N = 184$ suppressing matter flow beyond this waiting point [64]. We will know which of the two mass models is more realistic once masses of very neutron-rich nuclei will be measured; this is one of the important aims at radioactive ion-beam facilities like FAIR. It should be stressed that observations of r-process abundances in old halo stars in our galaxy show always the same pattern between mass numbers $A = 130$ and 195 (Fig. 9), while the patterns differ for $A < 130$. It is intriguing that these old stars in completely different locations of the Milky Way, which have witnessed only a few, and importantly not the same, supernova outbursts, show identical r-process patterns between the second and third peaks as obtained when averaged over the chemical history of the galaxy (solar r-process abundance). The fact that r-process simulations including fission and based on the FRDM mass model yield just such a behavior is certainly interesting. However, the nuclear and astrophysical input in these simulations is yet too uncertain to draw any conclusions. including fission and based on the FRDM mass model yield just such a behavior is certainly interesting. However, the nuclear and astrophysical input in these simulations is yet too uncertain to draw any conclusions.

4.4. The neutrino process

When neutrinos, produced in the hot supernova core, pass through the outer shells of the star, they can induce nuclear reactions and in this way contribute to the elementsynthesis (the ν-process, [65]). For example, the nuclides ^{11}B and ^{19}F are produced by $(\nu, \nu' n)$ and $(\nu, \nu' p)$ reactions on the quite abundant nuclei ^{12}C and ^{20}Ne. These reactions are dominantly induced by ν_μ and ν_τ neutrinos and their antiparticles (combined called ν_x neutrinos) which have larger

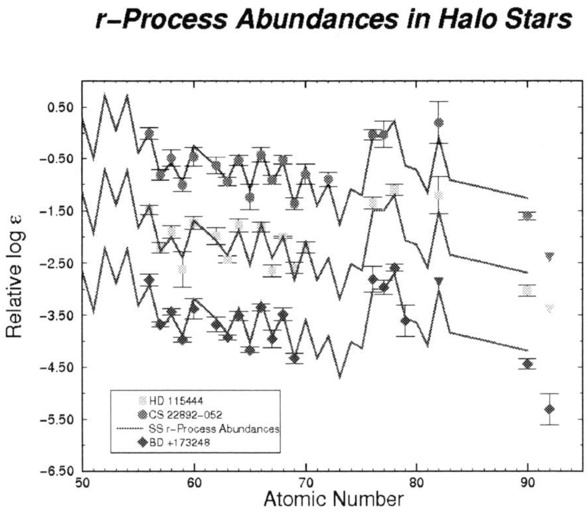

Figure 9. Observed r-process abundances in three old stars in the galactical halo, compared to the solar r-process abundance (blue line) which reflects the chemical history of the galaxy. For better viewing the different curves are off-set by arbitrary numbers. (courtesy John Cowan)

average energies (about 20 MeV) than ν_e and $\bar{\nu_e}$ neutrinos. As found in detailed stellar evolution studies [66] the rare odd-odd nuclides ^{138}La and ^{180}Ta are mainly made by the charged-current reaction ^{138}Ba$(\nu_e, e^-)^{138}$La and ^{180}Hf$(\nu_e, e^-)^{180}$Ta. Hence, the ν-process is potentially sensitive to the spectra and luminosity of ν_e and ν_x neutrinos, which are the neutrino types not observed from SN1987a. Both charged-current cross sections are dominated by the low-energy tail of the GT_- strength distribution which is notoriously difficult to simulate. As large-scale shell model calculations are yet not possible for ^{138}Ba or ^{180}Hf, the charged-current cross sections had been evaluated within the random phase approximation [66]. As a major improvement it has been possible last year to measure the GT_- strengths on ^{138}Ba and ^{180}Hf below the particle thresholds and to convert these data into the relevant (ν_e, e^-) cross sections [67]. It is found that the new cross sections are somewhat larger than the RPA predictions. Upon using these results in stellar evolution models, slightly larger ^{138}La and ^{180}Hf abundances are found, which agree reasonably well with observation [67].

We mention that neutrino nucleosynthesis studies are quite evolved requiring state-of-the-art stellar models with an extensive nuclear network [66]. In the first step stellar evolution and nucleosynthesis is followed from the initial hydrogen burning up to the presupernova models. The post-supernova treatment then includes the passage of a neutrino fluence through the outer layers of the star, followed by the shock wave which heats the material and also induces noticeable nucleosynthesis, mainly by photodissociation (the γ process). Modelling of the shock heating is quite essential as the associated γ process destroys many of the daughter nuclides previously produced by neutrino nucleosynthesis.

Neutrino nucleosynthesis has recently also been studied as a possible thermometer for supernova neutrinos, which is particularly interesting under the assumption of certain oscillation scenarios [68].

5. Conclusion

Supernova simulations are a quite interdisciplinary task. In recent years significant progress has been achieved due to improved astrophysical modelling and based on more reliable nuclear inputs. The latter is mainly concerned with weak-interaction processes like electron capture on nuclei, beta decays and neutrino-induced reactions. Here it has been possible to derive the relevant rates on the basis of modern nuclear many-body models, constrained and guided by recent measurements of nuclear Gamow-Teller and M1 strengths functions. The improved nuclear inputs have led to quite significant changes in the core-collapse dynamics and the supernova neutrino spectra. Nuclear network simulations, consistently coupled to trajectories of state-of-the-art supernova simulations, have yielded new insights in supernova nucleosynthesis, resulting for example to the discovery of the νp process which produces protonrich nuclides and might be the searched-for source of the light p-nuclides. Progress in the understanding of the r-process is expected from future radioactive ion-beam facilities which will help to remove the large uncertainties on the nuclear input data which still hamper current r-process simulations.

Acknowledgement

The work presented here has benefitted from fruitful discussions and collaborations with David Dean, Carla Fröhlich, Alexander Heger, Raphael Hix, Thomas Janka, Andrius Juodagalvis, Andreas Marek, Tony Mezzacappa, Bronson Messer, Matthias Liebendörfer, Peter von Neumann-Cosel, Achim Richter, Jorge Sampaio, F.-K. Thielemann and Stan Woosley. The work has been supported by the DFG grant SFB 634 and the Helmholtz Alliance EMMI.

[1] H.A. Bethe, Rev. Mod. Phys. **62** (1990) 801
[2] R. Buras, M. Rampp, H.-Th. Janka and K. Kifonidis, Phys. Rev. Lett. 90 (2003) 241101
[3] H.Th. Janka and M. Rampp, Astrophys. J. **539** (2000) L33
[4] A. Mezzacappa *et al.*, Phys. Rev. Lett. **86** (2001) 1935
[5] A. Burrows *et al.*, Astr. j. **640** (2006) 878
[6] A. Marek and H.-Th. Janka, Astr. J. **694** (2009) 664
[7] W.R. Hix, proceedings of the conference on Nucleus-Nucleus Collisions, Beijing (2009)
[8] E. Caurier, K. Langanke, G. Martínez-Pinedo and F. Nowacki, Nucl. Phys. A **653** (1999) 439.
[9] K. Langanke and G. Martínez-Pinedo, Nucl. Phys. **A673** (2000) 481
[10] D. Frekers, Prog. Part. Nucl. Phys. **57** (2006) 217
[11] K. Langanke and G. Martinez-Pinedo, Rev. Mod. Phys. **75** (2003) 819
[12] G.M. Fuller, Astr. J. **252** (1982) 741
[13] K. Langanke *et al.*, Phys. Rev. **C52** (1995) 718
[14] S.E. Koonin, D.J. Dean and K. Langanke, Phys. Rep. **278** (1997) 2
[15] K. Langanke *et al.*, Phys. Rev. Lett. 90 (2003) 241102
[16] A. Juodagalvis *et al*, submitted to Phys. Rev. C
[17] A. Heger, K. Langanke, G. Martinez-Pinedo and S.E. Woosley, Phys. Rev. Lett. **86** (2001) 1678
[18] A. Heger, S.E. Woosley, G. Martinez-Pinedo and K. Langanke, Astr. J. **560** (2001) 307
[19] R.W. Hix *et al.*, Phys. Rev. Lett. **91** (2003) 210102
[20] H.-Th. Janka *et al.*, Phys. Rep. **442** (2007) 38
[21] C. Bäumer *et al.*, Phys. Rev. **C68** (2003) 031303(R)
[22] A. Juodagalvis *et al.*, Nucl. Phys. **A747** (2005) 87
[23] J. Sampaio *et al.*, Phys. Lett. **B529** (2002) 19
[24] K. Langanke *et al.*, Phys. Rev. Lett. **93** (2004) 202501
[25] K. Langanke *et al.*, Phys. Rev. Lett. **100** (2008) 011101
[26] J. Pruet *et al.*, Astr. J. **623** (2005) 1
[27] C. Fröhlich *et al.*, Astr. J. **637** (2006) 415
[28] S. Wanajo, Astr. J. **647** (2006) 1323
[29] C. Fröhlich *et al.*, Phys. Rev. Lett. **96** (2006) 142502

[30] A. Frebel *et al.*, Nature (London) **434** (2005) 871
[31] J. Pruet, R. Hoffman, S. E. Woosley, H.T. Janka and R. Buras, Astr. J. **644** (2006) 1028
[32] E.M. Burbidge, G.R. Burbidge, W.A. Fowler and F. Hoyle, Rev. Mod. Phys. **29** (1957) 547
[33] A.G.W. Cameron, Chalk River Report CRL-41 (1957)
[34] J.J. Cowan, F.-K. Thielemann and J.W. Truran, Phys. Rep. **208** (1991) 267
[35] K.L. Kratz *et al.*, Astrophys.J. **402** (1993) 216
[36] S.E. Woosley *et al.*, Astr. J. **399** (1994) 229
[37] C. Freiburghaus, S. Rosswog and F.-K. Thielemann, Astr. J. **525** (1999) L121
[38] G.J. Wasserburg, M. Busso, R. Gallino and K.M. Nollett, Nucl. Phys. **A777** (2006) 5,
[39] C. Sneden *et al.*, Astr. J. **591** (2003) 936
[40] P. Möller, J.R. Nix and K.L. Kratz, At. Data Nucl. data Tables **66** (1997) 131
[41] I.N. Borsov, S. Goriely, and J.M. Pearson, Nucl. Phys. **A621** (1997) 307c
[42] M. Samyn, S. Goriely and J.M. Pearson, Nucl. Phys. **A725** (2003) 69
[43] S. Goriely, M. Samyn, J.M. Pearson and M. Onsi, Nucl. Phys. **A750** (2005) 425
[44] J.R. Stone, J. Phys. **G 31** (2005) 211
[45] K.L. Kratz *et al.*, J. Phys. **G24** (1988), S331
[46] K.L. Kratz *et al.*, Z. Phys. **A325** (1986), 489
[47] J. Lettry et al. Rev. Sci. Instr. **69** (1998), 761
[48] T. Kurtukian-Nieto *et al.*, submitted to Phys. Lett.
[49] I. N. Borzov, Phys. Rev. **C67** (2003) 025802
[50] J. Engel *et al.*, Phys. Rev. **C60** (1999), 014302
[51] I.N. Borzov *et al.*, Z. Phys. **A355** (1996) 117
[52] I.N. Borzov *et al.*, Nucl. Phys. **A814** (2008) 159
[53] G. Martinez-Pinedo and K. Langanke, Phys. Rev. Lett. **83** (1999), 4502
[54] G. Martinez-Pinedo, Nucl. Phys. **A688** (2001), 57c
[55] J.J. Cuenca-Garcia *et al.*, Eur. J. Phys. **34** (2007) 99
[56] G.M. Fuller and B.S. Meyer, Astrophys.J. 453 (1995) 792
[57] Y.-Z. Qian, W.C. Haxton, K. Langanke and P. Vogel, Phys. Rev. **C55** (1997) 1532
[58] W.C. Haxton, K. Langanke, Y.-Z. Qian and P. Vogel, Phys. Rev. Lett. **78** (1997) 2694
[59] F. Catara, E.G. Lanzo, M.A. Nagarajan and A. Vitturi, Nucl. Phys. **A624** (1997) 449
[60] S. Goriely, Phys. Lett. **436** (1998) 10
[61] E. Litvinova *et al*, Nucl. Phys. **A823** (2009) 26
[62] J. Benlliure *et al.*, Nucl. Phys. **A628** (1998) 458
[63] G. Martinez-Pinedo *et al.*, Prog. Part. Nucl. Phys. **59** (2007) 199
[64] G. Martinez-Pinedo *et al.*, to be published
[65] S.E. Woosley *et al.*, Astr. J. **356** (1990) 272
[66] A. Heger *et al.*, Phys. Lett. **B606** (2005) 258
[67] A. Byelikov it et al., submitted to Phys. Rev. Lett. **98** (2007) 082501
[68] T. Yoshida *et al.*, Phys. Rev. Lett. **96** (2006) 091101

26th Winter Workshop on Nuclear Dynamics IOP Publishing
Journal of Physics: Conference Series **230** (2010) 012029 doi:10.1088/1742-6596/230/1/012029

The Relationship Between the Soft and Hard Ridges

George Moschelli[1] **and Sean Gavin**[1]

[1]Department of Physics and Astronomy, Wayne State University, 666 W Hancock, Detroit, MI, 48202, USA

E-mail: moschelli@wayne.edu

Abstract. Two particle correlation measurements in relativistic heavy ion collisions show a 'ridge' of correlated particle pairs not seen in proton on proton collisions. The measured two-dimentional correlation landscape resembles a mountain ridge since correlated pairs are focused into a narrow regions of relative azimuthal angle, but can extend to a long range in relative pseudorapidity. Similar ridge structures are observed in correlations of particles associated with and without a jet trigger. Explanations, with and without a jet, are challenged to explain the long range pseudorapidity behavior of correlations where causality limits the influence of transport dynamics. Glasma initial conditions as predicted by the theory of Color Glass Condensate provide a and early stage correlation that naturally extends far in rapidity. We have previously shown that the soft ridge is a consequence of particles forming from an initial Glasma phase that experience a later stage transverse flow. We extend this work to study the ridge dependence on the p_t of the correlated pairs and determine the relationship between the triggered and untriggered measurements.

1. Introduction

Two particle correlation measurements have played an integral role in the most exciting discoveries at the Brookhaven Relativistic Heavy Ion Collider. Both jet quenching and elliptic flow measurements examine correlations in azimuth, but study distinctly different phenomenon. Jet quenching measurements find that when identifying a high p_t trigger particle, associated particles - expected to be jet fragments - are seen at small azimuthal differences from the trigger, but are suppressed at 180° where the away side jet would be expected. Elliptic flow measurements find that the azimuthal distribution of lower p_t bulk particles is related to the geometry of the collision. The success of these measurements inspired another correlation measurement that compares particles in both relative azimuth $\phi = \phi_1 - \phi_2$ and relative pseudorapidity $\eta = \eta_1 - \eta_2$ simultaneously. Triggered measurements find a narrow peak in peripheral (and pp) collisions centered at $\eta = \phi = 0$ that is attributed to a jet. Strikingly, in central collisions, even after subtraction of correlations due to elliptic flow, the jet is accompanied by a "ridge" that is similarly narrow in ϕ but broad in η [1]. Due to its association with a jet, it is sometimes called the "hard ridge". The most striking feature is that the ridge can extend to a long range in pseudorapidty [2]. A similar "soft ridge" structure is also seen in correlation measurements without a jet trigger [3], but due to the separation in p_t the soft and hard ridges have been thought of as separate phenomenon. Based on particle production in an early Glasma stage followed by later stage transverse flow, we propose a common explanation for these phenomena that addresses the long range behavior.

In [4] we extend the model of Ref.[5] to incorporate jet production and address the soft and hard ridges. Long range rapidity correlations and the insight they provide on early time dynamics are our driving concerns [6, 5]. We take particle production to occur through a Glasma state, and emphasize how computed Glasma correlations can affect ridge measurements. In Ref. [5] we found excellent agreement with the peak amplitude and azimuthal width shown in current Au+Au data. In the next section we extend this work here to include Cu+Cu systems. We then extend the model of Ref. [5] to address varying p_t ranges so that we may address the hard ridge. In Sec. 3 we add a contribution of jets following the model of Ref. [7]. We extend that model to compute both the strength and azimuthal dependence of the jet contribution to the hard and soft ridges. We then combine the flow and jet effects and find that correlations of thermally produced pairs constitute a significant contribution to the triggered measurement. We then discuss how experiments might distinguish the different contributions.

2. Glasma Correlations

The theory of Color Glass Condensate (CGC) predicts an early Glasma stage in a high energy collision in which particles are produced by strong longitudinal color fields. As nuclei collide, the transverse fields of each nucleus are instantaneously transformed into longitudinal fields that are approximately uniform in rapidity. These fields are essentially random over transverse distances r_t larger than the saturation scale Q_s^{-1}, where $Q_s \sim 1-2$ GeV. We can think of such field configurations as consisting of a collection of longitudinal flux tubes. Flux tubes are ubiquitous in QCD-based descriptions of high energy collisions. In the Glasma they are closely packed and not strictly distinct due to saturation. They are, however, uncorrelated for $r_t > Q_s^{-1}$. This is their essential feature for this work. The Glasma changes to plasma as particles form from the fields and thermalize.

In the saturation regime, the number of gluons in a rapidity interval Δy is

$$N = (dN/dy)\Delta y \sim {\alpha_s}^{-1} Q_s^2 R_A^2, \qquad (1)$$

where R_A is the nuclear radius and α_s is the strong coupling constant at the saturation scale Q_s [8]. We understand (1) as the number of flux tubes $K \sim (Q_s R_A)^2$ times the density of gluons per flux tube $\propto \alpha_s^{-1}$. The scale of correlations is set by

$$\mathcal{R} = \frac{\langle N^2\rangle - \langle N\rangle^2 - \langle N\rangle}{\langle N\rangle^2}, \qquad (2)$$

where the brackets denote an average over collision events. This quantity vanishes for uncorrelated gluons, for which multiplicity fluctuations are necessarily Poissonian. In Ref. [5] we argued that the Glasma correlation strength is $\mathcal{R} \propto \langle K\rangle^{-1} = (Q_s R)^{-2}$, and found that the Glasma contribution to correlations is

$$\mathcal{R} dN/dy \sim \alpha_s(Q_s)^{-1}, \qquad (3)$$

a result consistent with calculations of Dumitru et al. in Ref. [6]. Equations (1) and (3) constitute initial conditions for the hydrodynamic evolution of the system.

These Glasma initial conditions affect the final state correlations in several ways. First, particles emitted from the same tube share a common origin that is localized to a very small transverse area, since $Q_s \ll R_A$. Second, the flux tubes correlate particles over a large pseudorapidity range. Some of the flux tubes can stretch across the full longitudinal extent of the system at times < 1 fm. These flux tubes rapidly fragment. At later times, the particles they produce can be separated by large longitudinal distances depending on their momenta. Consequently, subsequent scattering and hydrodynamic evolution cannot erase their correlations

– they are causally disconnected. Third, the strength of the correlation depends on the number of flux tubes. The number of tubes depends on the centrality and energy of the collision, as well as the transverse area of the tube, all of which, in turn, depend on Q_s. Finally, as a result of the common origin, particles coming from the same tube must have the same initial radial position and feel the same effects from flow, independent if their rapidity.

During the Glasma phase, flux tubes thermalize into partons and pressure builds as the systems moves toward an equilibrated Quark Gluon Plasma. Partons initially localized in tubes are now localized in small fluid cells with a uniform azimuthal distribution of particles. Following a Hubble-like expansion of the system, the transverse fluid velocity takes the form $\gamma_t \mathbf{v}_t = \lambda \mathbf{r}_t$. All of partons in a fluid cell are boosted radially depending on their initial radial position. Consequentially, partons in any given fluid cell gain transverse momentum in the radial direction and the relative angle between any two momentum vectors in that cell becomes smaller. Furthermore, fluid cells at a larger radial position have a larger final transverse velocity and the relative angle between parton momentum vectors is narrower. This angular narrowing depends only on the initial radial position and the small transverse area of the flux tube source. In our simplistic view, all flux tubes are uniform in rapidity and extend to the same longitudinal length. This provides for a longitudinally uniform fireball that experiences the same radial flow at every longitudinal position. In this way, at every rapidity, angular correlations are enhanced in the same way and the initial state spatial correlations are both preserved through freeze out and represented in momentum correlations.

Following [5] we define the momentum space correlation function at freeze out as

$$r(\mathbf{p}_1, \mathbf{p}_2) = \rho_2(\mathbf{p}_1, \mathbf{p}_2) - \rho_1(\mathbf{p}_1)\rho_1(\mathbf{p}_2) \tag{4}$$

where $\rho_2(\mathbf{p}_1, \mathbf{p}_2) = dN/dy_1 d^2p_{t1} dy_2 d^2p_{t2}$ is the pair distribution, and $\rho_1(\mathbf{p}) \equiv dN/dyd^2p_t$ is the single particle spectrum. We describe the effect of flow using the familiar blast-wave model [9, 10, 11, 12, 13]. In this model, the single particle spectrum is $\rho_1(\mathbf{p}) = \int f(\mathbf{x}, \mathbf{p})\, d\Gamma$, where $f(\mathbf{x}, \mathbf{p}) = (2\pi)^{-3} \exp\{-p^\mu u_\mu / T\}$ is the Boltzmann phase-space density and $d\Gamma = p^\mu d\sigma_\mu$ is the differential element of the Cooper-Frye freeze out surface. We assume a constant proper time freeze out, so that $d\Gamma = \tau_F m_t \cosh(y - \eta) d\eta d^2 r_t$, where $\eta = (1/2)\ln((t+z)/(t-z))$ is the spatial rapidity.

We argue in Ref. [5] that the final-state momentum space correlation function is

$$r(\mathbf{p}_1, \mathbf{p}_2) = \int c(\mathbf{x}_1, \mathbf{x}_2) \frac{f(\mathbf{x}_1, \mathbf{p}_1)}{n_1(\mathbf{x}_1)} \frac{f(\mathbf{x}_2, \mathbf{p}_2)}{n_1(\mathbf{x}_2)} d\Gamma_1 d\Gamma_2, \tag{5}$$

where $n_1(\mathbf{x}) = \int f(\mathbf{x}, \mathbf{p}) d\Gamma$. The spatial correlation function $c(\mathbf{x}_1, \mathbf{x}_2)$ depends on the Glasma conditions as follows

$$c(\mathbf{x}_1, \mathbf{x}_2) = \mathcal{R}\, \delta(\mathbf{r}_t) \rho_{_{FT}}(\mathbf{R}_t), \tag{6}$$

where $\mathbf{r}_t = \mathbf{r}_{t1} - \mathbf{r}_{t2}$ is the relative transverse position, and $\mathbf{R}_t = (\mathbf{r}_{t1} + \mathbf{r}_{t2})/2$ is the average position. The delta function accounts for the fact that Glasma correlations are highly localized to $r_t < Q_s^{-1}$. The factor $\rho_{_{FT}}(\mathbf{R}_t)$ describes the transverse distribution of the flux tubes in the collision volume, which we assume follows the thickness function of the colliding nuclei.

We comment that the form of (6) holds as long as $\mathcal{R}$ is unmodified from its initial Glasma value by particle production and hydrodynamic evolution. This is only strictly true as long as a) subsequent evolution doesn't change the relative number of particles in the rapidity interval of interest; and b) the number of observed hadrons is proportional to the initial number of gluons. Causality prevents these effects from altering $\mathcal{R}$ for truly long range correlations, $|\eta_1 - \eta_2| > 1 - 2$. This would hold for smaller rapidities in Glasma theory as long as boost invariance is a reasonable approximation, since dN/dy is then a hydrodynamic constant of motion in each event. Moreover, assumption (b) is common in Glasma/CGC calculations. On the other hand, for $|\eta_1 - \eta_2| < 1-2$, the experimental dN/dy is not flat and will change with time due to particle diffusion and number changing processes; see [14]. For now, we will assume that $\mathcal{R}$ is constant and defer the hydrodynamic modification for later work.

The analysis in Ref. [5] focused on a measurement of the near side peak of the soft ridge in 200 GeV Au+Au using the observable $\Delta\rho/\sqrt{\rho} = (\rho_{sib} - \rho_{ref})/\sqrt{\rho_{ref}}$ [3]. The quantity $\rho_{sib}(\phi, \eta)$ represents the distribution of "sibling pairs" from the same event, as a function of relative pseudorapidity and relative azimuthal angle, and is comparable to our ρ_2. The quantity $\rho_{ref}(\phi, \eta)$ represents uncorrelated pairs from mixed events and is equivalent to the square of our ρ_1. The difference $(\rho_{sib} - \rho_{ref})$ is a measure of correlated pairs and is comparable to the integral of (5) over the transverse momenta and average azimuthal angle $\Phi = (\phi_1 + \phi_2)/2$. It is convenient to compute the quantity

$$\frac{\Delta\rho}{\rho_{ref}} = \frac{\int r(\mathbf{p}_1, \mathbf{p}_2) p_{t1} p_{t2} dp_{t1} dp_{t2} d\Phi}{\int \rho_1(p_{t1})\rho_1(p_{t2}) d^2p_{t1} d^2p_{t2}}, \tag{7}$$

which is independent of the overall scale of the multiplicity. We emphasize that this quantity includes correlated pairs in which both particles can have any momentum, while measurements of the hard ridge correlate particles from different p_t ranges. We will extend our approach to address such quantities below.

To construct the observed quantity $\Delta\rho/\sqrt{\rho_{ref}}$, we notice that $r(\mathbf{p}_1, \mathbf{p}_2)$ computed from eq. (7) is proportional to the correlations strength $\mathcal{R}$ (2). We can write

$$\frac{\Delta\rho}{\rho_{ref}} = \mathcal{R} F(\phi), \tag{8}$$

where $F(\phi)$ is normalized such that $\int_0^{2\pi} F(\phi) d\phi = 1$. The distribution $F(\phi)$ depends only on the blast-wave parameters $\gamma m/T$ and v_s, and represents the angular correlations of particles from flux tubes after hydrodynamic expansion. The factor $\mathcal{R}$ scales the strength of the correlations with both energy and centrality and determines the rapidity dependence (which is flat in this case).

We now combine (7) and (1) to obtain the observed quantity

$$\Delta\rho/\sqrt{\rho_{ref}} = \kappa \mathcal{R} dN/dy\ F(\phi), \tag{9}$$

where we equate the factor $\mathcal{R}dN/dy$ with (3), which accounts for all of the Glasma energy and centrality dependence. The scale constant κ is independent of energy. We adjust κ for Au+Au 200 GeV systems only. To highlight the energy dependence of the Glasma scale (3), in Ref.[5] we fix κ such that $F(\phi)$ for the most *central* collisions is aligned with the most central data point. Here we show that without readjustment of κ the blast wave formalism alone cannot account for the difference in correlation amplitude from in 200 and 62 GeV systems. This difference is governed by the magnitude, transverse size, and number of fluctuations in the initial system. In Ref.[4] we fix κ such that $F(\phi)$ for *peripheral* collisions is aligned with the most peripheral Au+Au 200 GeV data point. Here we emphasize that the centrality dependence of a wounded nucleon model based on *pp* like collisions does not describe the shape of the data. We show this situation here for Cu+Cu systems without readjustment of κ.

We now apply (9) and (3) to address new and forthcoming data for $\Delta\rho/\sqrt{\rho_{ref}}$ and the azimuthal width Cu+Cu for both 200 and 62 GeV as a function of centrality. The Q_s^2 has a dependence on the density of participants $\rho_{\rm part}$ that is determined in Ref. [8]. The saturation scale in central Cu+Cu is then

$$Q_s(\mathrm{Cu})^2 = Q_c(\mathrm{Au})^2 \frac{\rho_{\rm part}(\mathrm{Cu,\ central})}{\rho_{\rm part}(\mathrm{Au,\ central})}. \tag{10}$$

Journal of Physics: Conference Series **230** (2010) 012029 doi:10.1088/1742-6596/230/1/012029

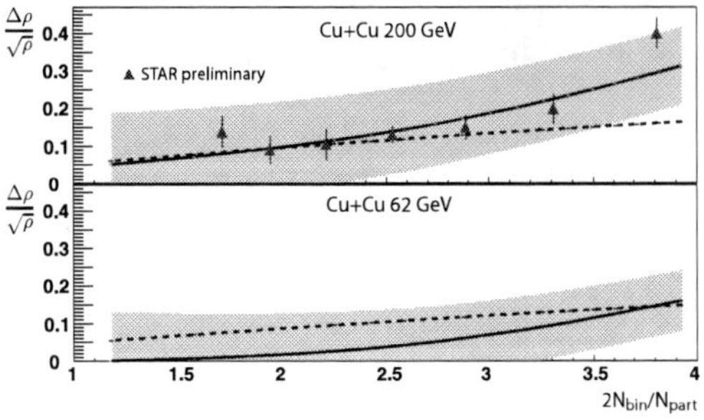

Figure 1. Cu+Cu soft ridge amplitude as a function of centrality for both 200 and 62 GeV. Dashed lines represent the blast wave blast-wave integrations $F(\phi)$ normalized to peripherial 200 GeV Au+Au.

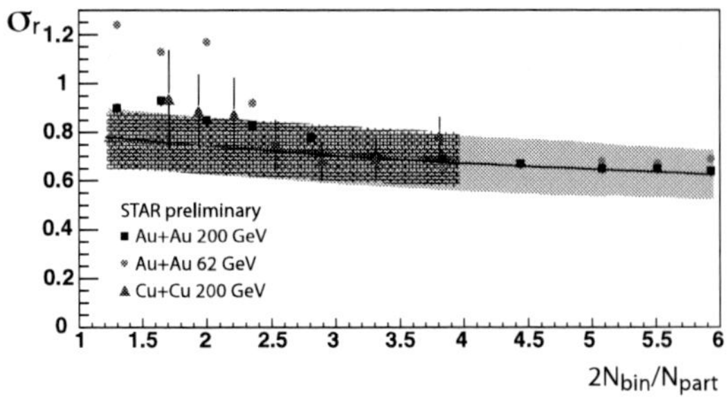

Figure 2. Comparison of the previously published angular width calculations for Au+Au (black line) with the angular widths for Cu+Cu systems (blue line, hatched error band).

After this scaling, we take the relative centrality dependence of Q_s to be the same in the Cu and Au systems. This assumption can be refined by measuring the centrality dependence of dN/dy in Cu+Cu. Similarly, we obtain the blast wave parameters T and v_s in Cu+Cu from the Au+Au values by assuming that they scale with the number of participants. These assumptions can be refined as single particle spectra measurements as a function of centrality become available.

Figure 1 shows our prediction for the soft ridge amplitude in Cu+Cu 200 and 62 GeV systems as a function of centrality compared to preliminary STAR data [15]. In both panels, the dashed line represents the blast-wave amplitudes $F(\phi = 0) - F(\phi = \pi)$ normalized to fit Au+Au at 200 GeV in Ref. [4]. The dashed lines are included to show how the energy and system-size dependence affects the calculation when the Glasma dependence (3) is omitted. The solid lines are the result of including the Glasma scaling (3) adjusted by (10). The error band represents a 10% uncertainty in the blast-wave parameters plus an additional uncertainty in the parameterization of Q_s that increases with decreasing centrality.

In Fig. 2 we show the soft ridge azimuthal width in Cu+Cu systems compared to preliminary STAR measurements as compared to previously published Au+Au result [5]. We have also included preliminary STAR measurement of the Cu+Cu width [15] to compare with other preliminary STAR measurements of the soft ridge in Au+Au width [3]. In [4, 5] we find that the azimuthal width of the near side peak of the soft ridge in Au+Au is due to radial flow, is constant with a change in energy, and is relatively uniform with change in centrality. The error band is representative of the uncertainty of fitting an offset gaussian to the angular calculation. Since the azimuthal width is completely determined by radial flow, which depends completely on the choice of centrality and blast-wave parameterizations, and all of those parameters have remained unchanged, we calculate the same enhancement in the width for Cu+Cu as Au+Au. The black line extending to $\nu = 6$ is the Au+Au result from [5], and the overlaid blue line extending to $\nu = 4$ with the hatched error band is the Cu+Cu result. Again, as with the Au+Au result, the Cu+Cu result is independent of energy since the measured transverse expansion does not depend on energy.

A key feature of the flow-based descriptions of the ridge is that it is the angular width, σ_r, of correlated pairs decreases as the mean p_t of the pair increases. The greater the radial boost given to a fluid cell, the narrower the relative angle between the momentum vectors of particles in that cell. A very high momentum correlated pair is more likely to have come from a fluid cell that received a very large transverse boost. To study whether this effect is present, we compute $\Delta\rho/\sqrt{\rho_{ref}}$ for pairs of $p_t > p_{t,\min}$. As $p_{t,\min}$ increases, we also expect the amplitude of the ridge to decrease, since it is more difficult to find higher p_t bulk particles. We therefore compute

$$\left(\frac{\Delta\rho}{\rho_{ref}}\right)_{p_t} = \frac{\int\limits_{p_{t2,min}}^{p_{t2,max}}\int\limits_{p_{t1,min}}^{p_{t1,max}} r(\mathbf{p}_1,\mathbf{p}_2)}{\int\limits_{p_{t1,\mathrm{min}}}^{p_{t1,max}}\rho_1(p_{t1})\int\limits_{p_{t2,min}}^{p_{t2,max}}\rho_1(p_{t2})} = \mathcal{R}F(\phi;p_{t1,min},p_{t1,max},p_{t2,min},p_{t2,max}), \tag{11}$$

where the integration measures are the same as in (7). To obtain the measured ratio we write

$$\frac{(\Delta\rho/\sqrt{\rho_{ref}})_{p_t}}{\Delta\rho/\sqrt{\rho_{ref}}} = \frac{F(\phi;p_{t,min},\infty,p_{t,min},\infty)}{F(\phi)}\frac{\int_{p_{t1,min}}^{\infty}\rho_1}{\int_0^{\infty}\rho_1}. \tag{12}$$

For increasing values of $p_{t,min}$ we calculate the correlation function as before using (11) and (12) and find that the azimuthal width does indeed decrease as shown in the upper panel of Fig. 4.

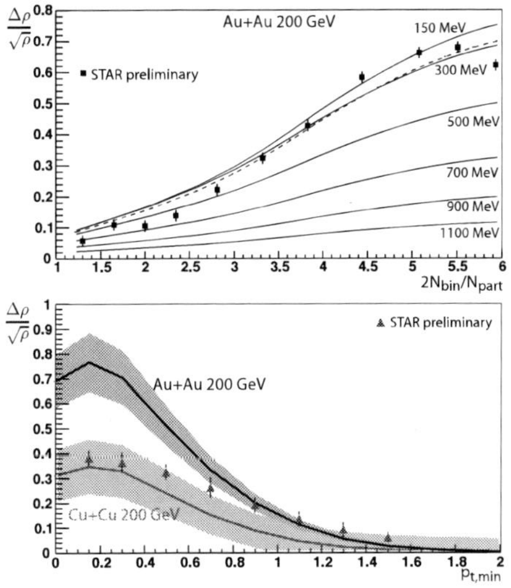

Figure 3. Top panel: Au+Au 200 GeV amplitude calculations for increasing $p_{t,min}$ limits. The dashed line is calculation with $p_{t,min} = 0$. Lower panel: the soft ridge amplitude for most central collisions vs the $p_{t,min}$ limit.

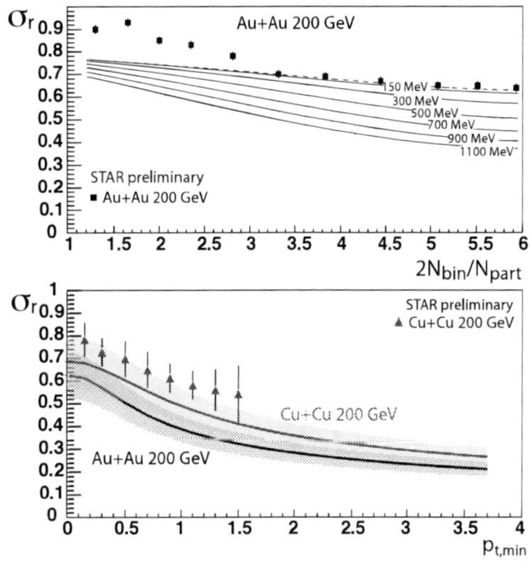

Figure 4. Top panel: Au+Au angular width for increasing $p_{t,min}$. The dashed line is the calculation $p_{t,min}$=0. Lower panel: The azimuthal width for most central collisions vs the $p_{t,min}$.

The upper panel of Fig. 3 shows the correlation amplitude vs. centrality for different choices of $p_{t,min}$. The amplitude decreases with increasing $p_{t,min}$ because the number of particles contributing to correlations is reduced. The lower panel of Fig. 3 shows the amplitude of $\Delta\rho/\sqrt{\rho_{ref}}$ for the most central collision as a function of the choice of $p_{t,min}$ for Au+Au and Cu+Cu at 200 GeV. We compare these calculations to preliminary Cu+Cu data [15]. Similarly, the blue curve in the lower panel of Fig. 4 represents the azimuthal width of the soft ridge in most central collisions as a function of $p_{t,min}$.

We see that the azimuthal width of the hard ridge is smaller than that of of the soft ridge, but as the $p_{t,min}$ limit of the soft ridge is increased, the amplitude of the correlations drops and the azimuthal width narrows. In the p_t range of the hard ridge, it appears that the azimuthal width could be narrow enough, but the amplitude is not directly comparable. To understand this difference, we must understand the differences in the two measurements. The most significant difference is the choice of the momentum range of the correlated particles. The hard ridge measurement analyzes the yield of associated particles per jet trigger where the associated particle p_t range and the trigger range do not overlap. The soft ridge measurement, however, finds the number of correlated pairs per particle where both particles are in the same range with p_t above minimum bias. The normalization of the soft ridge is found by taking the square root of the uncorrelated pair reference spectrum.

STAR measures the hard ridge, or yield of associated particles per jet trigger, for Au+Au 200 GeV for $3 < p_{t,trigg} < 4$ GeV with $2 < p_{t,assoc} < 3$ [1]. Identifying p_{t1} with the trigger range and the associated range with p_{t2}, we calculate $\Delta\rho/\sqrt{\rho_{ref}}$ and transform to yield by

$$\text{Yield} = \left(\frac{\Delta\rho}{\sqrt{\rho_{ref}}}\right)_{p_t} \left(\frac{\int \rho_1(p_{t2})}{\sqrt{\int \rho_1(p_{t1}) \int \rho_1(p_{t2})}}\right). \tag{13}$$

At higher ranges of $p_{t1,2}$ the contribution from jets should become more significant. It is important therefore to know the relative contribution of thermal particles and jet particles. As will be discussed in more detail later, we decompose the total particle spectrum into thermal bulk and jet fractions, and to obtain the contribution of bulk correlations to the hard ridge, we multiply (13) by the bulk fraction $\int \rho_1(p_{t1}) / \int \rho_{tot}(p_{t1})$ where $\int \rho_{tot}(p_{t1})$ is the total number of particles in the range of p_{t1}.

The blue curve in Fig. 5 represents the contribution to the hard ridge from only thermal bulk pairs. As can be seen on the figure, bulk-bulk correlations contribute significantly to the amplitude of the triggered measurement, but seems to have a somewhat narrow profile in azimuth. It was shown in [7] that a jet acquires angular correlations with flowing matter due to quenching, but the width of the correlation is wider than the data. The contribution of jet correlations with bulk particles could make up the difference between the blue curve in Fig. 5 and the data by increasing both the amplitude and the width of the calculation.

In the next section we combine a theory angular correlations from [7] with spatial correlations of jets and flux tubes to obtain a jet-bulk contribution to the hard ridge.

3. Jets, Glasma, and Correlations

As the p_t of correlated particles is increased, the contributions from jets should become prevalent, particularly for small η. At small rapidity differences, correlations of jet particles with fragments should be large, but restricted to the size of the jet cone, and the transfer of momentum from jet particles to bulk particles is causally limited to $\sim 1-2$ units in rapidity. The existence of correlations with jet particles at larger η would require a correlation early in the collision that remains through the longitudinal expansion of the system and is still present at freeze out. Both the hard collisions and flux tubes are made in the initial moments of the nuclear collision. Assuming that the entire overlap region of the colliding nuclei is in the saturation regime, flux tubes would fill the collision volume and a jet formed at any transverse position would be accompanied by a flux tube at the same position. Since the flux tube extends to large rapidities, the correlation of particles from the tube and the jet can extend to large η. Angular correlations arise since particles from the tube acquire a radial trajectory from flow as before, but the jet trajectory has a bias in the radial direction due to quenching [7].

We construct a distribution of jets as follows. We assume that jets are produced with a hard scattering rate $f_0(\mathbf{p}_1)$ that is independent of position, multiplied by a spatial profile $P_{rod} \propto (1 - r_1^2/R_A^2)$ that is roughly proportional to the density of binary collisions. The phase space density of jet particles is then

$$f_J(\mathbf{x}_1, \mathbf{p}_1) = f_0(\mathbf{p}_1) P_{prod}(r_1) S(r_1, \phi_1), \tag{14}$$

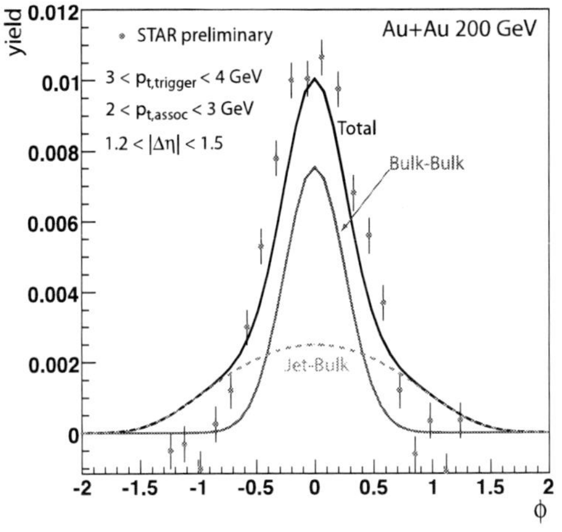

Figure 5. Angular profile of the jet triggered ridge in a rapidity range away from the jet peak. The solid black line combines of long range correlations. Bulk-bulk (blue line) and jet-bulk (dashed red line) contributions are shown separately. The jet fraction is determined by $p_s = 1.25$ GeV; bulk-bulk correlations make up $\sim$ 75% of the total amplitude.

where S is the survival probability of a jet due to jet quenching. In practice $f_0(\mathbf{p}_1)$ cancels in $\Delta\rho/\rho_{ref}$ so we need not specify it. We follow Ref. [7] and take

$$S(r_1, \phi_1) = exp(-L(r_1, \phi_1)/l_{abs}), \tag{15}$$

where $l_{abs} = 0.25fm$ is the jet attenuation length. The survival of the jet depends on the path it takes out of the medium

$$L(r_1, \phi_1) = \sqrt{R_A^2 - r_1^2 \sin^2(\phi_1)} - r_1 \cos(\phi_1). \tag{16}$$

The path (16) is the distance a jet would have to travel out of a circular transverse area at an angle ϕ_1 with respect to the radial vector pointing to its position of production r_1 [7]. In view of (16) and (15), the shortest path, which is the path a jet is most likely to survive, is one that is radially outward from its position of production. Although a jet production is at a minimum on the surface, $r \approx R_A$, with little material for the jet to pass through, the survival probability is maximum in any direction (not pointing into the volume). The resulting angular correlations are weak, since only a small fraction of phase space is restricted by quenching. The largest probability of production would occur at the center where $L \approx R_A$ in all directions, but there are no correlations in this case since quenching would be maximum. More concisely, jets are less likely to be produced at the surface and have a wide angular distribution. The most jets are produced the center and would have the narrowest correlation with radially flowing particles, but have the highest probability of being quenched. As the production point of the jet moves from the center toward the surface, the probability of production decreases, the probability of survival increases, and the angular correlation with radially flowing particles widens. Integration over all possibilities determines the width of the correlations.

The calculation follows the analysis of (5), but with the first particle from a jet and the second from a flux tube, so that

$$r_{JB}(\mathbf{p}_1, \mathbf{p}_2) = \int c_{JB}(\mathbf{x}_1, \mathbf{x}_2) \frac{f_J(\mathbf{x}_1, \mathbf{p}_1)}{n_{1J}(\mathbf{x}_1)} \frac{f_B(\mathbf{x}_2, \mathbf{p}_2)}{n_{1B}(\mathbf{x}_2)} d\Gamma_1 d\Gamma_2. \tag{17}$$

The correlation function $c_{JB}(\mathbf{x}_1, \mathbf{x}_2)$ requires that a jet and a bulk particle must come from the same radial position. The rest of the equation accounts for the different the spectra for jet and bulk particles. We write

$$c_{JB}(\mathbf{x}_1, \mathbf{x}_2) = \frac{\mathcal{R}_{JB}\langle N_J \rangle}{\mathcal{R}\langle N_B \rangle} c(\mathbf{x}_1, \mathbf{x}_2), \tag{18}$$

where c is given by (6).

To relate the correlation strength $\mathcal{R}_{JB}$ to the bulk correlation strength $\mathcal{R}$ discussed earlier, we assume that the hard scattering rate is independent of the flux tube dynamics. Recall that the bulk quantity $\mathcal{R}$ in (2) is related to the number of flux tubes; see (3). If we take the fraction of jet and bulk particles per flux tube to be independent of the number of flux tubes, we can write $\langle N_J\rangle = \alpha\langle N\rangle$, where α and β may depend on momentum, but do not vary event by event. We then follow [16] to find $\langle N_B\rangle = \beta\langle N\rangle$, and $\langle N_J N_B\rangle = \alpha\beta\langle N(N-1)\rangle$, so that

$$\mathcal{R}_{JB} = \frac{\langle N_J N_B\rangle - \langle N_J\rangle\langle N_B\rangle}{\langle N_J\rangle\langle N_B\rangle} = \frac{\alpha\beta\langle N(N-1)\rangle - \alpha\langle N\rangle\beta\langle N\rangle}{\alpha\langle N\rangle\beta\langle N\rangle} = \frac{\langle N(N-1)\rangle - \langle N\rangle^2}{\langle N\rangle^2} = \mathcal{R}. \quad (19)$$

Therefore, the addition of jets to the total multiplicity doesn't change the correlation strength. In essence, the beam jet associated with the hard process is just another flux tube in the high density Glasma state.

We can now rewrite (9) for jet-bulk correlations as

$$\Delta\rho_{JB}/\sqrt{\rho_{ref}} = \kappa\mathcal{R}dN_{jet}/dy\ F_{JB}(\phi). \quad (20)$$

Our calculation of yield and implementation of lower p_t limits follows (13) and (11) but, this time, we scale by the jet fraction $\int \rho_{1,J}(p_{t1})/\int \rho_{1,tot}(p_{t1})$.

In order to compute the amplitude and azimuthal width of the hard ridge in Fig. 5, we must determine the relative contributions from both bulk-bulk and jet-bulk correlations. The measured p_t spectrum follows an exponential behavior at low p_t and a power law behavior where jets play a larger role, see e.g. Ref. [17]. Our blast wave formulation describes the exponential behavior of the low p_t spectrum well. The scale p_s at which the spectrum begins to deviate from exponential behavior is proportional to Q_s in Glasma theory, but the proportionality constant is not known. This introduces a free parameter – p_s at $\sqrt{s} = 200$ GeV – that we fix below. We then find the number of jet particles by taking the difference between the total number of particles and the number of thermal particles $\rho_{1,J} = \rho_{1,tot} - \rho_{1,B}$. We take $\rho_{1,tot}$ from the measured spectrum in Ref. [17] and $\rho_{1,B}$ from the blast wave calculation with the appropriate normalization.

We now calculate the combined effect of Glasma, flow, and jets on the correlation function. Adding the bulk-bulk and jet-bulk contributions, we obtain

$$\frac{\Delta\rho}{\sqrt{\rho_{ref}}} = \kappa\mathcal{R}\frac{dN}{dy}F_{BB}(\phi)\frac{\int\rho_{1,B}(p_t)}{\int\rho_{1,tot}(p_t)} + \kappa\mathcal{R}\frac{dN_{jet}}{dy}F_{JB}(\phi)\frac{\int\rho_{1,J}(p_t)}{\int\rho_{1,tot}(p_t)}. \quad (21)$$

In order to compare to the yield of associated particles in the hard ridge, we combine (21) and (13) with the appropriate integration limits. We find that the agreement with the data in Fig. 5 requires $p_s = 1.25$ GeV. The dashed red curve in Fig. 5 represents the contribution to the yield from correlations of jet and bulk thermal particles. This contribution is too wide. On the other hand, the bulk-bulk correlation function describing the effect of flow alone, which is given by the blue curve, is too narrow and the computed peak height is too small. The combination of the two effects shown as the black curve in Fig. 5 gives nice agreement with both the amplitude and azimuthal width.

We now compare hard and soft ridge measurements directly by computing the momentum dependent correlation function (21). At low $p_{t,min}$ the contribution from jets is negligible, therefore the amplitude and width of (21) is determined by the bulk-bulk term. As the $p_{t,min}$ is increased, both the amplitude and the azimuthal width of the bulk-bulk term decreases, while the amplitude of the jet-bulk term increases. The azimuthal width of jet-bulk correlations is roughly independent of $p_{t,min}$; the growth in the figures for $p_{t,min} > 1$ GeV is due to the growth of the jet fraction. Jet-bulk correlations should become a more significant fraction of the total as $p_{t,min}$ is increased, and the azimuthal width of the ridge should increase toward the jet-bulk width. The trend shown in Figs. 4 follow preliminary data from Ref. [15]. The small difference may reflect the fact that our calculation omits the jet peak.

We emphasize that the decrease of the bulk-bulk contribution to the amplitude of $\Delta\rho/\sqrt{\rho}$ and σ_r with increasing p_t in Figs. 4 and 3 is a direct consequence of transverse flow and, consequently, is a firm prediction if the jet-bulk contribution is neglected. The role of jets and other phenomena like recombination are less clear. We have chosen the model of Ref. [7] because it relies only of the well-studied phenomena of jet quenching. Our calculations using this model predict that the width would increase for higher p_t ranges.

4. Summary

We have shown that long range Glasma correlations contribute significantly to both the hard and soft ridges. Bulk-bulk correlations arise due to particle production from a Glasma state followed by transverse flow [4, 5], [6]. The Glasma initial conditions allow us to extend to study of Cu+Cu systems at both 200 and 62 GeV. We see in Fig.1 and Fig.2, that we maintain good agreement with current data. To study the soft contribution to the hard ridge, we study the momentum dependence of bulk-bulk correlations and find that they are still significant, even in the p_t ranges of the hard ridge measurement. To explore the effect of jet production and quenching on the ridge we included a model of jet-bulk correlations following [7]. As seen in Fig.5 we find that we cannot explain the magnitude and width of the hard ridge without a substantial soft component.

Acknowledgments

We thank L. McLerran, R. Bellwied, C. De Silva, A. Timmins, A. Dumitru, A. Majumder, J. Nagel, P. Sorenson, P. Steinberg, R. Venugopalan, C. Pruneau, and S. Voloshin. This work was supported in part by U.S. NSF grants PHY-0348559 and PHY-0855369.

References

[1] Putschke J 2007 *J. Phys.* **G34** S679–684 (*Preprint* nucl-ex/0701074)
[2] Alver B *et al.* (PHOBOS) 2010 *Phys. Rev. Lett.* **104** 062301 (*Preprint* 0903.2811)
[3] Daugherity M (STAR) 2008 *J. Phys.* **G35** 104090 (*Preprint* 0806.2121)
[4] Moschelli G and Gavin S 2010 *Nucl. Phys.* **A836** 43–58 (*Preprint* 0910.3590)
[5] Gavin S, McLerran L and Moschelli G 2009 *Phys. Rev.* **C79** 051902 (*Preprint* 0806.4718)
[6] Dumitru A, Gelis F, McLerran L and Venugopalan R 2008 *Nucl. Phys.* **A810** 91 (*Preprint* 0804.3858)
[7] Shuryak E V 2007 *Phys. Rev.* **C76** 047901 (*Preprint* 0706.3531)
[8] Kharzeev D and Nardi M 2001 *Phys. Lett.* **B507** 121–128 (*Preprint* nucl-th/0012025)
[9] Schnedermann E, Sollfrank J and Heinz U W 1993 *Phys. Rev.* **C48** 2462–2475 (*Preprint* nucl-th/9307020)
[10] Kiyomichi A 2005 *Study of identified hadron spectra and yields at mid- rapidity in s(NN)**(1/2) = 200-GeV Au + Au collisions* Ph.D. thesis
[11] Barannikova O Y (STAR) 2004 (*Preprint* nucl-ex/0403014)
[12] Iordanova A, Barannikova O and Hollis R S (STAR) 2007 *Int. J. Mod. Phys.* **E16** 1800–1805 (*Preprint* 0705.0691)
[13] Retiere F and Lisa M A 2004 *Phys. Rev.* **C70** 044907 (*Preprint* nucl-th/0312024)
[14] Gavin S and Abdel-Aziz M 2006 *Phys. Rev. Lett.* **97** 162302 (*Preprint* nucl-th/0606061)
[15] De Silva L C (STAR) 2009 (*Preprint* 0910.5938)
[16] Pruneau C, Gavin S and Voloshin S 2002 *Phys. Rev.* **C66** 044904 (*Preprint* nucl-ex/0204011)
[17] Adams J *et al.* (STAR) 2003 *Phys. Rev. Lett.* **91** 172302 (*Preprint* nucl-ex/0305015)

Forward-Backward Multiplicity Correlations for Identified Particles in Au+Au 200 GeV Collisions

Michael Skoby (for the STAR Collaboration)

Department of Physics, Purdue University, 525 Northwestern Avenue, West Lafayette, IN 47907

E-mail: skoby@purdue.edu

Abstract. Preliminary STAR measurements of the forward-backward multiplicity correlation as a function of rapidity separation for pions and (anti)protons in Au+Au 200 GeV collisions are presented for 0-10% and 10-20% centralities. A strong, uniform correlation is shown for pions, and is smaller for (anti)protons. The Color String Percolation Model predicts the long-range correlation, and suggests a possible measurement to study a transition from percolation to the quark-gluon plasma.

1. Introduction

Kinematic observables of hadronic matter from relativistic heavy-ion collisions are measured at STAR in search of the so-called 'quark-gluon plasma'. Long-Range (rapidity separation > 1) Forward-Backward multiplicity correlations (LRC) may be a signal for multiple partonic interactions in dense matter, whereas short-range (rapidity separation < 1) correlations are due to independent sources [1,2]. Previously, strong LRC have been measured at STAR in 200 GeV central Au+Au collisions, and were shown to decrease with decreasing centrality [3]. To further investigate the origin of the forward-backward correlation, it can be studied with respect to its particle species dependence (pions, kaons, protons), and measured as a function of rapidity separation (Δy). The Color Glass Condensate (CGC) model, which describes sources as longitudinal flux tubes, predicts that the correlation will grow with centrality [4,5]. Furthermore, fluctuations in the number of gluons at early times will produce a long range correlation significantly larger for pions than (anti)protons [5]. Pions and (anti)protons are identified at STAR by measuring their specific ionization energy loss in the Time Projection Chamber (TPC) [6]. Preliminary results of the measured correlation are presented for pions, kaons, and protons for 0-10% and 10-20% centralities. As seen previously in hadron-hadron experiments, the average multiplicity of particles in the backward region can be related to the multiplicity in the forward region.

$$\langle N_B\rangle(N_F) = a + bN_F \tag{1}$$

Applying a linear regression one can obtain the correlation strength b.

$$b = \frac{<N_f N_b> - <N_f><N_b>}{<N_f^2> - <N_f>^2} = \frac{D_{bf}^2}{D_{ff}^2} \tag{2}$$

2. Data analysis

The data used in this analysis are from 2004 Au+Au 200 GeV collisions acquired at STAR. The correlation was measured for 0-10% and 10-20% centralities. Tracks in the TPC with rapidity < 1 and distance of closest approach from the primary vertex < 3 cm were used for the measurement. Previous measurements of forward-backward correlations were made as a function of pseudorapidity. However, once particle species are identified the measurement of the correlation as a function rapidity is possible. Particle identification is done by making transverse momentum and average energy loss cuts. Particle species are separated into bands when the average energy loss is plotted as a function of transverse momentum, as in figure 1.

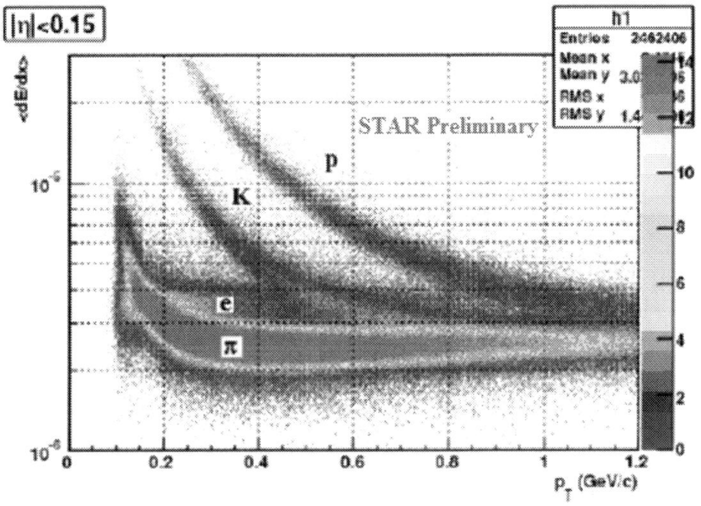

Figure 1. Average energy loss vs. p_T shows bands for various particle species.

The particle species bands overlap with each other in certain momentum ranges. The p_T cuts used in this analysis are 0.2-0.6 GeV/c for pions and kaons, and 0.4-1.0 GeV/c for (anti)protons. Background protons contaminate the proton sample by about 30% for p_T = 0.4 GeV/c and about 2% for p_T = 0.8 GeV/c. Also, the cuts in the number of standard deviations from the measured particle species' <dE/dx> to those obtained from theoretical expectations [7] were $n\sigma_\pi<2$, $n\sigma_K>2$, $n\sigma_e>2$ for pions, $n\sigma_\pi>3$, $n\sigma_K<1.5$ for kaons, and $n\sigma_\pi>2$, $n\sigma_K>2$, $n\sigma_p<2$ for (anti)protons. After particles are identified the rapidity region is divided into bins of width y = 0.16, so the correlation is measured across Δy = 0.16, 0.32, 0.48, 0.64, 0.80, 0.96 for kaons and (anti)protons. In addition to the aforementioned rapidity separations, the measurement is also done for Δy = 1.12, 1.28, 1.44, 1.60 for pions. The pseudorapidity acceptance in the TPC accounts for the limited rapidity measurement for kaons and (anti)protons. Particles used to determine the centrality of an event were counted in a rapidity region separate from the particles used to measure the correlation in order to avoid autocorrelations.

Particles were counted in the forward and backward regions on an event-by-event basis, and binned by event centrality. In order to reduce the effect of centrality fluctuations, a fit is made to the average values of N_f, N_b, $N_f N_b$, $N_b N_b$ as a function of the number of charged particles used to determine the centrality. After applying acceptance corrections and tracking efficiencies this fit gives the average values used to calculate the dispersions D^2_{ff} and D^2_{bf}, normalized by the total number of events in each bin.

3. Results and discussion

Figure 2 shows the forward-backward correlation for *(a)* (anti)protons, *(b)* kaons, and *(c)* pions in 0-10% and 10-20% centrality Au+Au 200 GeV collisions as a function of rapidity separation Δy. Only the short-range correlation is measured for (anti)protons and kaons due to the pseudorapidity acceptance at STAR. The correlation is small (~0.25) and uniform for kaons and (anti)protons in 0-10% collisions, and decreases to ~0.10 for 10-20% collisions. The decrease in the (anti)proton correlation strength from 0-10% to 10-20% is not expected in the context of the CGC picture, which says the baryon correlation should not grow with increasing centrality [5].

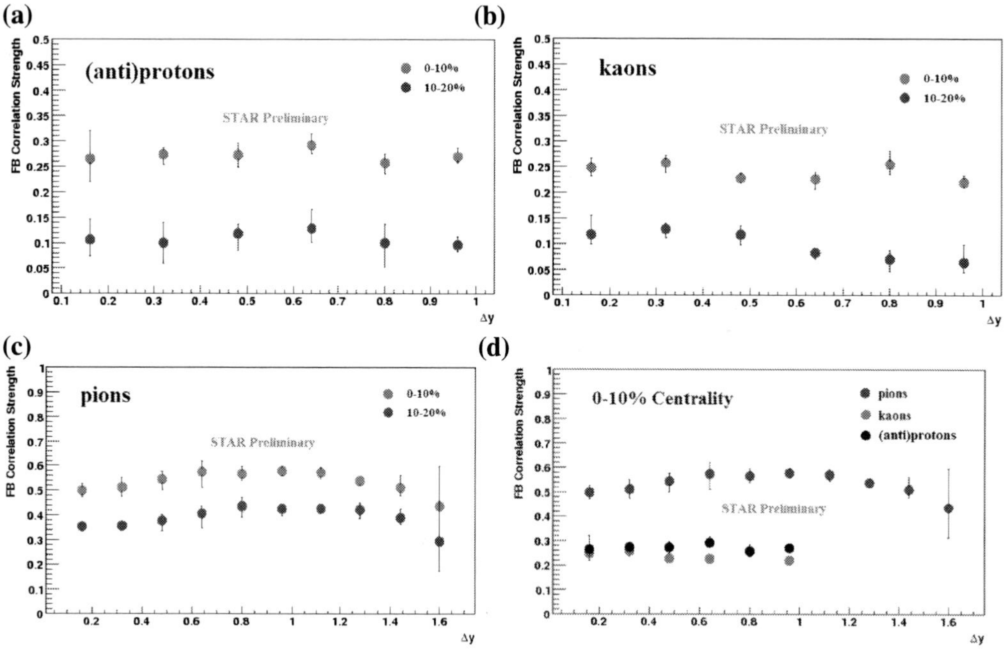

Figure 2. The FB correlation strength in Au+Au 200 GeV collisions for (a) (anti)protons, (b) kaons, and (c) pions for 0-10% and 10-20% centralities, and (d) a comparison of all species in central collisions.

Conversely, a strong (~0.50) uniform correlation is observed for pions in 0-10% collisions and decreases to ~0.40 in 10-20% collisions. The small short-range correlation for (anti)protons and kaons, compared to pions, suggests the LRC will also be small for these species. Figure 2 (d) shows a comparison of the correlation for all species in central collisions, where the LRC for pions is substantially larger than what is expected for (anti)protons. CGC predicts the LRC will be larger for pions, which presumably arise from gluons, than for (anti)protons due to fluctuations in the number of gluons in early times [5]. CGC also predicts measurements of the correlation for inclusive charged hadrons at high transverse momentum in central collisions may not be as large since contributions from baryons increase at high p_T. This will be studied in the future.

The observation of the LRC motivates a future investigation into other models that predict it. In the Color String Percolation Model (CSPM) color strings responsible for particle production overlap as the string density (□) increases-forming clusters [8]. At a critical string density these clusters form a connected system that extends across the medium (see figure 3). STAR can investigate a transition from percolation to QGP in the context of the CSPM by extracting the string density from the measured pT spectra for increasing centralities.

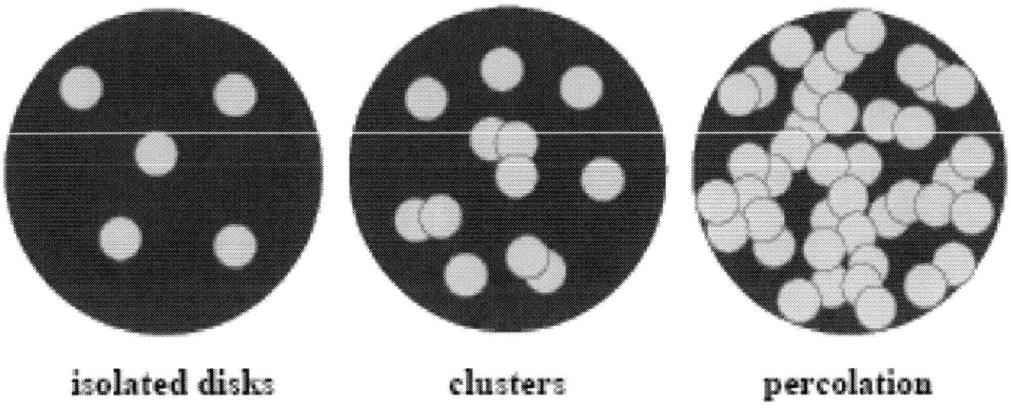

Figure 3. 2-D representation of percolation. The green disks are strings, which overlap to form clusters as the string density increases, and finally percolation is reached when the clusters span the across the medium as a connected system.

The CSPM predicts LRCs due to multiplicity fluctuations within overlapping strings, where the overlap region acts as an independent emitter. Color fields added in the overlap area reduce the effective color field, invoking an overlapping factor.

$$F(\eta) = \sqrt{\frac{1-e^{-\eta}}{\eta}} \tag{3}$$

Measuring the string density in p+p and Au+Au collisions can be done by extracting $F(\eta)$ from fits to the p_T distributions [9]. First, a fit to the p+p p_T distribution will yield fit parameters a, p_o, and m.

$$\frac{dN}{dp_t^2} = \frac{a}{(p_0 + p_t)^m} \tag{4}$$

For Au+Au collisions, adjust p_o to account for the percolation of strings.

$$p_0 \to p_0 \left(\frac{\langle nS_1 / S_n \rangle_{Au-Au}}{\langle nS_1 / S_n \rangle_{pp}} \right)^{1/4} \tag{5}$$

Here, n is the number of strings in a cluster, S_1 is the area of one string, and S_n is the area of a cluster. $F(\eta)$ can be determined using the relation

$$\sqrt{\frac{F(\eta)_{pp}}{F(\eta)_{Au-Au}}} = \left(\frac{\langle nS_1 / S_n \rangle_{Au-Au}}{\langle nS_1 / S_n \rangle_{pp}} \right)^{1/4} \tag{6}$$

Due to the low string overlap probability in p+p collisions $F(\eta)_{pp} \sim 1$.

4. Summary

Strong long-range correlations indicate the occurrence of multiple partonic interactions. Preliminary measurements show a strong, uniform LRC across Δy for pions in central Au+Au collisions at 200 GeV, which decreases from central to peripheral collisions. The small short-range correlation for (anti)protons, compared to pions, suggests the LRC will also be small for this species. CGC claims that the LRC is larger for pions than (anti)protons due to the fluctuation in the number of gluons, and can only be created at early times. A possible transition from percolation to QGP can be investigated by measuring the percolation string density in the context of the CSPM picture.

References

[1] A. Capella and A. Krzywicki, Phys. Rev. D **18**, 4120 (1978).
[2] A. Capella *et al.*, Phys. Rep. **236**, 225 (1994).
[3] B. I. Abelev *et al.* (STAR Collaboration), Phys. Rev. Lett. **103**, 172301 (2009).
[4] Y. V. Kovchegov, E. Levin, and L. McLerran, Phys. Rev. C **63**, 024903 (2001).
[5] N. Armesto, L. McLerran, and C. Pajares, Nucl. Phys. A **781**, 201 (2007).
[6] K.H. Ackermann *et al.*, Nucl. Instr. and Meth. A **499**, 659 (2003).
[7] B. I. Abelev *et al.* (STAR Collaboration), Phys. Rev. C **79**, 034909 (2009).
[8] M. A. Braun and C. Pajares, Eur. Phys. J. C **16**, 349 (2000).
[9] M. A. Braun and C. Pajares, Phys. Rev. C **65**, 024907 (2002).

Journal of Physics: Conference Series **230** (2010) 012031 doi:10.1088/1742-6596/230/1/012031

Studying the Jet Medium Interaction by Two Particle $\Delta\eta$-$\Delta\phi$ Correlations

Chin-Hao Chen for the PHENIX Collaboration

Department of Physics and Astronomy, Stony Brook University, Stony Brook, 11794-3400, New York

E-mail: chinchen@ic.sunysb.edu

Abstract. We present two dimensional $\Delta\eta$-$\Delta\phi$ inclusive photon-charged hadron correlations measured by the PHENIX experiment. Jet-like correlations are modified in central Au+Au compared to p+p, in both the trigger jet and opposing jet. The trigger jet is elongated in pseudo-rapidity (the "ridge"), while the opposing jet shows a double peak structure ("head" and "shoulder"). We decompose the structures by fitting the $\Delta\eta$ and $\Delta\phi$ correlations to disentangle contributions from the medium and the punch-through and trigger jets. The ridge exists for associated particle p_T below 3 GeV/c; it is broad in rapidity and narrow in $\Delta\phi$. The away side correlated particle yield is enhanced in central collisions. The yield of the ridge closely resembles the shoulder in the centrality dependence of particle yield and spectra.

The hot dense QCD medium created in high energy heavy ion collisions at RHIC has been shown to suppress the production high p_T hadrons which is indicative of parton energy loss in the medium [1, 2]. This phenomenon is known as "jet quenching". The single particle nuclear modification factors provide quantitative constraints to the properties of the medium, such as the initial color-charge density, dN^g/dy, or the medium transport coefficient $\langle\hat{q}\rangle$ to $\pm 20-25\%$ level [3]. Two particle correlations are a powerful tool to study the jet-medium interactions and provide further information on the jet-energy loss mechanisms and the medium response to the jets. Correlations in azimuthal angle, $\Delta\phi$, and in pseudorapidity, $\Delta\eta$, are constructed with a trigger particle" from a specified p_T range and all other particles in the event which fall in a given p_T range ("associated particle"). When triggering on high p_T particles, the away-side shape ($\Delta\phi \approx \pi$) depends on the p_T of the associated particles. For high p_T associated particles, the away-side shows a jet-like structure, that is suppressed compared to p+p collisions in which no hot medium in present [4] [5]. As the associated partner p_T is lowered, instead of a jet-like structure peaked at $\Delta\phi = \pi$, a double peak structure is observed [6]. The shift of the away-side peak to $\Delta\phi \approx \pi \pm 1$ is independent from colliding species and energies, when number of participants, N_{part}, is larger than 100 [7].

The medium not only modifies the away-side but also modifies the near-side ($\Delta\phi \approx 0$). When measuring $\Delta\eta$-$\Delta\phi$ correlations, a significant enhancement along the $\Delta\eta$ direction has been observed on the near-side in central collisions and is known as "the ridge". This feature has a strong associated p_T dependence, which is most notable at low associated p_T [8].

To investigate the medium response due to the jet-medium interaction, we are focusing on the low partner p_T correlations. We measured the two particle $\Delta\eta$-$\Delta\phi$ correlation functions of Au+Au and p+p collisions at $\sqrt{S_{NN}} = 200GeV$ in PHENIX at RHIC in Run 4 and Run 5. The trigger particles are inclusive photons with $p_T = 2-3$ GeV/c , and the associated particles are inclusive charged hadrons with $p_T = 1 - 5$ GeV/c. The inclusive photons at this p_T range are mainly from meson decays. The underlying combinatorial events are removed by assuming there is zero yield at the minimum in the $\Delta\phi$ correlation, which is known as the ZYAM method [6].

Fig. 1 shows the $\Delta\eta$-$\Delta\phi$ correlation function in p+p collisions. A di-jet structure is visible. In the near-side, the correlation function is peaked at $\Delta\phi = 0$. In the away-side, it is peaked at $\Delta\phi = \pi$ and extends along the $\Delta\eta$ direction, due to k_T smearing. In central Au+Au collisions (0-20%), as shown in Fig. 2, we can still see a peak in the near-side at $\Delta\phi = 0$, but there also exists a significant enhancement along $\Delta\eta$, which forms a "ridge". In the away-side, instead of a jet-like structure as shown in Fig. 1, a double-peak structure is observed. The two peaks are located at $\Delta\phi \approx \pi \pm 1$, which is consistent previous measurements [7]. This double peak structure also extends along $\Delta\eta$ direction.

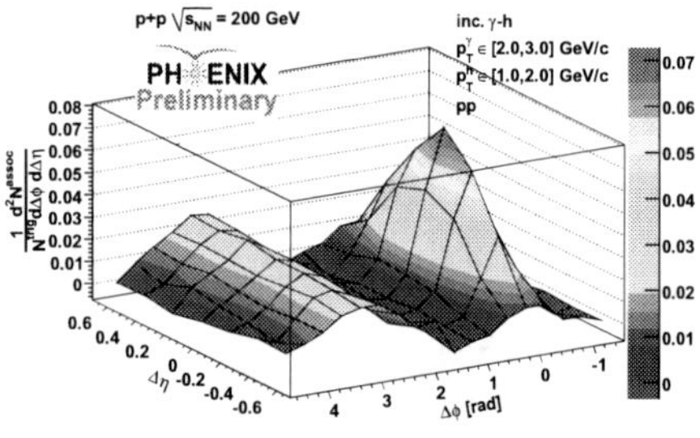

Figure 1. $\Delta\eta$-$\Delta\phi$ correlation function for p+p collisions.

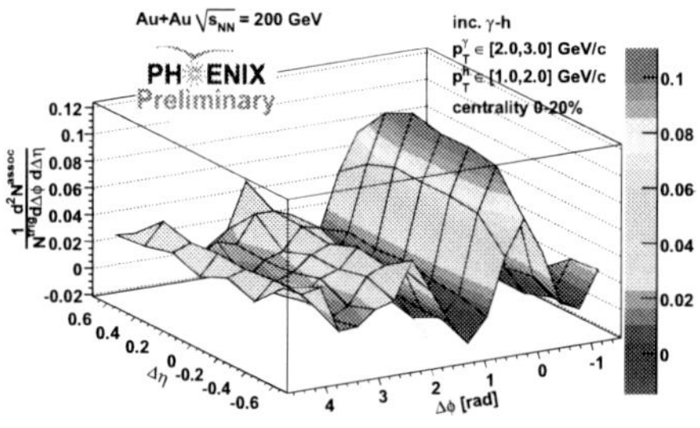

Figure 2. $\Delta\eta$-$\Delta\phi$ correlation function for Au+Au collisions in 0-20% centralites.

The near-side conditional yields are measured and plotted as a function of number of participants, N_{part}, in four different $|\Delta\eta|$ regions, from 0 - 0.1 up to 0.5 - 0.7, as shown in Fig. 3. The near-side yields in p+p collisions decrease with increasing $\Delta\eta$, which is consistent with a jet-like structure. For Au+Au collisions, as N_{part} increases, up to $N_{part} \approx 300$, the yield in all four $\Delta\eta$ regions increases. At the most central collisions, the near-side yield does not increase and may decrease. This is consistent with previous measurements [6, 9]. In the furtherest $\Delta\eta$ region, 0.5-0.7, the conditional yield in Au+Au collisions is significantly larger than in p+p. This is a clear sign of the existence of the long range rapidity correlations ("the ridge"), since there is very little jet-like yield in this $\Delta\eta$ region in the p+p collisions. Hence in this analysis, we will use the near-side yield at $0.5 < |\Delta\eta| < 0.7$ as the yield of the ridge, without removing the possible jet contamination.

In Fig. 3, the near side yield is significantly enhanced with increasing N_{part} due to the ridge. In heavy ion collisions, the multiplicity of the underlying event also increases with the number of participants, N_{part}. This poses the question if the ridge comes from the underlying combinatoric background. In the case the ridge only relates to the underlying events, the yield of the ridge should scale with the underlying events. We measured the ratio between the ridge yield and the yield of the underlying event in the near-side and plotted it as a function of number of participants and partner p_T as shown in Fig. 4. The yield of the ridge is always a few percent of the underlying events. At $N_{part} > 200$, the ratios show a decreasing trend. This means that the ridge does not increase as fast as the underlying combinatoric background and therefore the ridge can not be due to the background. We also observe that as the partner p_T increases, the ratio also increases. This is evidence that the ridge is harder than the underlying background.

The double-peak structure appears in the away-side in central Au+Au collisions as seen in

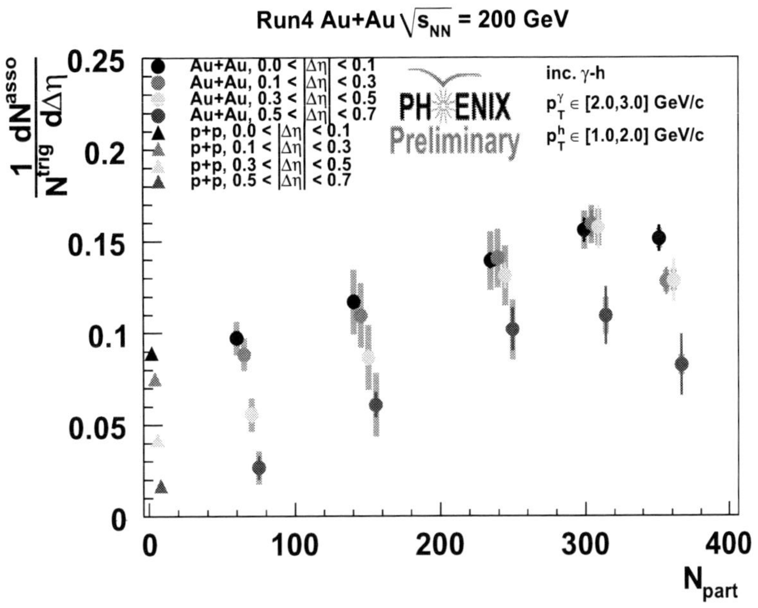

Figure 3. Near-side yield of different $\Delta\eta$ regions as a function of number of participants, N_{part}. N_{part}s in different $\Delta\eta$ are shifted slightly to avoid overlap.

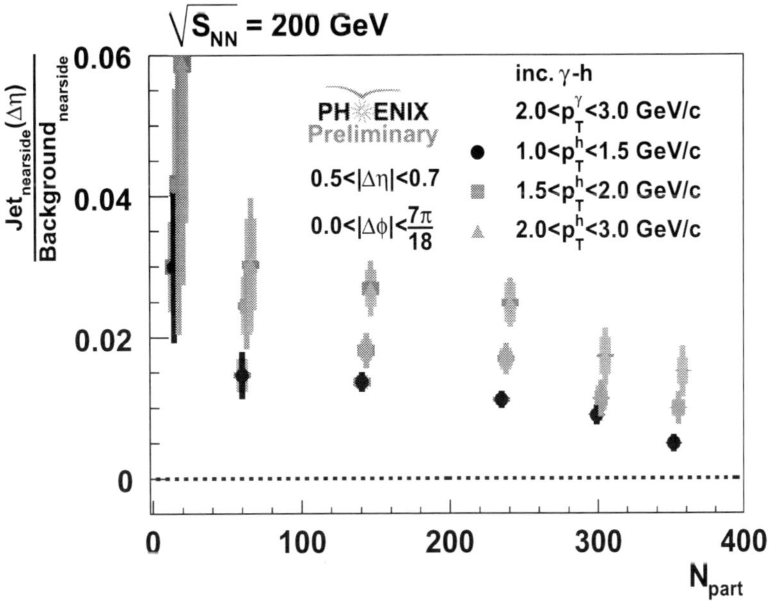

Figure 4. Ratio between the ridge yield and the underlying combinatorial background as function of N_{part} and partner p_T.

Fig. 2. We assume the away-side consists of two different components. The first component is the jet like structure, the "head", which is described by a Gaussian distribution peaked at $\Delta\phi = \pi$. The width of the head component is fixed to the away-side width of the p+p correlation function. The second component is referred to as "shoulder". A double Gaussian which peaks at $\Delta\phi \approx \pi \pm 1$ is used to describe this structure. We use a fitting method to separate the head and shoulder. The yield of the shoulder is the sum of the two peaks.

After decomposing the away-side, which presumably separates out the jet and medium components, we can compare the medium modification on both side: the ridge in near-side and the shoulder in away-side. The two yields are plotted as a function of the number of participants in Fig. 5. Both ridge and shoulder increase with N_{part}. Within errors, both yields are consistent up to $N_{part} \approx 300$. At the most central collisions where $N_{part} \approx 350$, the two yields do not agree. Here the yield of the shoulder is roughly constant while the yield of ridge is slightly decreasing.

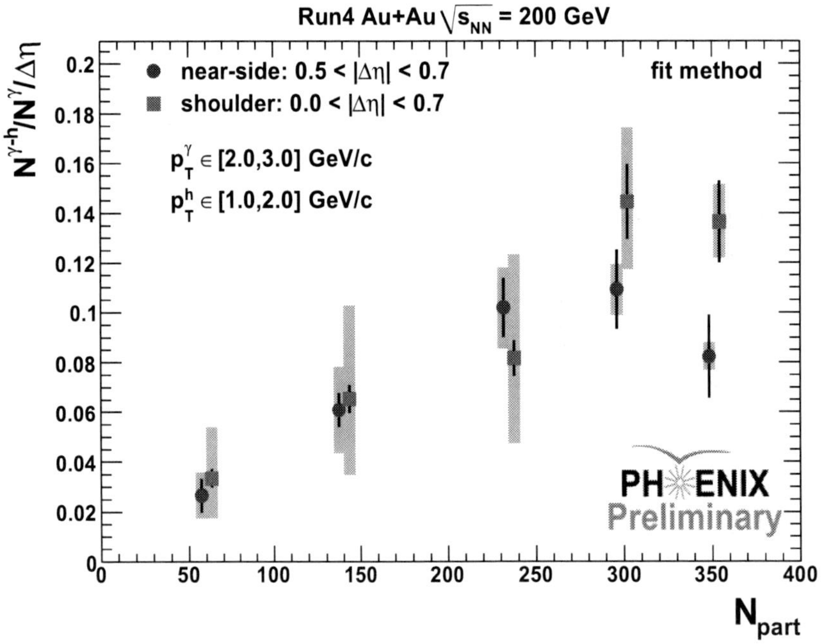

Figure 5. Yields of ridge and shoulder as a function of N_{part}.

The spectra of the ridge and shoulder are measured and are fitted with exponential function to extract the inverse slope. Fig. 6 shows the inverse slope of ridge and shoulder as a function of N_{part}. The inverse slope of ridge and shoulder are consistent in all centralities, which indicates that hadrons in both structures may come from a similar production mechanism. Both the ridge and shoulder are softer than p+p collisions. This suggests that the ridge and the shoulder do not come directly from hard processes. The spectra of the shoulder is consistent with the inclusive charged hadrons which come from the medium while the ridge is slightly harder. If the ridge and shoulder both come from the medium, there should be some mechanism to excite both features, which causes the spectra to be harder than the medium.

If the away-side jet is quenched in the medium, conservation of transverse momentum requires that the momentum lost by the jet must come out somewhere. On the other hand, the existence of the ridge means that some momentum is carried out along with the jet. In order to understand

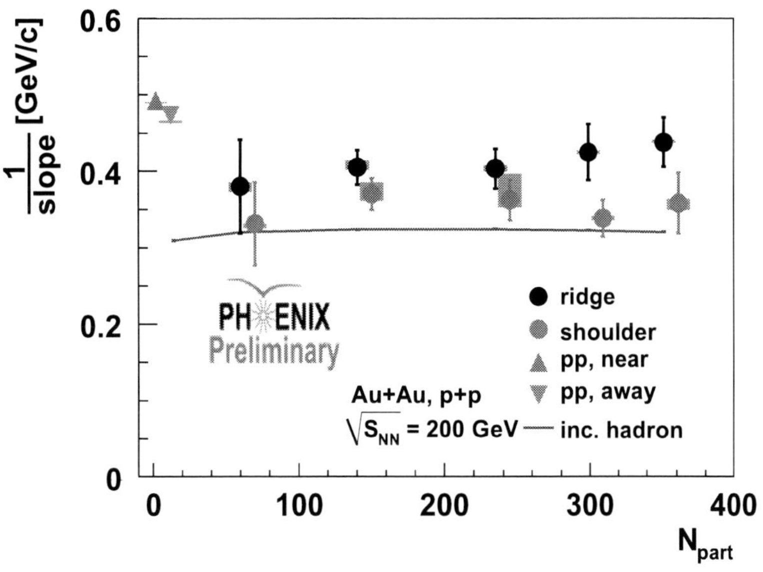

Figure 6. Inverse slope of ridge, shoulder, near and away side of p+p collisions. The magenta line is the inclusive charged hadrons.

how the transverse momentum is distributed, the conditional yields in both near and away side are weighted with the transverse momentum with respect to the direction of the trigger particle. The momentum of the trigger particle is not included.

We take the ratio of the away to near side p_T weighted yields as a function of N_{part} in different $\Delta\eta$ regions and plot this in Fig. 7. In right hand side of Fig. 7, at $0.0 < |\Delta\eta| < 0.1$, the near-side is predominantly from jet origin. The yields in both near and away side in Au+Au collisions increase with N_{part}. After weighting with p_T and taking the ratio between different away side components, such as head and shoulder, and the near side, we see a different trend for the head and shoulder as a function of N_{part}. As N_{part} increases, the ratio between head and near side decreases. This is consistent with the jet quenching scenario, where more medium medium is present, the more quenched are the jets. On the other hand, the ratio of shoulder to near side increases with N_{part}. This is consistent with Fig. 5, where the yield of the shoulder increases with N_{part}.

When the ratio is taken by the total away side over the near side, where the away side is the head + shoulder, we find that the ratios in Au+Au are consistent with p+p, where the ratio is about 0.55. This means that at this $\Delta\eta$ range, this ratio scales with N_{part}, and the p_T carried in the near and away side increase together by the same ratio with N_{part}. In addition, this indicates that the p_T lost in the medium from the scattered parton, which forms the head component is recovered in the shoulder.

The same method is also applied to $0.5 < |\Delta\eta| < 0.7$ and is shown in right hand side of Fig. 7. At this $\Delta\eta$ region, the near-side is mostly from the ridge. The ratio of the total away-side over near-side components increases when moving from central to peripheral collisions due to fewer near-side particles in peripheral Au+Au and p+p collisions. When looking only at the ratio of shoulder to the near-side, the ratio is about 0.6 and is independent of N_{part}. This tells us two things. First, the transverse momentum carried by ridge scales with the yield in shoulder, which is possibly populated by hadrons produced in a similar mechanism. Second, the ratio is less than one which means that there is more momentum carried by the ridge or the near-side than in shoulder.

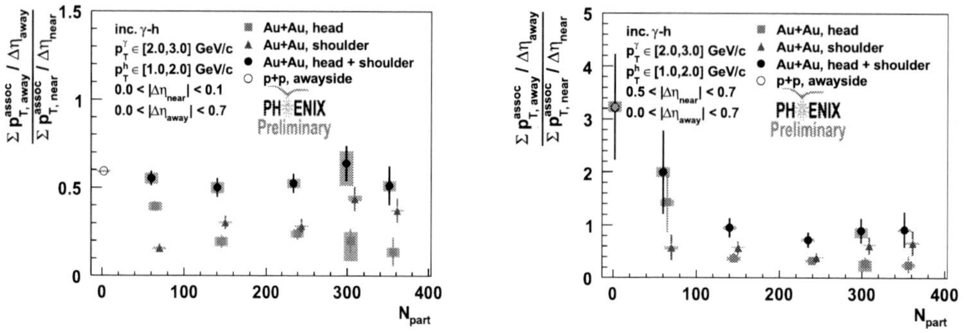

Figure 7. Ratio of away-side p_T over near-side p_T vs N_{part} at $0.0 < |\Delta\eta| < 0.1$ and $0.5 < |\Delta\eta| < 0.7$

In order to further investigate the relation between the ridge and the shoulder, more studies are needed. The conditional yields of near-side and the shoulder as a function of angle between reaction plane and the trigger particle, Φ_s, have been measured recently [10]. The results show that for the near-side, the yield increases when moving from in-plane ($\Phi_s \approx 0$), where the direction of the trigger particle is closer to the reaction plane, to out-of-plane, where the trigger is more perpendicular to the reaction plane or $\Phi_s \approx \pi/2$. On the other hand, the shoulder yield does not follow the same trend and the yield is at its strongest at $\Phi_s \approx \pi/4$. Such studies may allow to distinguish the production mechanism of these two structures.

In summary, in central Au+Au collisions, strong modification in the shape of two particle $\Delta\eta$-$\Delta\phi$ correlations has been observed in comparison to the correlations measured in p+p collisions. The modification in both sides, which is the ridge in the near side and the shoulder in the away side, are compared. The ridge and shoulder are similar in conditional yields. Both have similar inverse slopes, which may indicate that both have the same origin. The inverse slope of the ridge and shoulder indicates both are not produced directly from hard scattering. The ratio of the p_T carried by the near-side and away-side associated particles at $\Delta\eta \approx 0$ indicates that the p_T loss in the medium comes out in the shoulder. The new high-statistics data sets taken in Run 7 and Run 10 will allow us to perform a more detailed analysis.

References

[1] Adare A *et al.* (PHENIX) 2008 *Phys. Rev. Lett.* **101** 232301
[2] Adare A *et al.* (PHENIX) 2008 *Phys. Rev. Lett.* **101** 162301
[3] Adare A *et al.* (PHENIX) 2008 *Phys. Rev. C* **77** 064907
[4] Adler C *et al.* (STAR) 2003 *Phys. Rev. Lett.* **90** 082302
[5] Adare A *et al.* (PHENIX) 2010 (*Preprint* 1002.1077)
[6] Adler S S *et al.* (PHENIX) 2006 *Phys. Rev. Lett.* **97** 052301
[7] Adare A *et al.* (PHENIX) 2007 *Phys. Rev. Lett.* **98** 232302
[8] Adare A *et al.* (PHENIX) 2008 *Phys. Rev.* **C78** 014901
[9] Adler S S *et al.* (PHENIX) 2005 *Phys. Rev.* **C71** 051902
[10] Holzmann W G 2009 *Nuclear Physics A* **830** 781c – 784c

Journal of Physics: Conference Series **230** (2010) 012032 doi:10.1088/1742-6596/230/1/012032

Dileptons from the nonequilibrium Quark-Gluon Plasma

E. L. Bratkovskaya[a], **O. Linnyk**[a] **and W. Cassing**[b]

[a]Institut für Theoretische Physik, Universität Frankfurt, 60438 Frankfurt am Main, Germany
[b]Institut für Theoretische Physik, Universität Giessen, 35392 Giessen, Germany

Abstract. According to the dynamical quasiparticle model (DQPM) – matched to reproduce lattice QCD results in thermodynamic limit, – the constituents of the strongly interacting quark-gluon plasma (sQGP) are massive and off-shell quasi-particles (quarks and gluons) with broad spectral functions. In order to address the electromagnetic radiation of the sQGP, we derive off-shell cross sections of $q\bar{q} \to \gamma^*$, $q\bar{q} \to \gamma^* + g$ and $qg \to \gamma^* q (\bar{q}g \to \gamma^* \bar{q})$ reactions taking into account the effective propagators for quarks and gluons from the DQPM. Dilepton production in In+In collisions at 158 AGeV is studied by implementing these processes into the parton-hadron-string dynamics (PHSD) transport approach. The microscopic PHSD transport approach describes the full evolution of the heavy-ion collision: from the dynamics of quasi-particles in the sQGP phase (when the local energy density is above ~ 1 GeV/fm^3) through hadronization and to the following hadron interactions and off-shell propagation after the hadronization. A comparison to the data of the NA60 Collaboration shows that the low mass dilepton spectra are well described by including a collisional broadening of vector mesons, while the spectra in the intermediate mass range are dominated by off-shell quark-antiquark annihilation, quark Bremsstrahlung and gluon-Compton scattering in the nonperturbative QGP. In particular, the observed softening of the m_T spectra at intermediate masses (1 GeV $\leq M \leq$ 3 GeV) is approximately reproduced.

1. Introduction

While the properties of hadrons are rather well known in free space (embedded in a nonperturbative QCD vacuum) the masses and lifetimes of hadrons in a baryonic and/or mesonic environment are subject of current research in order to achieve a better understanding of the strong interaction and the nature of confinement. A modification of vector mesons has been seen experimentally in the enhanced production of lepton pairs above known sources in nucleus-nucleus collisions at Super-Proton-Synchroton (SPS) energies [1, 2]. This can be attributed to a shortening of the lifetime of the vector mesons ρ, ω and ϕ. The question arises if the enhancement might (in part) be due to new radiative channels [3] from the strong Quark-Gluon Plasma (sQGP). The answer to this question is nontrivial due to the nonequilibrium nature of the heavy-ion reactions and covariant transport models have to be incorporated to disentangle the various sources that contribute to the final dilepton spectra seen experimentally.

2. The PHSD approach

To address the vector meson properties in a hot and dense medium – as created in heavy-ion collisions – we employ an up-to-date relativistic transport model, i.e. the Parton Hadron String Dynamics model [4] (PHSD) that incorporates the relevant off-shell dynamics of the vector

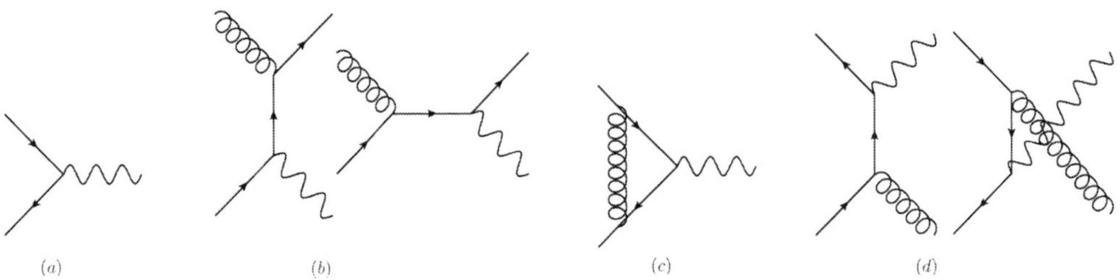

Figure 1. Diagrams contributing to the dilepton production from a QGP: (a) Drell-Yan mechanism, (b) gluon-Compton scattering (GCS), (c) vertex correction, (d) gluon Bremsstrahlung (NLODY), where virtual photons (wavy lines) split into lepton pairs, spiral lines denote gluons, arrows denote quarks. In each diagram the time runs from left to right.

mesons as well as the explicit partonic phase in the early hot and dense reaction region. PHSD consistently describes the full evolution of a relativistic heavy-ion collision from the initial hard scatterings and string formation through the dynamical deconfinement phase transition to the quark-gluon plasma (QGP) as well as hadronization and to the subsequent interactions in the hadronic phase.

In the hadronic sector PHSD is equivalent to the Hadron-String-Dynamics (HSD) transport approach [5, 6, 7] that has been used for the description of pA and AA collisions from SIS to RHIC energies and has lead to a fair reproduction of hadron abundances, rapidity distributions and transverse momentum spectra. In particular, HSD incorporates off-shell dynamics for vector mesons – according to Refs. [8] – and a set of vector-meson spectral functions [10] that covers possible scenarios for their in-medium modification. The transport theoretical description of quarks and gluons in PHSD is based on a dynamical quasiparticle model for partons matched to reproduce lattice QCD results in thermodynamic equilibrium (DQPM). The transition from partonic to hadronic degrees of freedom is described by covariant transition rates for the fusion of quark-antiquark pairs to mesonic resonances or three quarks (antiquarks) to baryonic states.

Various models predict that hadrons change in the (hot and dense) nuclear medium; in particular, a broadening of the spectral function or a mass shift of the vector mesons has been expected. Furthermore, QCD sum rules indicated that a mass shift may lead to a broadening and vice versa [9]; therefore both modifications should be studied simultaneously. Thus we explore three possible scenarios: (1) a broadening of the ρ spectral function, (2) a mass shift, and (3) a broadening plus a mass shift. The HSD (PHSD) off-shell transport approach allows to investigate in a consistent way the different scenarios for the modification of vector mesons in a hot and dense medium. In off-shell transport the hadron spectral functions change dynamically during the propagation through the medium and evolve towards the on-shell spectral function in the vacuum.

As demonstrated in Ref. [10] the off-shell dynamics is particularly important for resonances with a rather long lifetime in vacuum but strongly decreasing lifetime in the nuclear medium (especially ω and ϕ mesons) but also proves vital for the correct description of dilepton decays of ρ mesons with masses close to the two pion decay threshold. For a detailed description of the various hadronic channels included for dilepton production as well as the off-shell dynamics we refer the reader to Refs. [10, 11].

3. SPS energies

By employing the HSD approach to the low mass dilepton production in relativistic heavy-ion collisions, it was shown in [11] that the NA60 Collaboration data for the invariant mass spectra for $\mu^+\mu^-$ pairs from In+In collisions at 158 A·GeV favored the 'melting ρ' scenario [13]. Also

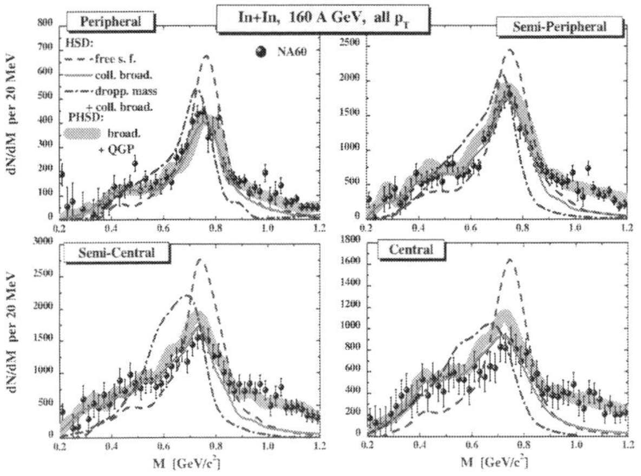

Figure 2. The HSD results for the mass differential dilepton spectra from $In + In$ collisions at 158 A·GeV in comparison to the excess mass spectrum from NA60 [13]. The actual NA60 acceptance filter and mass resolution have been incorporated [23]. The solid lines show the HSD results for a scenario including the collisional bradening of the ρ-meson whereas the dashed lines correspond to calculations with a 'free' ρ spectral functions for reference. The dash-dotted lines represent the HSD calculations for the 'dropping mass + collisional broadening' model. The bands represent the preliminary PHSD results incorporating direct dilepton radiation from the QGP in addition to a broadened ρ-meson.

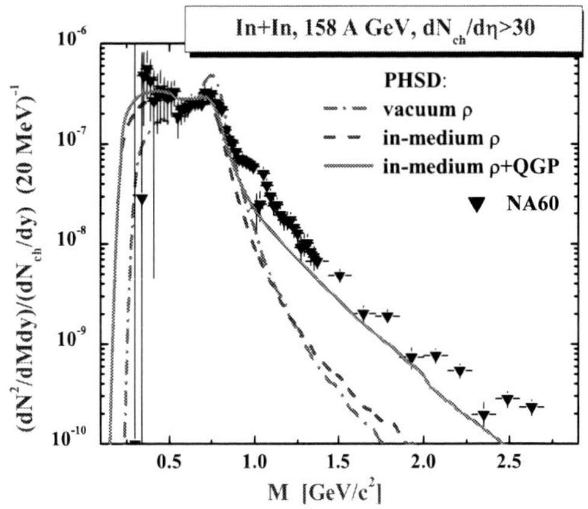

Figure 3. Acceptance corrected mass spectra of the excess dimuons from $In + In$ at 158 AGeV integrated over p_T in $0.2 < p_T < 2.4$ GeV from PHSD compared to the data of NA60 [15]. The green dash-dotted line shows the dilepton yield from the vacuum ρ meson. The blue dashed line is the contribution to the dilepton yield from the in-medium ρ with broadened spectral function. The red solid line presents the sum of the in-medium ρ and QGP dilepton radiation (the latter is calculated in the on-shell approximation in this work).

the data from the CERES Collaboration [14] showed a preference for the 'melting ρ' picture. On the other hand, the dilepton spectrum from In+In collisions at 158 A·GeV for $M > 1$ GeV could not be accounted for by the known hadronic sources (see Fig.2 of [11]). Also, hadronic models do not reproduce the softening of the m_T distribution of dileptons for $M > 1$ GeV [13]. This observation pointed towards a partonic origin.

Dilepton radiation by the constituents of the strongly interacting QGP proceeds via the elementary processes illustrated in Fig. 1: the basic Drell-Yan $q + \bar{q}$ annihilation mechanism, Gluon Compton scattering ($q + g \to \gamma^* + q$ and $\bar{q} + g \to \gamma^* + \bar{q}$), and quark + anti-quark annihilation with gluon Bremsstrahlung in the final state ($q + \bar{q} \to g + \gamma^*$). In the on-shell approximation, one uses perturbative QCD cross sections for the processes listed above. The obtained off-shell elementary cross sections have been implemented into the PHSD transport code. Note that in our calculations the running coupling α_S depends on the local energy density ϵ according to the DPQM [12], while the energy density is related to the temperature T by the lQCD equation of state. Numerically, we observe that α_S is of the order $O(1)$ and thus the contribution of the higher order diagrams in Fig. 1 is not subleading!

As we find in Fig. 3 the NA60 data favor the scenario of the in-medium broadening of vector mesons. Note that in the data - presented in this plot - the D-meson contribution has not been subtracted. The NA60 collaboration has published acceptance corrected data with subtracted

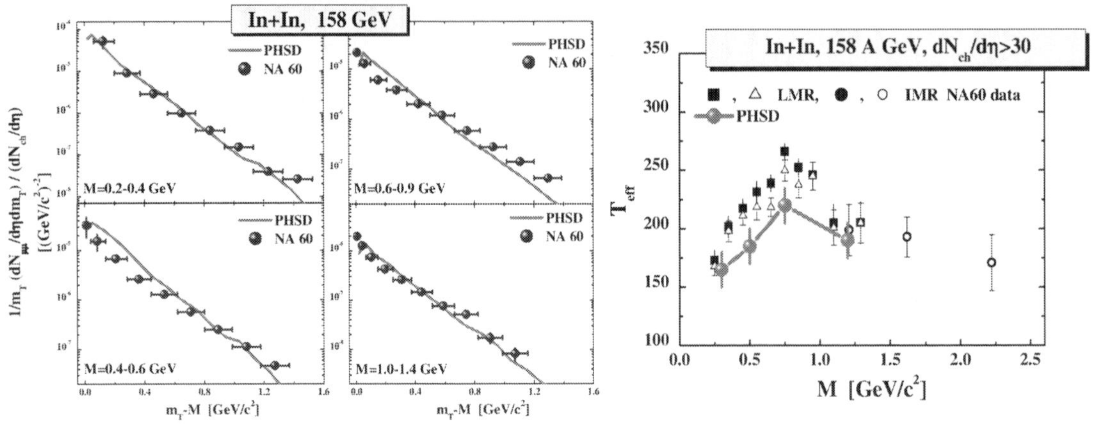

Figure 4. Left: Transverse mass spectra of dileptons for In+In at 158 A·GeV. **Right: Right:** The inverse slope parameter T_{eff} of the dimuon yield from In+In at 158 A·GeV as a function of the dimuon invariant mass in PHSD compared to the data of the NA60 Collaboration [13, 15].

charm contribution recently [15]. In Fig. 3 we present PHSD results for the dilepton spectrum excess over the known hadronic sources as produced in $In+In$ reactions at 158 AGeV compared to the acceptance corrected data. We find that the spectrum at invariant masses below 1 GeV is well reproduced by the ρ meson yield, if a broadening of the meson spectral function in the medium is assumed. On the other hand, the spectrum at $M > 1$ GeV is shown to be dominated by the partonic sources.

It is also interesting to note that accounting for partonic dilepton sources allows to reproduce in PHSD (cf. Fig. 3, rhs) the intriguing finding of the NA60 Collaboration [13, 15] that the effective temperature of the dileptons (slope parameters) in the intermediate mass range is lower than the T_{eff} of the dileptons from the hadronic phase. The softening of the transverse mass spectrum with growing invariant mass reflects that the partonic channels occur dominantly before the collective radial flow has developed. This feature of the data is also reproduced in PHSD (cf. Fig. 4, rhs). A detailed look at the PHSD results shows that in total we still underestimate the slope parameter T_{eff} which might be due to missing partonic initial state effects or an underestimation of partonic flow in the initial phase of the reaction.

4. RHIC energies

In 2008, the PHENIX Collaboration has presented first dilepton data from pp and $Au + Au$ collisions at Relativistic-Heavy-Ion-Collider (RHIC) energies of $\sqrt{s} = 200$ GeV [17] which show an even larger enhancement in $Au + Au$ reactions (relative to pp collisions) in the invariant mass regime from 0.15 to 0.6 GeV than the data at SPS energies [13, 14]. We recall that HSD provides a reasonable description of hadron production in $Au + Au$ collisions at $\sqrt{s} = 200$ GeV [19] such that we can directly continue with the results for e^+e^- pairs which are shown in Fig. 5 for $p + p$ collisions. We find that the dilepton radiation in the elementary channel is well under control in HSD, which is equivalent to PHSD for $p + p$ reactions.

We step on with the case of inclusive $Au + Au$ collisions in comparison to the data from PHENIX [17] (Fig 6). When including the in-medium modification scenarios for the vector mesons, we achieve a sum spectrum which is only slightly enhanced compared to the 'free' scenario. Whereas the total yield is quite well described in the region of the pion Dalitz decay as well as around the ω and ϕ mass, HSD clearly underestimates the measured spectra in the regime from 0.2 to 0.6 GeV by an average factor of ~ 3. The low mass dilepton spectra from

Journal of Physics: Conference Series **230** (2010) 012032 doi:10.1088/1742-6596/230/1/012032

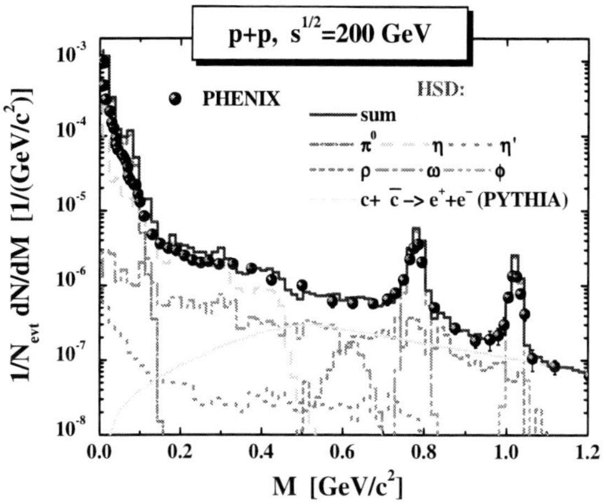

Figure 5. The HSD results for the mass differential dilepton spectra in case of pp collisions at $\sqrt{s}$ = 200 GeV in comparison to the data from PHENIX [16]. The actual PHENIX acceptance and mass resolution have been incorporated (see legend for the different color coding of the individual channels). Figure taken from [11].

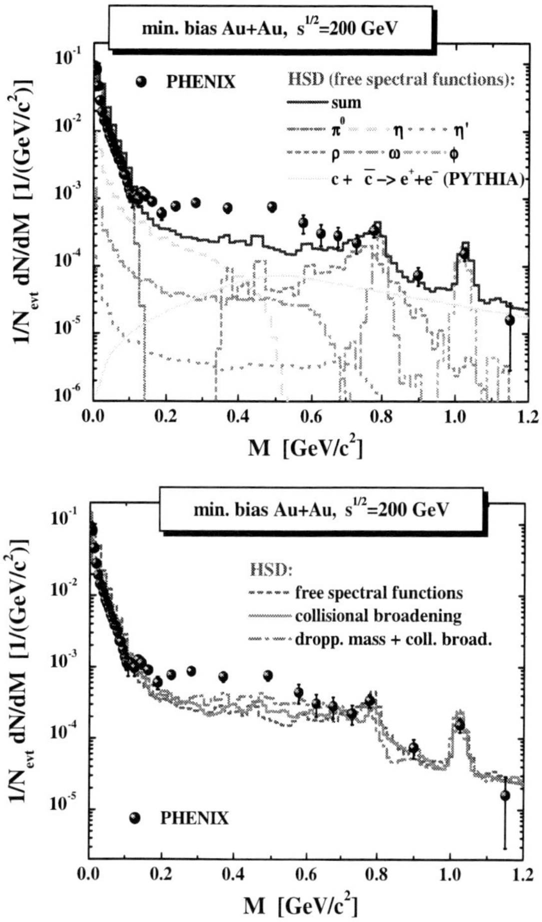

Figure 6. The HSD results for the mass differential dilepton spectra in case of inclusive $Au + Au$ collisions at $\sqrt{s}$ = 200 GeV in comparison to the data from PHENIX [17]. The actual PHENIX acceptance filter and mass resolution have been incorporated [18]. In the upper part the results are shown for vacuum spectral functions (for ρ, ω, ϕ) including the channel decompositions (see legend for the different color coding of the individual channels). The lower part shows a comparison for the total e^+e^- mass spectrum in case of the 'free' scenario (dashed line), the 'collisional broadening' picture (solid line) as well as the 'dropping mass + collisional broadening' model (dash-dotted line). Figure taken from [11].

$Au+Au$ collisions at RHIC (from the PHENIX Collaboration) are clearly underestimated in the invariant mass range from 0.2 to 0.6 GeV in the 'collisional broadening' scenario as well as in the 'dropping mass + collisional broadening' model, i.e. when assuming a shift of the vector meson mass poles with the baryon density. We mention that our results for the low mass dileptons are very close to the calculated spectra from van Hees and Rapp [20] as well as Dusling and Zahed [21] (cf. the comparison in Ref. [22]).

Consequently we attribute this additional low mass enhancement seen by PHENIX to non-hadronic sources. This assumption has to be tested quantitatively. However, in order to make quantitative predictions at experimentally relevant low dilepton mass ($\leq$ 0.6 GeV) and strong coupling ($\alpha_S \sim 0.5 \div 1$), we have to take into account not only the higher order pQCD reaction mechanisms, but also the non-perturbative spectral functions and self-energies of the quarks,

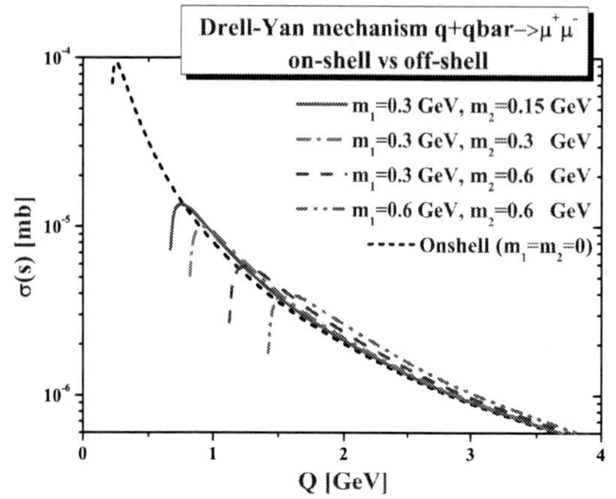

Figure 7. Dilepton production cross sections in the Drell-Yan channel ($q + \bar{q} \to \mu^+ + \mu^-$). The short dashes (black) line shows the on-shell, i.e. the standard pQCD result. The other lines show the off-shell cross section, in which the annihilating quark and antiquark have finite masses m_1 and m_2 with different values: $m_1 = 0.3$ GeV, $m_2 = 0.15$ GeV (solid magenta line), $m_1 = 0.3$ GeV, $m_2 = 0.3$ GeV (dash-dotted red line), $m_1 = 0.3$ GeV, $m_2 = 0.6$ GeV (dashed blue line), $m_1 = 0.6$ GeV, $m_2 = 0.6$ GeV (dash-dot-dot green line).

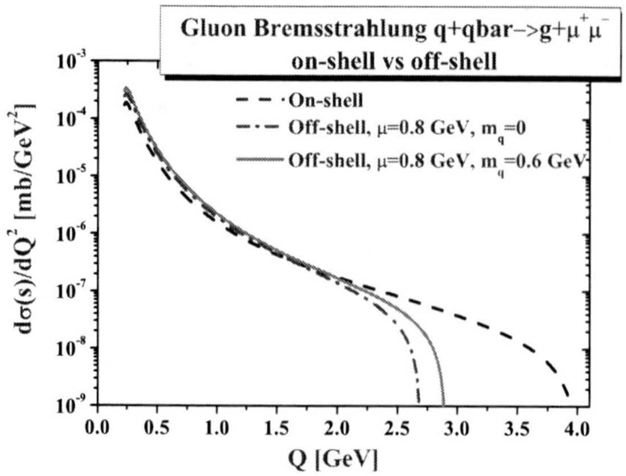

Figure 8. Comparison of off-shell and on-shell cross sections for dilepton production in the gluon Bremsstrahlung $q + \bar{q} \to g + \mu^+\mu^-$ channel at $\sqrt{s} = 4$ GeV. The dashed black line shows the on-shell (pQCD) cross section regularized by the cut on the gluon mass $\mu_{cut} = 0.206$ GeV, the blue dash-dotted line presents the off-shell cross section for the gluon mass fixed to $\mu = 0.8$ GeV and on-shell quark and anti-quark ($m_1 = m_2 = m_3 = 0$). The red solid line gives the off-shell result for $\mu = 0.8$ GeV, $m_1 = m_2 = m_3 = m_q = 0.6$ GeV.

anti-quarks and gluons thus going beyond the leading twist.

For this purpose, off-shell cross sections were derived in [24] for dilepton production in the reactions $q + \bar{q} \to l^+l^-$ (Drell-Yan mechanism), $q + \bar{q} \to g + l^+l^-$ (quark annihilation with the gluon Bremsstrahlung in the final state), $q(\bar{q}) + g \to q(\bar{q}) + l^+l^-$ (gluon Compton scattering), $g \to q+\bar{q}+l^+l^-$ and $q(\bar{q}) \to q(\bar{q})+g+l^+l^-$ (virtual gluon decay, virtual quark decay) in the sQGP in effective perturbation theory by dressing the quark and gluon lines in the diagrams in Fig. 1 with the DQPM propagators for quarks and gluons. The DQPM describes QCD properties in terms of single-particle Green's functions (in the sense of a two-particle irreducible approach) and leads to the notion of the constituents of the sQGP being effective quasiparticles, which are massive and have broad spectral functions (due to large interaction rates).

Dilepton production cross sections in the Drell-Yan mechanism are plotted on the l.h.s. of Fig. 7. The short dashes (black) line shows the on-shell, i.e. the standard perturbative result. The other lines show the off-shell cross section, in which the annihilating quark and antiquark have finite masses m_1 and m_2 with different values: $m_1 = 0.3$ GeV, $m_2 = 0.15$ GeV (solid magenta line), $m_1 = 0.3$ GeV, $m_2 = 0.3$ GeV (dash-dotted red line), $m_1 = 0.3$ GeV, $m_2 = 0.6$ GeV (dashed blue line), $m_1 = 0.6$ GeV, $m_2 = 0.6$ GeV (dash-dot-dot green line).

In Fig. 8, the off-shell cross sections for the quark annihilation with gluon bremsstrahlung channel at various values of quark and gluon off-shellnesses (masses) are compared to the on-shell (pQCD) result. The dashed black line shows the on-shell cross section regularized by a cut on gluon mass $\mu_{cut} = 0.206$ GeV, the red solid line presents the off-shell cross section for the gluon mass fixed to $\mu = 0.8$ GeV and on-shell quark and anti-quark ($m_1 = m_2 = m_3 = 0$). The blue dash-dotted line gives the off-shell result for $\mu = 0.8$ GeV, $m_1 = m_2 = m_3 = m_q = 0.6$ GeV.

One readily notices the shift of the maximum allowed mass of the lepton pair to a lower value (in order to produce a massive gluon in the final state).

By implementing the off-shell partonic processes into the PHSD transport approach, a consistent calculation of the dilepton production in heavy-ion collisions at RHIC energies will be performed in near future. The comparison to the dilepton data double differentially in mass and p_T will open the possibility to study the relative importance of different processes in the dilepton production and guide us towards a better understanding of the properties of matter at high densities and temperatures as created in heavy-ion collisions.

5. Summary

Dilepton production in relativistic heavy-ion collisions is a valuable probe of two phenomena of fundamental interest: the vector meson properties in a hot and dense medium and of the quark and gluon dynamics in the deonfined state of matter (sQGP). The Parton Hadron String Dynamics [4] (PHSD) transport approach incorporates the relevant off-shell dynamics of the vector mesons as well as the explicit partonic phase in the early hot and dense reaction region.

A comparison of the transport calculations to the data of the CERES and NA60 Collaborations points towards 'melting' modification of the ρ'-meson at high densities, i.e. the broadening of the vector meson's spectral function. On the other hand, the spectrum for $M > 1$ GeV is shown to be dominated by the partonic sources.

The low mass dilepton spectra from $Au + Au$ collisions at RHIC (from the PHENIX Collaboration) are clearly underestimated by the hadronic channels in the invariant mass range from 0.2 to 0.6 GeV in the 'collisional broadening' scenario as well as in the 'dropping mass + collisional broadening' model, i.e. when assuming a shift of the vector meson mass poles with the baryon density. We attribute this additional low mass enhancement seen by PHENIX to radiation from the strongly interacting QGP. This assumption has to be tested quantitatively. Strong interaction of partons - reflected in their high masses and broad widths - lead to higher-twist corrections to the perturbative cross sections especially at the experimentally relevant low dilepton masses (≤ 0.6 GeV). By implementing the off-shell partonic processes into the PHSD transport approach, a consistent calculation of the dilepton production in heavy-ion collisions at RHIC energies will be performed in near future that will allow us answer whether the observed access can be attributed to the radiation by off-shell partons.

Acknowledgments

OL acknowledges financial support within the "HIC for FAIR" center of the "LOEWE" program.

References

[1] G. Agakichiev *et al.*, CERES Collaboration, *Phys. Rev. Lett.* **75** (1995) 1272; Th. Ullrich *et al.*, *Nucl. Phys.* A**610** (1996) 317c; A. Drees, *Nucl. Phys.* A**610** (1996) 536c.
[2] M. A. Mazzoni, HELIOS Collaboration, *Nucl. Phys.* A**566** (1994) 95c; M. Masera, *Nucl. Phys.* A**590** (1995) 93c; T. Åkesson et al., *Z. Phys.* C**68** (1995) 47.
[3] O. Linnyk, E. L. Bratkovskaya and W. Cassing, *Nucl. Phys.* A **830** (2009) 491C.
[4] W. Cassing and E. L. Bratkovskaya, *Phys. Rev.* C **78** (2008) 034919; *Nucl. Phys.* A **831** (2009) 215.
[5] W. Cassing, E. L. Bratkovskaya, *Phys. Rept.* **308** (1999) 65.
[6] E. L. Bratkovskaya, W. Cassing, *Nucl. Phys.* A**619** (1997) 413.
[7] W. Ehehalt, W. Cassing, *Nucl. Phys.* A **602** (1996) 449.
[8] W. Cassing, S. Juchem, *Nucl. Phys.* A **665** (2000) 377; *ibid.* A **672** (2000) 417.
[9] J. Ruppert, T. Renk and B. Muller, *Phys. Rev.* C 73 (2006) 034907.
[10] E. L. Bratkovskaya, W. Cassing, *Nucl. Phys.* A **807** (2008) 214.
[11] E. L. Bratkovskaya, W. Cassing and O. Linnyk, *Phys. Lett.* B **670** (2009) 428.
[12] W. Cassing, *Nucl. Phys.* A 791 (2007) 365; *ibid.* A 795 (2007) 70.
[13] R. Arnaldi *et al.*, NA60 Collaboration, *Phys. Rev. Lett.* **96** (2006) 162302; J. Seixas *et al.*, *J. Phys.* G **34** (2007) S1023; S. Damjanovic *et al.*, *Nucl. Phys.* **A 783** (2007) 327c.

[14] D. Adamova *et al.* CERES Collaboration, *Nucl. Phys.* A **715** (2003) 262; *Phys. Rev. Lett.* **91** (2003) 042301; G. Agakichiev *et al.*, *Eur. Phys. J.* C **41** (2005) 475; D. Adamova *et al. Phys. Lett.* B **666** (2008) 425; A. Marin *et al.*; Proceedings of CPOD07, PoS 034 (2007).
[15] R. Arnaldi *et al.*, NA60 Collaboration, Eur. Phys. J. C **59**, 607 (2009)
[16] A. Adare *et al.*, PHENIX Collaboration, *Phys. Lett.* B **670** (2009) 313
[17] A. Toia *et al.*, PHENIX Collaboration, *Nucl. Phys.* A **774** (2006) 743; *Eur. Phys.* J **49** (2007) 243; S. Afanasiev *et al.*, PHENIX Collaboration, arXiv:0706.3034 [nucl-ex]; A. Adare *et al.*, PHENIX Collaboration, arXiv:0912.0244 [nucl-ex].
[18] A. Toia, private communication.
[19] E. L. Bratkovskaya, W. Cassing, H. Stöcker, *Phys. Rev.* **C 67** (2003) 054905; E. L. Bratkovskaya *et al. Phys. Rev.* **C 69** (2004) 054907.
[20] H. van Hees, R. Rapp, Phys. Rev. Lett. **97** (2006) 102301.
[21] K. Dusling and I. Zahed, *Nucl. Phys.* A **825** (2009) 212.
[22] A. Toia, *J. Phys.* G **35** (2008) 104037.
[23] S. Damjanovic, private communication.
[24] O. Linnyk, arXiv:1004.2591 [hep-ph].

Calculations of direct photon emission in Heavy Ion Collisions at $\sqrt{s_{\rm NN}} = 200$ GeV

B Bäuchle and M Bleicher

Frankfurt Institute for Advanced Studies, Ruth-Moufang-Straße 1, 60438 Frankfurt am Main, Germany

E-mail: baeuchle@th.physik.uni-frankfurt.de

E-mail: bleicher@th.physik.uni-frankfurt.de

Abstract. Direct photon emission in heavy-ion collisions is calculated within a relativistic micro+macro hybrid model and compared to the microscopic transport model UrQMD. In the hybrid approach, the high-density part of the collision is calculated by an ideal 3+1-dimensional hydrodynamic calculation, while the early (pre-equilibrium-) and late (rescattering-) phase are calculated with the transport model. We study both models with $Au + Au$-collisions at $\sqrt{s_{\rm NN}} = 200$ GeV and compare the results to experimental data published by the PHENIX collaboration.

1. Introduction

Creating and studying high-density and -temperature nuclear matter is the major goal of heavy-ion experiments. A state of quasi-free partonic degrees of freedom, the Quark-Gluon-Plasma (QGP) [1, 2] may be formed, if the energy density reached in the reaction is high enough. Strong jet quenching, large elliptic flow and other observations made at the Relativistic Heavy Ion Collider (BNL-RHIC) suggest the successful creation of a strongly coupled QGP (sQGP) at these energies [3, 4, 5, 6], and possible evidence for the creation of this new state of matter has also been put forward by collaborations at the Super Proton Synchrotron (CERN-SPS), as for instance the step in the mean transverse mass excitation function of protons, kaons and pions and the enhanced K^+/π^+-ratio [7].

Electromagnetic probes provide a unique insight into the early stages of heavy-ion collisions, since they have the advantage of negligible rescattering cross-sections. Therefore, they leave the production region without rescattering and carry the information from this point to the detector. Besides single- and dileptons, direct photon emission can therefore be used to study the early hot and dense, possibly partonic, stages of the reaction.

Unfortunately, most photons measured in heavy-ion collisions come from hadronic decays. The experimental challenge of obtaining spectra of only direct photons has been gone through by several experiments; WA98 (CERN-SPS) [8] and PHENIX (BNL-RHIC) [9] have published explicit data points for direct photons.

On the theory side, the elementary photon production cross-sections are known since long, see e.g. Kapusta *et al.* [10] and Xiong *et al.* [11]. The major problem is the difficulty to describe the time evolution of the produced matter, for which first principle calculations from Quantum

Chromodynamics (QCD) cannot be done. Well-developed dynamical models are therefore needed to describe the space-time evolution of nuclear interactions.

Among the approaches used are relativistic transport theory [12, 13, 14, 15, 16, 17, 18, 19, 20] and relativisitc fluid- or hydrodynamics [21, 22, 23, 24, 25, 26, 27, 28, 29, 30, 31, 32, 33, 34, 35, 36]. For both models, approximations have to be made, and in both models, the restrictions imposed by these approximations can be loosened. For transport theory, the necessary approximations include the restriction of scattering processes to two incoming particles, which limits the applicability to low particle densities. For hydrodynamics, on the other hand, matter has to be in local thermal equilibrium (for ideal, non-viscous hydrodynamic calculations) or at least close to it (for viscous calculations) [37, 38, 39].

From this, it is clear where the advantages for both models are: While in transport, non-equilibrium matter, which is present in the beginning of the heavy-ion reaction, and dilute matter, which is present in the late phase, can be described, hydrodynamics may be better suited to describe the intermediate stage, which is supposed to be dense, hot and thermalized. In addition, the transition between two phases of matter, such as Quark Gluon Plasma (QGP) and Hadron Gas (HG) can be easily described in hydrodynamics, while this is not (yet) possible for transport models, since the microscopic details of this transition are not known.

2. The Model

UrQMD v2.3 (Ultra-relativistic Quantum Molecular Dynamics) is a microscopic transport model [13, 14, 40]. It includes all hadrons and resonances up to masses $m \approx 2.2$ GeV and at high energies can excite and fragment strings. The cross-sections are either parametrized, calculated via detailed balance or taken from the additive quark model (AQM), if no experimental values are available. At high parton momentum transfers, PYTHIA [41] is employed for pQCD scatterings.

UrQMD differentiates between two regimes for the excitation and fragmentation of strings. Below a momentum transfer of $Q < 1.5$ GeV a maximum of two longitudinal strings are excited according to the LUND picture, at momentum transfers above $Q > 1.5$ GeV hard interactions are modelled via PYTHIA. For detailed information on the inclusion of PYTHIA, the reader is referred to Section II of [40]. In the UrQMD framework, propagation and spectral functions are calculated as in vacuum.

In the following, we compare results from this microscopic model to results obtained with a hybrid model description [42]. Here, the high-density part of the reaction is modelled using ideal 3+1-dimensional fluid-dynamics. The unequilibrated initial state and the low-density final state are described by UrQMD. In the hydrodynamic intermediate stage we use a Hadron Gas Equation of State (HG-EoS) which includes the same degrees of freedom as are present in the transport phase. This allows to explore the effects due to the change of the kinetic description.

To connect the initial transport phase with the high-density fluid phase, the baryon-number-, energy- and momentum-densities are smoothed and put into the hydrodynamic calculation after $t = 0.6$ fm. Temperature, chemical potential, pressure and other macroscopic quantities are determined from the densities by the Equation of State used in the current calculation. During this transition, the system is forced into an equilibrated state. Particles with high rapidity $y > 2$, however, are excluded from the hydrodynamic grid and propagated in the cascade without interaction to the hydrodynamic medium.

In non-central collisions, the spectators are propagated in the cascade. After the local rest frame energy density has dropped below a threshold value of $\epsilon_{\text{crit}} \approx 5\epsilon_0$, particles are created on a hyper-surface from the densities by means of the Cooper-Frye formula and propagation is continued in UrQMD.

The transition from hydrodynamic to cascade description used in the calculations presented here is gradual. I.e. each transverse slice (constant z) is transferred to the cascade at the same time, when the condition is met throughout that slice. This represents a pseudo-eigentime condition.

Photon emission is calculated perturbatively in both models, hydrodynamics and transport, because the evolution of the underlying event is not altered by the emission of photons due to their very small emission probability. The channels considered for photon emission may differ between the hybrid approach and the binary scattering model. Emission from a Quark-Gluon-Plasma can only happen in the hydrodynamic phase, and only if the equation of state used has partonic or chirally restored degrees of freedom. Photons from baryonic interactions are neglected in the present calculation.

For emission from the transport part of the model, we use the well-established cross-sections from Kapusta *et al.* [10], and for emission from the hydrodynamic phase, we use the parametrizations by Turbide, Rapp and Gale and Arnold *et al.* [43] (the latter for QGP-emission). For detailed information on the emission process, the reader is referred to Bäuchle and Bleicher [44].

3. Results

Photon emission has been calculated for minimum bias (0-92 %) and central (0-10 %) $Au+Au$-collions at $\sqrt{s_{\rm NN}} = 200$ GeV with both the pure cascade and the cascade-hydrodynamic hybrid models. We note that the high-$p_\perp$ data obtained by PHENIX [9] can be described very well with pQCD-calculations by Gordon and Vogelsang [45], therefore we investigate the contributions of our model to low-$p_\perp$ photons only.

In Figure 1, we show the direct photons emitted by the cascade- and hybrid models and compare them to the aforementioned pQCD-calculations and data from the PHENIX-collaboration.

While our hadronic model fails to predict the amount of direct photons emitted at intermediate transverse momentum, the disagreement is smaller at low $p_\perp$. The spectra obtained with an intermediate hydrodynamic part are significantly higher than those with pure cascade calculations. In any case, the spectra are negligible compared to the pQCD contributions predicted by Gordon and Vogelsang, therefore, an enhancement with respect to those data cannot be explained by hadronic sources.

In Figure 2, we explore the contribution of the different stages to the hybrid model spectra. The contribution of the intermediate hydro phase turns out to be dominant at low transverse momenta $p_\perp < 3$ GeV. The contributions of the initial and final stages are lower than the complete spectrum from the cascade calculations, so that we can assume the intermediate stage in the cascade calculation to be on the order of the final stage contribution in the hybrid model calculations.

The difference between the hydrodynamic and transport descriptions of the intermediate stage has not been observed for collisions with lower energies $E_{\rm lab} = 158$ AGeV (see [44]) and $E_{\rm lab} = 45$ AGeV (see [46]).

4. Summary

In this article, we have applied UrQMD and the UrQMD+Hydro hybrid model to calculate photon spectra from central and minimum bias Au+Au-collions at $\sqrt{s_{\rm NN}} = 200$ GeV. The comparison of the (hadronic) calculations to data from the PHENIX collaboration suggest that a significant contribution to the measured spectra comes from non-hadronic sources. The low-$p_\perp$-excess over pQCD-predictions seen by the PHENIX-collaboration [47] cannot be explained by hadronic sources; partonic sources such as a Quark-Gluon-Plasma are therefore very likely to be responsible for this excess.

The comparison of transport and hybrid calculations show that the conclusion drawn at lower energy, which suggested that there is no difference between the spectra obtained with or without intermediate hydrodynamic stage, is not valid at these high energies. The excess of the hybrid

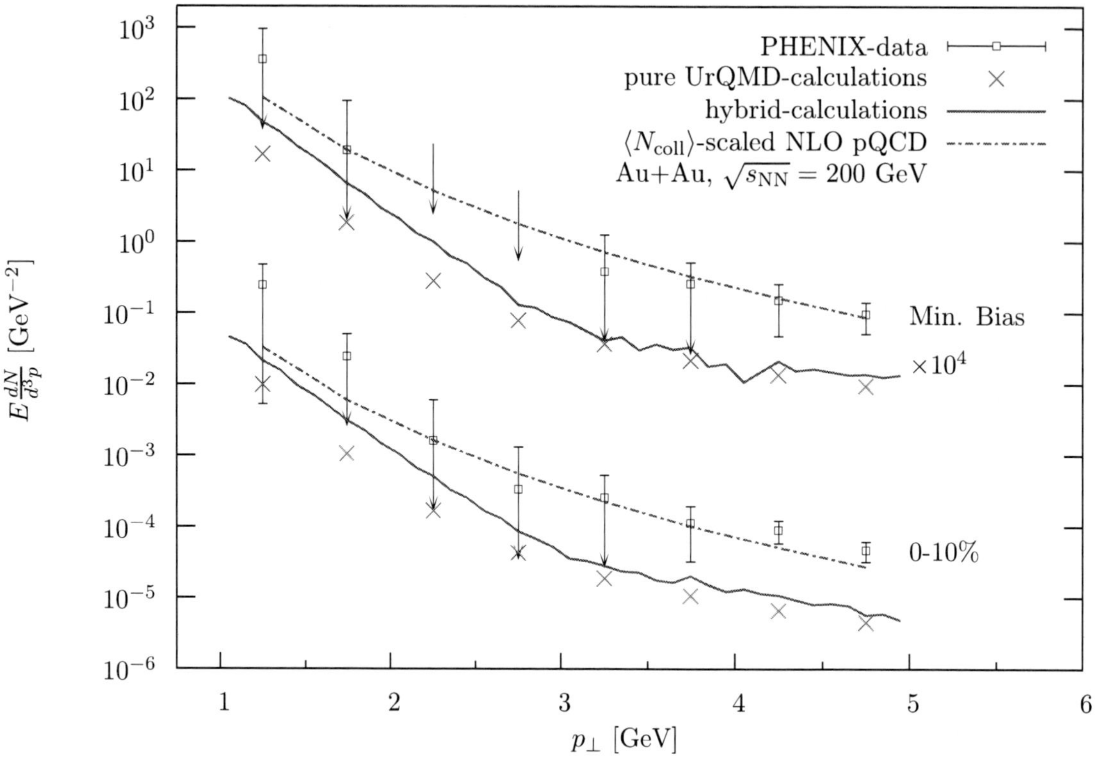

Figure 1. Comparison of the direct photon spectra from UrQMD cascade- (red crosses) and hybrid- (blue solid line) calculations to the spectra measured by the PHENIX collaboration (black symbols) and to the pQCD contribution predicted by Gordon and Vogelsang (violet dash-dotted line) for minimum bias and central Au+Au-collisions at $\sqrt{s_{NN}} = 200$ GeV.

model calculations over the transport calculations are visible in both central and minimum bias collision samples, and the magnitude of this excess is similar in both models. This suggests that the excess depends only the collision energy, not on the system size.

5. Outlook

The results shown here suggest the need for further studies. Calculations of direct photon spectra have to be done with the current model for other centrality selections, and with different Equations of State such as a Bag Model EoS and an EoS with a chirally restored phase. In order to investigate the difference between the hybrid model and the cascade model, an energy dependent investigation of the excess is advisable.

Different systems measured at the RHIC-facility, such as Cu+Cu-collisions and collisions at smaller center-of-mass energy, will also be calculated.

The parameters of the model – the conditions for switching to and from the hydrodynamic description and the scenario for the latter transition – will be investigated in the future.

Acknowledgments

This work has been supported by the Frankfurt Center for Scientific Computing (CSC), the GSI and the BMBF. The authors thank Hannah Petersen for providing the hybrid- and Dirk Rischke

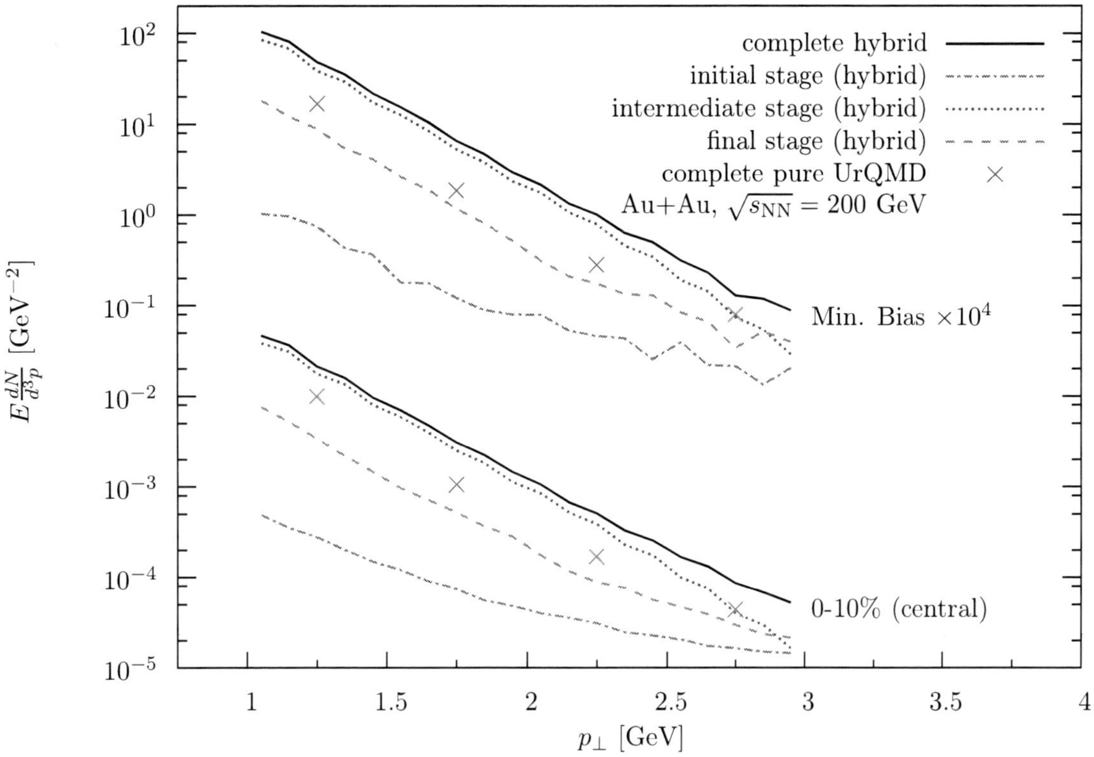

Figure 2. Comparison of the contribution of the different stages (initial stage, green dash-dotted lines; intermediate stage, blue dotted lines, final stage, red dashed lines) to the overall spectrum from the hybrid model (black solid line) and the overall spectrum from pure cascade calculations (red crosses) for minimum bias and central Au+Au-collisions at $\sqrt{s_{NN}} = 200$ GeV.

for the hydrodynamic code. B. Bäuchle gratefully acknowledges support from the Deutsche Telekom Stiftung, the Helmholtz Research School on Quark Matter Studies and the Helmholtz Graduate School for Hadron and Ion Research. This work was supported by the Hessian LOEWE initiative through the Helmholtz International Center for FAIR.

The authors thank Elvira Santini, Pasi Huovinen and Rene Bellwied for valuable discussions and Klaus Reygers for experimental clarifications.

References

[1] J. W. Harris and B. Muller, Ann. Rev. Nucl. Part. Sci. **46** (1996) 71 [arXiv:hep-ph/9602235].
[2] S. A. Bass, M. Gyulassy, H. Stoecker and W. Greiner, J. Phys. G **25** (1999) R1 [arXiv:hep-ph/9810281].
[3] J. Adams *et al.* [STAR Collaboration], Nucl. Phys. A **757** (2005) 102 [arXiv:nucl-ex/0501009].
[4] B. B. Back *et al.*, Nucl. Phys. A **757** (2005) 28 [arXiv:nucl-ex/0410022].
[5] I. Arsene *et al.* [BRAHMS Collaboration], Nucl. Phys. A **757** (2005) 1 [arXiv:nucl-ex/0410020].
[6] K. Adcox *et al.* [PHENIX Collaboration], Nucl. Phys. A **757** (2005) 184 [arXiv:nucl-ex/0410003].
[7] C. Alt *et al.* [NA49 Collaboration], Phys. Rev. C **77** (2008) 024903 [arXiv:0710.0118 [nucl-ex]].
[8] M. M. Aggarwal *et al.* [WA98 Collaboration], arXiv:nucl-ex/0006007.
[9] S. S. Adler *et al.* [PHENIX Collaboration], Phys. Rev. Lett. **94** (2005) 232301 [arXiv:nucl-ex/0503003].
[10] J. I. Kapusta, P. Lichard and D. Seibert, Phys. Rev. D **44** (1991) 2774 [Erratum-ibid. D **47** (1993) 4171].
[11] L. Xiong, E. V. Shuryak and G. E. Brown, Phys. Rev. D **46** (1992) 3798 [arXiv:hep-ph/9208206].
[12] K. Geiger, Comput. Phys. Commun. **104** (1997) 70 [arXiv:hep-ph/9701226].

Journal of Physics: Conference Series **230** (2010) 012033 doi:10.1088/1742-6596/230/1/012033

[13] S. A. Bass *et al.*, Prog. Part. Nucl. Phys. **41** (1998) 255 [Prog. Part. Nucl. Phys. **41** (1998) 225] [arXiv:nucl-th/9803035].
[14] M. Bleicher *et al.*, J. Phys. G **25** (1999) 1859 [arXiv:hep-ph/9909407].
[15] W. Ehehalt and W. Cassing, arXiv:hep-ph/9507274.
[16] D. Molnar and P. Huovinen, Phys. Rev. Lett. **94** (2005) 012302 [arXiv:nucl-th/0404065].
[17] Z. Xu and C. Greiner, Phys. Rev. C **71** (2005) 064901 [arXiv:hep-ph/0406278].
[18] Z. W. Lin, C. M. Ko, B. A. Li, B. Zhang and S. Pal, Phys. Rev. C **72** (2005) 064901 [arXiv:nucl-th/0411110].
[19] G. Burau, J. Bleibel, C. Fuchs, A. Faessler, L. V. Bravina and E. E. Zabrodin, Phys. Rev. C **71** (2005) 054905 [arXiv:nucl-th/0411117].
[20] S. A. Bass, T. Renk and D. K. Srivastava, Nucl. Phys. A **783** (2007) 367.
[21] J. Cleymans and K. Redlich,
[22] L. D. McLerran, M. Kataja, P. V. Ruuskanen and H. von Gersdorff, Phys. Rev. D **34** (1986) 2755.
[23] H. Von Gersdorff, L. D. McLerran, M. Kataja and P. V. Ruuskanen, Phys. Rev. D **34** (1986) 794.
[24] M. Kataja, Z. Phys. C **38** (1988) 419.
[25] D. K. Srivastava and B. Sinha, Phys. Lett. B **261** (1991) 1.
[26] D. K. Srivastava, B. Sinha and T. C. Awes, Phys. Lett. B **387** (1996) 21.
[27] D. K. Srivastava, B. Sinha, M. Gyulassy and X. N. Wang, Phys. Lett. B **276** (1992) 285.
[28] D. K. Srivastava, J. Alam, S. Chakrabarty, S. Raha and B. Sinha, Phys. Lett. B **278** (1992) 225.
[29] J. Cleymans and H. Satz, Z. Phys. C **57** (1993) 135 [arXiv:hep-ph/9207204].
[30] D. H. Rischke, Y. Pursun and J. A. Maruhn, Nucl. Phys. A **595** (1995) 383 [Erratum-ibid. A **596** (1996) 717] [arXiv:nucl-th/9504021].
[31] T. Hirano, Phys. Rev. C **65** (2002) 011901 [arXiv:nucl-th/0108004].
[32] P. Huovinen, P. V. Ruuskanen and S. S. Rasanen, Phys. Lett. B **535** (2002) 109 [arXiv:nucl-th/0111052].
[33] P. Huovinen, M. Belkacem, P. J. Ellis and J. I. Kapusta, Phys. Rev. C **66** (2002) 014903 [arXiv:nucl-th/0203023].
[34] P. F. Kolb and U. W. Heinz, arXiv:nucl-th/0305084.
[35] C. Nonaka and S. A. Bass, Phys. Rev. C **75** (2007) 014902 [arXiv:nucl-th/0607018].
[36] E. Frodermann, R. Chatterjee and U. Heinz, J. Phys. G **34** (2007) 2249 [arXiv:0707.1898 [nucl-th]].
[37] K. Dusling and D. Teaney, Phys. Rev. C **77** (2008) 034905 [arXiv:0710.5932 [nucl-th]].
[38] R. Baier, P. Romatschke and U. A. Wiedemann, Phys. Rev. C **73** (2006) 064903 [arXiv:hep-ph/0602249].
[39] H. Song and U. W. Heinz, Phys. Rev. C **78** (2008) 024902 [arXiv:0805.1756 [nucl-th]].
[40] H. Petersen, M. Bleicher, S. A. Bass and H. Stocker, arXiv:0805.0567 [hep-ph].
[41] T. Sjostrand, S. Mrenna and P. Z. Skands, JHEP **0605** (2006) 026 [arXiv:hep-ph/0603175].
[42] H. Petersen, J. Steinheimer, G. Burau, M. Bleicher and H. Stocker, Phys. Rev. C **78** (2008) 044901 [arXiv:0806.1695 [nucl-th]].
[43] S. Turbide, R. Rapp and C. Gale, Phys. Rev. C **69** (2004) 014903 [arXiv:hep-ph/0308085].
[44] B. Bauchle and M. Bleicher, Phys. Rev. C **81** (2010) 044904 [arXiv:0905.4678 [hep-ph]].
[45] L. E. Gordon and W. Vogelsang, Phys. Rev. D **48** (1993) 3136.
[46] B. Baeuchle and M. Bleicher, arXiv:1003.5454 [nucl-th].
[47] A. Adare *et al.* [PHENIX Collaboration], Phys. Rev. Lett. **104** (2010) 132301 [arXiv:0804.4168 [nucl-ex]].

Dilepton Continuum at PHENIX

Sarah Campbell for the PHENIX Collaboration
Department of Physics and Astronomy, Stony Brook University Stony Brook, NY 11794
E-mail: campbell@skipper.physics.sunysb.edu

Abstract. The dilepton mass spectrum consists of light vector meson decays in addition to decays from other hadronic and photonic sources. Thermal radiation and modification of light vector meson masses via chiral symmetry restoration may provide signals above known hadronic sources. The newly published PHENIX $\sqrt{s_{NN}} = 200$ GeV Au+Au dielectron continuum results are discussed. The PHENIX $\sqrt{s_{NN}} = 200$ GeV Cu+Cu results in centrality classes are shown, providing increased sensitivity to centrality dependent trends.

1. Introduction

Relativistic heavy ion collisions at RHIC produce a strongly-coupled medium, called quark gluon plasma (QGP) [1]. This state of matter is characterized by its high energy density and rapid thermalization time, its anisotropic or "elliptic" flow [2] and the energy loss experienced by light [3] and heavy quarks [4] in the medium. Dileptons are electro-magnetic probes of this matter and are produced throughout the entire time evolution of the collision system. They have no color charge and, therefore, no final state interactions, providing a clean measurement of the earliest stages of the collision.

The physics signals within the dilepton mass spectra include virtual photon emission, hadronic decays and semi-leptonic heavy flavor decays. Alterations of the mass spectrum due to interactions in the medium allow us to study additional properties of this matter. Sources for this modification include thermal radiation [5], chiral symmetry restoration producing a continuum enhancement due to broadening [6] and/or shifting [7] of the light vector meson masses, correlated charm modification due to suppression of high p_T non-photonic single electrons [4], and J/ψ suppression [8]. This proceeding focuses on the modifications in the low mass region, masses below that of the ϕ, primarily due to thermal radiation and enhancement via alteration of the light vector meson masses.

Earlier results from multiple SPS experiments have shown modification of the ρ meson in dimuon measurements [9], [10]. NA60's In-In results, in particular, strongly supported the mass broadening scenario over the purely dropping mass scenario; however the dropping mass scenario may still be a possibility when combined with broadening [11]. These proceedings highlight advancements in this subject from the PHENIX dielectron continuum results, including the Au+Au and p+p results recently accepted by Phys. Rev. C [12] and new Cu+Cu preliminary results in centrality bins.

2. Analysis

The PHENIX central arms, shown in Figure 1, track particles in the drift chamber (DC) and first pad chamber (PC1), providing a momentum resolution of $(0.7p\oplus)$ % [13]. Electron tracks

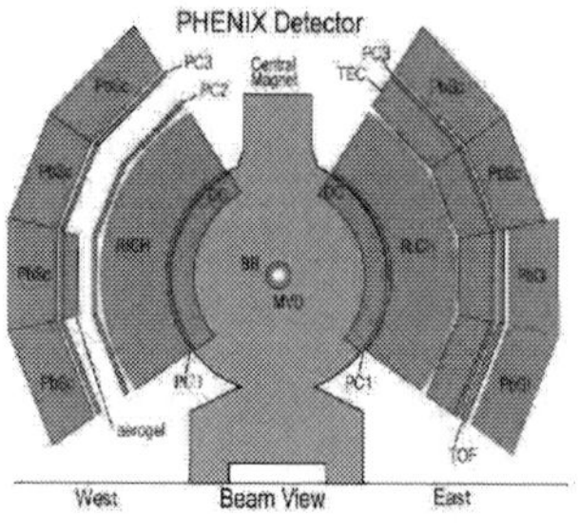

Figure 1. A beam view of the central arm detectors in PHENIX in the 2004 and 2005 run configuration. The relevant detectors for this analysis are the drift chamber (DC), the first pad chamber (PC1), the ring imaging cherenkov detector (RICH) and the two types of electromagnetic calorimeter (PbSc and PbGl).

are identified by a Cherenkov ring in the ring imaging cherenkov detector (RICH) [14] and an electromagnetic shower in the electro-magnetic calorimeters [15]. The centrality, a measure of the collision's nuclear overlap, is determined via coincidences between the beam beam counters and the zero degree calorimeter [16]. Once all electrons within an event are identified all possible dielectron pairs are made including like-sign and unlike-sign pairs. As a result many background pairs are generated. Removing background pairs is the crux of this analysis.

Two types of background pairs can be directly identified and rejected. These are pairs due to overlaps within the RICH and conversion pairs. Overlapping pairs within the RICH consist of pairs where the Cherenkov rings of the two electrons are not fully distinct. This results in a non-physical excess in both the like- and unlike-sign pair spectra which is removed by requiring a separation of twice the nominal ring size. Conversion pairs are opened by the central magnet's field and can be characterized by their mass and their orientation with respect to the magnetic field. The pair's mass is a conversion indicator because the electrons' momenta are misreconstructed, tracking back to the collision vertex not the conversion vertex. This gives the pair a mass that is roughly proportional to the distance between the conversion point and the collision vertex. Since the location of conversion material is well known, we are able to determine a range of masses in which these false pairs lie. As a result of being opened by the magnetic field, the charges of the conversion pair are ordered and the decay plane of the pair is perpendicular to the magnetic field. A cut on the mass, charge order and opening angle orientation relative to the magnetic field allows us to remove conversion pairs.

2.1. Statistical background removal

The remaining backgrounds are removed statistically. The like-sign pairs are used as a test to ensure that all backgrounds are removed. This test is possible because the relative rates of the like-sign and unlike-sign backgrounds are well known; for many backgrounds the relative rate is one. Since the like-sign pairs contain zero physics signals, if we describe all of the like-sign pairs as a combination of background sources then we can remove all backgrounds in the unlike-sign pairs.

The largest source of background in heavy ion collisions arises from combinatorial pairs. Combinatorial background increases as the square of the multiplicity, or number of electrons per event. We are able to determine the shape of the combinatorial background as a function of mass and pair transverse momentum, p_T, by creating non-physical pairings across events. The normalization of the combinatorial background is determined using the like-sign pairs in a region where the combinatorial background dominates. A ratio of all like-sign pairs over the like-sign combinatorial shows that there exist remaining correlated backgrounds at the lowest masses.

These correlated backgrounds are due to two sources, cross pairs and jet pairs. Cross pairs are generated by decays that produce multiple lepton pairs. An example of this is a dalitz decay

Journal of Physics: Conference Series **230** (2010) 012034 doi:10.1088/1742-6596/230/1/012034

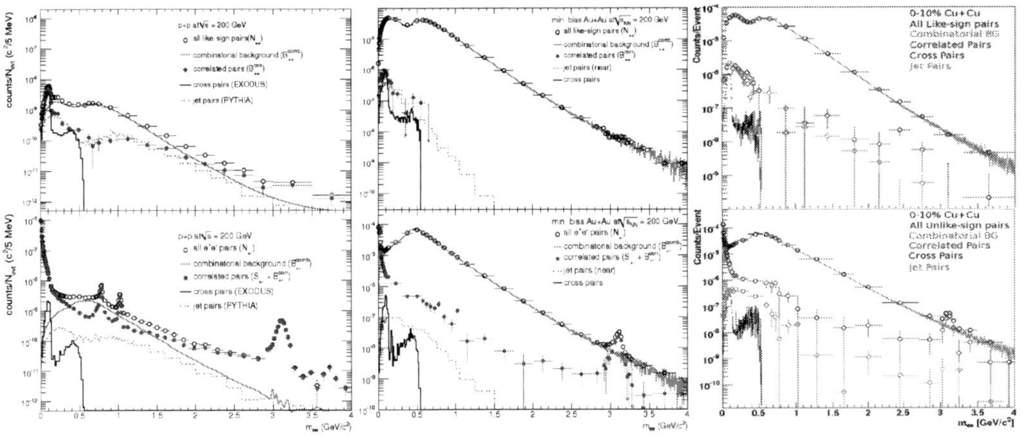

Figure 2. The dielectron counts per event in the p+p (left), minimum bias (MB) Au+Au (center) and 0 − 10% Cu+Cu (right) systems. The top row contains like-sign pairs and backgrounds; the bottom row contains unlike-sign pairs and backgrounds. All like(unlike)-sign pairs are shown in black circles, combinatorial background pairs are shown in red, correlated pairs (all pairs with the combinatorial background removed) are shown in blue, cross pairs are shown as a black histogram, and jet pairs are in green.

with a photon conversion where four unlike-sign pairs are created and two like-sign pairs are created. One of the unlike-sign pairs is kept as part of our standard signal as a dalitz pair included in our cocktail comparison. Another unlike-sign pair will be identified and removed as a photon conversion. This leaves two additional unlike-sign pairs and two like-sign pairs all of which are correlated through the original pion. In this example we see something that is observed generally for cross pair backgrounds. Like-sign pairs are produced at the same rate as the additional unlike-sign pairs. We are able to simulate this source of background using EXODUS, a PHENIX-made event generator.

Jet pairs are produced by dalitz decays of multiple pions within a jet, the resulting like- and unlike-sign electron pairs are correlated by the jet. This background source is simulated with PYTHIA. Here we see again that the like- and unlike-sign pairs due to this background are produced at the same rate. In the heavy ion case, the away-side ($d\phi > 90^o$) component of the jet background is reduced to account for jet modification [17], [18].

The like-sign spectra are used to set the normalization factors of the three different backgrounds. All of the like-sign pairs are well described by the three backgrounds; example spectra for each system are shown in Figure 2. In the unlike-sign spectra the backgrounds can then be removed considering the like-sign background rates. This results in the relevant unlike-sign signal also in Figure 2. The background subtracted unlike-sign spectra are corrected for efficiency and inactive regions up to the ideal PHENIX aperture and are compared to the cocktail of known sources.

2.2. Cocktail Generation

The cocktail of known sources contains a hadronic component and a open heavy flavor component. The hadronic component of the cocktail consists of the direct and dalitz hadronic decays and is modeled using EXODUS. The measured pion p_T spectra [19], [20], [22], [21] [23] is characterized by a modified Hagedorn function, $E\frac{d^3\sigma}{dp^3} = A\left(e^{-(ap_T+bp_T^2)} + \frac{p_T}{p_0}\right)^{-n}$, whose parameters (A, a, b, p_0, n) are input into EXODUS. Other mesons are extrapolated from this momentum information using m_T scaling ($p_T \rightarrow \sqrt{p_T^2 + m_{meson}^2 - m_\pi^2}$). In cases where the mesons' p_T distributions are measured the observed distributions are used [22-29].

The open heavy flavor component of the cocktail stems from two open charm (or open bottom) mesons that both semi-leptonically decay. The final state electron-positron pair, one from each meson, are correlated by the original $c\bar{c}$ (or $b\bar{b}$) production. These sources are simulated with PYTHIA. In the p+p case, we are able to measure the charm and bottom cross sections by subtracting the hadronic sources estimated with EXODUS and fitting the remaining pairs to the generated PYTHIA shapes. This results in the charm cross section, $\sigma_{c\bar{c}} = 544 \pm 39\,(stat) \pm 142\,(sys) \pm 200\,(model)\,\mu b$, and bottom cross section, $\sigma_{b\bar{b}} = 3.9 \pm 2.5\,(stat)^{+3}_{-2}\,(sys)\,\mu b$ [32]. The charm cross section agrees with the more precise PHENIX non-photonic single electron measurement's result, $567 \pm 57\,(stat) \pm 193\,(sys)\,\mu b$ [33]. In Au+Au and Cu+Cu, the p+p heavy flavor components are scaled by the average number of binary collisions, N_{coll}. The Drell Yan component is similarly generated in PYTHIA, included in the p+p fit, and scaled by N_{coll} for the heavy ion results.

3. Results

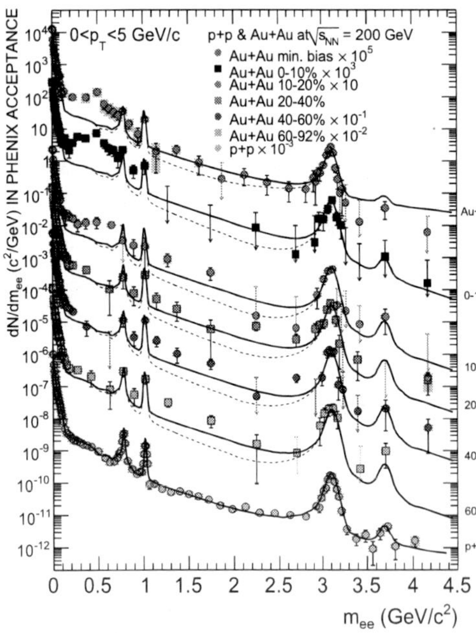

Figure 3. The dielectron mass spectra in MB, $0-10\%$, $10-20\%$, $20-40\%$, $40-60\%$, $60-92\%$ Au+Au and p+p collisions are scaled and overlaid with their respective cocktails, one with the charm component from PYTHIA (solid) and one with an uncorrelated charm contribution (dotted). The grey boxes represent the systematic errors.

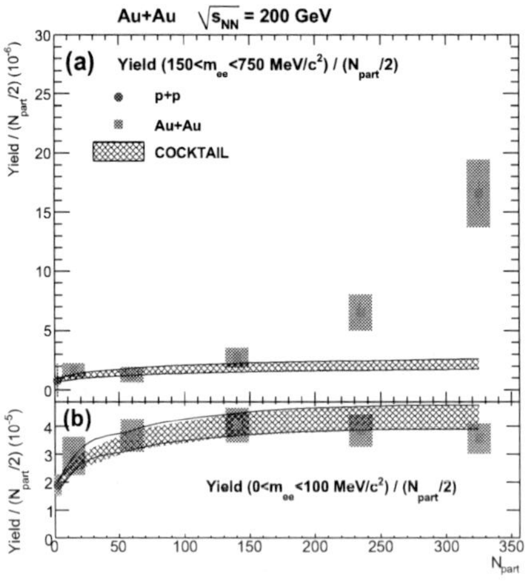

Figure 4. The dielectron yield scaled by $N_{Part}/2$ as a function of N_{Part} for the mass ranges a) 0.15 to 0.75 GeV$/c^2$ and b) 0 to 0.1 GeV$/c^2$. The hashed bands represent the expected yield with in the model's systematic error from the cocktail. The systematic error of the data is shown as grey boxes.

The measured Au+Au and p+p dielectron mass spectra are compared to their respective cocktails in Figure 3; the Au+Au minimum bias (MB), Au+Au in centrality bins and p+p result are shown. The MB, $0-10\%$ and $10-20\%$ Au+Au spectra have an excess above the cocktail in the region above the π and below the ω masses. In the case of the MB, this excess is remarkably large, a factor of $4.7 \pm 0.4(stat) \pm 1.5(sys) \pm 0.9(model)$ above the expected cocktail yield. The Au+Au excess is largest in the most central collisions and recedes as the centrality decreases. The integrated yield increases faster than the square of the average number of participating nucleons in the collision (N_{part}^2). This is seen in Figure 4 where the yield in the mass region 0.15 to 0.75 GeV/c^2 over $\frac{N_{part}}{2}$ increases faster than linear in N_{part}.

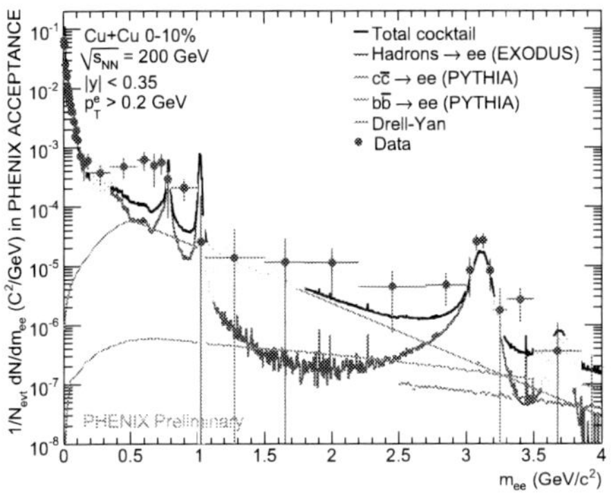

Figure 5. The dielectron mass spectra in $0-10\%$ Cu+Cu collisions overlaid with the cocktail. The yellow bars represent the systematic errors.

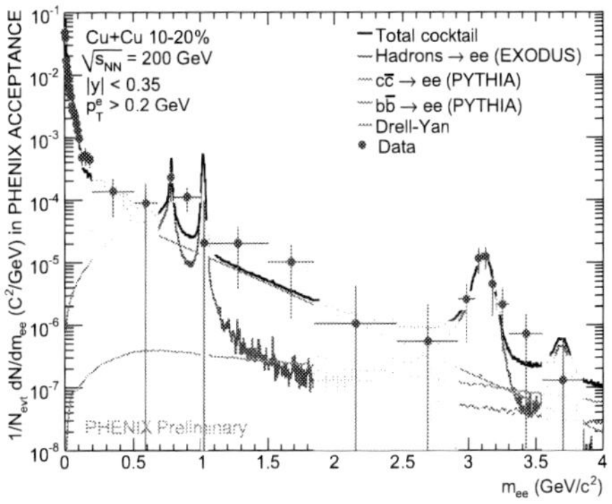

Figure 6. The dielectron mass spectra in $10-20\%$ Cu+Cu collisions overlaid with the cocktail. The yellow bars represent the systematic errors.

The Cu+Cu centrality data, first published here, allows us to gain insight into the onset of the excess. The Cu+Cu $0-10\%$ data contains an excess in the low mass region, as shown in Figure 5. The excess is not present in the $10-20\%$ data, in Figure 6, or in the more peripheral centrality bins. This is consistent with the centrality trend seen in Figures 3 and 4. The Cu+Cu results provide valuable additional information about the emergence of the enhancement. With these new results, theories should attempt to describe the enhancement as a function of centrality.

3.1. Virtual photons

The p+p and MB Au+Au mass spectra in different pair p_T slices are shown in Figures 7 and 8 respectively. In the p+p case the data matches the cocktail fairly well, particularly at low p_T; there is a small excess at high p_T. In the Au+Au case there exists an excess in all p_T slices with the largest portion in the low p_T ranges. This leads us to consider that the Au+Au excess has two sources, one that dominates at high p_T and another that dominates at low p_T.

One such source is direct virtual photons that convert to dielectron pairs. Virtual photons are produced by the same processes that generate real photons. A direct photon component, or thermal photon component, is described according to Equation 1 when the mass is much smaller than the p_T ($m < 0.3$ GeV/c^2 and $p_T > 1$ GeV/c)

$$\frac{d^2N_{ee}}{dm_{ee}dp_T} \simeq \frac{2\alpha}{3\pi}\frac{1}{m_{ee}}\frac{dN_\gamma}{dp_T} \text{ [34].} \qquad (1)$$

The photonic component is modeled in EXODUS, filtered by the acceptance and smeared by the detector resolution. This shape is shown in Figure 9. By looking at the ratio of the excess above the cocktail divided by the modeled direct photon shape, R, we can compare the mass

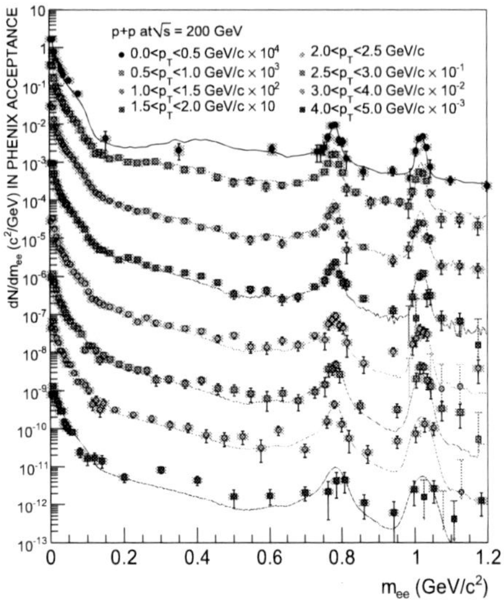

Figure 7. The dielectron mass spectra from p+p collisions separated into different p_T ranges. The data is compared to the cocktail including the PYTHIA charm component.

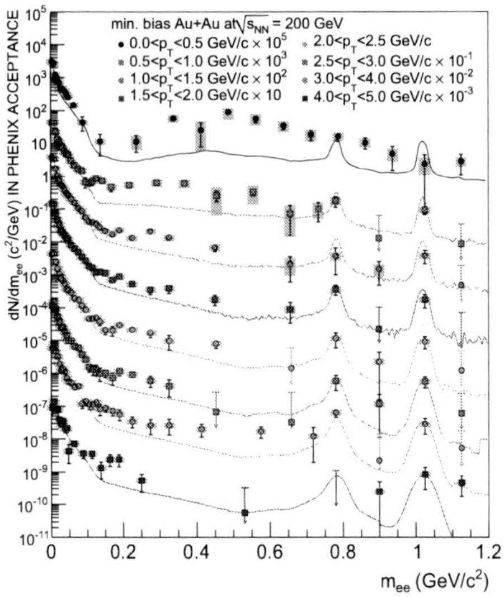

Figure 8. The dielectron mass spectra from MB Au+Au collisions separated into different p_T ranges. The data is compared to the cocktail including the N_{Coll} scaled PYTHIA charm component.

shape of the excess to the direct photon shape. When p_T is greater than one, as in Figure 10, R is flat as a function of mass; a direct photon source models this excess. For p_T less than one, R is not flat, as seen in Figure 11. At low p_T, there is more modification than a photonic source can describe. A second source of masses modification exists in the low p_T region. This is most likely due to alteration of the light vector mesons via chiral symmetry restoration. The small p+p mass excess at high p_T is fully described by a direct photon component.

The p+p data is well described by the cocktail plus a thermal photon component in both the mass spectra in p_T slices and 2π acceptance corrected p_T spectra in mass slices. The MB Au+Au 2π acceptance corrected p_T spectra in mass slices are compared to the cocktail plus the photonic component (not shown). In the high p_T regions, the spectra are well described at all masses. However, there is additional yield at low p_T in mass regions above 0.3 GeV/c^2, again indicating a second source. Thus, the p_T spectra measurements provide additional evidence that the low mass excess in Au+Au is produced by two sources. Direct (i.e. thermal) photons are the dominant source of the excess in the high p_T yield. There is a second modification at low p_T, presumably due to chiral symmetry restoration.

3.2. Temperature measurements

A detailed analysis of the photonic source in dielectrons at PHENIX is presented in the paper [34]. The MB Au+Au thermal photon measurement provides the most direct extrapolation of the initial temperature of the fireball, projecting values of 300 to 600 MeV for thermalization times of 0.6 to 0.15 fm/c respectively [33-38]. The fireball created at RHIC has "the hottest temperature ever reached in a laboratory, about 250,000 times hotter than the center of the Sun [41]."

The temperature of the fireball is not the only temperature information carried by the

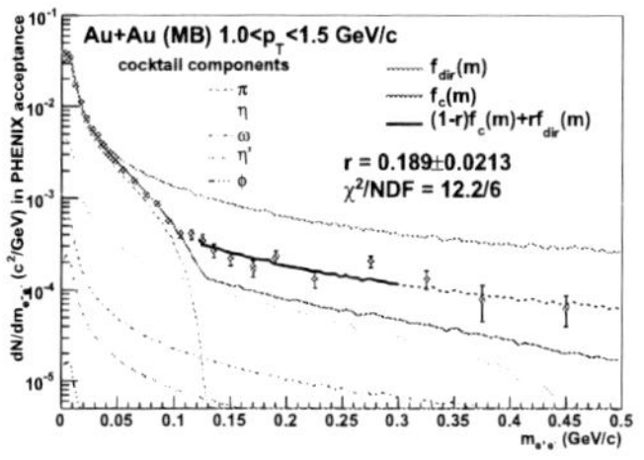

Figure 9. The MB Au+Au mass spectra in the p_T range 1 to 1.5 GeV/c overlaid with the cocktail (blue) and the direct photon shape (red). Additional cocktail components are also shown and the result of a fit (black) determining the relative size of the direct photon component.

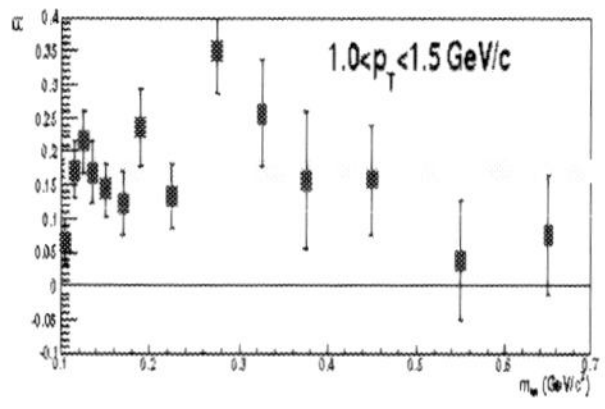

Figure 10. The ratio of the excess, data minus cocktail, over the direct photon shape (R) as a function of mass in the p_T range 1 to 1.5 GeV/c. The yellow band displays a one sigma range on the fit determining the direct photon component.

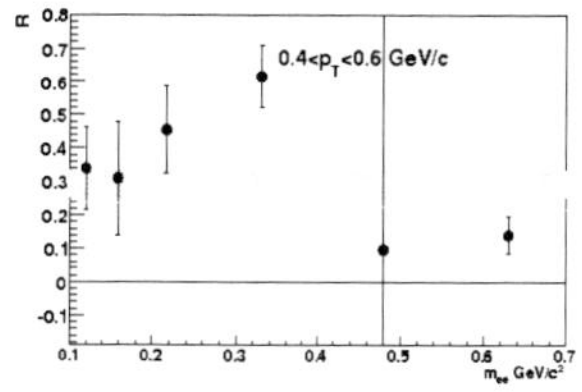

Figure 11. The ratio of the excess, data minus cocktail, over the direct photon shape (R) as a function of mass in the p_T range 0.4 to 0.6 GeV/c. The yellow band displays a one sigma range on the fit determining the direct photon component.

dielectron spectra. To measure the effective temperature of the low p_T excess, we plot the $m_T - m_0$ spectra in the mass region 0.3 to 0.75 GeV/c^2 corrected into the full 2π angular acceptance in Figures 12 and 13. This m_T spectrum has a kink structure with a much softer component at $m_T - m_0$ below 0.6 GeV/c^2 and a harder trend above 0.6 GeV/c^2 corresponding to the two sources of the excess in this mass region.

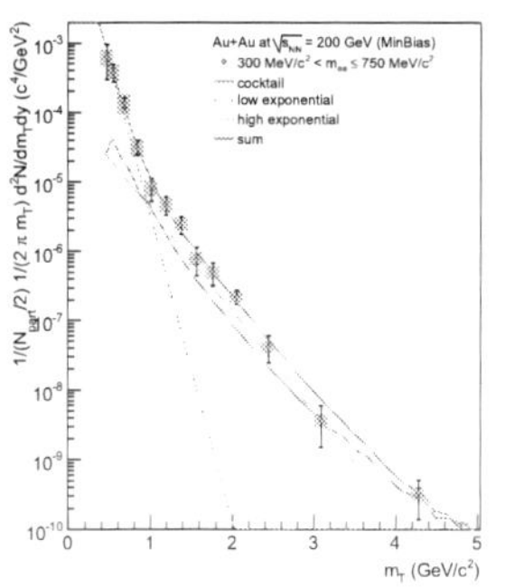

Figure 12. The MB Au+Au $m_T - m_0$ spectrum in the mass range 0.3 to 0.75 GeV/c^2. The spectrum is fit to the sum (solid) of two exponentials (dashed and dashed-dot) plus the cocktail.

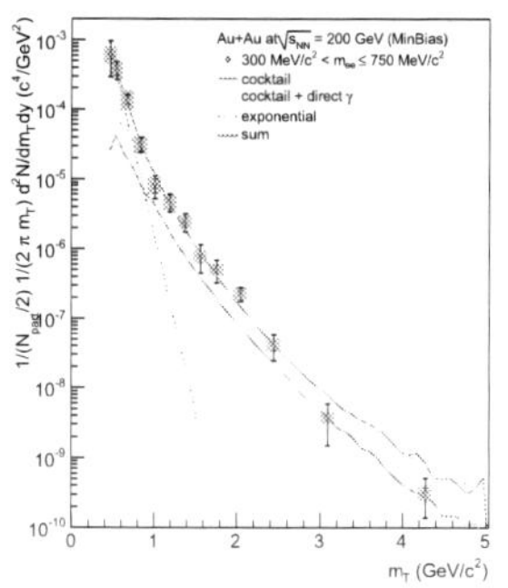

Figure 13. The MB Au+Au $m_T - m_0$ spectrum in the mass range 0.3 to 0.75 GeV/c^2. The spectrum is fit to the sum (solid) of an exponential (dashed), a thermal component (dashed-dot), and the cocktail.

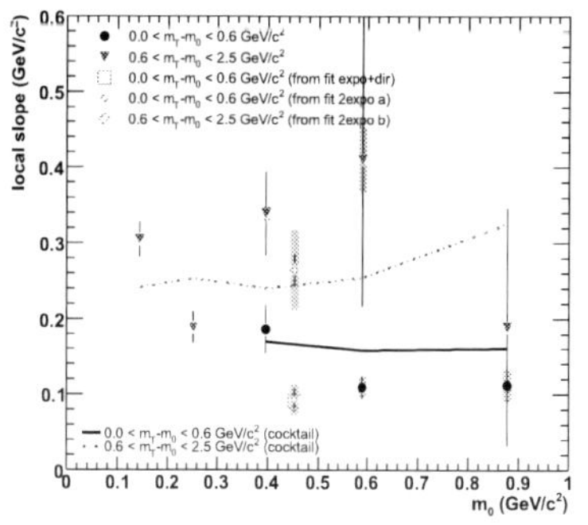

Figure 14. The local inverse slopes of the m_T spectra in Figures 12 and 13 in the regions above (blue) and below (black) 0.6 GeV/c^2. The results of the fits in Figures 12 and 13 are shown in red. The lines represent the local slope of the cocktail in the mass ranges above (blue dashed) and below (black solid) 0.6 GeV/c^2

The $m_T - m_0$ spectrum is fit twice, once with two exponentials in Figure 12 and again with an exponential plus a photonic component in Figure 13. This results in soft effective temperatures for the region below 0.6 GeV/c^2, $92.0 \pm 11.4 \pm 8.4$ MeV and $86.5 \pm 12.7^{+11}_{-28.4}$ MeV respectively, and harder effective temperatures for the region above 0.6 GeV/c^2, $258.4 \pm 37.7 \pm 9.6$ MeV and $221 \pm 19 \pm 19$ MeV respectively. Hydrodynamic models relate the photonic effective temperature to the larger initial temperature by a factor between 1.5 and 3 [35]. The effective temperature of the photonic component, $221 \pm 19 \pm 19$ MeV, is consistent with the initial temperatures from the thermal photon measurement described earlier. These effective temperatures are compatible with calculations of the local slope of the $m_T - m_0$ spectrum in the regions above and below 0.6 GeV/c^2, as seen in Figure 14. The expected local slopes from the hadronic sources are displayed as well, agreeing in the high $m_T - m_0$ region but overestimating the low $m_T - m_0$ region. This suggests that the excess below 0.6 GeV/c^2, which is the majority of the excess, is very soft with an effective temperature of around 100 MeV, much lower than one would expect from the hadronic component of the cocktail or from a flowing source.

3.3. Theoretical comparisons

A variety of theoretical comparisons (not shown) have been made to the MB Au+Au mass, p_T and $m_T - m_0$ spectra including Rapp and van Hees [42], Dusling and Zahed [43], and Cassing and Bratkovskaya [44]. Rapp and van Hees' theory describes the light vector mesons' spectral functions using the vector dominance model. In this theory, medium modifications are due to hadronic many-body interactions, and a shifting mass is modeled for comparison. Dusling and Zahed's theory uses a chiral reduction formalism to calculate the spectral functions in vacuum; medium modification occurs as a result of the hydrodynamic evolution of the system. Cassing and Bratkovskaya's theory applies a microscopic relativistic transport model (HSD) with the ρ spectral function modified via collisional broadening and a tunable mass shift.

None of the theories currently describe the MB Au+Au data, particularly in the mass region around 0.5 GeV/c^2 where the excess is the largest. Rapp and van Hees and Cassing and Bratkovskaya both are close to the data for masses above 0.6 GeV/c^2. Comparisons of the MB Au+Au mass spectra in p_T slices with the same theoretical calculations reveal that the difficulty lies in the low p_T ranges. Comparisons with the p_T and $m_T - m_0$ spectra produce similar discrepancies, particularly in the low p_T and $m_T - m_0$ ranges. In each case, m, p_T and $m_T - m_0$, the theories are close to the measured spectra at high p_T or m_T but a bit low. Including virtual photon processes, such as q + g $\rightarrow$ q + γ^*, could improve all of the calculations at high p_T or $m_T - m_0$. Resolving the discrepancy between the data and the theories at low p_T or $m_T - m_0$ may prove more difficult.

4. Conclusion

The p+p data is well described by the cocktail plus a virtual photon component. There is an excess in the low mass region of the dielectron spectra in the MB, 0 – 10% and 10 – 20% Au+Au and the 0 – 10% Cu+Cu data. The excess grows with centrality faster than N_{Part} squared. We have identified two sources of the excess in the low mass region, a photonic component at high p_T and a second modification at low p_T, possibly due to chiral symmetry restoration. The thermal photon component allows an extrapolation of the initial temperature of the fireball to between 300 and 600 MeV for thermalization times of 0.6 to 0.15 fm/c. The remainder of the excess, which is not described by thermal photons, contains the majority of the yield, dominates at low p_T and is very soft with an effective temperature of around 100 MeV.

The future of this subject at PHENIX lies in forthcoming analyses. A d+Au analysis, which is in progress, will allow us to identify cold nuclear matter effects in the heavy ion dielectron spectra. Low energy scan measurements will differentiate between the theoretical models. The hadron blind detector [45] has been successfully taking data since the 2009 p+p run and is currently taking Au+Au data. This detector reduces the combinatorial background, our largest background source, by allowing identification and removal of pion dalitz and conversion decay electrons from our electron sample before making pairs. The results shown in these proceedings, coupled with the future analyses and detector upgrades described above, move the discussion of the dilepton excess at PHENIX beyond the discovery phase and into a detailed study of its onset and characteristics, opening a new window onto the properties of the QGP medium and the fundamental nature of QCD.

References
[1] Adcox K et al. 2005 *Nucl. Phys. A* **757** 184-284
[2] Afanasiev S et al. 2007 *Phys. Rev. Lett.* **99** 052301
[3] Adler S S et al. 2006 *Phys. Rev. Lett.* **96** 202301
[4] Adler S S et al. 2006 *Phys. Rev. Lett.* **96** 032301
[5] Stankus P 2005 *Ann. Rev. Part. Sci.* **55** 517-54
[6] Rapp R and Wambach J 2000 *Adv. Nucl. Phys.* **25** 1-166
[7] Brown G E and Rho M 1991 *Phys. Rev. Lett.* **66** 2720
[8] Adare A et al. 2007 *Phys. Rev. Lett.* **98** 232301
[9] Agakichiev G et al. 1995 *Phys. Rev. Lett* **75** 1272
[10] Masera M et al. 1995 *Nucl. Phys. A* **590** 103-16
[11] Arnaldi R et al. 2006 *Phys. Rev. Lett.* **96** 162302
[12] Adare A et al. 2010 accepted to *Phys. Rev. C* [arXiv 0912.0244]
[13] Adcox K et al. 2003 *Nucl. Instrum. Meth. A* **499** 489-507
[14] Adcox K et al. 2003 *Nucl. Instrum. Meth. A* **499** 508-20
[15] Adcox K et al. 2003 *Nucl. Instrum. Meth. A* **499** 521-36
[16] Allen M et al. 2003 *Nucl. Instrum. Meth. A* **499** 549
[17] Adare A et al. 2008 *Phys. Rev. C* **78** 014901
[18] Jia J 2008 *J. Phys. G:Nucl. Part. Phys.* **35** 104033
[19] Adare A et al. 2007 *Phys. Rev. D* **76** 051106
[20] Adler S S et al. 2006 *Phys. Rev. C* **74** 024904
[21] Adler S S et al. 2004 *Phys. Rev. C* **69** 034909
[22] Adler S S et al. 2003 *Phys. Rev. Lett.* **91** 072301
[23] Adare A et al. 2008 *Phys. Rev. Lett.* **101** 162301
[24] Alder S S et al. 2007 *Phys. Rev. C* **75** 024909
[25] Alder S S et al. 2007 *Phys. Rev. C* **75** 051902
[26] Ryabov Y 2007 *J. Phys. G: Nucl. Part. Phys.* **34** S925
[27] Ryabov V 2009 *Nucl. Phys. A* **827** 395-7
[28] Alder S S et al. 2005 *Phys. Rev. C* **72** 014903
[29] Adare A et al. 2007 *Phys. Rev. Lett.* **98** 232002
[30] Adare A et al. 2008 *Phys. Rev. Lett.* **101** 122301
[31] daSilva C 2009 *Nucl. Phys. A* **830** 227-30
[32] Adare A et al. 2009 *Phys. Lett. B* **670** 313-20
[33] Adare A et al. 2006 *Phys. Rev. Lett.* **97** 252002
[34] Adare A et al. 2006 accepted to *Phys. Rev. Lett.* [arXiv 0804.4168]
[35] d'Enterria D and Peressounko D 2006 *Eur. Phy. J. C* **46** 451-64
[36] Turbide S, Rapp R and Gale C 2004 *Phys. Rev. C* **69** 014903
[37] Huovinen P, Ruuskanen P V and Rasanen S S 2002 *Phys. Lett. B* **535** 109-16
[38] Srivastava D K and Sinha B 2001 *Phys. Rev. C* **64** 034902
[39] Alam J, Sarkar S, Hatsuda T, Nayak T and Sinha B 2001 *Phys. Rev. C* **63** 021901
[40] Liu F M, Hirano T, Werner K and Zhu Y 2009 *Phys. Rev. C* **79** 014905
[41] BNL press release Feb 15,2010 http://www.bnl.gov/bnlweb/pubaf/pr/PR_display.asp?prID=1074
[42] Liu W and Rapp R 2007 *Nucl. Phys A* **796** 101-21
[43] Dusling K and Zahed I 2009 *Nucl. Phys. A* **825** 212 and Dusling K thesis 2008, Stony Brook Univeristy
[44] Bratkovskaya E L, Cassing W and Linnyk O 2009 *Phys. Lett. B* **670** 428-33
[45] Azmoun B, Caccavano A, Citron Z, Durham M, Hemmick T, Kamin J, Rumore M and Woody C 2009 *IEEE TNS* **56** 1544-9

Journal of Physics: Conference Series **230** (2010) 012035 doi:10.1088/1742-6596/230/1/012035

Jets associated with Z^0 boson production in heavy-ion collisions at the LHC

R. B. Neufeld
Los Alamos National Laboratory, Theoretical Division, MS B238, Los Alamos, NM 87545, U.S.A.

Abstract. The heavy ion program at the LHC will present unprecedented opportunities to probe hot QCD matter, that is, the quark gluon plasma (QGP). Among these exciting new probes are high energy partons associated with the production of a Z^0 boson, or Z^0 *tagged jets*. Once produced, Z^0 bosons are essentially unaffected by the strongly interacting medium produced in heavy-ion collisions, and therefore provide a powerful signal of the initial partonic energy and subsequent medium induced partonic energy loss. When compared with theory, experimental measurements of Z^0 tagged jets will help quantify the jet quenching properties of the QGP and discriminate between different partonic energy loss formalisms. In what follows, I discuss the advantages of tagged jets over leading particles, and present preliminary results of the production and suppression of Z^0 tagged jets in relativistic heavy-ion collisions at LHC energies using the Guylassy-Levai-Vitev (GLV) partonic energy loss formalism.

The suppression of high energy partons in the quark gluon plasma (QGP), that is *jet quenching*, is one of the most striking results from the heavy-ion program at RHIC [1, 2]. Much theoretical effort has been invested in understanding the mechanism of this suppression and in particular of partonic energy loss [3, 4]. Because of the limited center-of-mass energies available at RHIC, experimental measurements have thus far focused on leading particle suppression relative to $p+p$ collisions [5]. Unfortunately, leading particle suppression alone is not sufficient to discriminate between partonic energy loss formalisms or to extract quantitatively the jet quenching properties of the QGP [6].

In order to constrain the underlying QCD theory of jet quenching, new and more differential observables are needed. In particular, jet shapes [7, 8] and jets tagged with electromagnetic [9] or weakly interacting probes are especially promising. Jet reconstruction and analysis of the subsequent jet shapes provide an experimental handle on the medium-induced bremsstrahlung spectrum. This is demonstrated in Figure 1, which is taken from Ref. [8]. The Figure shows predictions for the sensitivity of jet R_{AA} on the Lorentz-invariant jet reconstruction cone size $R = \sqrt{(\Delta\phi)^2 + (\Delta y)^2}$ using the Guylassy-Levai-Vitev [4] (GLV) partonic energy loss formalism. The plot was done for impact parameter b = 3 fm in Au+Au collisions at $\sqrt{s} = 200$ GeV per nucleon. The dependence of the suppression on cone size R directly reflects the angular distribution of medium-induced radiation. The heavy-ion program at the LHC will enable a high statistics experimental measurement of this feature for the first time.

In addition to jet shapes, *tagged jets* provide an exciting opportunity to quantify the jet quenching properties of the QGP. Tagged jets refer to high energy partons which are produced in association with the tagging particle. Electromagnetic and weakly interacting particles are ideal tags to study partonic energy loss because, once produced, they have negligible medium induced

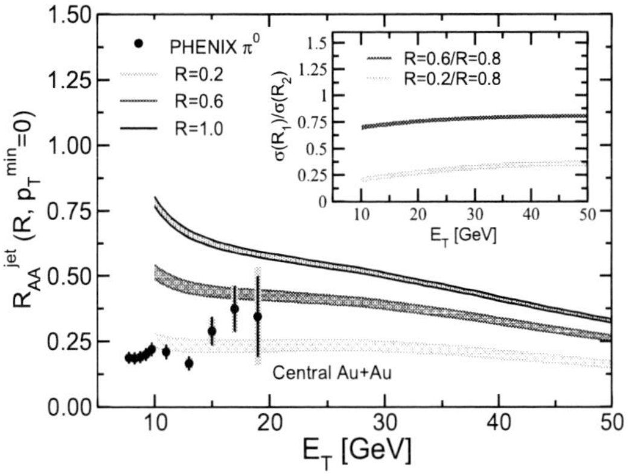

Figure 1. The figure, taken from Ref. [8], shows predictions for jet R_{AA} as a function of Lorentz-invariant jet reconstruction cone size $R = \sqrt{(\Delta\phi)^2 + (\Delta y)^2}$ using the Guylassy-Levai-Vitev [4] (GLV) partonic energy loss formalism. The plot was done for impact parameter b = 3 fm in Au+Au collisions at $\sqrt{s} = 200$ GeV per nucleon. The dependence of the suppression on cone size R directly reflects the angular distribution of medium-induced radiation. The insert shows the ratios of jet cross sections for selected values of R.

modifications. Consequently, the tagging particle serves as a signal for the initial associated jet energy. By reconstructing the tagged jet, one can obtain the amount and distribution of partonic energy loss. The leptonic final states of the Z^0 boson are an especially attractive jet tag because the large invariant mass of the Z^0 boson makes it easy to distinguish from the background generated in a heavy-ion collision. Again, the heavy-ion program at the LHC will enable a high statistics experimental measurement of this feature for the first time.

Consider the cross section for Z^0 tagged jet production in $p + p$ collisions. The leading-order (LO) production amplitude is represented by the Feynman diagrams in Figure 2. In each case, the Z^0 is produced either from quark radiation or quark/anti-quark annihilation. If one makes the approximation that the initial partons are collinear with the nucleon momenta, it is immediately clear that conservation of momentum requires the transverse (to the beam line) momentum (denoted $\mathbf{p}_T$ or $p_T = |\mathbf{p}_T|$ from here on) of the Z^0 and the scattered parton be equal. At mid-rapidity, the p_T of the associated Z^0 boson completely specifies the initial energy of the tagged jet at LO.

As mentioned above, I am here interested in the leptonic decay products of the Z^0 boson are. For this reason, one must also include the contribution of virtual photon tagged jets in the invariant mass range of the Z^0. Diagrammatically, this corresponds to making the replacement $Z^0 \leftrightarrow \gamma^*$ in Figure 2. In the results presented below, dilepton pairs in the invariant mass range $65 - 115$ GeV are considered (recall that $M_z = 91.1876$ GeV and $\Gamma_z = 2.4952$ GeV [11]). In this invariant mass range, the contribution from the Z^0 dominates that of the virtual photon by roughly two orders of magnitude.

Obtaining the LO result requires a lengthy but textbook level calculation, and the details will not be presented here. The LO result, in the notation of Field [14], for the matrix elements squared is

$$
\begin{aligned}
|M_g|^2 &= \frac{16\pi^2\,\alpha_s\,\alpha\,(R_q^2 + L_q^2)}{9x_w(1-x_w^2)}\left(\frac{\hat{t}^2 + \hat{u}^2 + 2\,\hat{s}\,m_z^2}{\hat{t}\hat{u}}\right) \\
|M_q|^2 &= -\frac{6\pi^2\,\alpha_s\,\alpha\,(R_q^2 + L_q^2)}{9x_w(1-x_w^2)}\left(\frac{\hat{s}^2 + \hat{t}^2 + 2\,\hat{u}\,m_z^2}{\hat{s}\,\hat{t}}\right).
\end{aligned}
$$

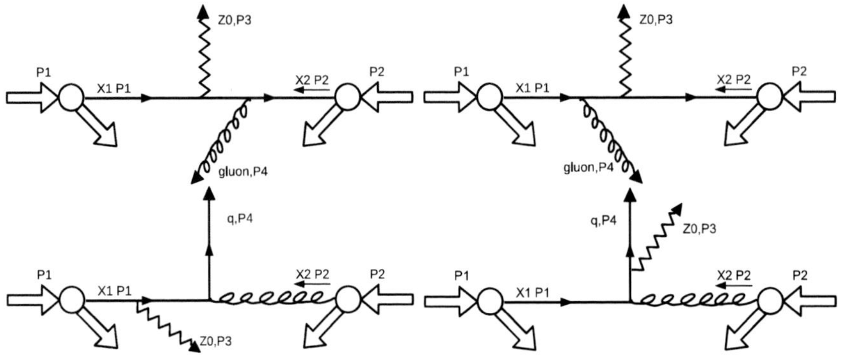

Figure 2. LO diagrams contributing to Z^0 tagged jet production. Curly lines denote gluons, straight lines quarks, and jagged lines Z^0 (labeled with $Z0$ in the figure). Two partons with momentum fraction x_1 and x_2 from nucleons with momenta P_1 and P_2, respectively, interact as seen above. At LO the contribution from quark and gluon jets are conveniently separable; however, at NLO real gluon emission complicates that distinction.

When the initial-state partons are interchanged the matrix elements can be trivially obtained with the substitution $\hat{t} \leftrightarrow \hat{u}$. In the above result, the subscript q or g refers to a quark or gluon jet, respectively, $\alpha_s = g_s^2/4\pi$, and the Lorentz-invariant Mandelstaam variables are defined such that

$$\begin{aligned} \hat{s} &= (p_1 + p_2)^2 \\ \hat{t} &= (p_1 - p_3)^2 \\ \hat{u} &= (p_2 - p_3)^2 \end{aligned}$$

where the notation of Figure 2 is used. Also, $x_w \equiv \sin^2\theta_w \approx 0.231$, $\alpha \approx 1/137.04$ [11] and

$$\begin{aligned} R_q^2 &= 4\,e_q^2\sin^4\theta_w, \\ L_q^2 &= \tau_q^2 - 4\,e_q\,\tau_q\sin^2\theta_w + 4\,e_q^2\sin^4\theta_w \end{aligned}$$

where τ_q is the weak isospin of quark q (that is, $\tau = 1$ for u, c, t and $\tau = -1$ for d, s, b) and e_q is the fractional electric charge of quark q. Modifying the above result to γ^* tagged jets can be done with the replacement

$$\frac{(R_q^2 + L_q^2)}{x_w(1 - x_w^2)} \rightarrow 2e_q^2 \qquad (1)$$

as well as letting $m_z^2 \rightarrow Q^2$, where Q is the photon mass. The total scattering probability requires summing $|M_g|^2 + |M_q|^2$ over quark flavors.

The leptonic final states from the LO Z^0/γ^* tagged jet cross section can be obtained by performing a Lorentz boost into the rest frame of the tagging particle. In the rest frame of the particle, a large number of isotropically distributed random decays are performed, which are then boosted back into the lab frame. The fraction of lepton pairs that are inside of the experimental limitations is then treated as a multiplicative factor, which, along with the appropriate branching ratio, is multiplied with the cross section.

Having the LO result is instructive and useful for understanding isospin effects when going from $p + p \rightarrow A + A$ or in obtaining the relative fraction of gluon versus quark jets. However, in order to study the suppression due to medium induced energy loss, it is necessary to have the

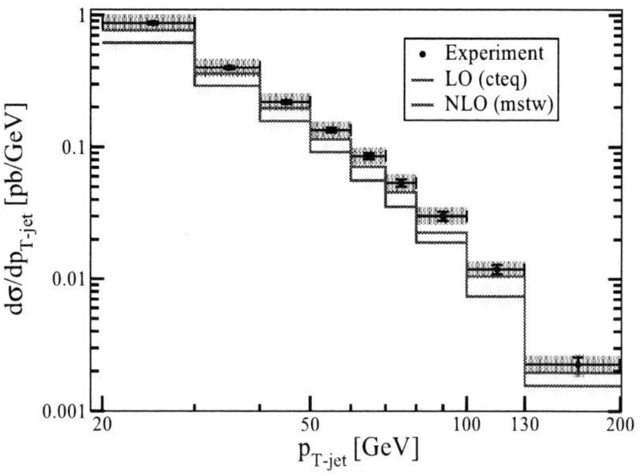

Figure 3. Experimental results from the FermiLab Tevatron Collider [13] of jets associated with $Z^0/\gamma^* \to \mu^+\mu^-$ in $p\bar{p}$ collisions at $\sqrt{s} = 1.96$ TeV. A comparison of the MCFM NLO result (blue line) and the LO result (within the yellow bands) is included. The experimental parameters are discussed in the text. The (yellow) bands around the LO result indicated uncertainty in the choice of scale.

next-to-leading-order (NLO) result. This is because, as mentioned above, at LO the p_T of the Z^0 and the scattered parton are exactly equal. Obviously medium induced energy loss in $A + A$ collisions will break this equality. In order to make a meaningful comparison for the case when the p_T of the Z^0 and the scattered parton are not equal (as is certainly the case in $A + A$), one must have the NLO result. At NLO, the p_T equality is smeared by real gluon emission, enabling a comparison between $p + p$ and $A + A$.

The NLO cross section is here obtained using the publicly available Monte Carlo for FeMtobarn processes (MCFM) developed by Campbell and Ellis [12], (available online at http://mcfm.fnal.gov/). MCFM provides NLO predictions for many processes at hadron colliders. The program has built-in options for a wide range of experimental acceptance cuts and run-time parameters. A comparison of the LO result and the MCFM NLO result for jets associated with $Z^0/\gamma^* \to \mu^+\mu^-$ in $p + \bar{p}$ collisions at $\sqrt{s} = 1.96$ TeV with experimental measurements at the FermiLab Tevatron Collider [13] is shown in Figure 3. The leading jet particle was required to have $p_T > 20$ GeV and rapidity $|y| < 2.8$, where

$$y = \frac{1}{2} \ln \frac{E - p_z}{E + p_z} \tag{2}$$

and $\hat{z}$ is oriented along the beam axis. Jets were reconstructed using a midpoint cone algorithm with Lorentz-invariant cone size $R = \sqrt{(\Delta\phi)^2 + (\Delta y)^2} = 0.5$. Additionally, the muons were required to have $p_T > 15$ GeV and $|y| < 1.7$. The yellow band around the LO result indicates the uncertainty due to the choice of scale in the problem. Notice that the results shown in Figure 3 are inclusive in the sense that the tagging particles are not constrained to lie within a certain p_T interval.

The real advantage of using tagged jets is that one can require the tagging particles to lie in a certain p_T range. Results for the suppression of tagged jets at NLO will not be presented in this proceedings, as they are still in progress and will be forthcoming soon. However, one can still obtain some information with LO $Z^0/\gamma^* \to \mu^+\mu^-$ tagged jets in $p + p$. For instance, in a probabilistic description of partonic energy loss, the probability for a parton to lose some fraction of energy ϵ is described by the normalized distribution $P(\epsilon)$. The probability for a parton not to lose energy, $P(0)$, is related to the average number of gluons emitted, $\langle N_g \rangle$. In

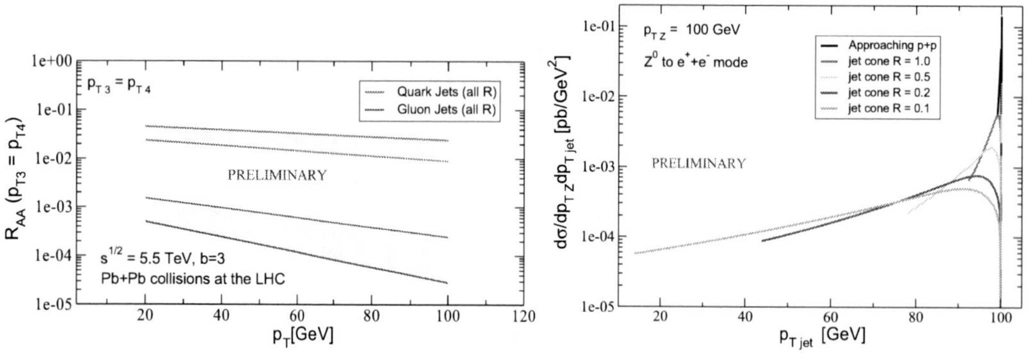

Figure 4. In a probabilistic description of partonic energy loss, the probability not to radiate, $P(0)$, can be directly related to the average number of gluons in the medium induced bremsstrahlung spectrum as a function of the p_T of the parent parton. This is the case considered in the left panel of the figure, $\ln R_{aa} = -\langle N_g \rangle$ (see discussion in text and equation (3)). In the right panel is plotted the modified nuclear cross section for different values of R as a function of jet p_T. As the jet cone radius, R, is increased the cross section approaches a δ function at $p_T = 100$ GeV, which is the p_T of the tagging particle. At NLO, even as $R \to \infty$, the cross section would still have some broadening. As R is decreased, the result smears out, reflecting the fact that less and less of the initial energy is recovered in the jet.

this specific case one has [15]

$$\ln R_{aa} = -\langle N_g \rangle . \tag{3}$$

This is precisely what is plotted in the left panel of Figure 4 for central $Pb + Pb$ collisions $\sqrt{S} = 5.5$ TeV per nucleon. The results were obtained using the GLV partonic energy loss formalism. From the Figure, one can determine the expected number of gluons emitted as a function of the p_T of the parent parton in this formalism.

One can also examine the sensitivity of the jet cross section in $Pb+Pb$ on cone radius. Again, using the LO result for $Z^0/\gamma^* \to \mu^+\mu^-$ tagged jets in $p+p$ and the GLV formalism, the modified nuclear cross section is plotted for different values of R as a function of jet p_T in the right panel of Figure 4. As the jet cone radius, R, is increased the cross section approaches a δ function at $p_T = 100$ GeV, which is the p_T of the tagging particle. At NLO, even as $R \to \infty$, the cross section would still have some broadening. As R is decreased, the result smears out, reflecting the fact that less and less of the initial energy is recovered in the jet.

In summary, I have argued that leading particle suppression alone is not sufficient to discriminate between partonic energy loss formalisms or to extract quantitatively the jet quenching properties of the QGP. New and more differential observables are needed to do this. In particular, jet shapes and jets tagged with weakly interacting probes are especially promising. The heavy-ion program at the LHC will enable high statistics experimental measurement of these observable for the first time. It is expected to usher in a new era in the study of QGP induced partonic energy loss and other related areas of QGP physics, such as medium response [16, 17] in the context of jets tagged with direct and virtual photons and electroweak bosons. I have here presented preliminary results for the production and nuclear modification of jets associated with Z^0/γ^* bosons at LHC energies using the GLV partonic energy loss formalism. The full results of this study are forthcoming.

Acknowledgments: I wish to thank my collaborators, Ivan Vitev and Ben-Wei Zhang, who have

Journal of Physics: Conference Series **230** (2010) 012035 doi:10.1088/1742-6596/230/1/012035

contributed equally in this work. This work was supported in part by the US Department of Energy, Office of Science, under Contract No. DE-AC52-06NA25396.

[1] M. Gyulassy, I. Vitev, X. N. Wang and B. W. Zhang, arXiv:nucl-th/0302077.
[2] P. Jacobs and X. N. Wang, Prog. Part. Nucl. Phys. **54**, 443 (2005).
[3] R. Baier, Y. L. Dokshitzer, A. H. Mueller, S. Peigne and D. Schiff, Nucl. Phys. B **484**, 265 (1997); B. G. Zakharov, JETP Lett. **65**, 615 (1997); X. f. Guo and X. N. Wang, Phys. Rev. Lett. **85**, 3591 (2000).
[4] M. Gyulassy, P. Levai and I. Vitev, Phys. Rev. Lett. **85**, 5535 (2000).
[5] S. S. Adler *et al.* [PHENIX Collaboration], Phys. Rev. Lett. **96**, 202301 (2006).
[6] S. A. Bass, C. Gale, A. Majumder, C. Nonaka, G. Y. Qin, T. Renk and J. Ruppert, Phys. Rev. C **79**, 024901 (2009) [arXiv:0808.0908 [nucl-th]].
[7] I. Vitev, S. Wicks and B. W. Zhang, JHEP **0811**, 093 (2008) [arXiv:0810.2807 [hep-ph]].
[8] I. Vitev and B. W. Zhang, Phys. Rev. Lett. **104**, 132001 (2010) [arXiv:0910.1090 [hep-ph]].
[9] D. K. Srivastava, C. Gale and T. C. Awes, Phys. Rev. C **67**, 054904 (2003) [arXiv:nucl-th/0212081].
[10] M. Gyulassy, P. Levai and I. Vitev, Phys. Rev. Lett. **85**, 5535 (2000) [arXiv:nucl-th/0005032].
[11] C. Amsler et al. (Particle Data Group), Physics Letters B667, 1 (2008) and 2009 partial update for the 2010 edition.
[12] J. M. Campbell and R. K. Ellis, Phys. Rev. D **65**, 113007 (2002) [arXiv:hep-ph/0202176].
[13] V. M. Abazov *et al.* [D0 Collaboration], Phys. Lett. B **669**, 278 (2008) [arXiv:0808.1296 [hep-ex]].
[14] R. D. Field, *Redwood City, USA: Addison-Wesley (1989) 366 p. (Frontiers in physics, 77)*
[15] I. Vitev, Phys. Lett. B **639**, 38 (2006) [arXiv:hep-ph/0603010].
[16] R. B. Neufeld, B. Muller and J. Ruppert, Phys. Rev. C **78**, 041901 (2008) [arXiv:0802.2254 [hep-ph]].
[17] R. B. Neufeld and B. Muller, Phys. Rev. Lett. **103**, 042301 (2009) [arXiv:0902.2950 [nucl-th]].

26th Winter Workshop on Nuclear Dynamics IOP Publishing
Journal of Physics: Conference Series **230** (2010) 012036 doi:10.1088/1742-6596/230/1/012036

Recent STAR results in high-energy polarized proton-proton collisions at RHIC

Bernd Surrow for the STAR Collaboration
Massachusetts Institute of Technology, 77 Massachusetts Avenue, Cambridge, MA 02139, USA
E-mail: surrow@mit.edu

Abstract. The STAR experiment at the Relativistic Heavy-Ion Collider at Brookhaven National Laboratory is carrying out a spin physics program in high-energy polarized $\vec{p}+\vec{p}$ collisions at $\sqrt{s} = 200 - 500\,\mathrm{GeV}$ to gain a deeper insight into the spin structure and dynamics of the proton.

One of the main objectives of the spin physics program at RHIC is the extraction of the polarized gluon distribution function based on measurements of gluon initiated processes, such as hadron and jet production. The STAR detector is well suited for the reconstruction of various final states involving jets, π^0, $\pi^{\pm}$, $e^{\pm}$ and γ, which allows to measure several different processes. Recent results will be shown on the measurement of jet production and hadron production at $\sqrt{s} = 200\,\mathrm{GeV}$.

The RHIC spin physics program has recently completed the first data taking period in 2009 of polarized $\vec{p}+\vec{p}$ collisions at $\sqrt{s} = 500\,\mathrm{GeV}$. This opens a new era in the study of the spin-flavor structure of the proton based on the production of $W^{-(+)}$ bosons. Recent STAR results on the first measurement of W boson production in polarized $\vec{p}+\vec{p}$ collisions will be shown.

1. Introduction

The spin structure and dynamics of the nucleon is one of the fundamental and unresolved questions in Quantum Chromodynamics (QCD). Contrary to atomic physics where the total angular momentum of an atom can be accounted for by basic quantum theory in terms of its underlying atomic constituents, a complete theoretical description of the proton spin decomposed in terms of contributions from quark and gluon angular momenta and spins is still missing. Various experimental programs have been conducted in the past to deepen our understanding on the proton spin. Deep-inelastic scattering (DIS) experiments have clearly established that the quark spin contribution is small and accounts for only $\approx 25\%$ of the proton spin [1, 2].

High energy polarized $\vec{p}+\vec{p}$ collisions at $\sqrt{s} = 200 - 500\,\mathrm{GeV}$ at RHIC provide a unique way to probe the proton spin structure and dynamics using hard scattering processes [3]. The production of jets and hadrons is the prime focus of the gluon polarization studies. The production of $W^{-(+)}$ bosons at $\sqrt{s} = 500\,\mathrm{GeV}$ provides an ideal tool to study the spin-flavor structure of the proton. This has been pointed out at the very early design stages of the polarized proton collider program at RHIC [4].

The first global analysis of polarized DIS data, as well as results obtained by the PHENIX [5] and STAR [6] experiments in polarized $\vec{p}+\vec{p}$ collisions at RHIC placed a strong constraint on the gluon spin contribution in the gluon momentum range of $0.05 < x < 0.2$, and suggested that the gluon spin contribution is not large in that range [7]. Constraining the polarized gluon

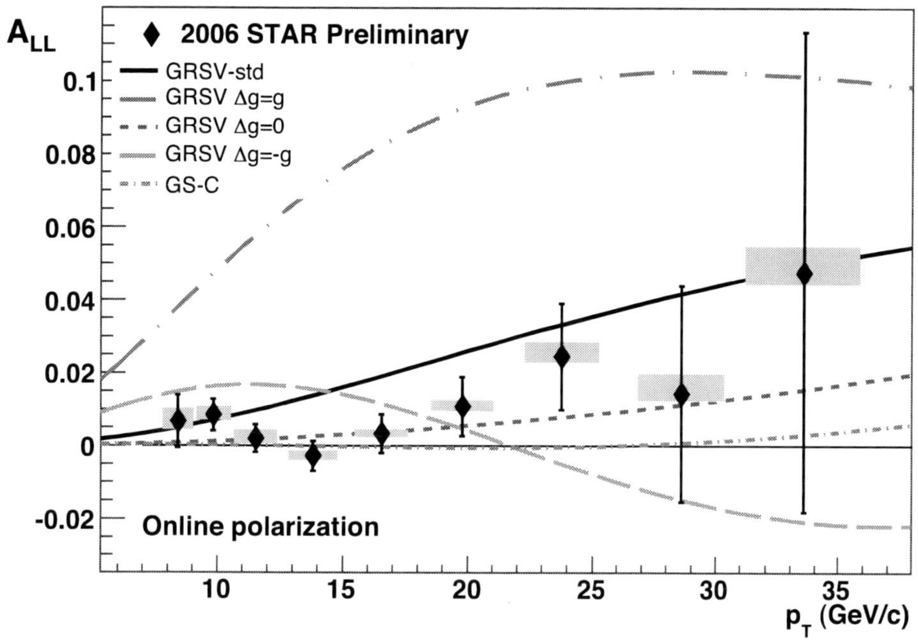

Figure 1. *Recent STAR 2006 A_{LL} inclusive jet result.*

distribution function, Δg, through inclusive measurements has been, so far, the prime focus of the STAR physics analysis program of the Run 3/4 [8], Run 5 [6] and Run 6 [9] data samples. Inclusive measurements, such as inclusive jet production, integrate over a fairly large x region for a given jet transverse momentum region. While those measurements provide a strong constraint on the value of Δg integrated over a range in x, those measurements do not permit a direct sensitivity to the actual x dependence. This motivates the need for correlation measurements in polarized $\vec{p}+\vec{p}$ collisions.

The STAR collaboration has presented a first measurement of the longitudinal spin transfer D_{LL} in inclusive Λ ($\Lambda \to p\pi^-$) and $\bar{\Lambda}$ ($\bar{\Lambda} \to \bar{p}\pi^+$) production in polarized proton-proton collisions at a center-of-mass energy of $\sqrt{s} = 200\,\mathrm{GeV}$ [10]. This measurement may provide constraints on strange (anti) quark polarization [11] and can yield new insight into polarized fragmentation functions.

The first data taking period in 2009 of polarized $\vec{p}+\vec{p}$ collisions at $\sqrt{s} = 500\,\mathrm{GeV}$ opens a new era in the study of the spin-flavor structure of the proton based on the production of $W^{-(+)}$ bosons. $W^{-(+)}$ bosons are produced predominantly through $\bar{u}+d$ ($u+\bar{d}$) collisions and can be detected through their leptonic decay. Quark and anti-quark helicity distribution functions are probed at large scales set by the mass of the W boson ($Q \sim m_W$) where theoretical calculations are well under control. The theoretical framework is well developed to describe the production of W bosons in high-energy polarized $\vec{p}+\vec{p}$ collisions including the description of the W decay leptons subject to experimental cuts, such as the transverse momentum p_T and pseudo-rapidity η of the final-state leptons [12, 13]. This development profits from a rich history of hadroproduction of weak bosons at the CERN SPS and the FNAL Tevatron and provides a firm basis to use W production as a new high-energy probe in polarized $\vec{p}+\vec{p}$ collisions [14].

The STAR Collaboration has provided critical measurements on the sudy of transverse spin effects in polarized p+ p collisions at RHIC. An overview of these results is discussed in a separate paper [15].

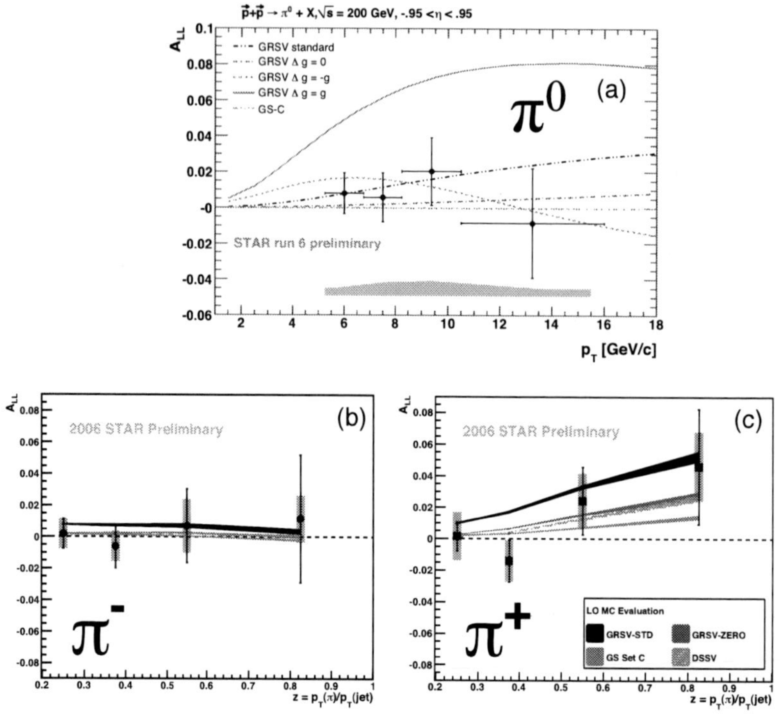

Figure 2. *Recent STAR 2006 A_{LL} neutral pion result (a) and STAR 2006 A_{LL} results for charged pion (π^- and π^+) / jet correlation measurements (b/c).*

2. Recent jet A_{LL} results

STAR reconstructs jets with the midpoint cone algorithm using clusters of charged track momenta measured with the STAR Time Projection Chamber (TPC) and tower energy deposits in the STAR Barrel Electromagnetic Calorimeter (BEMC) within a cone radius of $R \equiv \sqrt{\Delta\eta^2 + \Delta\phi^2}$ [9]. The jet sample was required to be within a fiducial range of $-0.7 < \eta_{\mathrm{Detector}} < 0.9$ for the 2006 data sample with a cone radius of $R = 0.7$. The dominant fraction of jet events in the 2006 data sample are based on a jet patch (JP) trigger that required a minimum energy deposition for a group of towers over a region of $\Delta\eta \times \Delta\phi = 1.0 \times 1.0$. This trigger was taken in coincidence with a minimum-bias condition using the STAR Beam-Beam Counter (BBC).

Throughout the following discussion, four gluon polarization scenarios have been used as input to NLO perturbative QCD calculations of A_{LL}. The GRSV standard case refers to a global analysis fit of polarized DIS data [16]. The case for a vanishing gluon polarization (GRSV-ZERO) and the case of a maximally positive (GRSV-MAX) or negative (GRSV-MIN) gluon polarization have also been considered. The GS-C [17] set of Δg is reflected by a large positive gluon polarization at low x, a node around $x \sim 0.1$ and a negative gluon polarization at large x at a scale of $Q^2 \sim 1\,\mathrm{GeV}^2$.

Figure 1 shows the most recent STAR preliminary result of A_{LL} for inclusive jet production as a function of p_T based on the 2006 data sample of $4.7\,\mathrm{pb}^{-1}$. Typical beam polarization values during the 2006 data-taking period were $55 - 60\%$. The leading systematic uncertainty comes from the trigger and jet reconstruction biases. Differences between observed and true jet p_T are estimated using PYTHIA and GEANT simulations and result in corrections being applied to the jet p_T values. Additionally, the trigger conditions at STAR can artificially bias the analyzed data

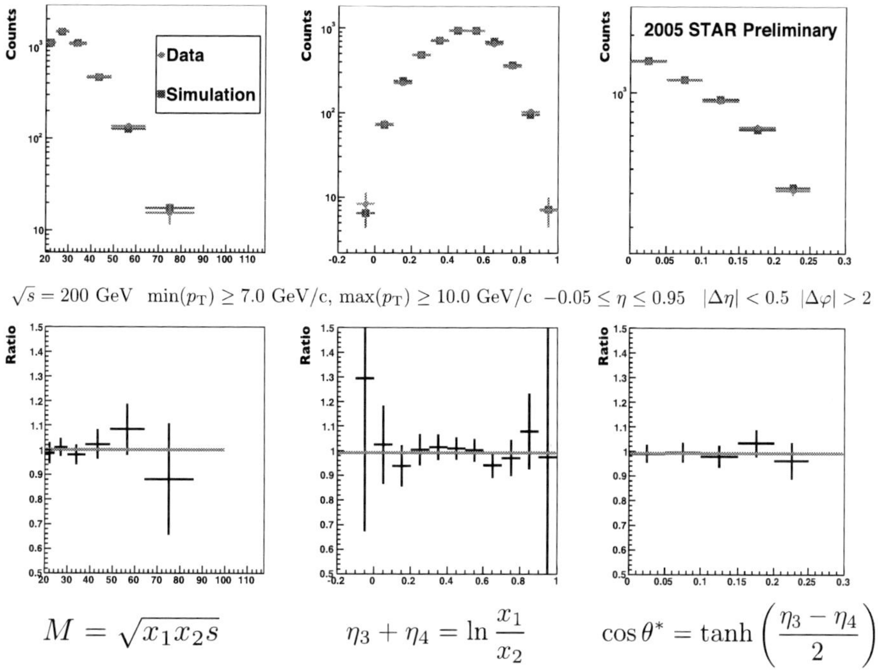

Figure 3. *Comparison of STAR 2005 di-jet data and a PYTHIA Monte-Carlo showing the yield at the top and the ratio at the bottom for three characteristic di-jet variables (See text for further details).*

sample towards particular flavors of partonic collisions. A conservative systematic uncertainty is evaluated to account for this effect which incorporates all allowable gluon polarization scenarios. The A_{LL} curves in Figure 1 are derived from NLO fits to world polarized DIS data by two separate theory groups, GRSV [16] and GS [17]. The currently measured result is consistent with a small gluon polarization scenario for the STAR x-range of $0.03 < x < 0.3$. This x-range accounts for $\sim 50\%$ of the total ΔG integral, i.e. $\Delta G = \int_0^1 \Delta g dx$, for GRSV-STD.

3. Recent neutral and charged pion A_{LL} results

Neutral pions are identified via the two-photon decay channel, which accounts for more than 98% of π^0 decays. Two primary sub-detectors, the STAR Barrel Electromagnetic Calorimeter (BEMC) and respective Shower Maximum Detector (SMD) are employed to reconstruct decay photons from mid rapidity neutral pions. Charged pion production is of particular interest since the difference of the longitudinal double-spin asymmetries for π^+ and π^- production, $A_{LL}(\pi^+)$ - $A_{LL}(\pi^-)$, tracks the sign of Δg, due to the opposite signs of the polarized distribution functions for up and down quarks. The STAR Time Projection Chamber (TPC) offers robust reconstruction and identification of charged pions up to a transverse momentum of 15 GeV/c. Particle identification is accomplished using measurements of ionization energy loss in the TPC.

Figure 2 (a) shows A_{LL} for neutral pion production at mid rapidity as a function of p_T based on the 2006 data sample [18]. This new measurement is shown in comparison to NLO calculations assuming different gluon polarization scenarios [9, 18]. This measurement reaches to higher p_T values compared to previous measurements by PHENIX [19] and STAR [20]. At the present level of precision, the data excludes extreme gluon polarization scenarios. The STAR collaboration has recently presented also results for neutral pion production at forward rapidity

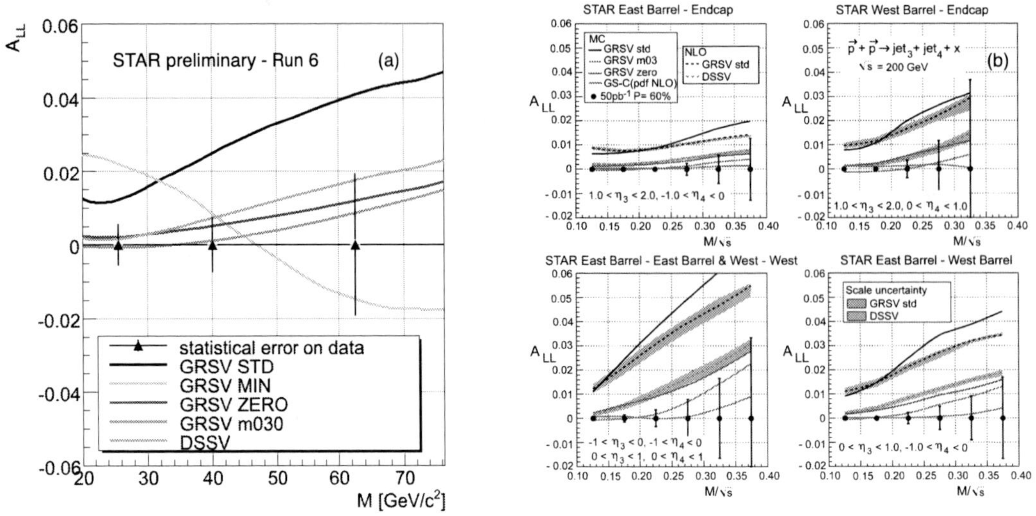

Figure 4. *(a) Statistical precision of the longitudinal double-spin asymmetry, A_{LL}, as a function of the di-jet invariant mass, M, for the 2006 RHIC data sample. (b) Longitudinal double-spin asymmetry, A_{LL}, for di-jet production as function of the ratio $M/\sqrt{s}$ for different topological combinations of the STAR BEMC and the STAR EEMC acceptance region.*

based on the STAR Electromagnetic Endcap Calorimeter (EEMC) and the STAR Forward Pion Detector (FPD) [21]. These measurements provide an important milestone for future photon-jet coincidence measurements at STAR.

Figure 2 (b) and (c) show STAR's new preliminary result of A_{LL} for charged pion (π^- and π^+) / jet correlation measurements at mid rapidity. Charged pions are reconstructed opposite a jet that triggered the experiment. The data are shown versus $z \equiv p_T(\pi)/p_T(jet)$ and were obtained in 2006 [22]. Full NLO calculations for this observable have recently been released [23]. These calculations will provide the basis for charged pion results to be included in a global analysis. The data shown for charged pion production are compared to a LO MC evaluation of A_{LL} excluding extreme scenarios of Δg. The measurement of $A_{LL}(\pi^+)$ is of particular interest, since its analyzing power is large because of the large ug scattering contribution to the production cross section and sizable $\Delta u/u$. In the future, more precise measurements of $A_{LL}(\pi^+)$ have a great potential for providing a better understanding of Δg.

4. Status and prospects of di-jet production

Correlation measurements such as those for di-jet production allow for a better constraint of the partonic kinematics and thus the shape of Δg. At LO, the di-jet invariant mass, M, is proportional to the product of the x values of the partons, $M = \sqrt{s}\sqrt{x_1 x_2}$, whereas the pseudo-rapidity sum of the final-state jets, $\eta_3 + \eta_4$, is proportional to the logarithm of the ratio of the x values, $\eta_3 + \eta_4 = \ln(x_1/x_2)$. Photon-jet coincidence measurements are expected to provide a theoretically clean way to extract Δg. A LO extraction of Δg alone would allow a model-independent way to constrain the x dependence, which would be an important contribution, without making an a priori assumption on the functional form of Δg as is currently required in a global analysis. Measurements at both $\sqrt{s} = 200\,\text{GeV}$ and $\sqrt{s} = 500\,\text{GeV}$ are preferred to maximize the kinematic reach in x and possibly provide a means to observe effects of scaling violations at fixed x by measuring different p_T values. The wide acceptance of the STAR experiment permits reconstruction of di-jet events with different topological configurations,

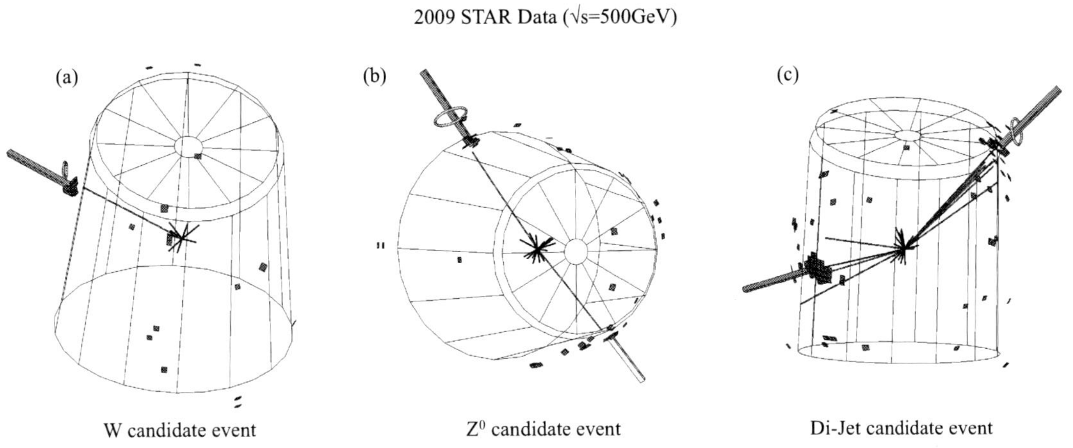

Figure 5. *Event displays from the 2009 data taking period at $\sqrt{s}$ = 500 GeV showing a W candidate event (a), a Z^0 candidate event and a di-jet candidate event (c).*

i.e. different η_3/η_4 combinations, ranging from symmetric ($x_1 = x_2$) partonic collisions to asymmetric ($x_1 < x_2$ or $x_1 > x_2$) partonic collisions. This, together with the variation of the center-of-mass energy, is expected to constrain Δg over a wide range in x of approximately $\sim 2 \cdot 10^{-3} < x < 0.3$ for di-jet and photon-jet events. The NLO framework for correlation measurements exists and therefore those measurements can be used in a global analysis [24].

Figure 3 shows a comparison of STAR 2005 di-jet data and a PYTHIA Monte-Carlo simulation with the yield at the top and the ratio at the bottom for the invariant mass M, the pseudo-rapidity sum $\eta_3 + \eta_4$ and the cosine of the partonic center-of-mass scattering angle $\cos\theta^*$ determined from the pseudo-rapidities of both final-state jets. The normalization between data and Monte-Carlo is fixed by the invariant mass distributions. The same normalization factor is then applied to both other di-jet distributions. Data and Monte-Carlo are found to be in good agreement.

Figure 4 (a) shows the statistical precision of the longitudinal double-spin asymmetry, A_{LL}, as a function of the di-jet invariant mass, M. Those uncertainties, extracted from the 2006 data sample, are compared to a LO MC evaluation of A_{LL} computed with a PYTHIA MC sample using different event weights to account for different polarized gluon distribution functions of GRSV [16] and DSSV [7]. The size of the statistical uncertainty at the highest invariant mass bin is at the level of the difference between GRSV-STD and DSSV.

Figure 4 (b) shows the expected precision for the longitudinal double-spin asymmetry, A_{LL}, for di-jet production as a function of $M/\sqrt{s}$ for different topological combinations of the STAR BEMC and the STAR Endcap Electromagnetic Calorimeter (EEMC) acceptance regions. Taking into account the different η ranges covered, and equivalently, the different $\cos\theta^*$ regions being probed, each panel represents a different range in x_1/x_2. At LO, $\cos\theta^*$ amounts to $\tanh\left(\frac{\eta_3-\eta_4}{2}\right)$. The upper left panel effectively probes asymmetric partonic collisions where predominantly a low-x gluon collides with a high-x quark at large invariant masses. The effective variation of x_1 and x_2 amounts to $0.2 < x_1 < 0.6$ and $0.07 < x_2 < 0.2$. In contrast, a kinematic region of larger x values is probed at predominantly symmetric partonic collisions such as the one shown in the lower right panel. The effective variation of x_1 and x_2 is roughly equal and given by the horizontal axis of the lower right panel. The projected uncertainties are shown for a luminosity of 50 pb^{-1} and a beam polarization of 60%. Those projected uncertainties are compared to a LO

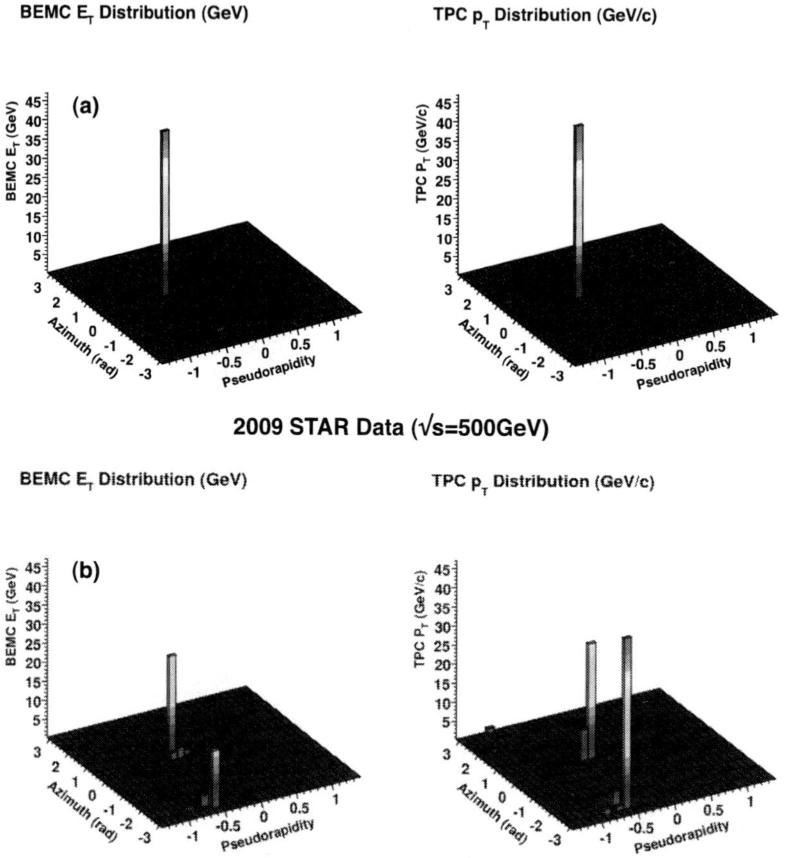

Figure 6. *Lego plot showing the BEMC E_T and the TPC p_T distribution for a W candidate event (a) and a di-jet candidate event (b).*

evaluation of A_{LL} and a full NLO A_{LL} calculation. Scale uncertainties are shown as a shaded band for DSSV and GRSV-STD reflecting a variation of the invariant mass M as a hard scale of $2M$ and $0.5M$. Asymmetric cuts are imposed for the LO MC and the NLO determination of $\min(p_T) \geq 7\,\text{GeV/c}$ and $\max(p_T) \geq 10\,\text{GeV/c}$. The result of a LO MC evaluation using GS-C [17] for Δg is also shown. GS-C is still consistent with the current inclusive jet results [9]. A cone radius of $R = 0.7$ has been used. Good agreement is found between a LO MC evaluation of A_{LL} and a full NLO calculation. Scale uncertainties are found to be small in comparison to the variation of the chosen polarized gluon distribution functions, in particular, at large values of M.

While inclusive measurements from STAR have provided important constraints on Δg, and will continue to do so, the impact of correlation measurements to the x dependence should greatly enhance our understanding of Δg.

5. Measurement of W boson production at STAR

The data used for the first W boson production analysis at STAR were collected in 2009 colliding polarized proton beams at 250 GeV. The STAR detector systems used in this analysis are the STAR Time Projection Chamber (TPC), the STAR Barrel Electromagnetic Calorimeter

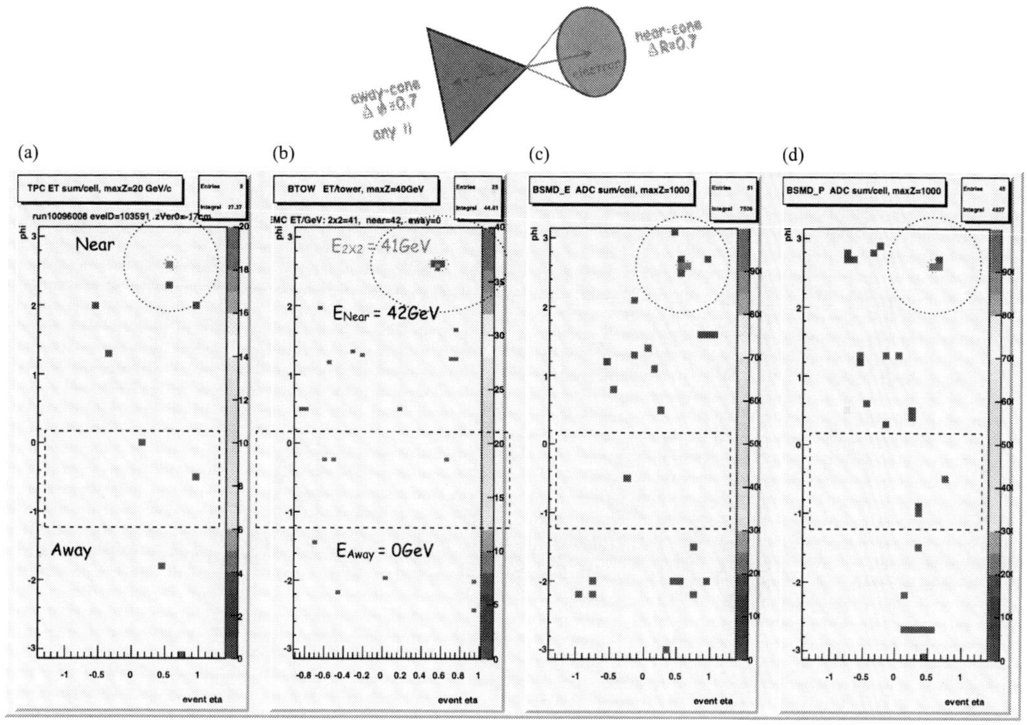

Figure 7. *Illustration of the W offline selection criteria (top) and the distribution of the TPC p_T (a) the BEMC E_T (b) and both BEMC SMD layers (c) / (d) in azimuth and pseudo-rapidity.*

(BEMC) and STAR Electromagnetic Endcap Calorimeter (EEMC). Only their tower response has been taken into account in this analysis. The BEMC was used to measure the transverse energy, E_T, of $e^{\pm}$. The suppression of the QCD background by several orders of magnitude was based on the TPC, BEMC and EEMC. Figure 5 shows event displays from the 2009 data taking period for a W candidate event (a), a Z^0 candidate event and a di-jet candidate event (c). A lego plot is shown in Figure 6 displaying the BEMC E_T and the TPC p_T distribution for a W candidate event at the top and a di-jet candidate event at the bottom.

Proton-proton collision events focusing on W production at $\sqrt{s} = 500$ were identified by a two-step energy requirement in the BEMC. Electrons and positrons from W production at mid rapidity are characterized by large E_T peaked at $\sim M_W/2$ (Jacobian peak). At the hardware trigger level (L0), a high tower (HT) calorimetric trigger condition required $E_T > 7.3\,\mathrm{GeV}$ in a single BEMC tower. At the software trigger level (L2), a dedicated trigger algorithm was developed that required that a 2×2 tower cluster E_T sum exceeds $13\,\mathrm{GeV}$. The offline selection of W candidate events is based on kinematical and topological difference between leptonic $W^{\pm}$ decays, and QCD background events as shown in Figure 5 and Figure 6. Events from $W^{\pm} \to e^{\pm} + \nu$ decays contain a nearly isolated $e^{\pm}$ with a neutrino in the opposite direction in azimuth. The neutrino escapes detection, leading to a large missing energy.

Figure 7 illustrates the basic idea of the selection of W candidate events. The 2D histograms show for one W candidate event the TPC p_T (a) and the BEMC tower E_T (b) distribution along with both BEMC shower-maximum detector (SMD) layers (c) / (d) as a function of the azimuthal angle ϕ and the pseudo-rapidity η. The approximate area of a 2×2 tower cluster is indicated by a small circle around the large energy deposition in a single BEMC tower. Also shown is the size of a cone (near-cone) around the electron candidate. The away-side area

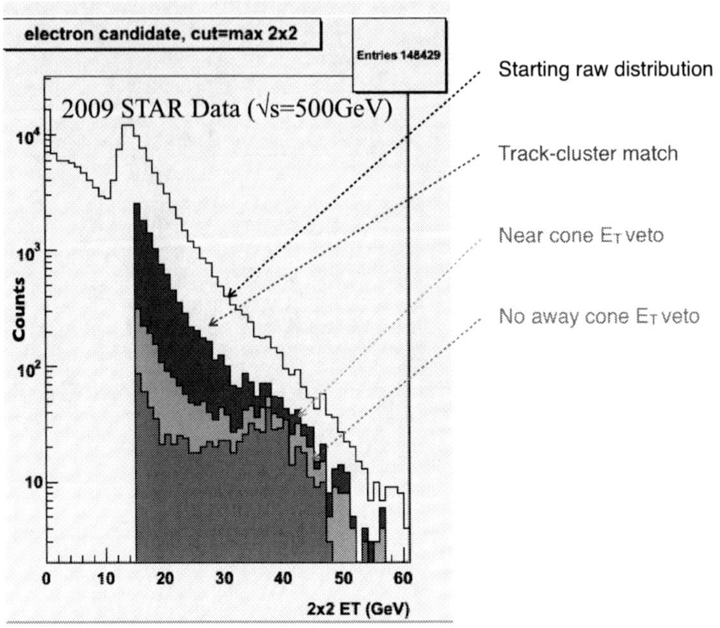

Figure 8. *Evolution of W offline selection cuts (See text for further details).*

is sketched by the black dashed rectangular region. The TPC high-p_T entry appears in the same $\eta - \phi$ region as the large BEMC E_T entry. Both BEMC SMD layers show a large energy deposition at the same location in the $\eta - \phi$ plane as the TPC high-p_T and the BEMC high-E_T entry. An electron candidate is defined to be any TPC track with $p_T > 10\,\mathrm{GeV}/c$ that is associated with a primary vertex with $|z| < 100\,\mathrm{cm}$, where z is the distance along the beam direction. A 2×2 BEMC tower cluster E_T sum is required to be larger than 15 GeV. The excess BEMC E_T sum in a 4×4 tower cluster centered around the respective 2×2 tower cluster is required to be below 5%. In addition, the distance between the 2×2 cluster tower centroid and the TPC track is required to be less than 7 cm. A near-cone is formed around the electron candidate direction with a radius in η-ϕ space of $R = 0.7$. The excess BEMC and TPC E_T sum is required to be less than 12% of the 2×2 cluster E_T. The away side requirement is based on a cut on the BEMC and TPC E_T sum to be less than 8 GeV. This sum is extended over the full η range of the BEMC and EEMC and $\Delta\phi = 0.7$ as shown in Figure 7. Figure 8 shows the evolution of all cuts for the BEMC 2×2 E_T distribution. A clear Jacobian peak emerges characteristic for W production in contrast to QCD background dominating the low E_T region.

A data-driven procedure was applied to estimate the level of QCD background events. The first part consists of an estimate of the impact of the missing calorimeter coverage for $-2 < \eta < -1$ on the cluster E_T distribution. This estimate was obtained by performing a parallel analysis without the EEMC as an active detector. The difference in yield of accepted events as a function of E_T provides an absolute estimate for the missing calorimetric coverage. The remaining background is estimated by fitting a separate QCD shape distribution in E_T to the region of $E_T < 19\,\mathrm{GeV}$ shown as the black histogram in Figure 9. The fitted QCD background distribution is shown by the blue histogram in Figure 9. The background subtracted distribution is shown by the yellow histogram. Figure 9 also shows the distribution from a PYTHIA Monte-Carlo simulation. Good agreement in yield and shape is found between the data and Monte-Carlo distribution for the charge-sum BEMC E_T distribution. In the meantime, a considerable

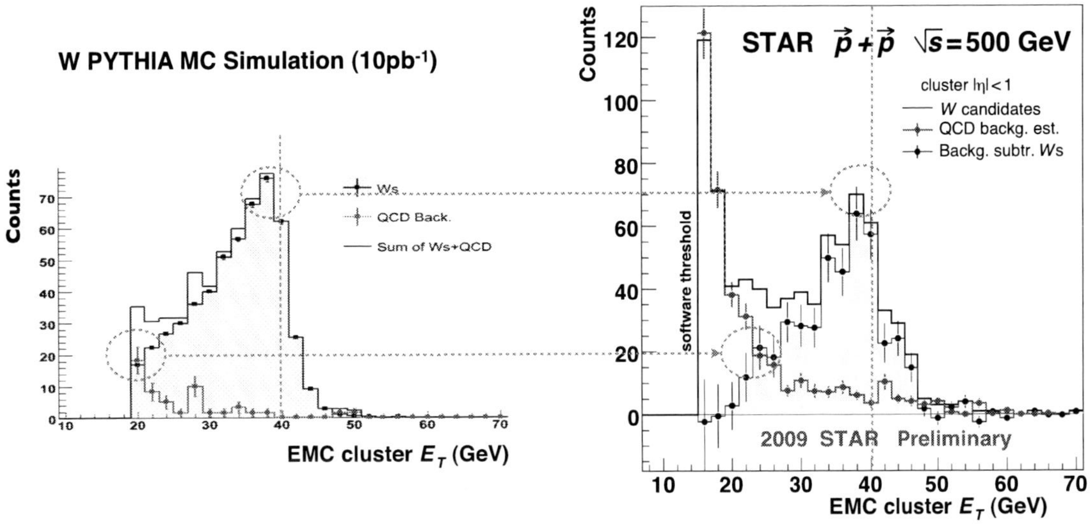

Figure 9. *Reconstructed BEMC lepton transverse energy for W events.*

effort has been performed on the TPC calibration to allow a full charge-sign separation. This and in particular the first measurement of the respective charge-separated cross-sections and parity-violating singe-spin asymmetries A_L at mid rapidity will be discussed in a forthcoming publication.

Future planned high-statistics measurements at forward and backward pseudo-rapidities will focus on constraining the polarization of $\bar{d}$ and $\bar{u}$ quarks, respectively to deepen our understanding of the spin structure of the QCD sea.

References

[1] Ashman J *et al.* (EMC) 1989 *Nucl. Phys.* **B328** 1
[2] Filippone B W and Ji X D 2001 *Adv. Nucl. Phys.* **26** 1
[3] Bunce G, Saito N, Soffer J and Vogelsang W 2000 *Ann. Rev. Nucl. Part. Sci.* **50** 525
[4] Underwood D *et al.* 1992 *Part. World.* **3** 1–12
[5] Adare A *et al.* (PHENIX) 2007 *Phys. Rev.* **D76** 051106
[6] Abelev B I *et al.* (STAR) 2008 *Phys. Rev. Lett.* **100** 232003
[7] de Florian D, Sassot R, Stratmann M and Vogelsang W 2008 *Phys. Rev. Lett.* **101** 072001
[8] Abelev B I *et al.* (STAR) 2006 *Phys. Rev. Lett.* **97** 252001
[9] Sarsour M *et al.* (STAR) 2008 *Proceedings of the 18th International Spin Physics Symposium*
[10] Abelev B I *et al.* (STAR) 2009 *Phys. Rev.* **D80** 111102
[11] Xu Q h, Liang Z t and Sichtermann E 2006 *Phys. Rev.* **D73** 077503
[12] Nadolsky P M and Yuan C P 2003 *Nucl. Phys.* **B666** 31–55
[13] de Florian D and Vogelsang W 2010 (*Preprint* 1003.4533)
[14] Kotwal A V and Stark J 2008 *Ann. Rev. Nucl. Part. Sci.* **58** 147–175
[15] Eun L *et al.* (STAR) 2010 *Proceedings of the 26th Winter Workshop on Nuclear Dynamics*
[16] Gluck M, Reya E, Stratmann M and Vogelsang W 2001 *Phys. Rev.* **D63** 094005
[17] Gehrmann T and Stirling W J 1996 *Phys. Rev.* **D53** 6100–6109
[18] Hoffman A *et al.* (STAR) 2008 *Proceedings of the 18th International Spin Physics Symposium*
[19] Adare A *et al.* (PHENIX) 2009 *Phys. Rev. Lett.* **103** 012003
[20] Abelev B I *et al.* (STAR) 2009 *Phys. Rev.* **D80** 111108
[21] Wissink S *et al.* (STAR) 2008 *Proceedings of the 18th International Spin Physics Symposium*
[22] Kocoloski A *et al.* (STAR) 2008 *Proceedings of the 18th International Spin Physics Symposium*
[23] de Florian D 2009 *Phys. Rev.* **D79** 114014
[24] de Florian D, Frixione S, Signer A and Vogelsang W 1999 *Nucl. Phys.* **B539** 455–476

Journal of Physics: Conference Series **230** (2010) 012037 doi:10.1088/1742-6596/230/1/012037

A fundamental test of the Higgs Yukawa coupling at RHIC in A+A collisions

M. J. Tannenbaum

Physics Department, 510c, Brookhaven National Laboratory, Upton, NY 11973-5000, USA

E-mail: mjt@bnl.gov

Abstract. Searches for the intermediate boson, $W^{\pm}$, the heavy quantum of the Weak Interaction, via its semi-leptonic decay, $W \rightarrow e+\nu$, in the 1970's instead discovered unexpectedly large hadron production at high p_T, notably π^0, which provided a huge background of $e^{\pm}$ from internal and external conversions. Methods developed at the CERN ISR which led to the discovery of direct-single-$e^{\pm}$ in 1974, later determined to be from the semi-leptonic decay of charm which had not yet been discovered, were used by PHENIX at RHIC to make precision measurements of heavy quark production in p-p and Au+Au collisions, leading to the puzzle of apparent equal suppression of light and heavy quarks in the QGP. If the Higgs mechanism gives mass to gauge bosons but not to fermions, then a proposal that all 6 quarks are nearly massless in a QGP, which would resolve the puzzle, can not be excluded. This proposal can be tested with future measurements of heavy quark correlations in A+A collisions

1. Introduction

In September 1991 the PHENIX experiment was born from the rejection of the Dimuon, Oasis and TALES/Sparhc Letters of Intent (LoI) which were combined into a new collaboration to build "a detector designed to study electrons and photons emerging from the QGP." Any collaborators "primarily interested in hadron physics will be welcomed by the STAR collaboration which has been empowered to build a large TPC detector" [1].

2. PHENIX is not your father's solenoid collider detector

The central spectrometer of PHENIX was designed to trigger on $J/\Psi \rightarrow e^+e^-$ at rest, i.e. $p_T^{J/\Psi} \approx 0$ at mid-rapidity, based on experience from the CERN ISR as proposed in the TALES [2] and later TALES/SPARHC [3] LoI. The CCRS experiment at the CERN ISR had discovered direct-single $e^{\pm}$ in 1974 [4] at a level of 10^{-4} of charged pion production at all 5 c.m. energies measured (Fig. 1). The same methods—i) $\geq 10^5$ charged hadron rejection; ii) minimum of material in the aperture to avoid external conversions; iii) zero magnetic eld on the axis to avoid de-correlating conversion pairs; iv) precision measurement of π^0 and η, the predominant background source; v) precision background determination in the direct-single-$e^{\pm}$ signal channel by adding external converter—were used in design of the highly non-conventional PHENIX mid-rapidity spectrometer, as I discussed in a previous winter workshop [6].

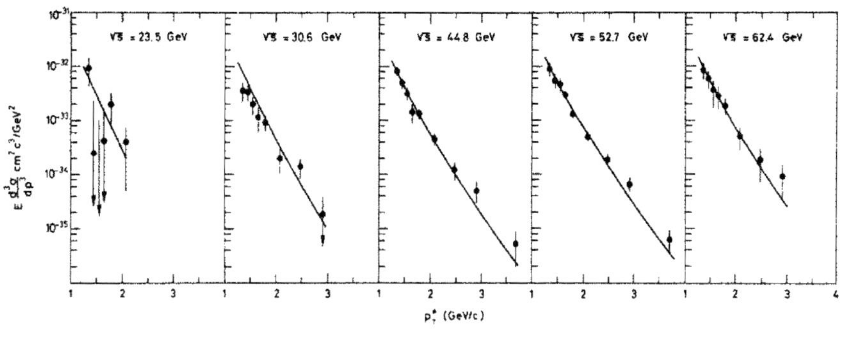

Figure 1. Invariant cross sections at mid-rapidity for 5 values of $\sqrt{s}$ in p-p collisions at the CERN-ISR: $(e^+ + e^-)/2$ (points); $10^{-4} \times (\pi^+ + \pi^-)/2$ (lines) [5]

Figure 2. PHENIX on cover of Physics Today, October 2003.

The key detector elements in achieving these goals were a fine grained Electromagnetic Calorimeter and a Ring Imaging Cerenkov counter for use in achieving the charged hadron rejection and providing an electron trigger at level 1; and an axial field Helmholtz coil spectrometer magnet, rather than the typical solenoid detector generally used at colliders, with two sets of coils which could be run with opposite currents to achieve zero field on the axis. It is no surprise that PHENIX made the cover of Physics Today instead of the more conventional STAR detector because PHENIX is a special purpose detector. In fact when Jack Steinberger came to visit BNL and I was showing and explaining PHENIX to him, he frowned more than usual and said, "Mike, is there a 'real collider detector' at RHIC?" So I took him to STAR for a minute and he was happy.

3. A charming surprise

We designed PHENIX specifically to be able to detect charm particles via direct-single $e^\pm$ since this went along naturally with $J/\Psi \to e^+ + e^-$ detection and since the single particle reaction avoided the huge combinatoric background in Au+Au collisions. We thought that the main purpose of open charm production, which corresponds to a hard-scale ($m_{c\bar{c}} \gtrsim 3$ GeV/c^2), would be a check of our centrality definition and $\langle T_{AA} \rangle$ calculation since the total production of c quarks should follow point-like scaling. In fact, our first measurement supported this beautifully [7]. However, our subsequent measurements proved to be much more interesting and even more beautiful. Figure 3a shows our direct-single-$e^\pm$ measurement in p-p collisions at $\sqrt{s}$ = 200 GeV [8] in agreement with a QCD calculation of c and b quarks as the source of the direct-single-$e^\pm$ (also called non-photonic $e^\pm$ at RHIC). The total yield of direct-single-$e^\pm$ for $p_T > 0.3$

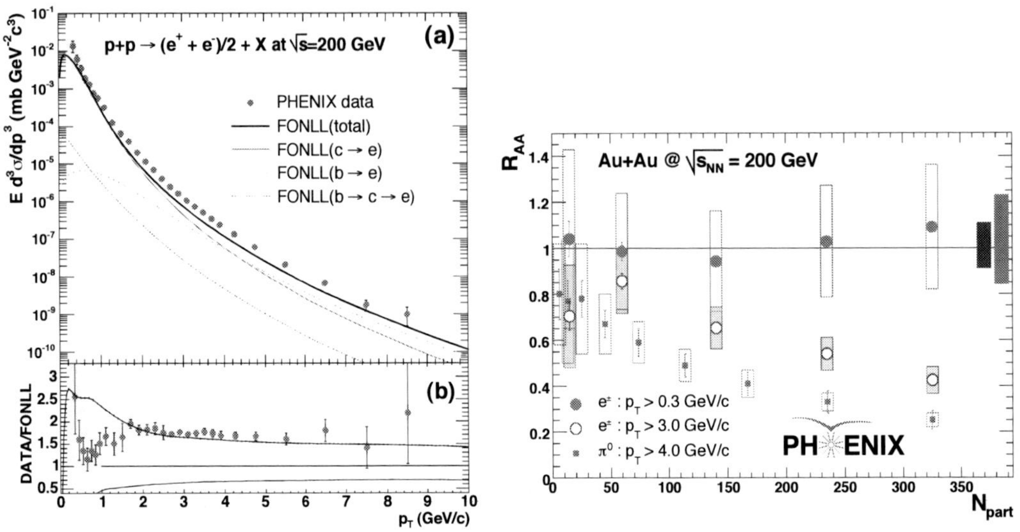

Figure 3. a),b) (left) Invariant cross section of direct $e^{\pm}$ in p-p collisions [8] compared to theoretical predictions from c and b quark semileptonic decay. c) (right) R_{AA} as a function of centrality ($N_{\rm part}$) for the total yield of $e^{\pm}$ from charm ($p_T > 0.3$) GeV/c, compared to the suppression of the $e^{\pm}$ yield at large $p_T > 3.0$ GeV/c which is comparable to that of π^0 with ($p_T > 4$ GeV/c) [9]

GeV/c was taken as the yield of c-quarks in p-p and Au+Au collisions [9]. The result, $R_{AA} = 1$ as a function of centrality (Fig. 3c), showed that the total $c - (\bar{c})$ production followed point-like scaling, as expected. The big surprise came at large p_T where we found that the yield of direct-single-$e^{\pm}$ for $p_T > 3$ GeV/c was suppressed nearly the same as the π^0 from light quark and gluon production. This strongly disfavors the QCD energy-loss explanation of jet-quenching because, naively, heavy quarks should radiate much less than light quarks and gluons in the medium; but opens up a whole range of new possibilities including string theory [10].

4. A History Lesson

"Those who cannot remember the past are condemned to repeat it", wrote George Santayna in 1905. Incredibly, the history of the discovery of direct-single-$e^{\pm}$ at the ISR by CCRS in 1974, which led to a spate of conflicting measurements by experiments not designed for the purpose, was repeated in 2005 at RHIC: STAR and PHENIX measurements of direct-$e^{\pm}$ in p-p collisions disagreed [11]. I will not discuss this further except to insist that PHENIX was specifically designed for this measurement. However, I do think that it is worthwhile to review the history at the CERN-ISR where direct-single-$e^{\pm}$ were discovered before the J/Ψ or charm particles were known and presented at the ICHEP in London in July 1974 [12].

The first interpretation of the CCRS discovery was by Farrar and Frautschi [13] who proposed that the direct-single-$e^{\pm}$ were due to the internal conversion of direct photons with a ratio γ/π^0 ~10-20%. CCRS was able to cleanly detect both external and internal conversions since there was zero magnetic field on the axis (Fig. 4) and set limits excluding this explanation [5].

The first correct explanation of the CCRS direct-single-$e^{\pm}$ (prompt leptons) was given by Hinchliffe and Llewellyn-Smith [14] as due to semi-leptonic decay of charm particles. Open charm was discussed at the 1975 Lepton-Photon conference at SLAC, the first major conference after the discovery of the J/Ψ and Ψ', and the paper submitted to Physical Review Letters

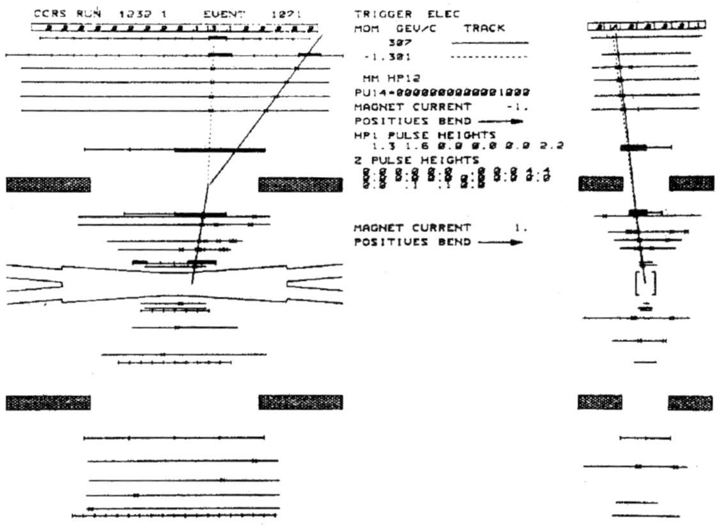

Figure 4. CCRS [5] identified e^+e^- pair which opens up in the magnet

in June 1975, but was not published until August 1976 [16]. The CCRS data submitted to the SLAC conference [15] and the prediction from charm decay [14] are shown in Fig. 5. A similar explanation was offered by two experimentalists, Maruice Bourquin and Jean-Marc

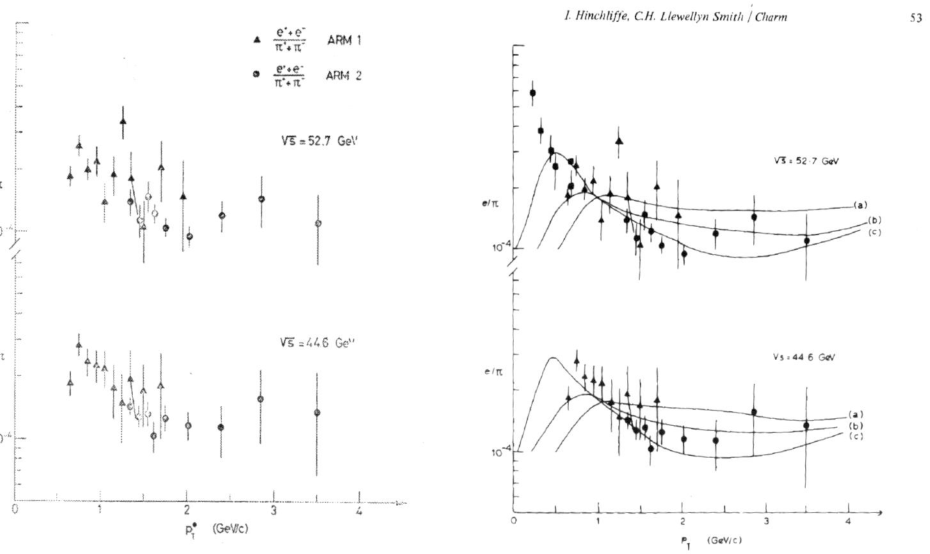

Figure 5. a)(left) CCRS e/π in two spectrometer arms at two values of $\sqrt{s}$ [15] b)(right) same data with prediction, see Ref. [14] for details

Gaillard [17], who compared the measured e/π ratios to a cocktail of all known leptonic decays including *"Possible contributions from the conjectured charm meson..."* (Fig. 6). The usage of "conjectured" charm is notable due to the delayed publication of reference [16]. Another notable point about both papers [14, 17] is that neither could fit the data points for $p_T < 1$ GeV/c from

Nuclear Physics B114 (1976) 334–364
© North-Holland Publishing Company

A SIMPLE PHENOMENOLOGICAL DESCRIPTION OF HADRON PRODUCTION

M. BOURQUIN
University of Geneva, Switzerland

J.-M. GAILLARD
Laboratoire de l'Accélérateur Linéaire, Orsay, France
CERN, Geneva, Switzerland

Received 13 May 1976

The inclusive production distributions of the non-leading particles in pp collisions are fitted by a simple general formula. The contributions from the ρ^0, ω, ϕ, and J/ψ decays to the direct lepton spectra are calculated. Possible contributions from the semi-leptonic decay of the conjectured charmed meson are discussed.

M. Bourquin, J.-M. Gaillard / Hadron production

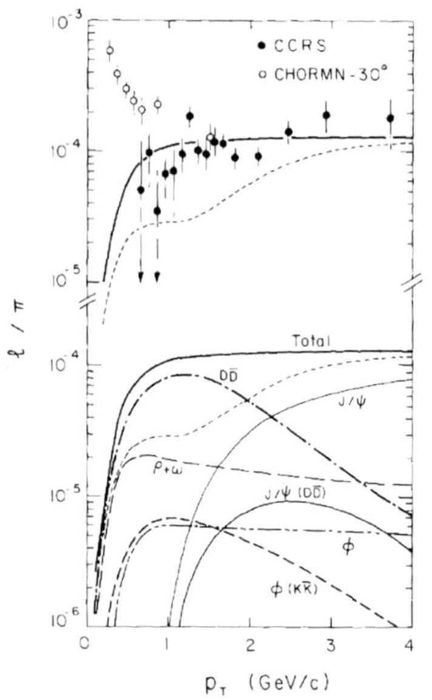

Figure 6. (left) Bourquin-Gaillard title page [17]; (right) Contribution of a cocktail of decays to the e/π ratio.

Volume 60B, number 5 PHYSICS LETTERS 16 February 1976

EXPERIMENTAL OBSERVATION OF A COPIOUS YIELD OF ELECTRONS WITH SMALL TRANSVERSE MOMENTA IN pp COLLISIONS AT HIGH ENERGIES

L. BAUM, M.M. BLOCK, B. COUCHMAN, J. CRAWFORD, A. DEREVSHCHIKOV, D. DIBITONTO, I. GOLUTVIN, H. HILSCHER, J. IRION, A. KERNAN, V. KUKHTIN, J. LAYTER, W. MARSH, P. McINTYRE, F. MULLER, B. NAROSKA, M. NUSSBAUM*, A. ORKIN-LECOURTOIS**, L. ROSSI, C. RUBBIA, D. SCHINZEL, B. SHEN, A. STAUDE, G. TARNOPOLSKY and R. VOSS
University of California at Riverside, Riverside, Cal., USA
CERN, Geneva, Switzerland
Harvard University, Cambridge, Mass., USA
Sektion Physik der Universität, Munich, Germany
Northwestern University, Evanston, Ill., USA

Received 15 January 1976

Inclusive electron and positron emission have been observed for $\theta_{cm} = 30°$ and $s = 2800$ GeV^2 at the CERN Intersecting Storage Rings (ISR). Over the transverse momentum interval 0.2 GeV/c $< p_T <$ 1.5 GeV/c, electrons and positrons, which are equal in number within the experimental accuracies, appear to grow with respect to other particles (pions) approximately like $1/p_T$. We are unable to explain their number and p_T-dependence in terms of "conventional" mechanisms.

Figure 7. One of the stars of 1976, measurement at 30° and 0.2 GeV/c $< p_T <$ 1.5 GeV/c

the CHORMN measurement [18] (Fig. 7). The important point here is that many experiments not designed for the purpose wanted to get into the prompt-lepton act and this is only one example. See Ref. [17] for more details. This led to the unfortunate situation as exemplified in Fig. 8 from a typical paper [19] about charm c. 1990 which didn't even give citations for the many

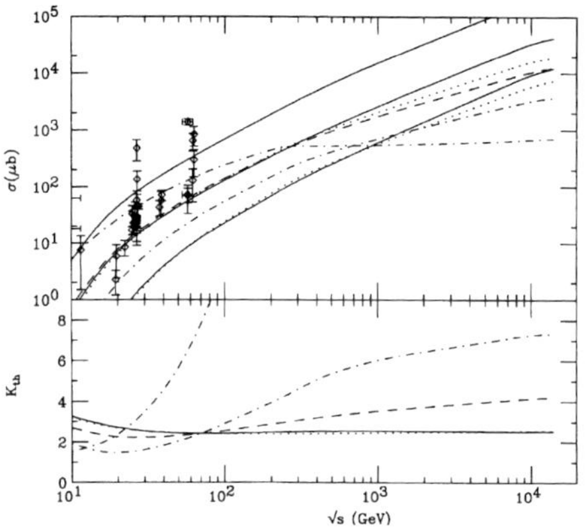

Figure 8. Predictions and measurement of open charm cross section [19]

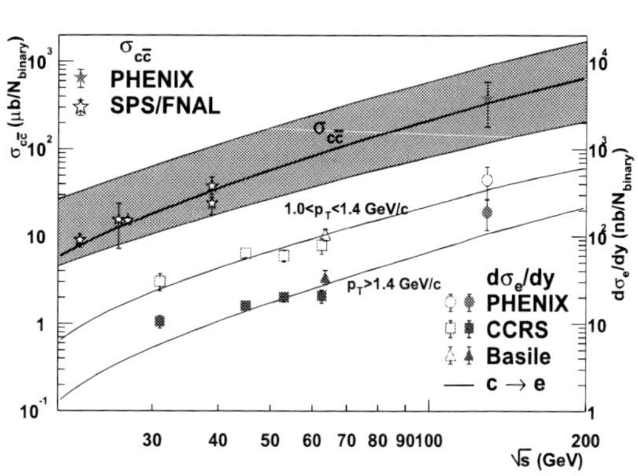

Figure 9. Predictions and measurements [21] of charm cross section $\sigma_{c\bar{c}}$

disagreeing ISR publications of the direct-single-$e^{\pm}$ from charm which were essentially ridiculed or, in the case of the original CCRS discovery [15], ignored [20] because it was published before either charm or the J/Ψ were discovered so there is no reference to the word "charm" in the publication. A fairer comparison of the ISR and fixed target measurements [22] of open charm is given in Fig. 9 from the first measurement of direct-single-$e^{\pm}$ at RHIC by PHENIX [21]. The

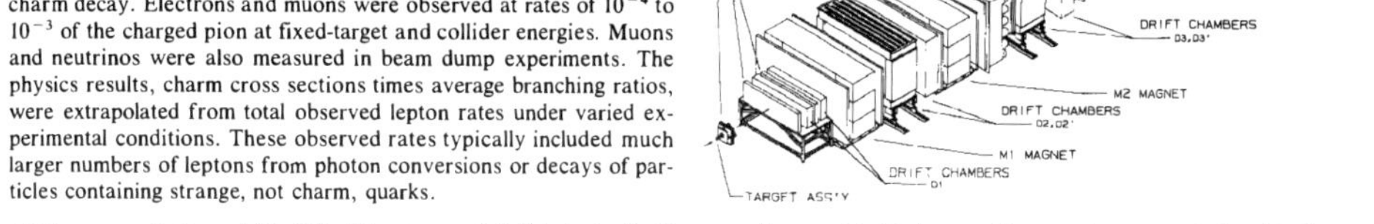
2. A BRIEF HISTORY OF MEASUREMENTS

The early years of open charm hadroproduction were limited by the capability of detectors in the face of difficult experimental conditions. Among the difficulties (still faced today) are (*a*) the small fractional charm production cross section (one $c\bar{c}$ pair event per 10^3 interactions, typically), (*b*) the high multiplicity of particles in the charm events, and (*c*) the small branching ratios to specific final states (typically 1–10%). As it turned out, many of the more reliable early measurements were indirect. Among these were the observations of prompt leptons resulting from the semileptonic decays of charm particles. Most of these early experiments had goals other than charm production and decay as their primary motivation. Nevertheless, leptons with intermediate transverse momentum have been interpreted to come from charm decay. Electrons and muons were observed at rates of 10^{-4} to 10^{-3} of the charged pion at fixed-target and collider energies. Muons and neutrinos were also measured in beam dump experiments. The physics results, charm cross sections times average branching ratios, were extrapolated from total observed lepton rates under varied experimental conditions. These observed rates typically included much larger numbers of leptons from photon conversions or decays of particles containing strange, not charm, quarks.

HADROPRODUCTION OF CHARM 375

Figure 10. a)(left) Text and b)(right) figure from Jeff Appel's review article [22].

fixed target measurements, which like to claim the discovery of hadroproduction of charm, all used Silicon Vertex Detectors, which is the first tiny element (SMD), actually an upgrade, in an otherwise giant experiment (Fig. 10b). At least, Jeff Appel in his review article [22] (Fig. 10a) did acknowledge the much earlier prompt lepton results.

4.1. Contemporary (c. 1977) controversies

To be fair, the direct-single-$e^{\pm}$ measurements as observations of semi-leptonic decays of open charm were not universally accepted by the experimental community in the late 1970's; and there were heavy hitters on both sides of the argument. First of all, after the discovery of the

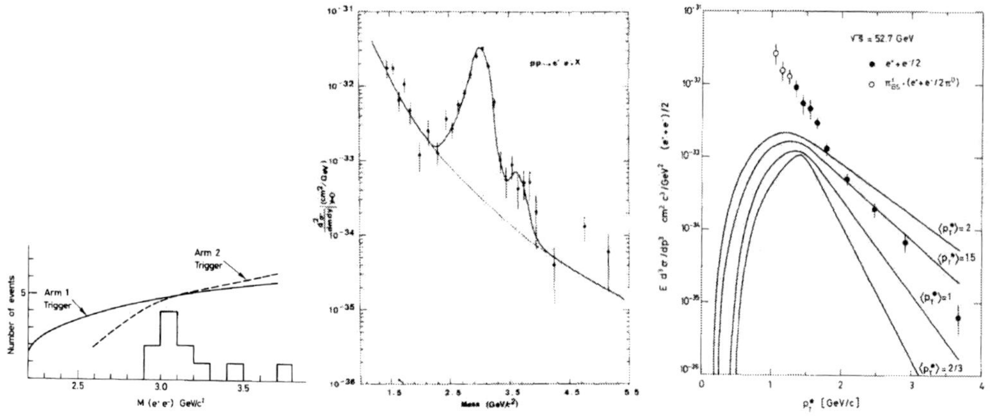

Figure 11. a)(left) First J/Ψ at ISR [23]; b) (center) Best $d\sigma_{ee}/dm_{ee}dy|_{y=0}$ [24]; c)(right) direct-$e^{\pm}$ data at $\sqrt{s}$ = 52.7 GeV (Fig. 1) with calculated $e^{\pm}$ spectrum for J/Ψ for several values of $\langle p_T \rangle$ [5].

J/Ψ in November 1974 and, in particular, with the near miss of this discovery at the CERN-ISR, there was concern that the $J/\Psi \rightarrow e^{+} + e^{-}$ decay could be the source of the single $e^{\pm}$. Fortunately, CCRS quickly demonstrated that the J/Ψ was not the source of the single-$e^{\pm}$ (Fig. 11). Fig. 11a [23] shows the first J/Ψ at the ISR [23], Fig. 11b shows the best J/Ψ measurement at the ISR [24] while Fig. 11c [5] shows that the direct electrons (Fig. 1) are not the result of J/Ψ decay since $\langle p_T \rangle = 1.1 \pm 0.05$ GeV/c [24].

Sam Ting, in his Nobel Lecture [25] (Fig. 12) also noted that the J meson could not explain the prompt leptons and indicated that he actually delayed announcing the J discovery at Viki Weisskopf's retirement ceremony in mid-October 1974 in order to investigate the prompt leptons in his AGS experiment.

On the other side of this argument was Jim Cronin, another Nobel Laureate (but incorrect on this issue), who prominently claimed in his plenary talk at the 1977 Lepton Photon Symposium in Hamburg [26] that "The origin of direct single leptons is principally due to the production of lepton pairs" (Fig. 13). Also, as one can see from the discussion, a young (at the time) Associate Professor from the Rockefeller University vehemently disagreed with Cronin's explanation. In private, I also said, with evidence [27], that "people who try to measure prompt leptons become the world's experts on η Dalitz decay".

5. Back to the "Heavy Quark Suppression Crisis" at RHIC

The suppression of direct-single-$e^{\pm}$ in Au+Au collisions at RHIC (Fig. 3) is even more dramatic as a function of $p_T \gtrsim 5$ GeV/c (Fig 14a) which indicates suppression of heavy quarks as large as that for π^0 (from light quarks and gluons) in the region where the $m \gtrsim 4$ GeV b-quarks dominate. Figure 14b shows that heavy quarks exhibit collective flow (v_2), another indication of a very strong interaction with the medium. This observation calls into question the QCD radiative energy-loss explanation of jet-quenching [28] because, naively, heavy quarks should radiate much less than light quarks and gluons in the medium. This issue has attracted much theoretical attention [29].

Journal of Physics: Conference Series **230** (2010) 012037 doi:10.1088/1742-6596/230/1/012037

V. I was considering announcing our results during the retirement ceremony for V. F. Weisskopf, who had helped us a great deal during the course of many of our experiments. This ceremony was to be held on 17 and 18 October 1974. I postponed the announcement. for two reasons. First, there were speculations on high mass e^+e^- pair production from proton-proton collisions as coming from a two-step process : $p+N\rightarrow\pi+...$, where the pion undergoes a second collision $\pi+N\rightarrow e^++e^-+....$ This could be checked by a measurement based on target thickness. The yield from a two-step process would

S. C. C. Ting 335

increase quadratically with target thickness, whereas for a one-step process the yield increases linearly. This was quickly done, as described in point (iv) above.

Most important, we realized that there were earlier Brookhaven measurements [24] of direct production of muons and pions in nucleon-nucleon collisions which gave the μ/π ratio as 10^{-4}, a mysterious ratio that seemed not to change from 2000 GeV at the ISR down to 30 GeV. This value was an order of magnitude larger than theoretically expected in terms of the three known vector mesons, p, ω, φ, which at that time were the only possible "intermediaries" between the strong and electromagnetic interactions. We then added the J meson to the three and found that the linear combination of the four vector mesons could not explain the μ^-/π^- ratio either. This I took as an indication that something exciting might be just around the corner, so I decided that we should make a direct measurement of this number. Since we could not measure the μ/π ratio with our spectrometer, we decided to look into the possibility of investigating the e^-/x^- ratio.

We began various test runs to understand the problems involved in doing the e/π experiment. The most important tests were runs of different e^- momenta as a function of incident proton intensities to check the single-arm backgrounds and the data-recording capability of the computer.

On Thursday, 7 November, we made a major change in the spectrometer (see Fig. 13) to start the new experiment to search for more particles. We began by measuring the mysterious e/π ourselves. We changed the electronic logic and the target, and reduced the incident proton beam intensity by almost two orders of magnitude. To identify the e^-background due to the decay of π^0 mesons, we inserted thin aluminium converters in front of the spectrometer to increase the $\gamma\rightarrow e^++e^-$ conversion. This, together with the C_B counter which measures the $\pi\rightarrow\gamma+e^++e^-$ directly, enabled us to control the major e^-background contribution.

Figure 12. Excerpt from Ting Nobel Lecture [25]

5.1. *Zichichi to the rescue?*

In September 2007, I read an article by Nino Zichichi, "Yukawa's gold mine", in the CERN Courier taken from his talk at the 2007 International Nuclear Physics meeting in Tokyo, Japan, in which he proposed: "We know that confinement produces masses of the order of a giga-electron-volt. Therefore, according to our present understanding, the QCD colourless condition cannot explain the heavy quark mass. However, since the origin of the quark masses is still not known, it cannot be excluded that in a QCD coloured world, the six quarks are all nearly massless and that the colourless condition is 'flavour' dependent."

HADRON INDUCED LEPTONS AND PHOTONS

James W. Cronin
Enrico Fermi Institute, University of Chicago
Chicago, Illinois, 60637 U.S.A.

ABSTRACT

A review of direct production of leptons and photons in hadron-hadron collisions is presented. Production of lepton pairs with large mass is well accounted for by the Drell-Yan process. The origin of direct single leptons is principally due to the production of lepton pairs. A dominant source of lepton pairs is at low effective mass, m < 600 MeV/c.

III. SINGLE LEPTON PRODUCTION

Since 1974 direct single lepton production has been the subject of intensive experimental investigation. The important qualitative observation has been that direct electron or muon production for $p_\perp$ > 1 GeV/c and large c.m. angle is approximately 10^{-4} of pion production. We shall argue that a very large fraction of the observed single muons have their source as one member of a lepton pair. These sources are best studied in experiments in which both members of the pair can be observed, or if a neutrino accompanies the lepton that the neutrino is detected indirectly through an absence of total energy balance. We believe that in the future, experiments which observe only a single lepton are going to yield little new information. We have come to understand the single lepton problem only through the study of lepton pairs.

V. CONCLUSIONS

1. The Drell-Yan mechanism provides a qualitative and nearly quantitative description for lepton pair production with mass > 4 GeV/c^2.

2. Direct single lepton production is consistent with a source which is lepton pair production. An important ingredient in this discussion is the existence of a significant low mass dilepton continuum which is an order of magnitude larger than expected from the Drell-Yan process.

3. A dedicated experiment will be required to establish definitively a significant direct photon yield induced by hadrons.

M.J. Tannenbaum, Rockefeller University: With reference to explaining ISR single electrons from Fermilab data at x_F > .15: First of all, the CCRS experiment made a great effort to eliminate low mass electron pairs (M_{ee} < 0.50 GeV/c^2) but allowed a continuum of dσ/dm 1/m or flatter to explain all the single electrons observed; secondly, the crucial parameters in the lepton pair cross section relevant to explaining the ISR(CCRS) single electrons are the width of the rapidity plateau and $d^2\sigma/dmdy|_{y=0}$, the value of the cross section on the plateau. The Fermilab measurements at x_F > .15 do not give these values and thus are irrelevant in trying to explain the ISR data. Thus, while it might seem reasonable that the dilepton continuum could explain all of the single electron signal, it is by no means proved.

J.W. Cronin: While technically it is true that the measurements of the CP(II) group do not correspond to the kinematic region observed in the CCRS experiment (Ref. 24), we do have some experience with continuity. The extrapolation of the CP(II) group was based on a constant cross section in rapidity between x_F = 0.15 and x_F = o. I insist on my main point, however, that in fact we learn little from the single lepton measurements alone. To "prove" that pairs explain the CCRS signal, pair measurements will have to be made at x_F ~ o.

Figure 13. Excerpt from Cronin talk and discussion at Lepton-Photon 1977 [26]

Nino's idea really excited me even though, or perhaps because, it appeared to overturn two of the major tenets of the Standard Model since it seemed to imply that: QCD isn't flavor blind; the masses of quarks aren't given by the Higgs mechanism. Massless b and c quarks in a color-charged medium would be the simplest way to explain the apparent equality of gluon, light quark and heavy quark suppression indicated by the equality of R_{AA} for π^0 and direct single-$e^\pm$ in regions where both c and b quarks dominate. Furthermore RHIC and LHC-Ions are the only place in the Universe to test this idea.

It may seem surprising that I would be so quick to take Nino's idea so seriously. This confidence dates from my graduate student days when I checked the proceedings of the 12th ICHEP in Dubna, Russia in 1964 to see how my thesis results were reported [30] and I found several interesting questions and comments by an "A. Zichichi" printed in the proceedings (Fig. 15). One comment about how to find the W boson in p+p collisions [31] deserves a verbatim

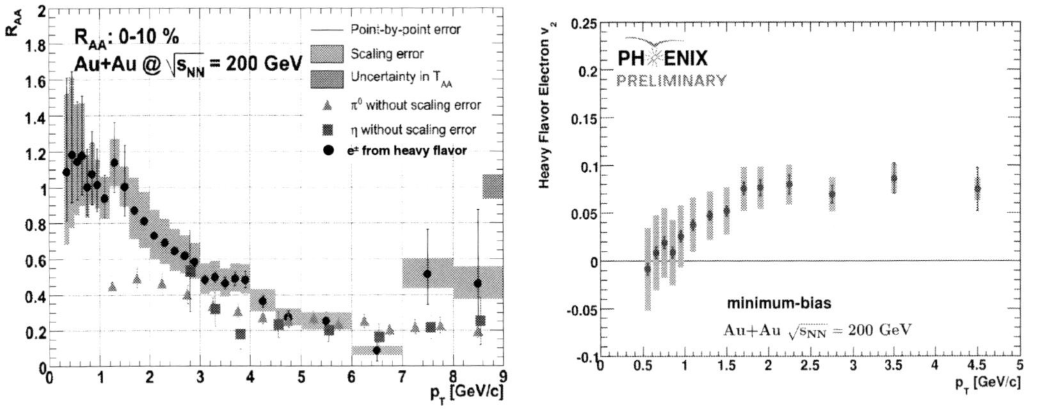

Figure 14. a) (left) R_{AA} (central Au+Au) b) (right) v_2 (minimum bias Au+Au) as a function of p_T for direct-$e^{\pm}$ at $\sqrt{s_{NN}} = 200$ GeV [9].

quote because it was exactly how the W was discovered at CERN 19 years later [32, 33]: "We would observe the μ's from W-decays. By measuring the angular and momentum distribution at large angles of K and π's, we can predict the corresponding μ-spectrum. We then see if the μ's found at large angles agree with or exceed the expected numbers."

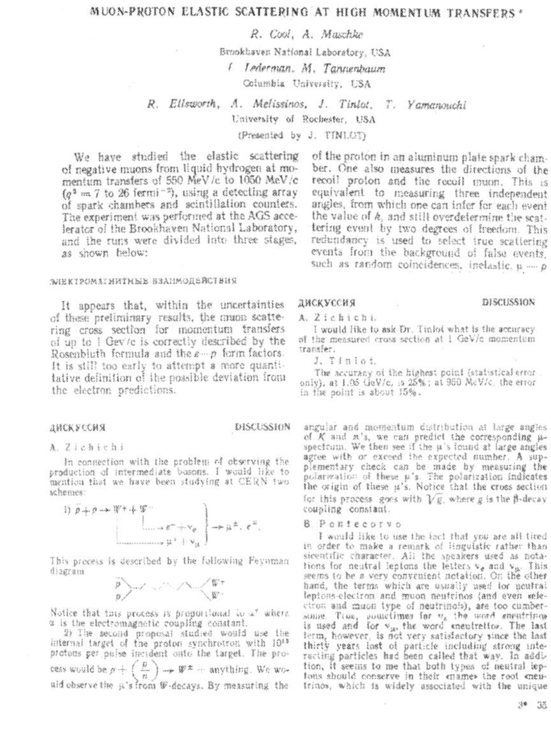

MUON-PROTON ELASTIC SCATTERING AT HIGH MOMENTUM TRANSFERS *

R. Cool, A. Maschke
Brookhaven National Laboratory, USA
L. Lederman, M. Tannenbaum
Columbia University, USA
R. Ellsworth, A. Melissinos, J. Tinlot, T. Yamanouchi
University of Rochester, USA
(Presented by J. TINLOT)

We have studied the elastic scattering of negative muons from liquid hydrogen at momentum transfers of 550 MeV/c to 1050 MeV/c (q^2 = 7 to 26 fermi^{-2}), using a detecting array of spark chambers and scintillation counters. The experiment was performed at the AGS accelerator of the Brookhaven National Laboratory, and the runs were divided into three stages, as shown below:

of the proton in an aluminum plate spark chamber. One also measures the directions of the recoil proton and the recoil muon. This is equivalent to measuring three independent angles, from which one can infer for each event the value of k, and still overdetermine the scattering event by two degrees of freedom. This redundancy is used to select true scattering events from the background of false events, such as random coincidences, inelastic $\mu - p$

ЭЛЕКТРОМАГНИТНЫЕ ВЗАИМОДЕЙСТВИЯ

It appears that, within the uncertainties of these preliminary results, the muon scattering cross section for momentum transfers of up to 1 Gev/c is correctly described by the Rosenbluth formula and the $e-p$ form factors. It is still too early to attempt a more quantitative definition of the possible deviation from the electron predictions.

ДИСКУССИЯ DISCUSSION

A. Zichichi.

I would like to ask Dr. Tinlot what is the accuracy of the measured cross section at 1 GeV/c momentum transfer.

J. Tinlot.

The accuracy of the highest point (statistical error only), at 1.05 GeV/c, is 25%; at 950 MeV/c, the error in the point is about 15%.

ДИСКУССИЯ DISCUSSION

A. Zichichi

In connection with the problem of observing the production of intermediate bosons. I would like to mention that we have been studying at CERN two schemes:

1) $\bar{p} + p \to W^+ + W^-$; $W \to e^- + \nu_e$, $\mu^+ + \nu_\mu$ $\to \mu^\pm, e^\mp$.

This process is described by the following Feynman diagram

Notice that this process is proportional to α^2 where α is the electromagnetic coupling constant.

2) The second proposal studied would use the internal target of the proton synchrotron with 10^{13} protons per pulse incident onto the target. The process would be $p + \left(\frac{p}{n}\right) \to W^\pm +$ anything. We would observe the μ's from W-decays. By measuring the angular and momentum distribution at large angles of K and π's, we can predict the corresponding μ-spectrum. We then see if the μ's found at large angles agree with or exceed the expected number. A supplementary check can be made by measuring the polarization of these μ's. The polarization indicates the origin of these μ's. Notice that the cross section for this process goes with $\sqrt{g}$, where g is the β-decay coupling constant.

B. Pontecorvo

I would like to use the fact that you are all tired in order to make a remark of linguistic rather than scientific character. All the speakers used as notations for neutral leptons the letters ν_e and ν_μ. This seems to be a very convenient notation. On the other hand, the terms which are usually used for neutral leptons-electron and muon neutrinos (and even «electron and muon type of neutrinos»), are too cumbersome. True, sometimes for ν_e the word «neutrino» is used and for ν_μ, the word «neutretto». The last term, however, is not very satisfactory since the last thirty years lost of particle including strong interacting particles had been called that way. In addition, it seems to me that both types of neutral leptons should conserve in their «name» the root «neutrino», which is widely associated with the unique

3* 35

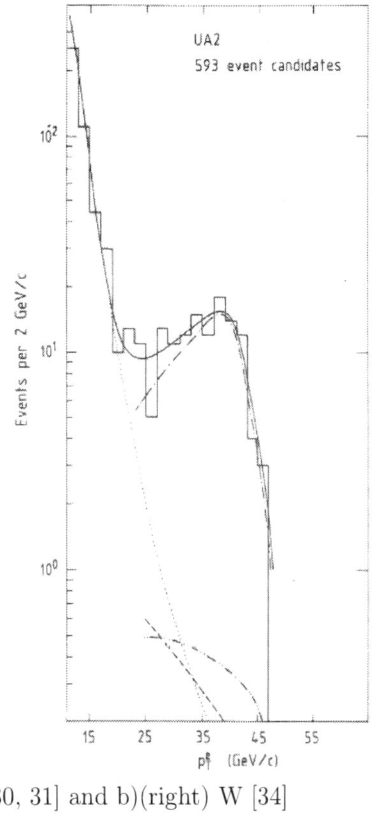

Figure 15. a)(left)Zichichi ICHEP 1964 [30, 31] and b)(right) W [34]

Nino's idea seems much more reasonable to me than the string theory explanations of heavy-quark suppression (especially since they can't explain light-quark suppression). Nevertheless, just to be safe, I asked some distinguished theorists what they thought, with these results:

- Stan Brodsky:"Oh, you mean the Higgs field can't penetrate the QGP."
- Rob Pisarski: "You mean that the propagation of heavy and light quarks through the medium is the same."
- Chris Quigg (Moriond 2008): "The Higgs coupling to vector bosons γ, W, Z is specified in the standard model and is a fundamental issue. One big question to be answered by the LHC is whether the Higgs gives mass to fermions or only to gauge bosons. The Yukawa couplings to fermions are put in by hand and are not required." "What sets fermion masses, mixings?"
- Bill Marciano:"No change in the t-quark, W, Higgs mass relationship (Fig. 16) if there is no Yukawa coupling: but there could be other changes."

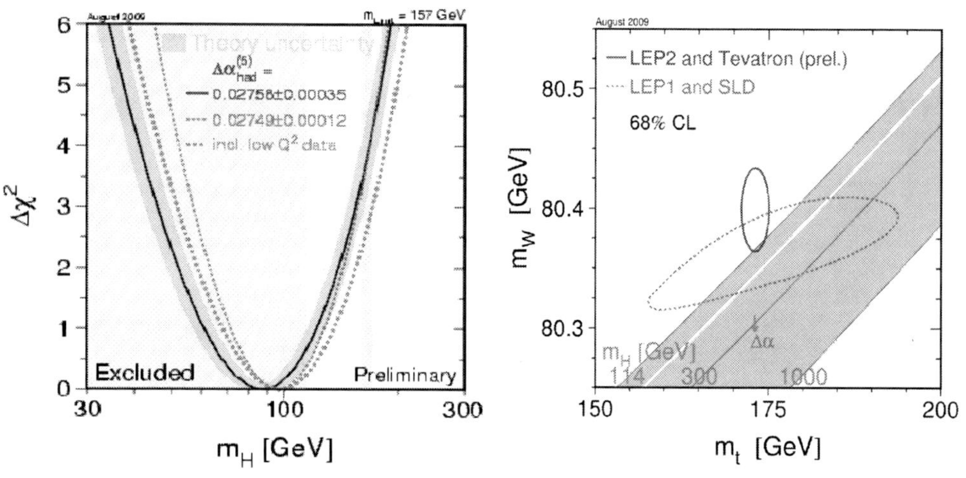

Figure 16. Limits on Higgs mass from Tevatron and LEP, Summer 2009 [35]

Nino proposed to test his idea by shooting a proton beam through a QGP formed in a Pb+Pb collision at the LHC and seeing the proton 'dissolved' by the QGP. My idea is to use the new PHENIX Silicon VTX detector, to be installed in 2010, to map out, on an event-by-event basis, the di-hadron correlations from identified $b - \bar{b}$ di-jets, identified $c - \bar{c}$ di-jets, which do not originate from the vertex, and light quark and gluon di-jets, which originate from the vertex and can be measured with π^0-hadron correlations. A steepening of the slope of the x_E distribution of heavy-quark correlations as in π^0-hadron correlations [36] will confirm in detail (or falsify) whether the different flavors of quarks behave as if they have the same energy loss (hence mass) in a color-charged medium. If Nino's proposed effect is true, that the masses of fermions are not given by the Higgs particle, and we can confirm the effect at RHIC or LHC-Ions, this would be a case where we Relativistic Heavy Ion Physicists may have something unique to contribute at the most fundamental level to the Standard Model, which would constitute a "transformational discovery." Of course the LHC could falsify this idea by finding the Higgs decay to $b - \bar{b}$ at the expected rate in p-p collisions. Clearly, there are exciting years ahead of us!

Acknowledgments
Research supported by U. S. Department of Energy, DE-AC02-98CH10886.

References

[1] Letter from Mel Schwartz, Associate Director High Energy and Nuclear Physics at BNL to spokespersons of the three competing experiments, TALES/SPARHC, OASIS, Dimuon, September 9. 1991
[2] Hayano R S *et al* RHIC LoI #2, September 27, 1990
[3] Zhang P H *et al* RHIC LoI #12, July 15, 1991
[4] Büsser F W *et al* CCRS Collaboration 1976 *Phys. Lett. B* **53** 212–216
[5] Büsser F W *et al* CCRS Collaboration 1976 *Nucl. Phys. B* **113** 189–245
[6] Tannenbaum M J 2005 *Proc. 20th Winter Workshop on Nuclear Dynamics, Acta Phys. Hung. A* **24** 259–267 (*Preprint* arXiv:nucl-ex/0406024v3)
[7] Adler S S *et al* PHENIX Collaboration 2005 *Phys. Rev. Lett.* **94** 082301
[8] Adare A *et al* PHENIX Collaboration 2006 *Phys. Rev. Lett.* **97** 252002
[9] Adare A *et al* PHENIX Collaboration 2007 *Phys. Rev. Lett.* **98** 172301
[10] e.g. see Ref. [9] for a list of references
[11] Abelev B I *et al* STAR Collaboration 2007 *Phys. Rev. Lett.* **98** 192301
[12] Segler S *et al* CCRS Collaboration 1974 *Proc. XVII Int. Conf. on High Energy Physics (London)* (Chilton UK: The science Research Council) pp V41–V43
[13] Farrar G R and Frautschi S C 1976 *Phys. Rev. Lett.* **36** 1017–1020
[14] Hinchliffe I and Llewellyn Smith C H 1976 *Nucl. Phys. B* **114** 45-60
[15] Büsser F W *et al* CCRS Collaboration 1975 *Proc. Intl. Symp. Lepton and Photon Interactions at High Energies, August 21-27* (Stanford, CA: SLAC) paper 202
[16] Goldhaber G *et al* 1976 *Phys. Rev. Lett.* **37** 255–259
[17] Bourquin M and Gaillard J-M 1976 *Nucl. Phys. B* **114** 334-364
[18] Baum L *et al* CHORMN Collaboration 1976 *Phys. Lett. B* **60** 485–490
[19] Vogt R 1996 *Z. Phys. C* **71** 475–481
[20] Tavernier S P K 1987 *Rep. Prog. Phys.* **50** 1439–1489
[21] Adcox K *et al.* PHENIX Collaboration 2002 *Phys. Rev. Lett.* **88** 192303
[22] Appel J A 1992 *Ann. Rev. Nucl. Part. Sci.* **42** 367–399
[23] Büsser F W *et al* CCRS Collaboration 1976 *Phys. Lett. B* **56** 482–486
[24] Clark A G *et al* CSZ Collaboration 1978 *Nucl. Phys. B* **142** 29-52
[25] Ting Samuel C C 1976 http://nobelprize.org/nobel_prizes/physics/laureates/1976/ting-lecture.html
[26] Cronin J W 1977 *Proc.Intl. Symp. Lepton and Photon Interactions at High Energies, August 25-31* (Hamburg, Germany: DESY) pp 579–598
[27] Jane M R *et al* 1978 *Phys. Lett. B* **73** 503
[28] Baier R, Schiff D and Zakharov B G 2000 *Ann. Rev. Nucl. Part. Sci.* **50** 37–69
[29] e.g. see Gyulassy M and Lévai P 2010 *Nucl. Phys. A* **834** 271c–222c and references therein
[30] Cool R *et al* 1964 *Proc. 12th Intl. Conf. High Energy Physics August 5-15* (Moscow: Atomizdat) v.1 pp 852–856
[31] Zichichi A 1964 *Proc. 12th Intl. Conf. High Energy Physics August 5-15* (Moscow: Atomizdat) v.2 p 35
[32] Arnison G *et al* UA1 Collaboration 1983 *Phys. Lett. B* **122** 102–116
[33] Banner M *et al* UA2 Collaboration 1983 *Phys. Lett. B* **122** 476–485
[34] Appel J A *et al* UA2 Collaboration 1986 *Z. Phys. C* **30** 1–22
[35] http://lepewwg.web.cern.ch/LEPEWWG/plots/summer2009/
[36] Adare A *et al* PHENIX Collaboration 2010 Trends in Yield and Azimuthal Shape Modication in Dihadron Correlations in Relativistic Heavy Ion Collisions *Preprint* arXiv:1002.1077v2 [nucl-ex]

Journal of Physics: Conference Series **230** (2010) 012038 doi:10.1088/1742-6596/230/1/012038

Muon Trigger Upgrade at PHENIX

I. Nakagawa for the PHENIX Collaboration

RIKEN institute, Hirosawa, Wako, 351-0198, Saitama, Japan

E-mail: itaru@bnl.gov

Abstract. The Relativistic Heavy Ion Collider (RHIC) at BNL provides a unique opportunity to collide polarized protons. One of the hightlight of the spin program at $\sqrt{s} = 500$ GeV is the direct measurement of sea quark contribution to the proton spin via W-boson production by measuring parity violating single spin asymmetry. A new trigger on forward muons in PHENIX identifies and triggers on high momentum Ws suppressing a large number of low momentum muons coming from hadronic decays. Since the current muon trigger will fire on any muon above 2 GeV/c, it will not provide the required rejection factor for 500 GeV running, which is about 7500. Two major upgrade projects are parts of this Muon Trigger; 1) Resistive Plate Chambers, 2) the muon tracker front end electronics upgrades. The status of these upgrade projects is discussed in this document as well as physics goal of W program at RHIC.

1. INTRODUCTION

Parity-violating production of the W boson with longitudinally polarized protons at RHIC provides a direct measure of the individual polarizations of the quarks and anti-quarks in the colliding protons. The high energy scale set by the W-mass makes it possible to extract quark and anti-quark polarizations from inclusive lepton spin asymmetries in W-production with minimal theoretical uncertainties. Sub-leading twist and higher-order terms in the perturbative QCD expansion are strongly suppressed, and a direct extraction without additional assumptions becomes possible[1]. This program thus will break new ground in our detailed understanding of the proton's structure. Especially, our present knowledge for the anti-quark contribution to the proton spin is limited by SIDIS measurements. In addition to forementioned naturally set high energy scale, the sea quark measurement via W has certain advantage with respect to SIDIS in following sense: 1) The W boson couples with the quark flavor selectively and quarks (antiquarks) with negative (positive) helicity can participate in the reaction due to the V-A structure of the weak interaction, 2) free from fragmentation function which is the unavoidable dominant source of model uncertainties for SIDIS on their way to extract the sea quark polarization from observed asymmetries. Due to unbalanced x_1 and x_2, the measurement of W asymmetry in forward/backward angle further enhances the sensitivity to the sea-quark polarization.

The W program is initiated by the first operation of RHIC polarized proton beams at its highest operational energy $\sqrt{s} = 500$ GeV in 2009 (Run9). This W program is the highlight of RHIC spin project for the next 5 years and thus following upgrade projects to PHENIX muon arms have been pursued to prepare for the rare probe measurement under high-background rate circumstances.

2. MUON TRIGGER AT PHENIX

A new trigger on forward muons in PHENIX identifies and triggers on high momentum muons from W decay suppressing a large number of background low momentum muons coming from hadronic decays. Since the current muon trigger will fire on any muon above 2 GeV/c, it will not provide the required rejection factor for 500 GeV running, which is about 7500. Two major upgrade projects are parts of this Muon Trigger; 1) Resistive Plate Chambers[2], 2) the muon tracker front-end-electronics (MuTr-FEE) upgrades[3]. Each projects are discussed in detail hereafter.

2.1. MuTr-FEE Upgrade

The MuTr-FEE upgrade features additional fast readouts to the existing MuTr-FEE, which provides momentum sensitivity to the existing muon trigger beyond 2 GeV/c. There are three tracking stations of cathode strip (5mm pitch) chambers in the muon tracker for measuring the trajectory of particles in a magnetic field (with an integrated $B - dl$ of approximately 0.75 Tesla-meters). A new amplifier discriminator (MuTRG-ADTX) board[4] divides the signal into two paths; existing read out path and the digitization path for the fast trigger.

Figure 1 visualizes the principle of the momentum sensitive trigger. The hit signals from three tracking chambers (named station-1, 2, and 3, respectively from the collision point) are immediately digitized and processed by MuTRG-ADTX and are sent to new merger boards (MuTRG-MRG)[5]. The MuTRG-MRG board collects hit signals from multiple MuTRG-ADTXs and formats the data to meet the input format of the Local Level-1 (LL1) trigger module. The LL1 module receives data from MuTRG-MRG and then select the events which have ± 1 strip sagitta at station-2 from the linear interpolation between station-1 and station-3 hits, providing the trigger of only high transverse momentum track.

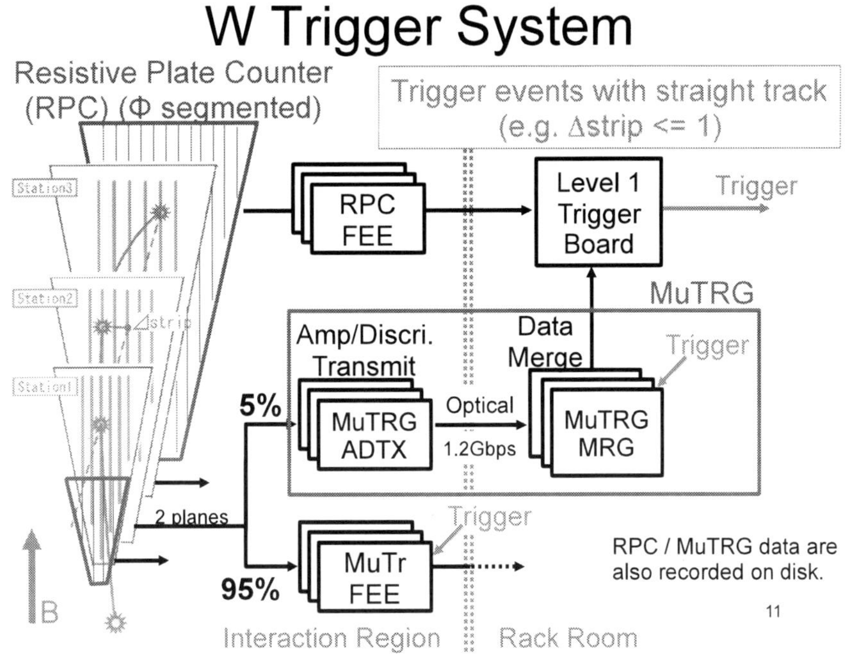

Figure 1. The concept of momentum sensitive trigger using muon tracker cathode strip readouts.

The R&D, production of these new boards were completed by the end of 2008. The installation to North muon arm was completed in 2008 and they were commissioned in Run9, whereas South muon arm was completed in 2009 and were commissioned in Run10 Au-Au collision data as well.

Shown in Fig. 2 is the turn on curve of the trigger efficiency plotted as a function of track momentum evaluated using reconstructed tracks in South and North MuTr. The trigger threshold is certainly pushed higher to around 5.1 and 7.6 GeV (parameter p1 in Fig. 2 fit) for South and North, respectively compared to the maximum threshold limit of about 2 GeV given by MuID. The plateau saturates around the efficiency of 0.93. This is the consequence of the product of the individual efficiencies in each station (about 0.98) and the vertex cut efficiency. Better efficiency can be achieved by relieving the operating conditions of the trigger electronics such as threshold, acceptance range of track sagitta, with or without strip clustering, AND or OR logic selection of MuTr redundant planes in each stations (MuTr chambers of Station-1, 2, and 3 consisted of 3, 3, and 2 gaps, respectively. New trigger electronics were implemented to 3, 2, and 2 non-stereo planes of these stations, respectively for redundancy.)

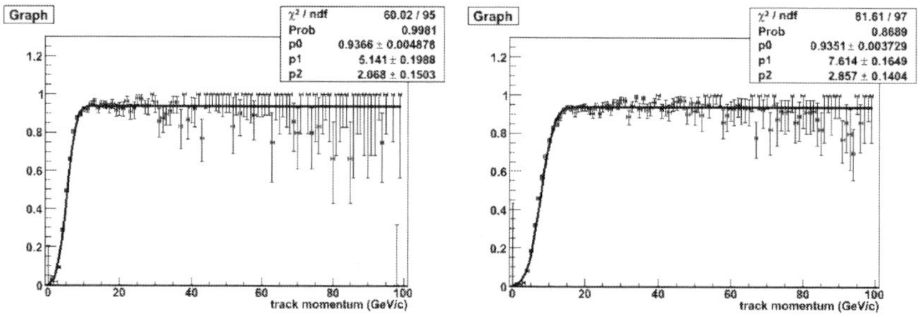

Figure 2. Turn on curve of the muon trigger as a function of track momentum evaluated from Run10 data.

On the other hand, the higher efficiency would be the trade off between the rejection power. Shown in Fig. 3 is the correlation between the trigger efficiency versus the total rejection power (BBC×MuID×MuTrig) evaluated at the BBC rate of 1.5MHz. Plotted data points are the performance of different operating conditions of new trigger electronics as described previously. The achieved efficiency of > 0.9 is satisfactory at this BBC rate, which requires only the total rejection power of 750. Projected efficiency is about 0.8 at the rejection factor of 3000, which is the required factor at the projected luminosity of 1×10^{32} (expected BBC rate of about 6MHz) for Run11.

Note this performance may be degraded as a function of the luminosity, particularly for rejection performance due to increasing accidental hits. Nevertheless, these results are still preliminary stage and there are still plenty of rooms to improve the trigger performance. Following items are some ideas to be addressed in future study to retrieve the performance:

- More efficient MuID trigger algorithm
- Track Matching with MuID
- Timing cut by RPC
- Track matching with RPC
- Tighter background shields
- Quenching cross talks in MuTr chamber
- Etc.

Journal of Physics: Conference Series **230** (2010) 012038 doi:10.1088/1742-6596/230/1/012038

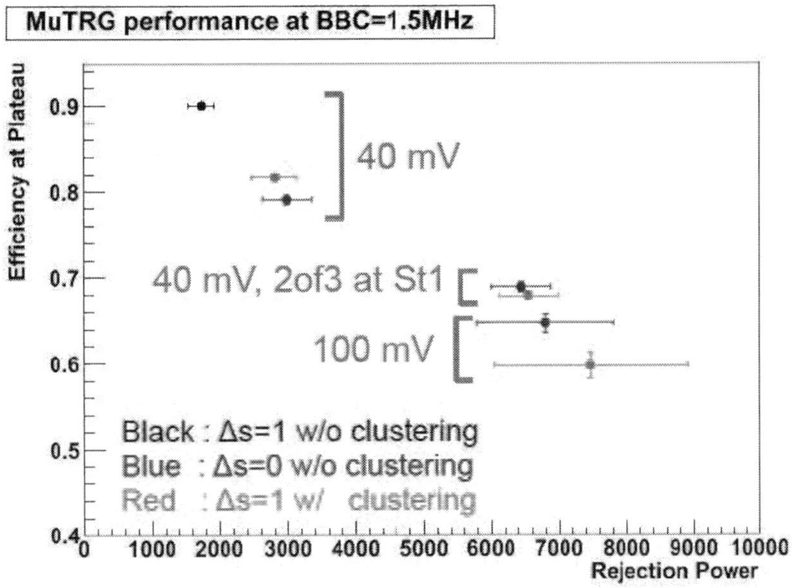

Figure 3. Correlation between the trigger efficiency vs. rejection power observed at BBC rate of 1.5MHz. 40mV and 100mV are threshold condition applied to the cathode signal at new trigger boards. $\Delta s=1$ and 0 represent the acceptance of sagitta range of strip ± 1 and ± 0 from the central strip of the trajectory.

2.2. RPC

In addition to lack of rejection power, another critical aspect of the muon trigger at high luminosity in PHENIX forward/backward arm is the timing resolution. The timing information from BBC would not work at the maximum luminosity at 500 GeV, because BBC is anticipated to be fired every single beam crossing. Given the fact Muon ID has the timing resolution of 2 beam clocks, we would not be able to determine which beam crossing the observed track comes from. This is the fatal situation for polarized proton experiments at its full intensity operation because polarization vector, e.g. + or - is altered every bunch in RHIC on purpose. Thus new timing device in muon arm was introduced called RPC.

Shown in Fig. 4 is the cross section structure of the RPC detector (left panel). The design of PHENIX RPC is based on precedent RPC detector developed for Compact Muon Solenoid (CMS) experiment in CERN. The RPCs are built with two parallel plates of high resistive material, i.e. bakelite as electrodes. The plates have a resistivity of the order of $1 \sim 10\times10^{10}\Omega$cm. The sensitive gas volume gap is typically 2 mm thick between the two plates. The outside surfaces of the RPC plates are coated with graphite for distributing high voltage on one side and the ground on the other in order to establish a strong electric field in the gas gap. The signal readout is made of copper strips and is located outside of the sensitive gas volume. This is one of the very attractive features of using RPCs. Ionizing particles create electron-ion clusters in the gas, where an intense constant electric field is present between the two parallel electrode plates. Strips are running along the radial direction to measure an azimuthal position of a trajectory which is the orientation of momentum kick in the magnetic field of MuTr. Taking advantage of the long level arm of RPC3 (behind MuID), additional requirement of track matching between MuTRG-FEE and RPC hits is expected to boost the rejection power against low momentum tracks.

Journal of Physics: Conference Series **230** (2010) 012038 doi:10.1088/1742-6596/230/1/012038

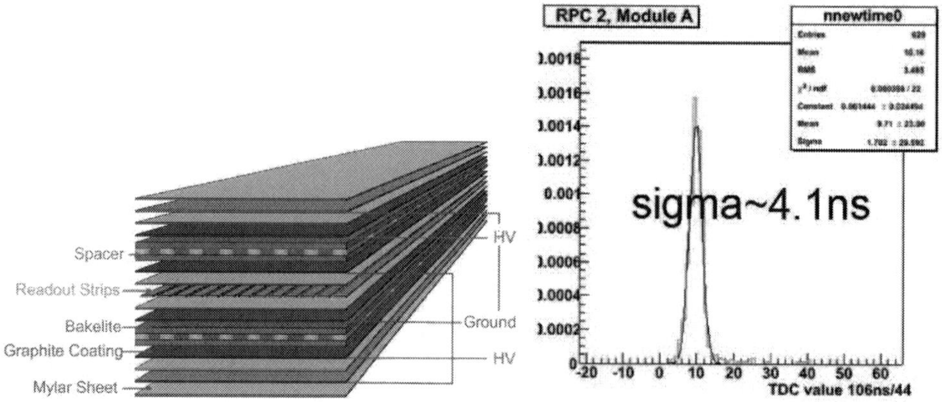

Figure 4. Structure of RPC (Left) and ToF peak observed by RPC2 prototype in Run9.

Table 1. Required performance of RPC

Efficiency	$> 95\%$
Time Resolution	≤ 3 ns
Average Cluster Size	≤ 2 strips
Rate Capability	0.5 kHz/cm^2
Operation Plateau	> 300 V

Considering overall forementioned demands, required performance for the PHENIX RPC is summarized in Table. 1. Shown in Fig. 4 (right panel) is the observed timing resolution of RPC2 prototype using the actual beam during Run9. As can be seen in the figure, satisfactory timing resolution of $\sim$ 4.4 ns was observed. Also presented in Fig. 5 are the efficiency (left) and the size of the cluster (right) as a function of operating HV observed with cosmic rays in test bench. These test bench results demonstrated high efficiency $> 95\%$, plateau region > 300 V, and small enough cluster size < 2 strips at reasonable operating voltage.

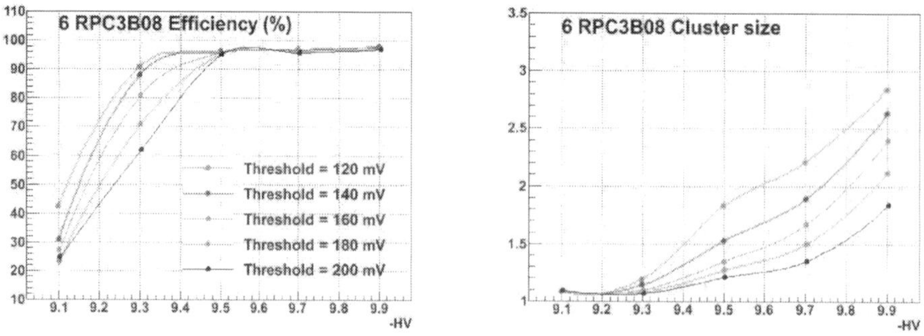

Figure 5. The efficiency of RPC (Left) and the size of the cluster (right) as a function of operating HVobserved with cosmic rays in test bench.

3. TOWARDS COMPLETION OF UPGRADE

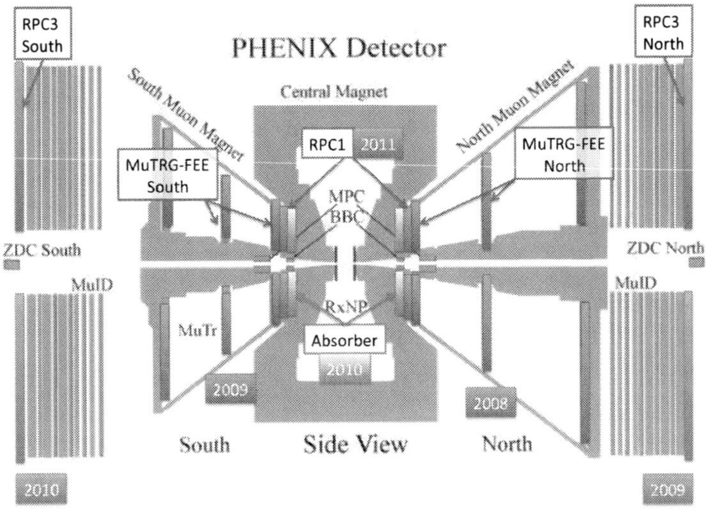

Figure 6. Installation schedule of PHENIX muon arm upgrade projects. Year in solid square is the scheduled year of installation.

The PHENIX muon arm upgrade project continues to improve the performance of the muon detection system as shown in Fig.6. An additional hadron background absorbers are in preparation in front of the MuTr. The absorber supposed to be installed in post Run10 shutdown period getting ready for the first production run of polarized proton-proton collision at 500 GeV in Run11. While implementing new devices as an upgrading performance of muon arms, efforts to improve performance of existing detectors are also underway in parallel. The momentum resolution of MuTr is one of the major concern, since it has been demonstrated only factor of 2 to 3 worse resolution in the past than its intrinsic one (100μm). One of the claim is the day-by-day motion of MuTr chambers which deteriorates the alignment precision. The relative alignment between MuTr chambers is carried out using straight particle trajectories in the zero magnetic field volume. Indeed we observed the chamber movement in the order of 100 μm over the course of months during Run9 using optical alignment system (OASys) implemented to MuTr chamber frames. The OASys consisted of light sources, lenses, and CCD cameras attached to Station-1, Station-2 and Station-3 chambers, respectively and tracks chambers' relative motion by continuously monitoring the light spot in the CCD camera[8]. The chamber displacement predicted by the OASys can be cross checked by the zero field data which are taken months a part. This consistency analysis is now underway.

[1] C. Burrell and J. Soffer, Phys. Lett. B314, 132 1993.
[2] Conceptual Design Report for a Fast Muon Trigger (2007).
[3] Technical Design Report on Amplifier-Discriminator board and Data Transfer board for the MuTr FEE upgrade (2008).
[4] Y. Fukao, *et. al.*, RIKEN Accel. Prog. Rep. **42** (2009).
[5] T. Mibe, *et. al.*, RIKEN Accel. Prog. Rep. **42** (2009).
[6] Y. Fukao, *et. al.*, RIKEN Accel. Prog. Rep. **43** (2010).
[7] K. Nakamura, *et. al.*, RIKEN Accel. Prog. Rep. **43** (2010).
[8] H. Akikawa, *et. al.*, NIM A499 537 (2003).

26th Winter Workshop on Nuclear Dynamics IOP Publishing
Journal of Physics: Conference Series **230** (2010) 012039 doi:10.1088/1742-6596/230/1/012039

Highlights of the PHENIX Transverse Spin Physics

Ming X. Liu

P-25 MS H846, Los Alamos National Laboratory, Los Alamos, NM 87545

mliu@lanl.gov

Abstract. The Relativistic Heavy Ion Collider (RHIC) at Brookhaven National Laboratory delivers the world highest energy polarized proton-proton collisions at a center of mass energy up to 500GeV and provides a unique opportunity to study the spin structure of the proton at high-energy scale. In recent years, there has been tremendous experimental and theoretical progress toward understanding the physics with transversely polarized beams (or targets). The scope of research in transverse physics involves from measurements of nucleon structure and parton distribution functions to the studies of QCD dynamics in high-energy polarized DIS and p+p collisions.

During the 2006 and 2008 RHIC runs, the PHENIX experiment took a significant amount of transversely polarized p+p collision data at the center of mass energies of 62 and 200 GeV, with integrated luminosity of 8 pb-1 at 200GeV and beam polarization of 45%(run8) and 57%(run6). In this report, I highlight the latest results from the PHENIX experiment, followed by a brief discussion of the future prospects of transverse physics with the PHENIX upgrade detectors, particularly on the importance of the unique measurements of open charm, J/Psi and Drell-Yan productions.

1. Introduction

Spin has been playing a key role in our understanding of fundamental interactions since the day it was postulated. Today, we are facing with two major puzzles in spin physics: (1) the challenge of "too small", i.e., the nucleon spin puzzle. It was believed that the spin was carried by the quarks that make up the proton. However, experiments in the 1980's led to the startling discovery that quarks contribute very little to the proton spin, setting off the "proton spin crisis". Where the rest of the proton spin coming from remains a major challenge to our understanding of the nucleon structure. Currently, there are several candidates that could make up the difference, such as the gluons' spin, and/or the quark and gluon's orbital angular momentum, and we need to determine them experimentally; (2) the challenge of "too big", i.e., the unexpectedly large transverse left-right single spin asymmetries (TSSA, A_N) in transversely polarized high energy $p^\uparrow$+p (and e+$p^\uparrow$) collisions. Prior to the early Fermilab E704 experimental measurements, it was believed based on pQCD arguments that transverse single spin asymmetry should be very small in high-energy p+p collisions, $A_N \sim \frac{m_q}{\sqrt{s}} \sim 10^{-3}$, while the experimental data showed asymmetry as large as 40%. Spin measurements have historically produced surprising results and are a stringent test to theories as spin is an intrinsically relativistic and quantum mechanical aspect of particle interactions. This "too big" challenge is the focus of my discussion here.

Journal of Physics: Conference Series **230** (2010) 012039 doi:10.1088/1742-6596/230/1/012039

Since the beginning of RHIC-Spin, very interesting results on transverse single spin asymmetries and cross sections have been measured by the RHIC experiments (BRAHMS, PHENIX and STAR) in a very wide kinematic range in polarized p+p collisions. All three experiments have observed and confirmed significant non-zero TSSAs in inclusive single particle productions in the forward rapidities. In the middle and backward rapidities, TSSAs are consistent with zero within the experimental uncertainty. The BRAHMS experiment has measured TSSAs for charged hadrons, including pion, Kaon, proton and antiproton at very forward rapidities (y~3.0 and 3.3). The STAR collaboration has measured π^0 TSSA as a function of xF and pT, up to pT ~ 4.5GeV. The PHENIX experiment has measured TSSA in pion and inclusive charged hadron production as a function of xF and pT in the central and forward rapidities and confirmed previous results from other experiments. PHENIX also carried out the first unique measurements of TSSAs in open heavy quark and J/Psi production at $\sqrt{s}$=200 GeV.

At RHIC energies, the NLO pQCD calculations have been applied successfully to describe the unpolarized cross section of pion production. For the TSSA physics, two different approaches have been used to explain the observed large TSSAs in the forward rapidity: one is based on the postulated transverse momentum dependent (TMD) functions, such as the nucleon spin and parton transverse momentum correlation functions (Sivers functions) and spin dependent fragmentation functions (Collins functions), to generate the asymmetries; the other one utilizes the collinear factoraization scheme and the asymmetry is generated from the initial and/or final state soft gluon interactions, the so called Twist-3 approach. Both models are used successfully to fit the observed pion TSSA v.s. xF at RHIC. However, the observed somewhat flat pT dependence of TSSA A_N at high pT remains as a challenge for both TMD and Twist-3 models, as both models expect smaller A_N at larger pT in the high pT region. At present, serious doubts are still lingering over the issues in QCD spin physics, such as the applicability of QCD-factorization under the current experimental conditions and the completeness of our understanding of the mechanisms that generate the transverse single-spin asymmetries. We expect future high statistic precision data from RHIC will provide a definite answer to many puzzles we have now.

2. The PHENIX experiment

The PHENIX experiment is designed for optimal measurements of rare probes [1]. In this report, we present the highlights of transverse spin asymmetry measurements using the PHENIX detectors at RHIC and also discuss the future plans.

2.1. PHENIX Highlights – A_N at central and forward rapidity in light hadron production

PHENIX central arm detectors have excellent capability of tracking, particle identification and photon and electron energy measurements in the pseudo-rapidity range of $|\eta| < 0.35$. In 2006 and 2007, the forward Muon Piston Calorimters (PMC) were installed to both south and north muon magnets, expanding the coverage of photon measurement to forward pseudo-rapidity $3.1 < |\eta| < 3.7$ with full azimuth. Figure 1 shows the π^0 and inclusive charged hadron TSSAs measured in the central rapidity $|\eta| < 0.35$. The results could indicate that the gluon's contribution is small in the probed kinematic range[2].

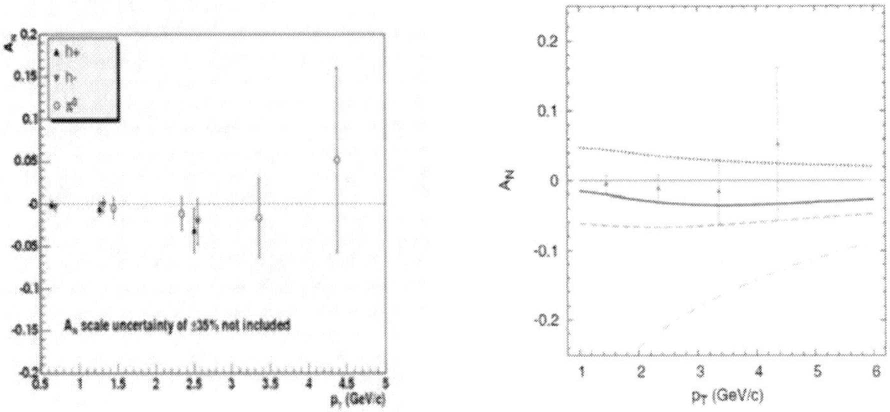

Figure 1. Left: Measured TSSAs A_N in inclusive π^0 and charged hadron production as a function of pT at central rapidity |eta| < 0.35. Right: A_N of π^0 from TMD pQCD model calculations with various assumptions about quark and gluons' Sivers contributions.

The new MPC detectors measure inclusive photons in the very forward rapidity where large TSSAs were observed at lower energies and also by other RHIC experiments (BRAHMS and STAR) in light hadron productions. Very significant none-zero TSSAs have been measured in π^0 production by the MPC at lower energy of $\sqrt{s}$ =62 GeV. The results are consistent with BRAHMS and STAR observations, see Figure 2.

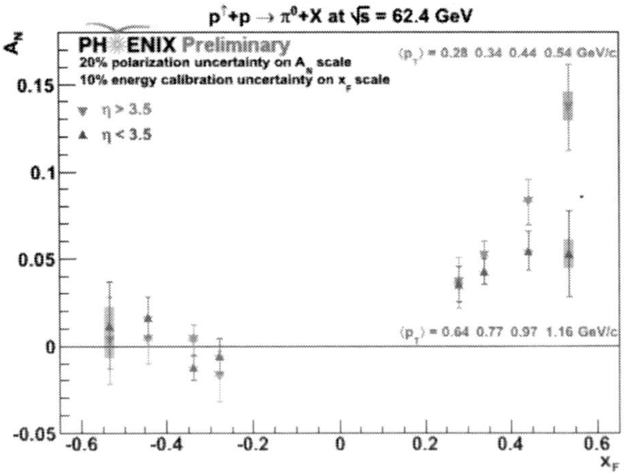

Figure 2. The measured A_N as a function of xF in pi0 production in the forward rapidity at the center of mass energy of $\sqrt{s}$ =62 GeV.

2.2. PHENIX Highlights – A_N in heavy quark production and the role of gluons

Besides the traditional light hadron probes, RHIC-Spin opens up new channels, such as heavy quark and Drell-Yan, that had never been possible in previous fixed target experiments. In particular, the PHENIX forward muon arms provide a unique probe to study the TSSA in open heavy quark and heavy quarkonium production, and in the future, the Drell-Yan production in the forward rapidity.

At RHIC energies, heavy quarks are predominantly produced via gluon-gluon interaction. Since gluon has zero transversity thus no Collins effect, we expect heavy quark asymmetries are sensitive to the gluon's Sivers function, which is currently not well constrained by DIS experiments. So open charm and beauty through high pT single muon production in the forward rapidity could be an excellent probe of the gluon's Sivers function at RHIC. Figure 3 shows the first measurement of TSSA in inclusive prompt muon production as a function of muon pT in the forward rapidity $1.4 < \eta < 1.9$. Although statistically limited, the data already put a constraint on the gluon's Sivers function and excluded the maximum gluon Sivers function scenario. The blue cure corresponding to the open charm prompt muon A_N with maximum gluon Sivers function. It should be noted that the measured prompt muons come from both open charm and open beauty contributions. With upgrade silicon vertex detectors, it is possible to separate these two contributions in the future.

Transverse spin physics could also provide new information about QCD dynamics. It is proposed recently that the TSSA of J/psi is sensitive to J/Psi's production mechanisms[3]. In p+p collsions, if J/Psi is produced via color octet channel, the initial and final states effects will cancel each other, as a results, the TSSA will be zero. However, if the production is dominated by color singlet channel, we could have none zero A_N. The situation is opposite in the polarized DIS process. It is nice to see that transverse spin physics could have an impact on a much broader aspects in QCD study. The PHENIX preliminary results from Run6 are statistically limited to draw any strong conclusion, but these combined with Run8 results will provide much improved constraint.

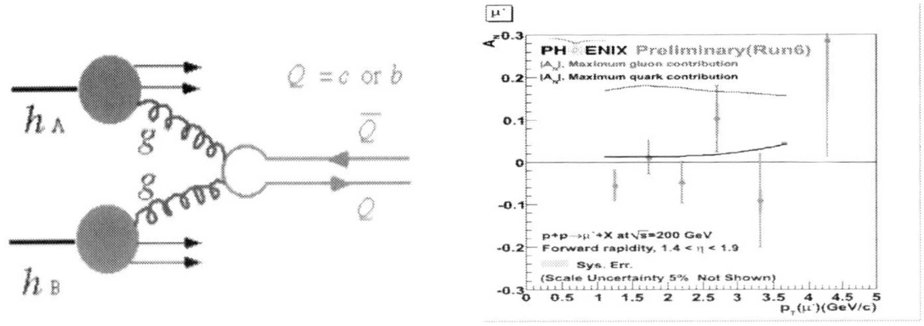

Figure 3. Left: At RHIC, heavy quarks are predominantly produced by gluon-gluon interaction. Right: measured prompt muon A_N as a function of pT in the forward rapidity $1.4 < \eta < 1.9$.

It is also important to note that the physics of TSSA in the gluonic sector could be more complicated and interesting. In the TMD approach, there is only one gluon Sivers function that could contribute to TSSA, thus, in the open charm production, both charm and anti-charm are expected to have identical TSSA A_N. However, the Twist-3 approach, there are two independent tri-gluon functions that could generate TSSAs, and depending on how these two functions coupled to quark and anti-quark, TSSAs could be very different in open charm and anti-charm production[4], see Figure 4.

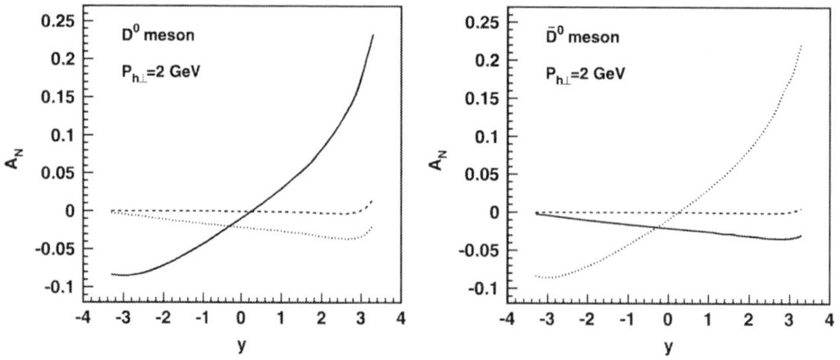

Figure 4. Twist-3 model calculations of the TSSA as a function of rapidity for D^0 and D^0-bar production in p+p collisions at $\sqrt{s}$=200 GeV. Different from TMD model where is only one gluon Sivers function, in the Twist-3 approach, the two independent tri-gluon functions could yield quite different TSSAs in charm and anti-charm produciton, a distinct experimental signature that can be tested at RHIC in the future.

2.3. Future Transverse Physics with PHENIX upgrade detectors

With upcoming detector upgrades for both PHENIX and STAR experiments and expected much improved accelerator performance, RHIC-Spin program will provide a great opportunity for new studies of polarized nucleon structure and the QCD dynamics. The silicon vertex detectors in PHENIX will allow us to cleanly identify and separate leptons from heavy quark and Drell-Yan decays. A lepton from a heavy quark (or light hadron) decay normally has a significantly displaced secondary vertex from the primary interaction point, while leptons from Drell-Yan process are prompt, see Figure 4.

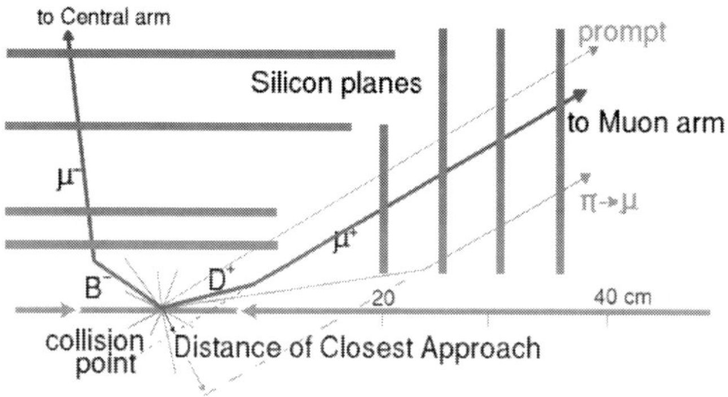

Figure 4. PHENIX Silicon vertex detectors will precisely measure the impact parameter of a lepton to the primary collision vertex point and identify a lepton from long-lived light hadron (pion), short-lived open heavy quark and prompt Drell-Yan decay.

The future Drell-Yan A_N measurement is particularly important at RHIC[5]. There is a fundamental prediction of QCD that A_N of Drell-Yan should have an opposite sign of the $A_N(\pi^+)$ observed in the polarized DIS experiment. See Figure 5. Its verification or disproval will be an important milestone in our understanding of QCD spin physics.

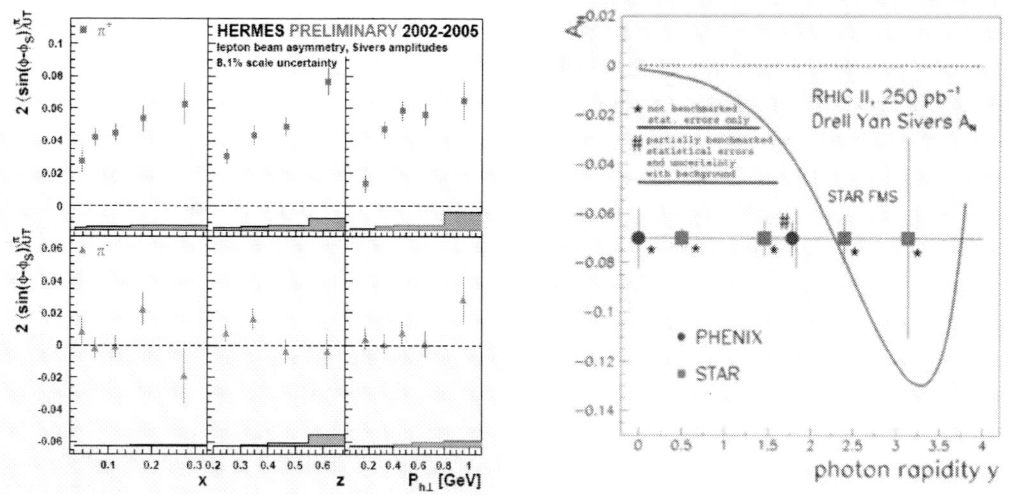

Figure 5. Left: Charged pion Sivers asymmetries measured by the HERMES experiment.
Right: Expected Drell-Yan TSSA A_N in p+p at RHIC (based on HERMES fit), the sign of the asymmetry should be opposite to the one observed in π^+ production in the polarized DIS process.

We expect the future high precision measurements of open heavy quark, J/Psi and Drell-Yan transverse SSAs will shed new light on our understanding of the origin of TSSAs, nucleon structure as well as the QCD dynamics, and could have a strong impact on the future EIC physics program.

3. Conclusion and outlook

The PHENIX experiment at RHIC has produced exciting physics results with transversely polarized proton beams. Significant none-zero asymmetries have been observed in the forward rapidity in π^0 production; in the middle and backward rapidities, the asymmetries are consistent with zero within experimental uncertainty in light hadron productions. PHENIX has also produced the first measurements of open heavy quark and J/Psi TSSAs in the electron and muon channels, and much improved measurements are expected with the upgrade silicon vertex detectors in the near future. Currently the PHENIX silicon vertex detector upgrade projects are in good progress, and part of the detectors will be installed as early as 2011. PHENIX upgrades also open up new physics opportunity, such as the Drell-Yan TSSA measurement that could provide a crucial test of a new fundamental prediction of QCD.

With upcoming detector upgrades for both PHENIX and STAR experiments and expected much improved accelerator performance, RHIC-Spin program will provide a great opportunity for new studies of polarized nucleon structure and the QCD dynamics.

References

[1] PHENIX Overview, NIM A499 469-479 (2003).
[2] Anselmino et al., Phys. Rev. D74, 094011 (2006).
[3] F. Yuan, Phys. Rev. D78 (2008).
[4] Kang, Qiu, Vogelsang and Yuan, Phys. Rev. D 78,114013(2008)
[5] RHIC Drell-Yan whitepaper, http://spin.riken.bnl.gov/rsc/write-up/dy_final.pdf

Longitudinal spin physics with the PHENIX detector at RHIC

David Kawall, for the PHENIX Collaboration

Department of Physics, University of Massachusetts, Amherst, MA 01003, USA, and RIKEN-BNL Research Center, Brookhaven National Laboratory, Upton, NY 11973, USA

E-mail: kawall@physics.umass.edu

Abstract. The Relativistic Heavy Ion Collider (RHIC) at Brookhaven National Laboratory has demonstrated the unique ability to collide beams of polarized protons at center of mass energies from $\sqrt{s}$=62.4 GeV to 500 GeV. The spin-dependent gluon distribution function of the proton, $\Delta g(x)$, can be determined from such collisions, through processes such as $p + p \to \pi + X$, $p + p \to \gamma + X$, $p + p \to \eta + X$, and others. The results of analyses for these kinds of processes in PHENIX will be presented, some of which provide strong constraints on the contribution of gluon spin to the spin of the proton.

Flavor-separated spin-dependent quark and antiquark distribution functions can also be extracted in polarized p+p collisions, from parity-violation in the production of W bosons, which we observe indirectly from their decay into leptons. Our progress in determining the W-production yield from our selection of $p + p \to e + X$ events will be presented.

1. Introduction

Polarized deep-inelastic scattering experiments conducted at SLAC, CERN, and DESY, revealed that quark and antiquarks carry roughly 25% of the spin of the proton [1]. This fraction is unexpectedly small, and understanding the remainder, which must come from the intrinsic spin of the gluons $\Delta g(x)$, and orbital angular momentum of the quarks L_q, and gluons L_g, presents an outstanding challenge to both theorists and experimentalists.

The Relativistic Heavy Ion Collider (RHIC) at Brookhaven National Laboratory has the unique ability to collide beams of highly polarized protons. This has provided experimentalists with a new tool to access, with leading order sensitivity, the gluon contribution to the spin of the proton [2]. In addition, in 2009 RHIC reached $+pp$ center of mass energies of 500 GeV, where measurements of Δu, $\Delta \bar{u}$, Δd, and $\Delta \bar{d}$ are possible by exploiting parity-violation in the process $pp \to W^{\pm} + X$ [2]. This ability is important since $\Delta \bar{u}$ and $\Delta \bar{d}$, like Δg, are poorly known.

2. RHIC : The polarized proton collider

RHIC has two rings, each containing up to 120 bunches of polarized protons (though usually not all "buckets" are filled). In the interaction regions, a bunch from one ring collides with a bunch from the other ring, then these same bunches circulate in opposite directions 3833 m around the RHIC rings and collide again about 13 μs later [3].

To preserve the spin during acceleration and store, a pair of Siberian snakes are used in RHIC. Spin rotators can be used to rotate the spins from the stable vertical direction to the

Table 1. Summary of polarized proton runs at RHIC and integrated luminosity recorded by PHENIX.

Run	Year	$\sqrt{s}$ (GeV)	Polarization	Recorded Luminosity
1,2	2001,2002	200	14%, Transverse	0.15 pb^{-1}
3	2003	200	34%, Longitudinal	0.35 pb^{-1}
4	2004	200	46%, Longitudinal	0.12 pb^{-1}
5	2005	200	47%, Longitudinal	3.4 pb^{-1}
			47%, Transverse	0.4 pb^{-1}
6	2006	200	55%, Longitudinal	7.5 pb^{-1}
		62.4	50%, Longitudinal	0.08pb^{-1}
8	2008	200	44%, Transverse	5.2 pb^{-1}
9	2009	500	34%, Longitudinal	11 pb^{-1}
		200	56%, Longitudinal	16 pb^{-1}

radial or longitudinal directions at the interaction region. In the most recent $p+p$ run in 2009, RHIC ran at $\sqrt{s}$=500 GeV for 30 days, and achieved 35% polarization and a peak luminosity of 8.5 × 10^{31} $cm^{-2}s^{-1}$, and store average of 5.5 × 10^{31} $cm^{-2}s^{-1}$. RHIC also ran at $\sqrt{s}$=200 GeV, at 56% polarization. The polarization of the beams was measured every few hours using proton-carbon CNI polarimeters, which are calibrated using a polarized hydrogen gas jet target to better than 5% absolute accuracy. A summary of the polarized proton runs at RHIC is given in Table 1.

3. The PHENIX detector

The PHENIX detector (see Fig. 1) comprises two forward detectors and four spectrometer arms [4]. The forward detectors (beam-beam counter (BBC) and zero-degree calorimeter (ZDC)) determine the collision vertex position and time, are calibrated to determine the proton beam luminosity and polarization (ZDC plus some scintillators are used for this), and can form a minimum bias trigger. The four spectrometers have specialized functions. The east and west "central" arms have tracking, particle identification, and an electromagnetic calorimeter at pseudorapidity $|\eta| < 0.35$. The muon spectrometer arms are specialized for the detection of muons.

The east and west central arms are in a solenoidal field along the beam axis, and can identify electrons, photons, and charged hadrons in a pseudorapidity range $-0.35 < \eta < 0.35$ over $|\Delta\phi| = 90°$ in azimuth. Tracking is provided by drift chambers (DC), time expansion chamber (TEC), and pad chambers (PC), while particle identification is possible with a time-of-flight system, (TOF) and ring-imaging Cerenkov (RICH). Of particular importance is the electromagnetic calorimeter (EMCal). Individual calorimeter towers are made of lead scintillator (PbSc) or lead glass (PbGl) and subtend $\Delta\eta \times \Delta\phi \approx 0.01 \times 0.01$. This fine segmentation allows the two photons from π^0 decay to be resolved up to p_T of 12 GeV/c. Shower shape analysis extends this range beyond 20 GeV/c, and allows the two photon invariant mass peak from π^0 decay to be used for energy calibration. Fine segmentation is also important for distinguishing rare direct photon events from the background from photons originating from neutral meson decay. This allows the measurement of the theoretically clean (but statistically limited) direct-photon asymmetry [2].

The north and south muon arms subtend a pseudorapidity interval $1.2 < |\eta| < 2.4(2.2)$

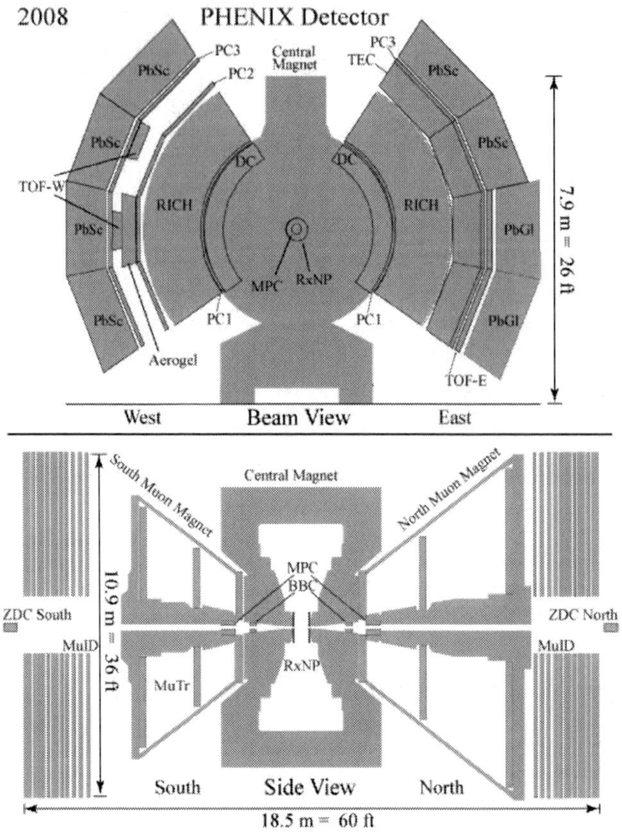

Figure 1.
Beam and side views of the PHENIX detector

respectively, with full azimuthal coverage. They have a hadron absorber, radial magnetic field, and cathode strip tracking chambers. Trigger electronics identify muons using layers of proportional tubes interspersed with hadron absorber. Of great interest to the spin program is the ability to trigger on muons from J/ψ decay, and high energy muons from W decay. The latter channel will allow the flavor separated measurement of the u and d polarized quark and antiquark distributions when an upgrade involving new planes of RPCs and new trigger electronics is complete.

In addition to its well-understood detectors, PHENIX has selective triggers for final states of interest to the spin program, including high p_T neutral pions, electrons, and photons. These events can be recorded at rates above 5 kHz with the PHENIX data acquisition system. Data analysis is done in parallel at the RHIC Computing Facility and the CCJ facility at RIKEN in Wako, Japan.

4. Physics Results

PHENIX is sensitive to Δg through measurements of the double helicity asymmetries in the production of particular final states, such as π^0. The asymmetries are defined :

$$A_{LL}^{\pi^0} = \frac{d\sigma_{++}/dp_T - d\sigma_{+-}/dp_T}{d\sigma_{++}/dp_T + d\sigma_{+-}/dp_T}$$

Journal of Physics: Conference Series **230** (2010) 012040 doi:10.1088/1742-6596/230/1/012040

where $\sigma_{++}(\sigma_{+-})$ is the production cross section for π^0 from $p+p$ collisions with like (unlike) proton helicities. In leading order this can be factorized as the sum of all partonic subprocesses $ab \to cX$ where parton c fragments into the detected π^0. In this framework, A_{LL} can be understood as the convolution :

$$A_{LL}^{\pi^0} = \frac{\sum_{abc} \Delta f_a \Delta f_b \hat{\sigma}(ab \to cX) \hat{a}_{LL}(ab \to cX) D_c^{\pi^0}}{\sum_{abc} f_a f_b \hat{\sigma}(ab \to cX) D_c^{\pi^0}},$$

where $f_a(\Delta f_a)$ are the unpolarized (polarized) parton distribution functions, and D_c^{π} is the fragmentation function of c into π. The spin-averaged partonic scattering cross-section for $ab \to cX$ is denoted by $\hat{\sigma}$, and $\hat{a}_{LL}$ is the analyzing power; both of which are calculable at next-to-leading order (NLO) in perturbative QCD (pQCD) [2].

The $A_{LL}(pp \to \pi^0 X)$ analyses have been PHENIX' most sensitive probes of Δg since we can trigger on, and reconstruct $\pi^0 \to 2\gamma$ with high efficiency. Also, at $\sqrt{s} = 200$ GeV, the cross-section is dominated by qg scattering for $p_{T\pi} > 5$ GeV, so the process is directly sensitive to Δg.

In the analyses, the invariant mass spectrum is formed from all pairs of photons, and the asymmetry is formed around the π^0 mass peak in a window from 112-162 MeV. The background fraction, and asymmetry in the background, are estimated from sidebands around the mass peak. Background fractions vary from 16% in the π^0 p_T range of 2-3 GeV, to 6% in the 9-12 GeV range at $\sqrt{s} = 200$ GeV.

In 2006, PHENIX recorded 0.08 pb^{-1} of $p+p$ collisions at $\sqrt{s} = 62.4$ GeV. This run was intended primarily to improve upon ISR results on inclusive pion cross-sections. At these low center of mass energies, NLO pQCD predictions of inclusive pion cross-sections tend to under-predict the observed yields. Including threshold resummation at next-to-leading-logarithm (NLL) accuracy improves the agreement with fixed target [5] and collider data [6]. The PHENIX measurement of the inclusive π^0 cross-section confirmed the necessity of such corrections [7]. In addition, a measurement of $A_{LL}(pp \to \pi^0 X)$ at $\sqrt{s} = 62.4$ GeV [7], made with one week of data, excluded the GRSV scenarios $\Delta g(x) = \pm g(x)$ [8] (see Fig. 2). The NLL predictions for $A_{LL}^{\pi^0}$ are slightly smaller than for NLO pQCD. The asymmetry measurement also extended our sensitivity to $\Delta g(x)$ to higher x in comparison to the measurements at $\sqrt{s} = 200$ GeV. While the collider luminosity is significantly worse at these low energies; the fact that the data can be used reliably in a global analysis, and the fact that the asymmetries become easier to measure (due to the approximate x_T scaling of the cross-sections and asymmetries) the author feels it can be advantageous to spend more running time at $\sqrt{s}$ =62.4 GeV in order to reduce the uncertainties on Δg.

From the 2005 and 2006 runs at $\sqrt{s} = 200$ GeV, PHENIX published results on A_{LL} of π^0 [9] (see Fig. 3). The asymmetries are consistent with zero, and lie below the GRSV standard model (which reflected the best fit to polarized deep inelastic scattering (DIS) data). This favors a smaller $\Delta g(x)$, and smaller integral Δg. An estimate of the integral of $\Delta g(x)$ was made from these data in the framework of the GRSV model. We use the notation $\Delta g^{[a,b]}$ to denote the integral of $\Delta g(x)$ from $a < x < b$. The polarized DIS data were refit using a constraint on the value of $\Delta g^{[0,1]}$, at a scale $\mu^2 = 0.4$ GeV^2, to extract a set of polarized parton distribution functions (PDFs) using the GRSV parametrization. From the resulting polarized PDFs, a prediction for $A_{LL}^{\pi^0}$ is made and compared with the measurement. The resulting χ^2 is plotted as a function of $\Delta g_{GRSV}^{[0.02,0.3]}$, evolved to $\mu^2 = 4$ GeV^2, in Fig. 3. The curve is not parabolic since π^0 can be produced from gg, qg, and qq' scattering. The former has quadratic sensitivity to Δg, leading to two minima in the χ^2 curve. The qg contribution is linear in Δg, breaking the symmetry. Including the statistical uncertainty, the results suggest $\Delta g_{GRSV}^{[0.02,0.3]} = 0.2 \pm 0.1(1\sigma)$ and $0.2^{+0.2}_{-0.8}$ (3σ) with an additional systematic uncertainty of

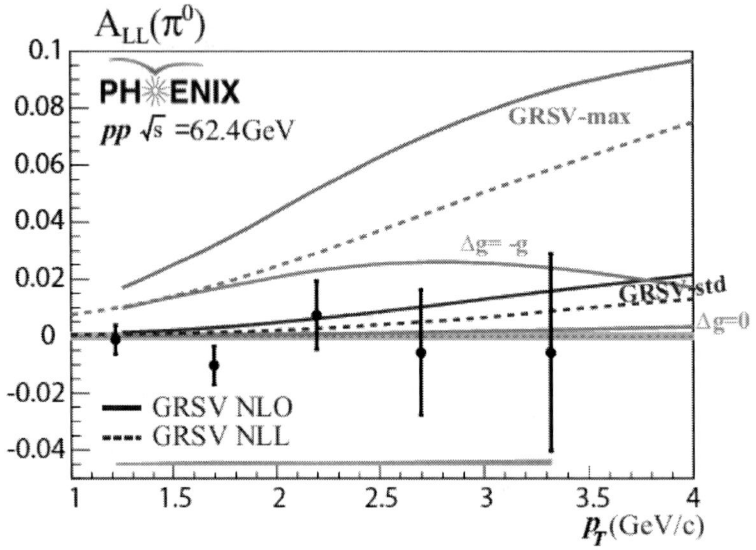

Figure 2. A_{LL} for π^0 at $\sqrt{s} = 62.4$ GeV as a function of p_T, showing statistical uncertainties. (14% uncertainty from beam polarization not shown.) Four GRSV NLO pQCD (solid curves) and NLL pQCD (dashed curves) are shown for comparison. Details are discussed in the text.

± 0.1 arising from the experimental uncertainty in beam polarization and relative luminosity. Using other parametrizations yields similar results, suggesting the asymmetry data are sensitive primarily to the integral $\Delta g^{[0.02,0.3]}$ and not to the specific shape of $\Delta g(x)$. A similar analysis showed that changes of a factor of 2 in the renormalization, factorization, and fragmentation scales used in the NLO pQCD prediction, lead to changes in the value of $\Delta g_{GRSV}^{[0.02,0.3]}$ extracted of $\pm 0.1(^{+0.1}_{-0.4})$ at $\Delta\chi^2 = 1(9)$. This suggest that calculations beyond NLO, while extraordinarily difficult, would be of great value in reducing the uncertainty on Δg.

The above results are consistent with a comprehensive global analysis of almost all polarized DIS and RHIC data [1, 10]. The RHIC data now pose the tightest constraint on $\Delta g(x)$ in the range $0.05 < x < 0.2$. The global analysis suggests a truncated first moment $\Delta g^{[0.001,1]} = 0.013^{+0.702}_{-0.314}$ at $Q^2 = 10$ GeV^2, and full first moment $\Delta g^{[0,1]} = -0.084$. These results are very interesting. First, they suggest that large, anomaly-inspired values for Δg are very unlikely. Second, the difference between the truncated and full first moment highlights the importance of extending the measurements at RHIC to lower x. This may be possible in the near future as PHENIX' barrel and forward silicon vertex detectors are installed, and a proposed forward calorimeter (covering $1 < \eta < 3$) is developed. These will increase the acceptance of the PHENIX detector substantially, and will allow new, correlated asymmetry measurements of γ-jet, hadron-jet, and jet-jet. These new observables will allow us to constrain the kinematics of the hard-scattered partons, and improve our sensitivity to $\Delta g(x)$ at lower momentum fraction x, perhaps to a few$\times 10^{-3}$.

The PHENIX collaboration also has preliminary asymmetry results on several other inclusive channels. The A_{LL}^{η} measurement has a slightly different sensitivity to Δg than π^0 since the flavor content of the η is different. The asymmetry measured is consistent with 0, and is shown in Fig. 4 along with GRSV model predictions [8]. PHENIX' measurement of A_{LL} of direct photons is shown also in Fig. 4. This asymmetry too is consistent with 0, and is heavily statistics limited. The virtues of this measurement are that it is dominated by qg scattering, and so is linear in

Journal of Physics: Conference Series **230** (2010) 012040 doi:10.1088/1742-6596/230/1/012040

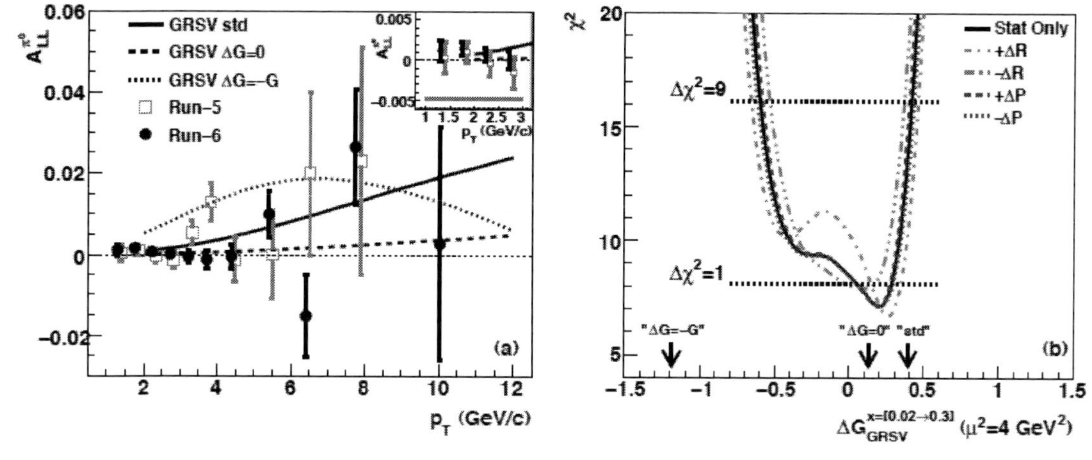

Figure 3. Left : A_{LL} for π^0 at $\sqrt{s} = 200$ GeV as a function of p_T, showing statistical uncertainties (8.3% uncertainty from beam polarization not shown.) Right : χ^2 profile as function of $\Delta g_{GRSV}^{[0.02,0.3]}$ using 2005 and 2006 $A_{LL}^{\pi^0}$ data.

Δg, and that the photon carries information about the hard scattering kinematics and is not the result of a fragmentation process.

In Fig. 5 is shown the $A_{LL}(pp \to h^+X)$ measured at $\sqrt{s} = 62.4$ GeV (where h^+ denotes a non-identified charged hadron (π,K,p)). These data were taken with the minimum bias trigger, with the momentum of the hadron measured using the drift chamber (DC). The PHENIX RICH has a threshold for $\pi^\pm$ of 4.7 GeV, so a RICH veto ensures the events are $h^\pm$ with $p_T < 4.7$ GeV. The GRSV model curves are shown on the plot. The GRSV max curve, corresponding to $\Delta g(x) = g(x)$ is not shown as it is clearly excluded by this data. Similar data exist for h^-.

At $\sqrt{s} = 200$ GeV, $A_{LL}(pp \to \pi^\pm X)$ has been measured at $p_T > 5$ GeV. Here we trigger on the EMCal, and look for associated tracks, measuring the particle momentum in the DC. The hadrons can be identified as pions as now they are above the RICH threshold. The charged pions are not as well measured as π^0 since PHENIX lacks a dedicated charged hadron trigger. Still, the channel is important because it has high analyzing power for Δg. This is apparent from two observations. First, at $p_T > 5$ GeV, π production is dominated by qg scattering, giving leading order sensitivity to Δg. Second, we can write $A_{LL}^{\pi^+} \propto \Delta g(x_1) \otimes (\Delta u(x_2)D_u^{\pi^+} + \Delta\bar{d}(x_2)D_{\bar{d}}^{\pi^+})$. Since Δu is large and positive, if Δg is positive, we expect $A_{LL}^{\pi^+} > 0$. Similarly, $A_{LL}^{\pi^-} \propto \Delta g(x_1) \otimes (\Delta d(x_2)D_d^{\pi^-} + \Delta\bar{u}(x_2)D_{\bar{u}}^{\pi^-})$. Since Δd is negative, this yields smaller asymmetries than for π^+. The asymmetry $A_{LL}^{\pi^0}$ will lie in-between. Thus we predict an ordering of the asymmetries $A_{LL}^{\pi^+} > A_{LL}^{\pi^0} > A_{LL}^{\pi^-}$ if $\Delta g > 0$. We also observe that $A_{LL}^{\pi^+}$ has the largest analyzing power for Δg of these channels.

Finally we note with great excitement that the W programs at PHENIX and STAR have their first data from the RHIC run in 2009 at $\sqrt{s} = 500$ GeV. PHENIX accumulated roughly 10 pb^{-1} of data during this 4 week run. The goals of the program (see [2, 11] for details), are to extract the flavor-separated spin-dependent quark distribution functions, in particular $\Delta\bar{u}$ and $\Delta\bar{d}$. These have been measured in semi-inclusive DIS experiments (SMC, HERMES, and COMPASS) but with limited accuracy. The virtue of the RHIC program is that these quantities will be be measured at very high momentum scales, without uncertainties due to fragmentation functions or possible higher-twist contributions.

With the recorded luminosity, PHENIX expects (before acceptance and efficiency cuts) $\approx$ 200 e^+ from $pp \to W^+ \to e^+\nu_e$ with $p_{Te} > 25$ GeV in the central arms and $\approx$ 35 e^-

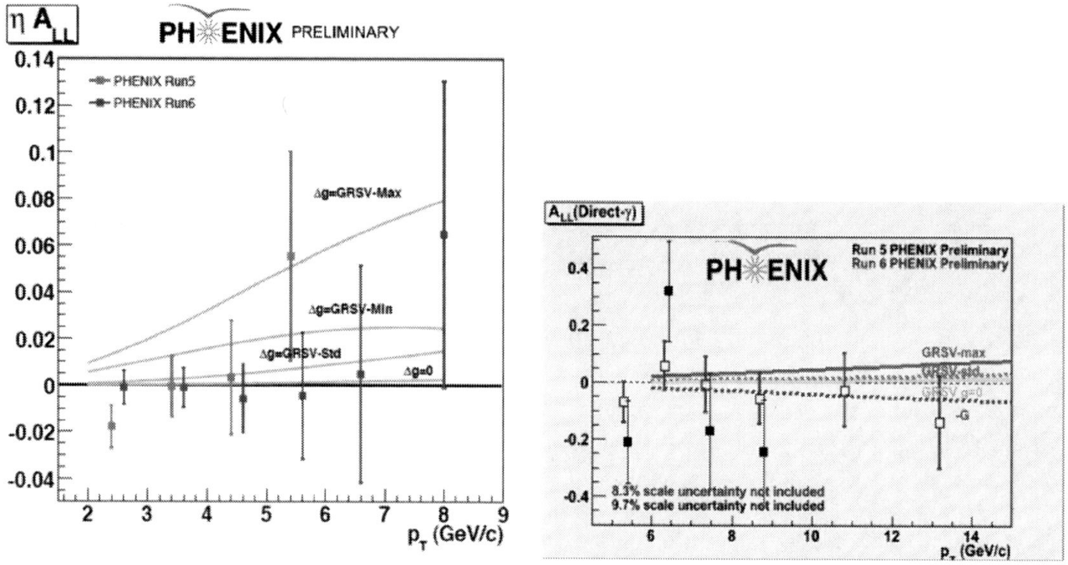

(a) (b)

Figure 4. **(a)** A_{LL} for η at $\sqrt{s} = 200$ GeV as a function of p_T, showing statistical uncertainties and GRSV model predictions (8.3% uncertainty in beam polarization not shown). **(b)** A_{LL} for direct photons at $\sqrt{s} = 200$ GeV as a function of p_T, showing statistical uncertainties and GRSV model predictions.

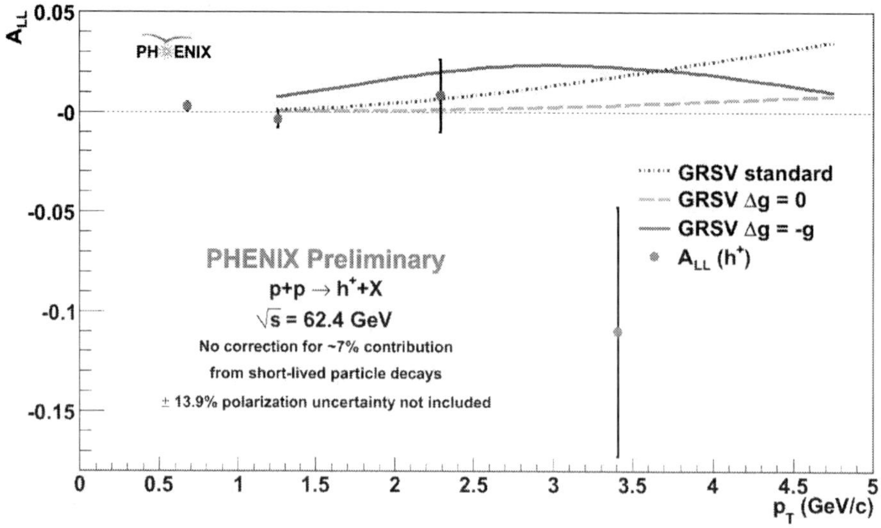

Figure 5. A_{LL} for h^+ versus p_T measured at $\sqrt{s} = 62.4$ GeV, showing statistical uncertainties, and GRSV model predictions.

with $p_{Te} > 25$ GeV from $pp \to W^- \to e^-\bar{\nu}_e$. In the asymmetry analysis, we look for > 25 GeV in the calorimeter with an isolated, high-momentum charged track passing a time-of-flight cut. The resolution of the DC is sufficient to distinguish e^+ from e^- even at $p_T > 40$ GeV, allowing us to identify the parent as W^+ or W^-. The main background is from $e^\pm$ from photon conversion from π^0 decay (plus a contribution from converted direct photons), with additional contributions from energetic charged hadrons showering in the EMCal. Other backgrounds include $e^\pm$ from heavy quark decay, cosmics, and accidentals. The process $pp \to Z \to e^+e^-$ constitutes $< 10\%$ background for W^+, and about 30% for W^-. We have clear evidence for a W signal above these backgrounds. After all cuts, and considering detectors inefficiencies and dead regions, we expect an uncertainty on the single spin asymmetry $\delta A_L^{W^+} \approx 0.3$. With a few more successful runs, this uncertainty should reach a few percent.

5. Summary

Results from the RHIC spin experiments, PHENIX and STAR, now pose the tightest constraints on $\Delta g(x)$, which is smaller than anticipated from polarized DIS experiments. There is still considerable uncertainty in $\Delta g(x)$ in the low-x ($x < 0.02$) region, which will be addressed by both experiments, though continued running and through upgrades in the future. The process $pp \to W \to e\nu$ has been observed at PHENIX, and cross-section and single-spin asymmetry A_L^W analyses are nearing completion. These are very exciting and productive times for the spin program.

References

[1] D. de Florian, R. Sassot, M. Stratmann, and W. Vogelsang, Phys. Rev. D **80**, 034030 (2009).
[2] G. Bunce, N. Saito, J. Soffer, and W. Vogelsang, Ann. Rev. Nucl. Part. Sci. **50**, 525 (2000).
[3] I. Alekseev *et al.*, Nucl. Instrum. Methods Phys. Res. Sect. A **499**, 392 (2003).
[4] K. Adcox *et al.*, Nucl. Instrum. Methods Phys. Res. Sect. A **499**, 469 (2003).
[5] D. de Florian and W. Vogelsang, Phys. Rev. D **71**, 114004 (2005).
[6] D. de Florian, W. Vogelsang, and F. Wagner, Phys. Rev. D **76**, 094021 (2007).
[7] A. Adare *et al.* (PHENIX), Phys. Rev. D **79**, 012003 (2009).
[8] M. Glück, E. Reya, M. Stratmann, and W. Vogelsang, Phys. Rev. D **63**, 094005 (2001).
[9] A. Adare *et al.* (PHENIX), Phys. Rev. Lett. **103**, 012003 (2009).
[10] D. de Florian, R. Sassot, M. Stratmann, and W. Vogelsang, Phys. Rev. Lett. **101**, 072001 (2008).
[11] P.M. Nadolsky and C.-P. Yuan, Nucl. Phys. B**666**, 31 (2003) (hep-ph/0304002).

26th Winter Workshop on Nuclear Dynamics IOP Publishing
Journal of Physics: Conference Series **230** (2010) 012041 doi:10.1088/1742-6596/230/1/012041

Transverse Spin Physics at STAR

Len K. Eun for the STAR Collaboration

Pennsylvania State University, University Park, PA 16802

Abstract. The large values of the Transverse Single Spin Asymmetry, A_N, seen in forward π^0 production from polarized proton collisions have stimulated important questions and have been studied in many QCD based transverse spin models. Previously, STAR has reported large A_N for forward π^0 production in the kinematic region where the cross-section is well described by pQCD. More recently, STAR reported the first RHIC measurement of A_N for forward η-meson production. The current analysis suggests that in this kinematic region, the η-meson asymmetry is larger than the already large π^0 asymmetry. Continued upgrades to the STAR forward calorimetry, leading to the commissioning of Forward Meson Spectrometer in 2008, provides STAR with greatly improved electromagnetic coverage, and opens up exciting future possibilities for the transverse spin program.

Transverse Single Spin Asymmetry

Large transverse single spin asymmetry (A_N) for forward meson production has been observed in previous experiments, most notably FNAL E704 [1,2] with its center of mass energy of 19.4 GeV. Over the years, several theoretical models were developed to explain the phenomenon in the framework of perturbative QCD, the best known of which are the Collins effect [3,4] and the Sivers effect [5]. Both models introduce small spin dependent transverse momentum (k_T) that generates the observed spin asymmetry. For the Sivers effect, the extra k_T comes from the spin dependent initial state parton distribution function, while for the Collins effect, it comes from the transversity dependence of the jet fragmentation. Consequently for the Collins effect, the asymmetry is in the hadronization relative to the jet axis, and full jet reconstruction should suppress this type of asymmetry. For the Sivers effect, the final state particle carries the asymmetry, and di-jets and γ-jets would not be exactly back to back. Unlike the Collins effect, the Sivers effect can generate spin asymmetry for prompt photons.

STAR Transverse Spin Physics

STAR Forward Pion Detector (FPD) is a modular lead glass calorimeter positioned at very forward regions of the STAR wide angle hall. Until RHIC run 6, the FPD was placed at both east and west ends of the interaction point. The main modules for each side were two 7x7 lead glass arrays positioned on both sides of the beam pipe. The distance from the modules to the beam was varied to sample a range of pseudo-rapidity points from 3.3 to 4.1.

STAR has reported both cross-section [6] and transverse single spin asymmetry (A_N) for forward neutral pion production [7]. Cross-section was measured at pseudo-rapidity of 3.3, 3.8. and 4.0. A_N was measured at pseudo-rapidity of 3.3 and 3.7. The data were collected during RHIC run 3, 5 and 6 polarized proton runs with $\sqrt{s}$=200 GeV.

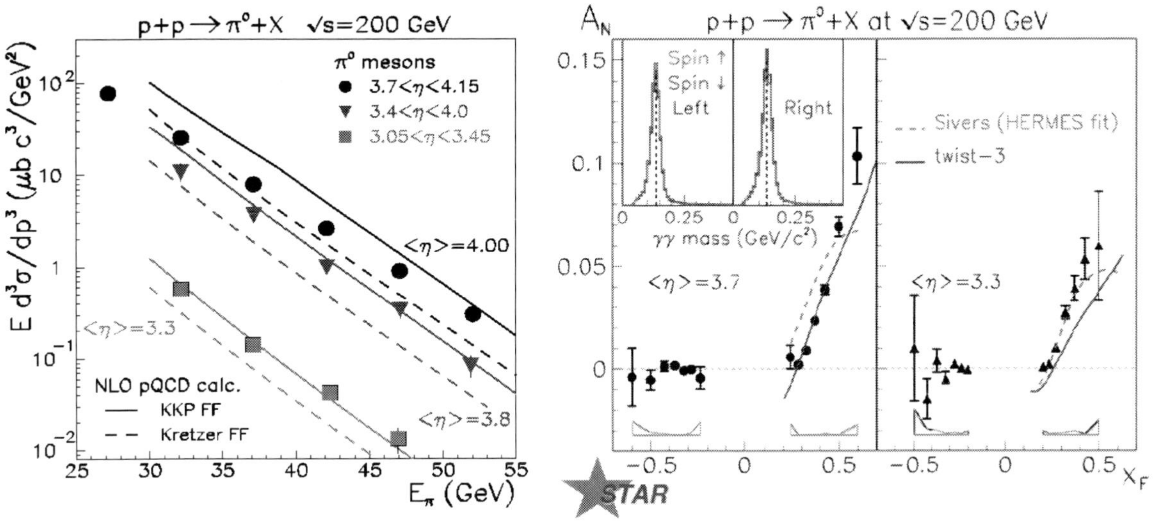

FIGURE 1. LEFT: π^0 Invariant Cross-section at $\sqrt{s}$=200 GeV [4]. RIGHT: Transverse Single Spin Asymmetry (A_N) vs. x_F for forward p0 production [7].

As shown in the left-hand panel of figure 1, the π^0 cross-section at η=3.3 and 3.8 were found to be in good agreement with the NLO pQCD predictions. In the same kinematic region, A_N was found to be large and positive as illustrated in the right-hand panel of figure 1. The result confirmed that the previously observed large transverse spin effect persists to the RHIC energy where pQCD describes the cross-section well. Furthermore, the existing QCD based theoretical models can also explain the x_F dependence of A_N reasonably well. However, the p_T dependence of A_N, shown in figure 2, was found to be at odds with all current theoretical predictions. While theory in general expects A_N to fall with p_T, we found that A_N rises with the increasing p_T. This trend was confirmed by the more recent STAR preliminary result from RHIC run 8, which is also included in figure 2.

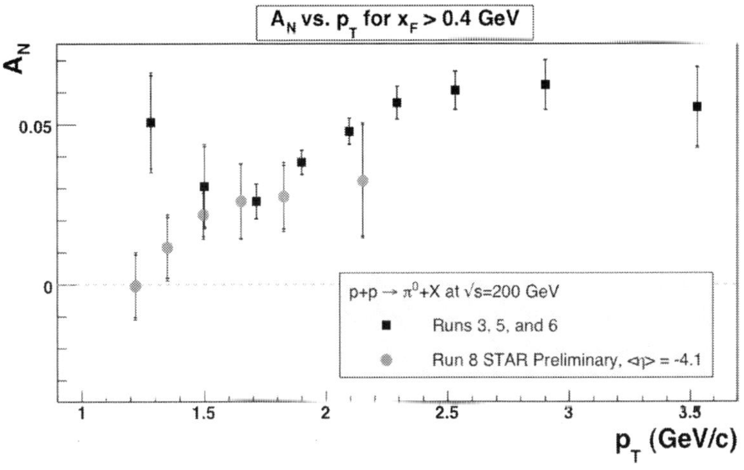

FIGURE 2. Transverse Single Spin Asymmetry (A_N) vs. Transverse Momentum (p_T) for forward p0 production. Black points have been published [7], while red points are preliminary [9]. Inner error bars are statistical errors, while outer error bars include both statistical and systematic errors.

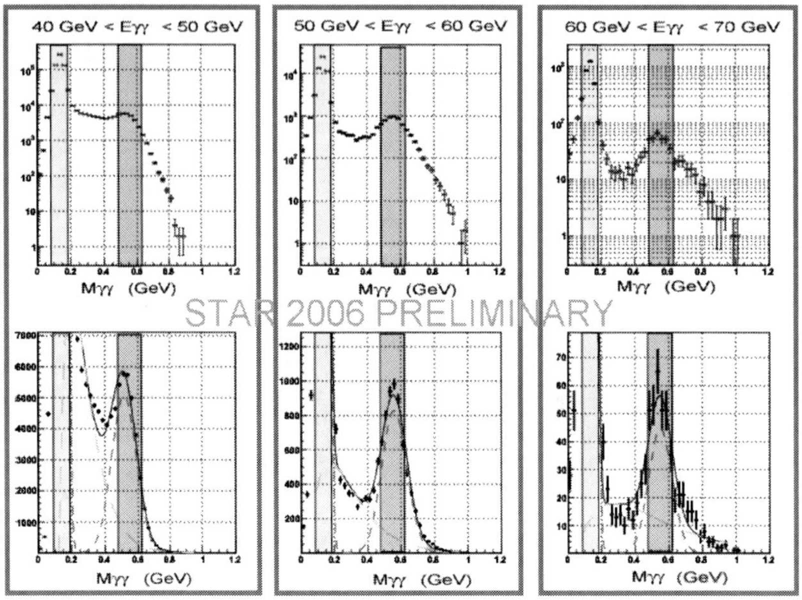

FIGURE 3. Di-photon Invariant Mass Distributions in Energy Bins. The upper three panels are in log scale. The bottom three panels show the corresponding distribution in linear scale to emphasize the η-meson signal.

In addition to the π^0, we observed η-mesons in the east FPD during RHIC run 6. Due to the relatively small physical size of the main FPD modules, (~27 cm square) we had very limited acceptance for the η-mesons at the range of x_F where previous π^0 measurements were performed. ($x_F<0.5$) However, at higher x_F, ($x_F>0.5$) we found sizeable η-meson signal in the FPD around the pseudo-rapidity of 3.7 through $\eta \rightarrow \gamma\gamma$ channel. This allowed us to measure the single spin asymmetry for the forward η-meson production for the first time at $\sqrt{s}$=200 GeV, up to x_F of 0.7.

During RHIC run6, the east FPD was positioned so that the average pseudo-rapidity of the π^0 was around 3.8. Events were triggered if the energy summed over any one of the two 7x7 arrays was nominally greater than 30 GeV. Due to the relatively small size of the FPD, the photon pairs from η-meson decay had a much greater chance of being contained in the FPD if the η-meson was heading towards the center of the detector. Consequently, we were able to enhance the η-meson signal by looking at photon pairs whose center of mass fell within a circle of radius 0.15 on the pseudo-rapidity and tan(ϕ) plane centered at (3.65,0). The contamination from the decays of heavier particles was minimal, as the opening angle for such decays would be too large for the FPD . Figure 3 shows the di-photon invariant mass distribution in three energy bins from 40 GeV to 70 GeV, with the aforementioned event cuts. We see the emergence of the η-mass peak as the energy increases. The data were collected from most of the RHIC run 6 transversely polarized proton collisions, which amounted to the integrated luminosity of about 6.8pb^{-1}. The average polarization for the projectile proton beam was 56%.

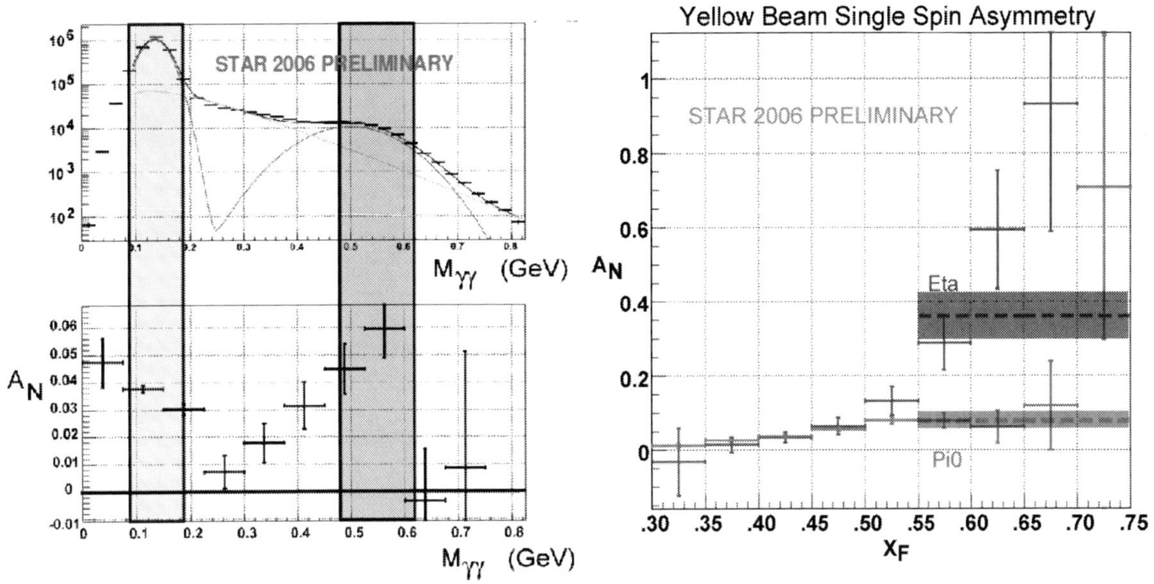

FIGURE 4. LEFT: A_N vs. Di-photon Invariant Mass. Shaded regions indicate the mass cuts for π^0 and η-meson. RIGHT: A_N vs. x_F for the π^0 and η mass regions, with 56% beam polarization. The dotted lines indicate weighted mean A_N for x_F>0.55, with the corresponding shaded area indicating the error of the mean.

The left-hand panel of figure 4 shows the preliminary result for transverse single spin asymmetry (A_N) as a function of the di-photon invariant mass. There is a clear structure in A_N that resembles the mass spectrum, with an "asymmetry valley" separating the two mass resonances. We then looked at the A_N as a function of x_F with mass cuts around the π^0 and η mass regions, as shown in the right-hand panel of figure 4. Despite the large statistical errors for high x_F bins, the surprisingly large A_N within the η mass region makes the measurement significant. We found that the average asymmetry for the η-mass region above x_F=0.55 is 4 standard deviation greater than that of the π^0. Similar measurement had been made by FNAL E704, which found the A_N for η-meson to be greater than π^0 for x_F>0.4, albeit with little statistical significance [8]. The result does not include background corrections, and therefore should be interpreted as average A_N in the given mass cut region. Systematic errors have not been calculated, but it is expected to be smaller than the statistical errors.

Since RHIC run6, the west FPD has received significant upgrades, leading eventually to the commission of the Forward Meson Spectrometer (FMS) during RHIC run 8. The FMS is a lead glass calorimeter similar to the FPD in components, but it consists of 1264 cells rather than 98 for the west FPD. Unlike the FPD, which only covered small regions on both sides of the beam pipe, the FMS covers full azimuth between η of 2.5 and 4.0. Along with the existing Barrel and Endcap calorimeters, STAR now has almost complete electromagnetic coverage from η of -1.0 to 4.0. The η–ϕ coverage map for STAR is shown in the left-hand panel of figure 5. The right-hand panel of figure 5 shows preliminary result for π^0 A_N, based on the data taken by the FMS during run 8. The result is consistent with the previously published measurements [7].

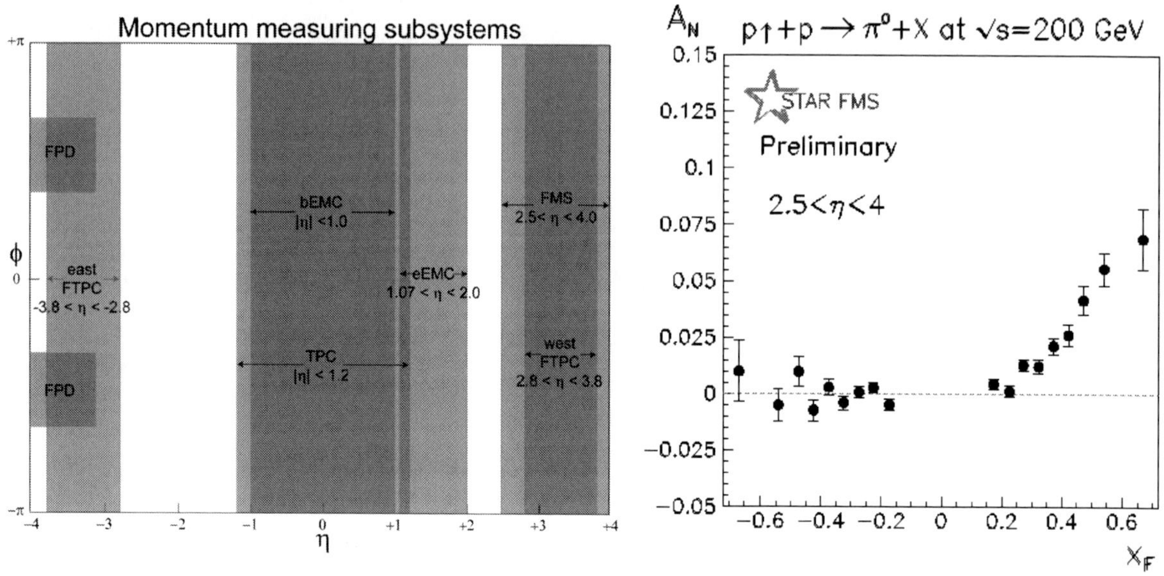

FIGURE 5. LEFT: Momentum Measuring Subsystems of STAR in $\eta-\phi$ space. From η of -1.0 to 4.0, the only gap in electro-magnetic coverage is between η of 2.0 and 2.5. RIGHT: Preliminary A_N for Forward π^0 Production measured with the FMS during RHIC run 8. The result is consistent with the previous measurements [7].

The dramatically improved coverage of the FMS opens up new opportunities for forward spin physics. With much larger area to serve as veto, FMS has the potential to measure forward prompt photons. In conjunction with the central detectors, this can also lead to the measurement of the Sivers type asymmetry for γ-jets. Another possibility is to disentangle the relative contributions from Collins and Sivers mechanism for the observed A_N for the π^0 and η. Furthermore, the FMS has the acceptance for neutral mesons heavier than the η, which provides the possibility of measuring spin asymmetry and cross-section for wider range of final states. With the potential future upgrades of charged particle tracking and/or hadron calorimetry in the forward region, full forward jet reconstruction and Drell-Yan are also future possibilities.

Summary

STAR has produced exciting results for the forward transverse spin physics. We have shown that A_N for forward π^0 production continues to be large at RHIC energy, where we can rely on pQCD to describe the cross-section. Our results suggest that while QCD based models can describe the x_F dependence of A_N, they fail to predict the observed p_T dependence. We have also reported the first measurement of A_N for forward η-meson production at RHIC energy, and found evidence that it is significantly larger than that of the π^0 around pseudo-rapidity of 3.7. With the vastly improved acceptance provided by the new Forward Meson Spectrometer, there are many new physics possibilities for the STAR transverse physics.

Journal of Physics: Conference Series **230** (2010) 012041 doi:10.1088/1742-6596/230/1/012041

REFERENCES

1. *D. L. Adams et al., Phys. Lett. B 261, 201 (1991).*
2. *D. L. Adams et al., Phys. Lett. B 264, 462 (1991).*
3. *J. Collins et al., Nucl. Phys. B 420, 565 (1994)*
4. *J. Collins, Nucl. Phys. B 396, 161 (1993).*
5. *D. Sivers, Phys. Rev. D 41, 83 (1990).*
6. *J. Adams et al., Phys. Rev. Lett. 97, 152302 (2006).*
7. *I. Abelev et al., Phys. Rev. Lett. 101, 222001 (2008).*
8. *D. L. Adams et al., Nucl. Phys. B 510, 3 (1998).*
9. *AIP Conf. Proc. 1149 517 (2009).*

26th Winter Workshop on Nuclear Dynamics IOP Publishing
Journal of Physics: Conference Series **230** (2010) 012042 doi:10.1088/1742-6596/230/1/012042

Quarkonia Measurements with ALICE

Frederick Kramer[1] and Christoph Blume[1] for the ALICE Collaboration

[1] Institut für Kernphysik, Johann Wolfgang Goethe-Universität Frankfurt, Max-von-Laue-Str. 1, D-60438 Frankfurt am Main, Germany

E-mail: kramer@ikf.uni-frankfurt.de

Abstract.
ALICE is the dedicated heavy-ion physics experiment at LHC. It is designed to provide excellent means to study the quark-gluon plasma, a deconfined state of matter assumed to be created under extreme conditions of temperature and/or baryonic density. A very promising observable is the measurement of quarkonia, including all the J/ψ, ψ' and Υ states. We will review the capabilities of ALICE to measure quarkonia and it's associated physics program. Furthermore we will present the current status of the analysis techniques and the results which can be obtained with the data collected during the first year of running.

1. Introduction

Bound hadronic states made of heayy quark-anti-quark pairs are called *quarkonia*. Due to their high masses, heavy quark pairs are expected to be created predominantly in the early stage of the collision of two heavy nuclei. Thus, quarkonia can be utilized as a very interesting probe to study the quark-gluon plasma (QGP).
The evolution of this state of matter is expected to take place in later stages of the collision and is therefore believed to modify the measured rates of quarkonia. Before being able to carry out such measurements in A-A, the elementary production processes have to be understood.

1.1. Elementary Reactions

Various theoretical models are supposed to destcribe the production of quarkonia in elementary reactions such as proton-proton collisions. Three of the most popular ones are the Colour Singlet Model (CSM), the Colour Octet Model (COM) and the Colour Evaporation Model (CEM). One of the main difference between the models is the way the necessity of coulor-neutrality of the final state is dealt with. This is also reflected by the models' names: In the CSM for example the bound state has the same quantum numbers as the initial $Q\bar{Q}$ pair; hence, only pairs that lead to a colour singlet state are allowed to bind [1, 2]. So far the theoretical predictions do not yet fully reproduce the experimental observations such as differential cross sections or polarization, or many free parameters limit their predictive power. The latter is the case e.g. for the CEM [2]. The fact that the CSM strongly underpredicts the measured yields, especially for the ψ'[3, 4] (see fig. 1), seem to indicate that colour octet states cannot be neglected [5].

Precise measurements at a new energy regime will give important constraints for predictions. ALICE [6] at LHC will contribute to such measurements.

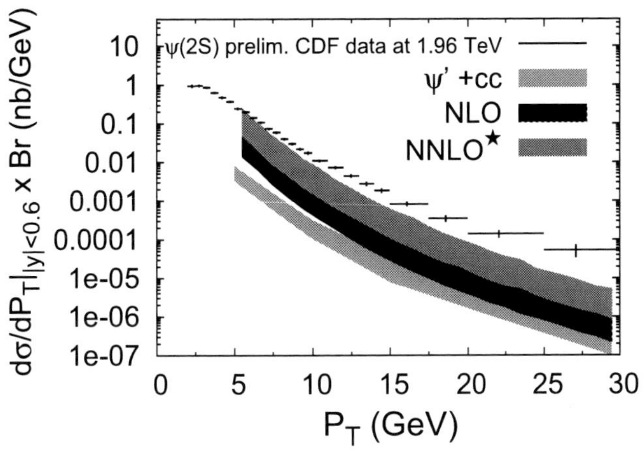

Figure 1. Comparison between CDF measurements of the cross sections of ψ' at $\sqrt{s} = 1.96$ TeV with predictions at various accuracies [3]

The above mentioned models describe the direct production of quarkonia. However, there are further sources for bound $Q\bar{Q}$ states and especially the J/ψ:

- Feed-down from higher mass quarkonium states, such as $\chi_c \to J/\psi + \gamma$
- b-hadron decays such as B $\to J/\psi + X$ or $\psi' + X$

For the understanding of quarkonia production these sources have to be taken into account. This subject will be addressed in 3.2.

1.2. Heavy Ions

In the collision of two heavy ions two further groups of effects might influence the measured rates of quarkonia and have to be taken into account: Cold nuclear matter effects (CNM) and QGP induced effects. The first group includes nuclear absorption, i.e. the breakup of pre-resonant $Q\bar{Q}$-pairs due to multiple scattering with nuclear matter surrounding the hot core of the collision scene at SPS energies.
The parton distribution function (PDF) of a free nucleon differs from the one of a nucleon bound in a nucleus. Depending on whether the PDF is suppressed or enhanced inside the nucleus at a given x this effect is called shadowing or anti-shadowing. Since the cross sections of quarkonia production directly depend on these distributions these modifications have to accounted for when comparing different collision systems as p-p and A-A. Such cold nuclear matter effects can be unveiled by measuring collisions in which no hot matter is created, such as p-A collisions.

Of most interest are the effects induced by the creation of a quark-gluon plasma. The first prediction of a modification of J/ψ yields in heavy ion collisions was put forward by T. Matsui and H. Satz [7]. The basic idea is the breakup of $Q\bar{Q}$ pairs in the hot environment of a quark-gluon plasma via Debye screening by free color charges, analogous to the well-known QED process. This effect would depend on the distance of the Q and the $\bar{Q}$ in the bound state which can be calculated from the binding energy of that particle. Depending on the temperature of the quark-gluon plasma and the corresponding Debye length the formation of such a state of matter would lead to a suppression of quarkonia rates in heavy ion collisions. By this the different quarkonia states can reflect the temperature of the quark-gluon plasma [8].

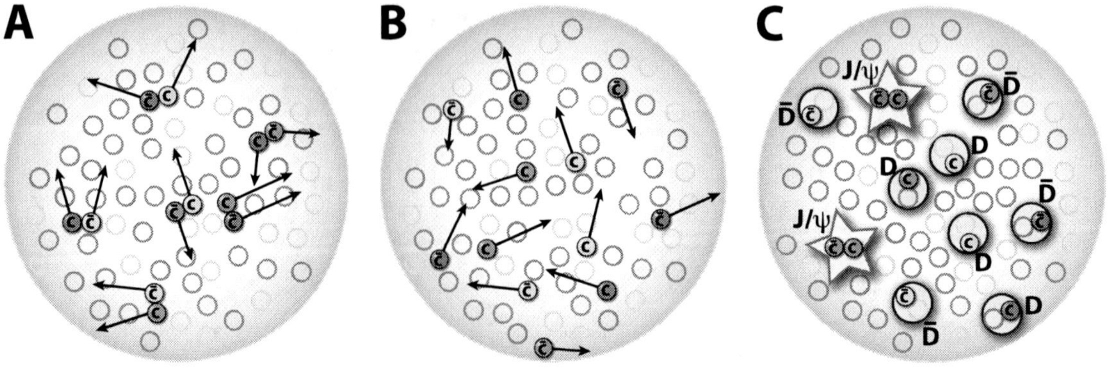

Figure 2. Sketch of the dissociation of correlated $c\bar{c}$ pairs (A), their diffusion (B) and the statistical production (C) of uncorrelated $c\bar{c}$ pairs in the quark-gluon plasma in heavy ion collisions

	$\sqrt{s_{NN}}$ (GeV)	$N_{c\bar{c}}$/ev. cent. AA
SPS	17.3	0.2
RHIC	200	10
LHC	5500	115

Table 1. Average numbers of produced $c\bar{c}$ pairs per central A-A collision for different accelerator energies (centre of mass)

A suppression of J/ψ yields has indeed been found at SPS and RHIC. But so far the interpretation of the results is not as unambigous as theoretically predicted, mainly because the CNM effects have not been fully understood yet [9].

Moreover, when going to higher accelerator energies, an additional effect may set in. The more $c\bar{c}$ pairs are created in one collision (see average values in table 1), after their dissociation, depicted in panel A of figure 2, and diffusion through the medium, panel B, the higher the probability for statistical production of uncorrelated c and $\bar{c}$ becomes. This effect might lead to an enhancement of the yields, especially of the J/ψ as depicted in panel C, fig. 2 [10, 11].

1.3. LHC - New Perspectives

The design values of the LHC collision energies are $\sqrt{s} = 14$ TeV for proton-proton collisions and $\sqrt{s_{NN}} = 5.5$ TeV for Pb-Pb, respectively. One nominal year of LHC data taking comprises approx. 10^7 s of proton-proton and 10^6 s of Pb-Pb running time. The maximum luminosities of the proton and the Pb beams in the ALICE experiment of $\mathcal{L}_{pp} \approx 5 \cdot 10^{30}$ cm^{-2}s^{-1} and $\mathcal{L}_{PbPb} \approx 5 \cdot 10^{26}$ cm^{-2}s^{-1}, 10^9 p-p and $2 \cdot 10^8$ Pb-Pb minimum bias events correspond to one nominal LHC running year at ALICE.

The LHC experiments will highly benefit from the dependence of the $Q\bar{Q}$ production cross sections on the collision energy. NLO calculations [12] as shown in fig. 3 predict a strong increase with the energy. Average expected numbers of produced $c\bar{c}$ pairs in central nucleus-nucleus collisions are summarized in table 1. Compared to the top RHIC energy for Au-Au, an order of magnitude higher rates are expected at the nominal LHC Pb-Pb collision energy. An even stronger increase with the energy is expected for $b\bar{b}$ production, see the right panel of fig. 3. Thus, the LHC will deliver excellent conditions for quarkonia measurements.

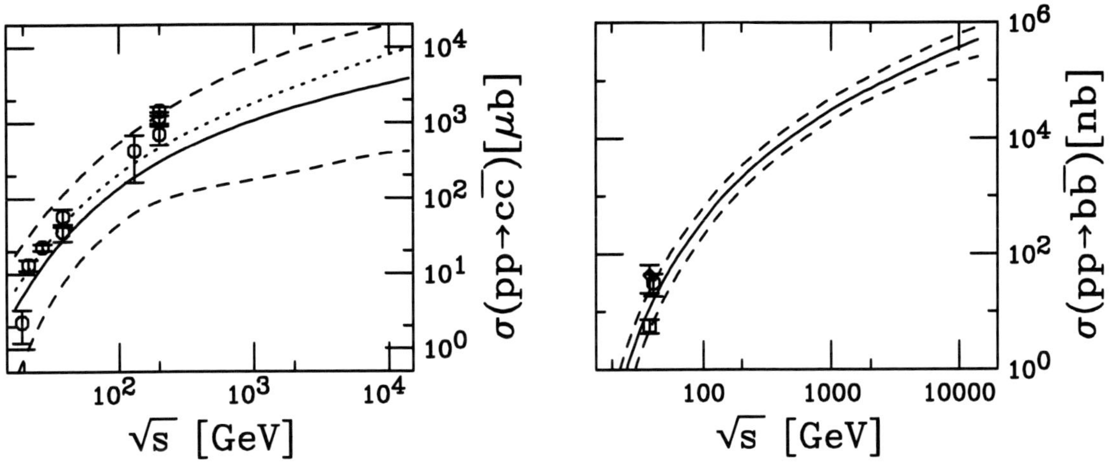

Figure 3. Extrapolations of the NLO total $c\bar{c}$ (left panel) and $b\bar{b}$ (right panel) production cross sections versus the collision energy in proton-proton collisions compared to measured data [12]. The solid curve is the central result, the dashed curves represent the uncertainty band. The dotted curve on the left panel is a calculation with different input parameters.

2. The Experiment

2.1. ALICE

ALICE is the dedicated heavy ion experiment at LHC [6]. Its general setup can be divided into two parts: The central barrel and the forward muon spectrometer.

The latter consists of a 10 plane tracking system with high granularity, partially inside a large area dipole magnet for momentum determination. Four planes of trigger chambers behind a filter wall provide a fast trigger decision on events with one or more opposite sign muon pairs with high p_t. The muon system is protected from hadrons and photons by a large front absorber and a beam shield. [13, 14] The muon system will measure J/ψ, ψ', Υ, Υ' and Υ'' in their dimuon decay channel. The longitudinal acceptance of the system is $-4.0 < \eta < -2.4$. Quarkonia will be measured over the whole range of transverse momenta down to $p_t = 0$.

The various detectors of the central barrel part are placed inside the L3 solenoidal magnet. In the following only the systems of importance for quarkonia measurements are discussed. All these cover the full azimuth and the longitudinal range of $|\eta| < 0.9$. Starting from the interaction vertex, the innermost detector is the Inner Tracking System (ITS). Consisting of two layers each of silicon pixel, silicon strip and silicon drift detectors it provides high precision primary and secondary vertex reconstruction, tracking and particle identification (PID). Surrounding the ITS, the cylindrical Time Projection Chamber (TPC) serves as the main tracking and particle identification detector. The Transition Radiation Detector (TRD) is the next subdetector which embraces the TPC. One of its main purposes is the identification of electron tracks. Since this is of main importance for the measurement of quarkonia the TRD will be introduced in more detail in the following.

2.2. Transition Radiation Detector

The two main purposes of the TRD [15] are PID in terms of electron-pion separation and triggering on high momentum electrons or electron pairs. The TRD will also be incorporated in the ALICE High Level Trigger which will allow full online data analysis, event selection and data compression. Therein complex trigger scenarios such as a trigger on a given e^+e^- invariant mass will be implemented.

The TRD is segmented in 18 super modules out of which seven are currently installed and operational. Each supermodule is further divided into five stacks with six layers. The left panel of fig. 4 shows a side view on one layer with a pion and an electron track and illustrates the working principle of the TRD. The particles have to traverse a radiator. Therein, particles with γ factors > 1000 are likely to emit transition radiation (TR). Electrons are fast enough already around 1 GeV/c whereas the pions, which are the main source of background, do not emit TR at momenta around a few GeV/c. The TR photon is absorbed in the Xe detector gas mixture and

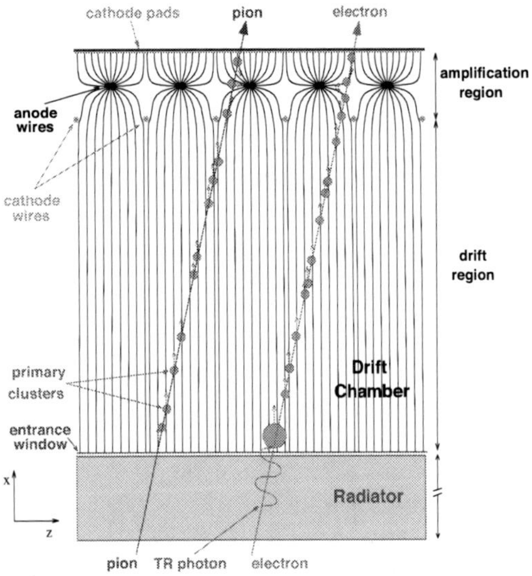

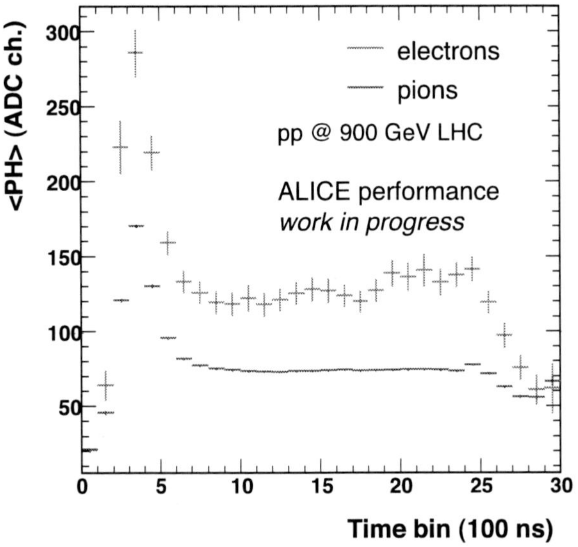

Figure 4. Left panel: setup and working principle of a ALICE TRD readout chamber. Traversing charged particles ionize the detector gas. Electrons create additional ionization clusters by transition radiation produced inside the radiator. After the drift of the electrons to the cathode wires, signals are created by their amplification at the anode wire plane [15]. Right panel: TRD signal (average pulse height) vs. drift time. This very first sample from 900 GeV p-p collisions contains mostly electrons (red) around 1 GeV/c which emit only litte TR. However, a clear rise at late drift times can already be seen.

leads to an enhanced signal at late drift times in the drift chamber of the TRD module (see right panel of fig. 4). By this the TRD reaches a factor of 100 in pion rejection (above $p > 1$ GeV/c) leading to an enormous reduction of the background in quarkonia measurements.

3. Performance Studies

3.1. Direct Quarkonia

ALICE will measure all quarkonia with the quantum numbers of the photon and mass below the $D\bar{D}/B\bar{B}$ thresholds by their dileptonic decay.

The main advantages of the muon system are the low background since only few other species than muons punch through the absorber. Further, the tracking and triggering stations are very fast detectors that can operate at high rates, triggered by dimuon pairs above a given momentum. Fig. 5 shows the expected invariant mass distribution of opposite sign muon pairs in one nominal year of p-p data taking at 14 TeV. In this simulation only the combinatorial background has been investigated. Very high signal rates bring along small statistical errors, the little combinatorial

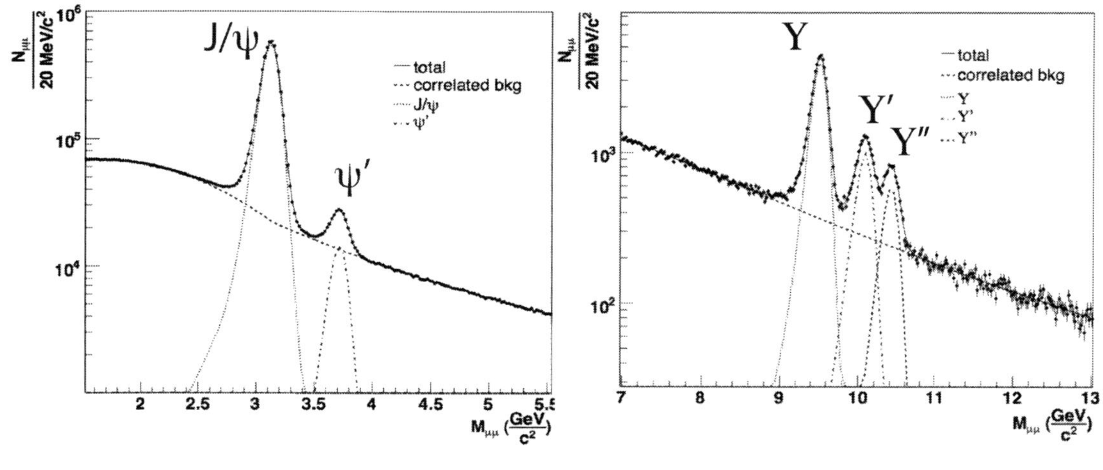

Figure 5. Simulated unlike-sign dimuon mass spectra for one nominal LHC p-p year at 14 TeV. The dashed lines represent the correlated background. The J/ψ and Υ mass regions are shown on the left and right panels, respectively [6].

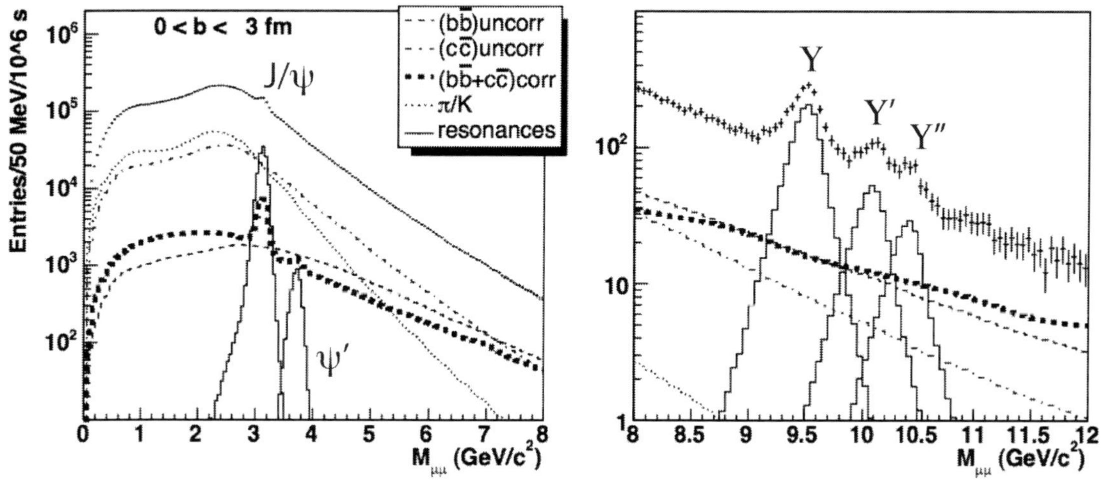

Figure 6. Simulated unlike-sign dimuon mass spectra of central events for one nominal LHC Pb-Pb year at 5.5 A·TeV. Points represent the total number of dimuons, the lines disentangle the different sources to the spectrum. The J/ψ and Υ mass regions are shown on the left and right panels, respectively [6].

background in p-p events result in good signal to background ratios. This holds for all analyzed quarkonia species, the numbers are summarized for J/ψ and Υ in table 2.

Analogous distributions for the most central ($0 < b < 3$ fm) Pb-Pb events (one nominal year of data taking) are shown in fig. 6. Less events will be recorded in comparison to proton-proton. Still, large signal rates are expected. Combinatorics and the larger underlying heavy ion event multiplicity lead to much more prominent background compared to p-p. Also, in this study more background sources have been taken into account and are represented by the dashed lines in fig. 6. Good measurements of the both ground states J/ψ and Υ will be possible already in one

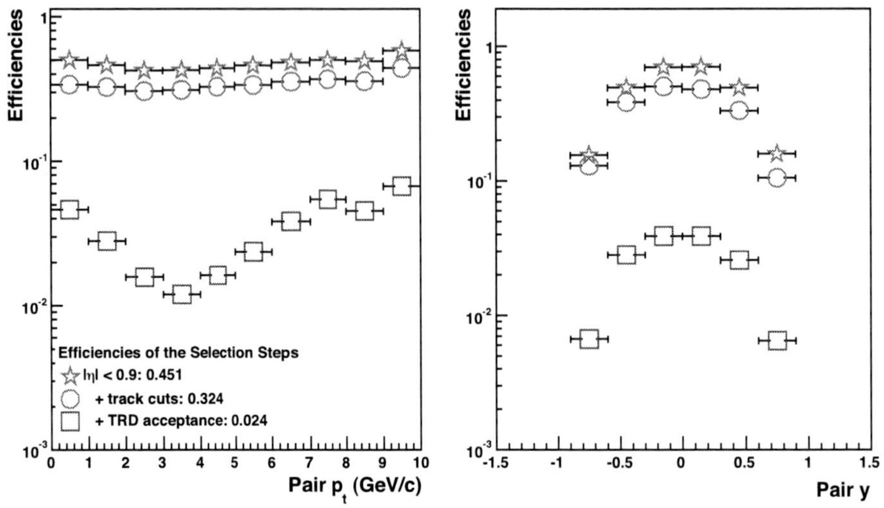

Figure 7. The efficiencies of J/ψ reconstruction of various reconstruction steps versus p_t (left panel) and y (right panel)

data taking period, with significances of around 150 and 30 respectively. Two or three data taking periods lead to comparable significances of the higher Υ states. For the ψ' the situation is worse, the signal is of the same order of magnitude as the the one of the Υ but the background is expected to be only a factor of two smaller compared to the J/ψ.

Since the detector does not cover the whole phase space and each step of the reconstruction performs only with a certain efficiency, simulations have to be carried out to correct for losses due to acceptance and reconstruction efficiency. Fig. 7 shows an example of the efficiency maps of various different reconstruction steps. Red stars represent the efficiency of measuring both J/ψ daughter tracks within the acceptance $|\eta| < 0.9$. For the distribution of the green circles the daughters also have to pass track quality cuts. The blue squares correspond to the additional requirement that both daughters hit the TRD. Since the TRD is not completely installed yet the efficiency drops about one order of magnitude when giving this requirement. Such maps are later applied to the measured yields. Signal to background ratios are under investigation both with TPC and TRD alone as well as both detectors combined.

In Pb-Pb collisions the situation in the dielectron channel is comparable to the one in the dimuon channel. The results of the simulation corresponding to $2 \cdot 10^8$ central events are shown in fig. 8 and summarized in table 2. In contrast to that study in the first year of Pb-Pb data taking the TRD will not be fully installed yet (see 2.2). Assuming that further the nominal luminosity will not be reached in the first running year the expected statistics therein are expected to be at least one order of magnitude lower than for one year at nominal conditions discussed here.
Due to the very little material budget of the central barrel detectors the mass resolution is even better, especially in the J/ψ mass region. On the other hand, the distributions have a long tail towards low masses due bremsstrahlung. The TRD drastically reduces the number of misidentified pions and with this the background of the measurement. The main source of the remaining background are semi-electronic decays of open charm and beauty particles.

3.2. Secondary J/ψ

As mentioned already in 1.1 a detailed investigation of all sources for quarkonia has to be carried out. Two substantial contributions for the total J/ψ yield are the radiative decay of χ_c and the

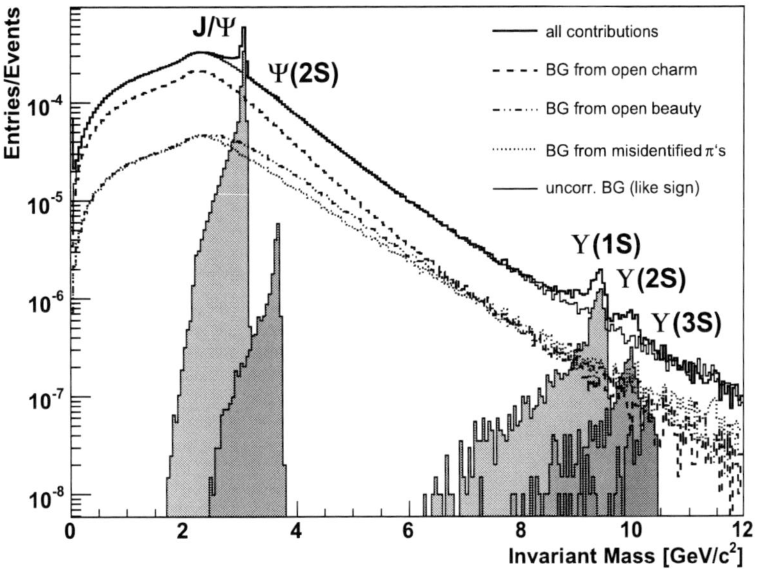

Figure 8. Simulated dielectron mass spectrum of 10% most central events for one nominal LHC Pb-Pb year at 5.5 A·TeV. The bold line represents the total number of opposite-sign dielectrons, the dashed lines disentangle the different sources to the spectrum, the thin line shows the distribution of the like-sign spectrum [16].

	$J/\psi \to \mu^+\mu^-$ dimuon triggered				$J/\psi \to e^+e^-$ centrality triggered	
	p-p		Pb-Pb		Pb-Pb	
	J/ψ	Υ	J/ψ	Υ	J/ψ	Υ
S	4.7M	44.7k	130k	1300	120k	900
S/B	12.6	5.8	0.2	1.7	1.2	1.1
$S/\sqrt{S+B}$	2081	195	150	29	245	21
Resolution (MeV/c^2)	~70	~100	~70	~100	~30	~90

Table 2. Summary of the expected signals, signal to background ratios, significances and mass resolutions for J/ψ and Υ in both dileptonic channels at top LHC energies with a fully installed TRD

weak decay of B mesons. Due to the excellent momentum resolution of the ALICE central barrel detectors and their acceptance down to very low p_t a total signal of around 7400 χ_c are assumed to be measurable in one nominal year of LHC p-p data taking [17]. Note that this estimation still assumes a perfect trigger, the real value will be reduced by the efficiency of that trigger. As depicted in the event display in fig. 9 the decay photon will be reconstructed by its conversion into an electron-positron pair in the detector material (mostly ITS). The probability of this process is expected in the order of 8.3%. The J/ψ is as well reconstructed by its dielectronic decay, the χ_c are finally identified in the resulting invariant mass spectrum. However, the difference in invariant mass $\Delta M = M(e^+e^-\gamma) - M(e^+e^-)$ provides a better resolution since the uncertainty of $M(e^+e^-)$ cancels out. As can be seen in fig. 9, the precision of the measurement is expected to be high enough to clearly separate the χ_{c1} and χ_{c2} states.

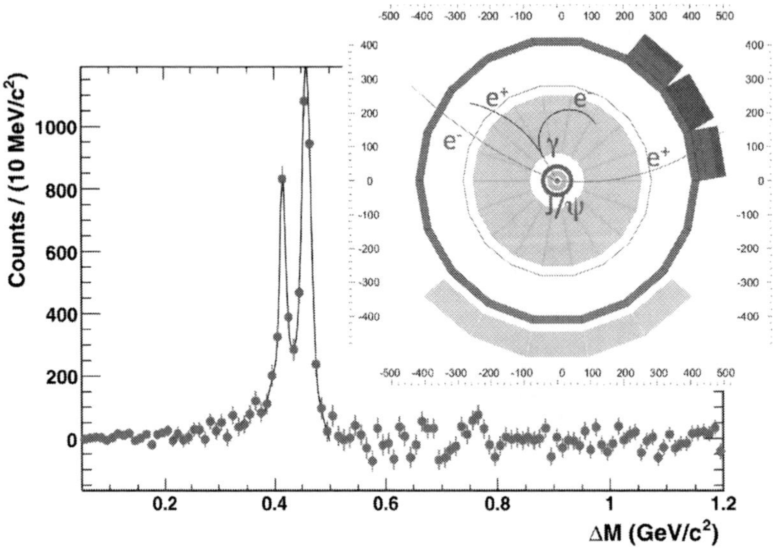

Figure 9. The $\Delta M = M(e^+e^-\gamma) - M(e^+e^-)$ invariant mass distribution with sharp χ_{c1} and χ_{c2} signals. The insert figure shows a typical event topology used for this kind of analysis [17].

Measurements of the fraction of J/ψ coming from b hadron decay will exploit the time scale of its weak decay process: With a $c\tau$ of 491.1 μm of the $B^{\pm}$, for example, the daughter J/ψ is likely to have a vertex displaced from the primary one. The measurement of that displaced vertex is carried out via the pseudo proper decay time $x = L_{\rm xy}\frac{M_{J/\psi}}{p_{\rm t}}$, where $L_{\rm xy} = \vec{L}\frac{\vec{p_{\rm t}}}{|p_{\rm t}|}$, the projection of the flight distance of the J/ψ, $\vec{L}$, on its transverse momentum [18, 19]. Secondary J/ψ lead to a asymmetric distribution of x as can be seen in fig. 10. The B fraction is then deduced from a simultaneous fit of the invariant mass and the pseudo proper decay time.

4. Summary

The LHC provides a very good environment for quarkonia measurements. ALICE will measure quarkonia of the J/ψ and the Υ family via their dielectron (around midrapidity) as well as dimuon (in forward direction) decay channel. It will be possible to analyze various different sources of secondary J/ψ. ALICE has a large acceptance in transverse momentum and will measure quarkonia down to $p_{\rm t} = 0$. Dedicated triggers will significantly enhance the signals in proton-proton and especially the Υ states in Pb-Pb. The very good mass resolution of the ALICE detectors will allow to clearly separate all different $Q\bar{Q}$ states.

References

[1] Berger E L and Jones D 1981 *Phys. Rev. D* **23** 1521–1530
[2] Lansberg J P 2006 *Int. J. Mod. Phys.* **A21** 3857–3916 (*Preprint* hep-ph/0602091)
[3] Lansberg J P 2009 *Eur. Phys. J.* **C61** 693–703 (*Preprint* 0811.4005)
[4] E Braaten, M A Doncheski, S Fleming and M L Mangano 1994 *Phys. Lett.* **B333** 548 (*Preprint* hep-ph/9405407)
[5] Cho P and Leibovich A K 1996 *Phys. Rev. D* **53** 150–162
[6] The ALICE Collaboration 2005 Physics Performance Report Volume II CERN/LHCC 2005-030
[7] Matsui T and Satz H 1986 *Phys. Lett. B* **178** 416
[8] Satz H 2006 (*Preprint* hep-ph/0602245)
[9] Arnaldi R (NA60) 2009 *Nucl. Phys.* **A830** 345c–352c (*Preprint* 0907.5004)

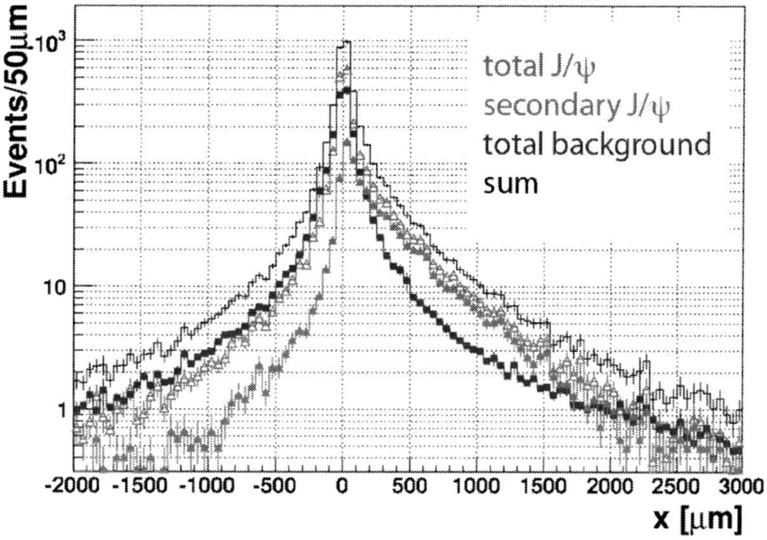

Figure 10. Distributions of the x variable, scaled from a larger statistical sample to the number of events corresponding to one nominal year of LHC p-p data taking. The total sum (black line) is composed of the background (blue squares) and signal (red open triangles) contributions. The latter is further consisting of secondary (magenta triangles) and prompt J/ψ [19].

[10] Andronic A, Braun-Munzinger P, Redlich K and Stachel J 2003 *Phys. Lett.* **B571** 36–44 (*Preprint* nucl-th/0303036)

[11] Thews R L 2005 *Eur. Phys. J.* **C43** 97–102 (*Preprint* hep-ph/0504226)

[12] Vogt R 2008 *Eur. Phys. J. ST* **155** 213–222 (*Preprint* 0709.2531)

[13] The ALICE Collaboration 1999 Dimuon Forward Spectrometer Technical Design Report CERN/LHCC 99-22

[14] The ALICE Collaboration 2000 Dimuon Forward Spectrometer Technical Design Report Addendum CERN/LHCC 2000-046

[15] The ALICE Collaboration 2001 TRD Technical Design Report CERN/LHCC 2001-021

[16] W Sommer, C Blume, F Kramer, J F Grosse-Oetringhaus 2007 *IJMPE* **16** 2484–2490 (*Preprint* nucl-ex/0702045v1)

[17] Gonzalez P, Ladron de Guevara P, Lopez Torres E, Marin A and Serradilla E (ALICE Collaboration) 2009 *Eur. Phys. J.* **C61** 899–903 and Erratum 915 (*Preprint* 0811.1592)

[18] D Acosta et al (CDF Collaboration) 2005 *Phys. Rev. D* **71** 032001

[19] Bruno G E (ALICE Collaboration) 2009 *J. Phys. G: Nucl. Part. Phys.* **36** 064053

26th Winter Workshop on Nuclear Dynamics IOP Publishing
Journal of Physics: Conference Series **230** (2010) 012043 doi:10.1088/1742-6596/230/1/012043

Forward Calorimeter Upgrades in PHENIX: Past and Future

Richard S. Hollis (for the PHENIX Collaboration)
University of California, Riverside, CA, 92521, USA
E-mail: rhollis@ucr.edu

Abstract. Over the past few years, the PHENIX detector has undergone several upgrades in the forward region ($1<|\eta|<4$). Initially covered only by the muon arms, the addition of the MPC and the future FOCAL expands on the physics capabilities of the PHENIX detector. The focus of these upgrades is toward a better understanding of the Color-Glass Condensate (CGC) and the interplay between the different components of the proton's spin valence/sea quark and gluon contributions. These proceedings highlight the latest results from the MPC related to the CGC as well the newly proposed forward calorimeter detector, FOCAL. Both detectors aim to constrain the current view of gluon saturation at small x in the Color-Glass condensate framework, through correlations using π^0's (MPC and FOCAL) and isolated direct photons (FOCAL) at high-p_T over a broad range of pseudorapidity.

1. Introduction

Over the course of the next decade, the Relativistic Heavy Ion Collider (RHIC) will study in detail several physics signals in order to understand the role of gluons within the nucleus. At the forefront of these tasks are the measurements of the gluon density at low-x in cold nuclear matter, the proton spin contribution from gluon polarization, to measure γ-jet correlations in Au+Au collisions, and test predictions for the relation between single-transverse spin in pp and those in DIS. The forward calorimeter upgrades will contribute towards these measurements through π^0 and direct-γ identification, at forward angles, with excellent angular and energy resolution. In these proceedings, the first of these challenges is discussed in detail, although all may be possible measurements with the forward detectors.

2. The PHENIX Detector

Currently, the PHENIX detector is comprised of two broadly defined regions: the central arms and the muon arms. The central arm measurements focus on charged hadrons and electromagnetic showers but have limited angular coverage in both ϕ and η, see Fig. 1. The focus of the forward region is μ (prominently toward J/ψ measurements and other similarly muon-decaying channels). For this, the coverage is symmetric in ϕ and extensive in η (Fig. 1). A series of recently implemented detectors (MPC), the in-progress forward silicon vertex detector, and planned FOCAL detector are paving the way *forward* to perform the measurements we desire to make over the next decade.

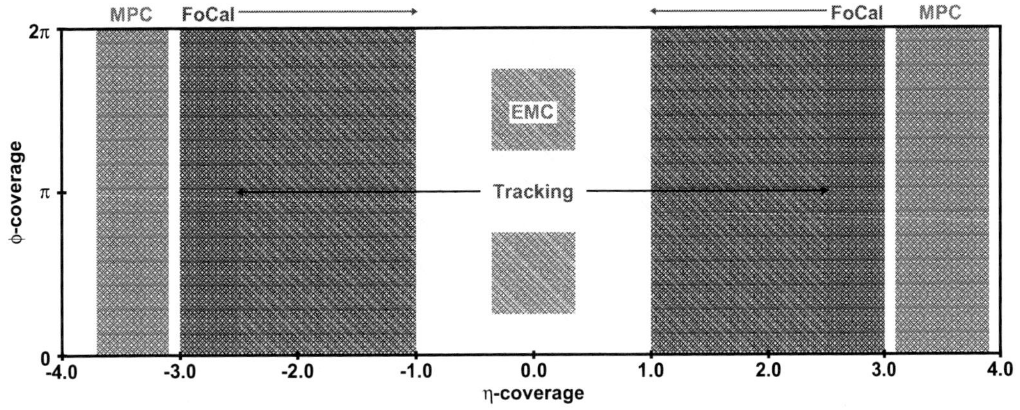

Figure 1. Future coverage of the PHENIX detector after upgrades. The actual coverage of the MPC (red) and EMC (green) are augmented by the proposed FOCAL (blue) and the vertex trackers (yellow hash).

3. The physics challenge

From the d+Au collision program at RHIC, several intriguing results have led theorists to new physics models. In the central region, a region well covered by PHENIX, an enhancement of intermediate-p_T particles (in central collisions relative to peripheral — or relative to pp collisions) is observed, associated to multiple-scattering of soft partons, prior to a hard interaction, giving rise to the *Cronin effect* [1]. In the more forward region a suppression is observed, associated to the *color-glass condensate* (CGC) [2]. The CGC is theorized to arise due to low-x partons, within the gold nucleus, where the nuclear wavefunctions overlap giving rise to saturation and the resultant suppression. This picture is consistent with observations made in electron-proton collisions at HERA [3]. The gold nucleus in d+Au collisions acts as an amplifier to this effect, owing to it's thickness. This allows an observation at a larger x than the former $e+p$ collisions, but still of the order of $x \sim 10^{-2} - 10^{-3}$. However, the observation still has to be made at "low-x", meaning that at RHIC this measurement will only be accessible at forward angles.

4. MPC contribution to physics goals

For the first collisions of d+Au, the PHENIX contribution to the CGC was limited in that the (then) detector measured charged hadrons and photons in the central region [4], later followed by punch-through hadrons [5] and J/ψ [6] measurements at forward angles. In 2006, a new detector, the Muon-Piston Calorimeter was installed. This calorimeter measures energy clusters which may be identified as π^0's through the direct reconstruction of the invariant mass.

The MPC is a Lead-Tungstate ($PbWO_4$) crystal calorimeter, approximately 18 cm ($20X_0$) long. In the transverse direction, the ~200 crystals are segmented into 2.2×2.2 cm blocks. The detector reconstructs π^0 in the low-p_T region (up to 1.7GeV/c). In reality, owing to the longitudinal boost, this is up to a total energy of about 17 GeV. Above this, as the photons from the π^0 decay merge into a single (inseparable) cluster, only single clusters are measured.

Once the clusters are calibrated and identified, each π^0 is correlated against a trigger π^0 (or charged hadron) from the central arms. Forming the correlation restricts the x range for the hard process to the range $0.006<x<0.1$. The *conditional* yields in the MPC are then compared to those from a similar analysis in pp interactions. Figure 2 illustrates the final ratio. For peripheral collisions, at a low number of binary (nucleon-nucleon) collisions, the forward yields are found to be similar to that in pp data (i.e. a ratio close to unity is observed). As expected

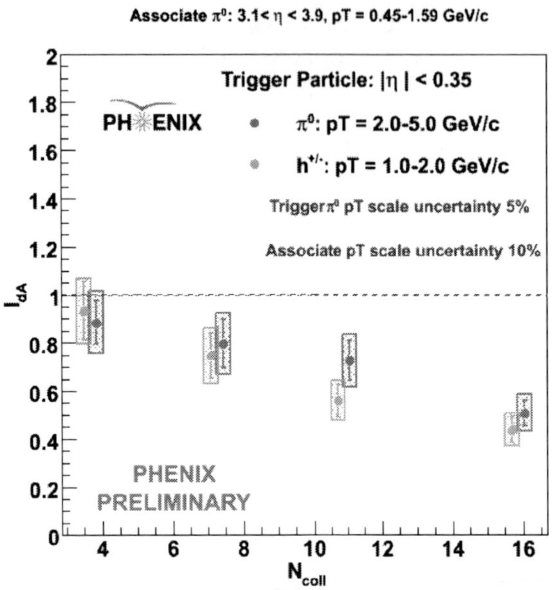

Figure 2. Ratio of conditional yields measured in d+Au collisions divided by that of pp collisions and scaled by the expected number of collisions from the d+Au data.

from the CGC, traversing a thicker nucleus – or thicker part of the nucleus – leads to a larger suppression. More central data (more binary collisions) is suppressed by a factor of two more than the peripheral. This suppression pattern is also observed by other experiments, whereby the central data (those in which the d undergoes the most collisions) are suppressed the most. More details can be found in Ref. [7].

5. A new FOward CALorimeter: FOCAL

To further our understanding of particle production at forward angles, indeed to access a more direct measure of the color-glass condensate, a new detector (FOCAL) has been proposed. It will measure direct photons at forward angles, and distinguish them from π^0's. Such a forward measurement, with identified γ, would be unique to the RHIC experiments.

5.1. The FOCAL Detector

FOCAL is a Tungsten-Silicon sampling calorimeter, approximately 16 cm in depth ($\sim 21X_0$) and located $\approx$40 cm from the nominal interaction point. The transverse direction is broken into small (6.2×6.2 cm) *supertowers* which are in turn subdivided into 4×4 pads on the read-out silicon wafers. Longitudinally, the supertowers are split into three segments of seven tungsten-silicon sandwiches (which are summed and read-out as a single signal). In the first segment, eight additional silicon strip wafers are included to facilitate the identification of the direct-γ. These are pairs of x- or y-oriented strips with 128×1 read-outs per wafer (compared to 4×4 in the pads). These strips are located at in the first segment of FOCAL – after $2X_0$, $3X_0$, $4X_0$, and $5X_0$ – in the region where the e.m. showers are in their infancy. Starting the read-out of the strips after two – rather than one – radiation lengths was chosen to optimize the conversion probabilities. The possibility of adding further strip read-out pairs (for example after $6X_0$ or $7X_0$) would not aid in the discrimination of high-p_T π^0 from γ, as the two γ showers from the π^0 begin to spread and merge into each other.

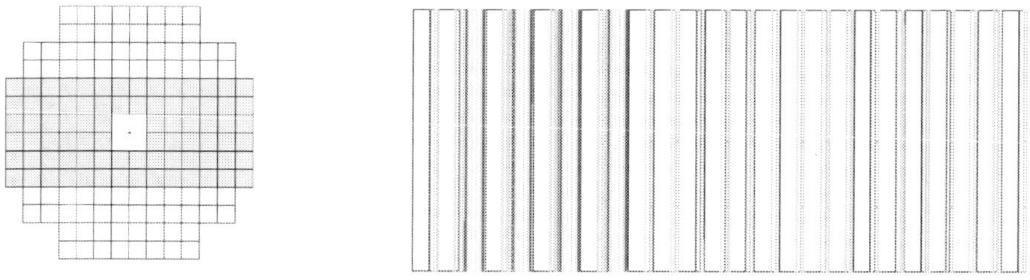

Figure 3. Geometrical layout of FOCAL. The left figure shows the beam-view of the detector, showing the brick structure. Each small square is read-out as a subdivided 4×4 pad. The right figure shows the longitudinal segmentation. The blue represents the tungsten, green shows the position of the pads silicon, and the red shows the strips.

5.2. FOCAL Acceptance

FOCAL will be positioned in the region $1<\eta<3$, with the possibility of full azimuthal acceptance. At present, it is proposed to only partially instrument the detector with silicon read-out, shown as yellow in Fig. 3. For reference, from this configuration, for $1.9<\eta<3.0$ ($1.5<\eta<1.9$) the full azimuth (0.8π) is covered.

The coverage of FOCAL in terms of the fraction of momentum carried by the interacting partons (x) is shown in Fig. 4. The leftmost panel shows the p_T dependence, which is weak, implying that differential measurements in p_T may not further restrict the measured x. The center panel of Fig. 4 shows the η dependence of x. A clear correlation of the gluon-momentum x (labeled x_2) with pseudorapidity is observed. By using FOCAL to probe different photon pseudorapidities, a narrow region of x can be probed. This is further shown in the rightmost panel, where the corresponding x_2 values from narrow slices of η are made. With this detector it is possible to probe x values as low as 10^{-3}. With the MPC (not shown) a slightly lower x

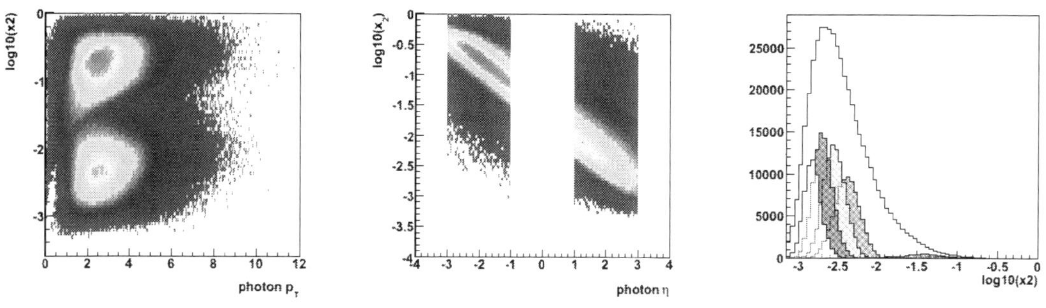

Figure 4. The x coverage of FOCAL. The leftmost panel shows x_2 versus p_T for forward (lower) and backward (upper concentration) photons. The center panel shows the η dependence and the rightmost panel shows x_2 for slices of η in FOCAL.

may be reached owing to it proximity at higher pseudorapidity.

5.3. FOCAL Reconstruction Methods

Reconstruction of showers in FOCAL is divided into two parts: (1) a global reconstruction of the energy, direction, and electromagnetic (e.m.)/hadronic shower discrimination followed by (2) an algorithm to identify π^0 and direct-γ signals. Although FOCAL does not have a mechanically projective geometry, the transversely segmented readout design allows for the tracking of particle showers through the detector in a projective manner: FOCAL is a tracking device. The first part of the reconstruction exploits this fact and utilizes all information available from the whole calorimeter. Clusters in pads and strip layers are used to determine a regression line through seven points in space, corresponding to each pad segment (three) and strip layer (four). A comparison of the candidate track to the features of single and multiple hits in that area completes this part of the reconstruction. However, *this* portion of the reconstruction cannot discriminate two close proximity γ's (for example from a high energy π^0) and single γ showers.

Isolated showers are expected to reach an energy resolution of $\Delta E/E \sim 23\%/\sqrt{E}$ and an angular resolution of better than 50 μrad. Discrimination between the e.m. and hadronic showers is made via a χ^2/NDF analysis of the energy deposition and widths in each of the segments, as well as the width in the strips. The expected mean and width is calibrated on a sample of "signal" (see below) simulations which are projected through a full GEANT description of the PHENIX detector and FOCAL. Once established, the same calibrations are applied to all tracks, whether e.m. or hadronic. A larger χ^2/NDF is found for hadronic showers, which either leave a MIP signal (low energy in a given layer) or have a wider shower profile (than e.m.) if a hadronic shower develops.

The second part of the reconstruction concentrates solely on the identification of direct-γ through the rejection of π^0's. A simplified Hough tracking algorithm is used to track straight line paths through the strip layers of the FOCAL. The x- (or y-) position of all strips in the

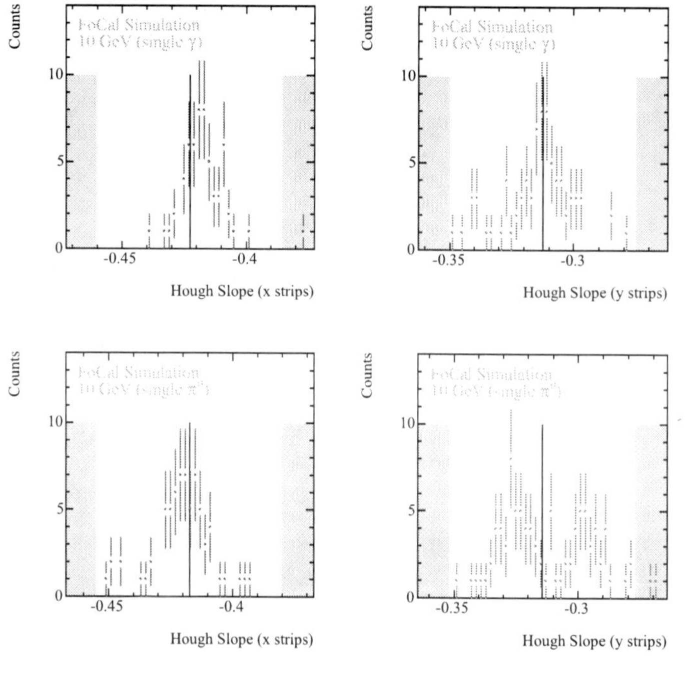

Figure 5. Strip reconstruction of single γ's (top) and single π^0's (bottom) for the x/y-direction (left/right). The line in the center represent the reconstructed center of gravity from the pads, the shaded region is the region excluded in the analysis of the strips.

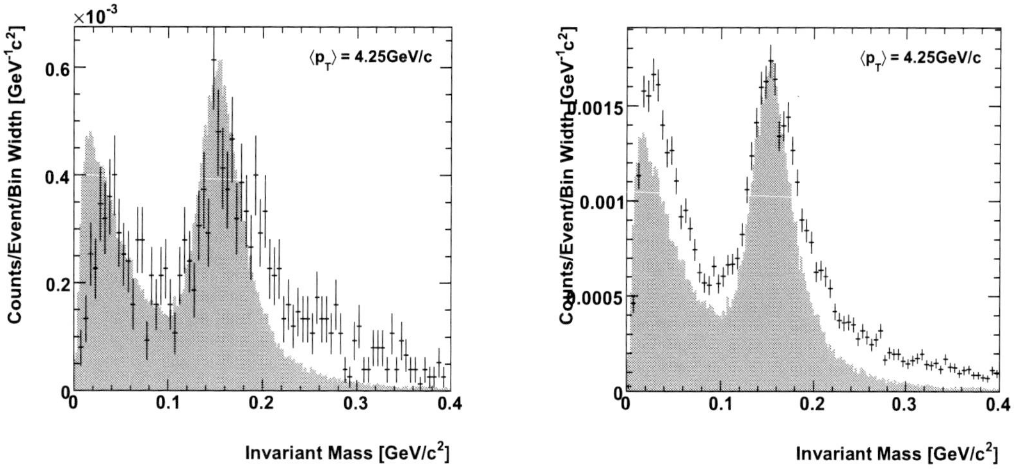

Figure 6. Invariant mass reconstruction of background events (black histogram) in d+Au (left) and pp (right) collisions both at $\sqrt{s_{NN}} = 200\,\text{GeV}$. For comparison, the grey histogram represents single-γ and π^0 simulations at the same energy. In each sample, the reconstructed transverse momentum is $4.0<p_T<4.5\,\text{GeV}/c$. This corresponds to a total energy of ~6.5 (at η=1) and ~26 (at η=2.5).

vicinity of a reconstructed cluster (in the pads as described above) are divided by their z-position to form a Hough parameter (in this case the slope). Note that only one Hough parameter is calculated, as the collision vertex position is explicitly used in the determination of the first parameter. All Hough parameters (slopes) are histogrammed ready to analyze. Figure. 5 shows the final strip-histograms used in the discrimination of γ and π^0. The top panels are single-input γ, the bottom are π^0. The left (right) panels show the x (y) strip layers, all x (y) layers are summed together. The line at the center represents the center-of-gravity reconstructed with the pads (first part described above) and the grey bands represent the edge of the excluded region in the strips analysis, which helps to reduce fake peaks in the full multi-particle HIJING/PYTHIA analysis. Clearly, a two-peak structure is visible in the π^0 (y strips) and is absent in the γ (x and y strips). In such cases, in conjunction with the reconstructed energy, an invariant mass is calculated. Invariant masses reconstructed close to (or above) the π^0 mass are rejected in the analysis. A secondary analysis uses the reconstructed masses close to the expected π^0 mass to directly reconstruct π^0's.

In the analysis of γ-jet, it is our aim to reject as much of the (predominantly) π^0 background as possible. To fully evaluate this rejection power of the FOCAL for direct-γ analysis, a sample of minimum bias HIJING was used as a background sample, and a second sample of minimum bias HIJING plus an embedded PYTHIA signal (γ+jet) was used to form a "signal" sample in $\sqrt{s_{NN}}$=200GeV d+Au collisions. Particle distributions resultant from both samples were propagated through full GEANT description of the PHENIX detector, including FOCAL. To evaluate the signal to background in d+Au collisions, the number of signal events were scaled to the level of that expected in data by a factor of $N_{coll}\frac{\sigma_{\gamma-jet}}{\sigma_{inel.pp}} \approx 5 \times 10^{-4}$. At such high E_T in HIJING, the background sample is mostly formed from fragmented jets, embedded γ+jets into the HIJING event are easily visible over the background at modest energies – from about E=10 GeV ($E_T \sim 2.7\,\text{GeV}$ at η=2) – and above this are always selected as the highest energy track (which is used in the analysis).

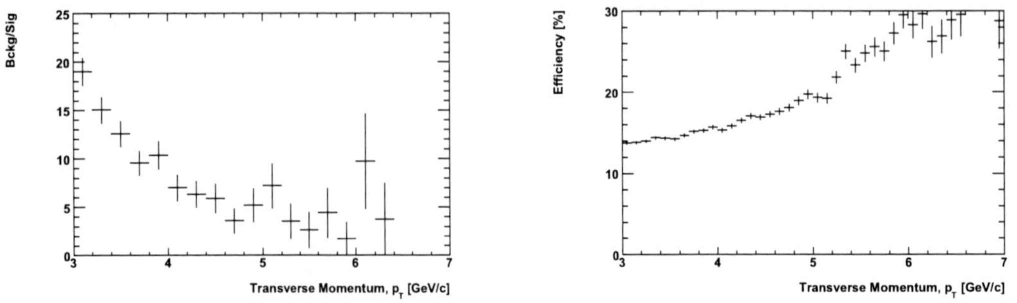

Figure 7. The left panel shows the ratio of background (all particles) to signal (direct-γ) reconstructed tracks, corrected for cross-section in d+Au collisions, on the deuteron-going side. The expected ratio is 5:1 for p_T>5 GeV/c. For the same p_T, the efficiency (shown on the right panel) is expected to be ~20%.

Figure 6 shows a comparison between the reconstructed background (from d+Au and pp collisions) and single-input π^0. The reconstructed invariant mass, even in the higher multiplicity d+Au background are only slightly altered, with a clear π^0 peak visible. The level at which the algorithm fails to reconstruct the π^0 mass is similar in both cases. In the final analysis, the candidate γ track will be correlated against tracks (as a proxy for jets) in the central arm and in FOCAL. This final candidate sample will contain a significant fraction (~80%) π^0 background, see Figure 7. Note that the correlation in this analysis is explicitly needed to calculate the momentum fraction x associated with the gluon. To evaluate this background, identified π^0's (for example those seen in Fig. 6) from FOCAL are also correlated to tracks in the central arm or FOCAL, in the same way as for the γ candidates. Analysis is underway to fully evaluate the expected systematic error on the π^0 measurement.

In summary, over the past few years PHENIX has made great strides toward forward measurements to study the properties of the collisions produced at RHIC. With the MPC, we observe a suppression of high-p_T particles (jets) as a function of nuclear thickness traversed in d+Au collisions. With FOCAL, we expect to expand on this by identifying direct-γ to further explore the color-glass condensate. This FOCAL measurement of direct γ's in d+Au collisions has been shown to reduce the background-to-signal of γ+jet events to a manageable level (5:1).

Acknowledgements

We thank the staff of the Collider-Accelerator and Physics Departments at BNL for their vital contributions. We acknowledge support from the Office of Nuclear Physics in DOE Office of Science and NSF (U.S.A.), MEXT and JSPS (Japan), CNPq and FAPESP (Brazil), NSFC (China), MSMT (Czech Republic), IN2P3/CNRS and CEA (France), BMBF, DAAD, and AvH (Germany), OTKA (Hungary), DAE and DST (India), ISF (Israel), NRF (Korea), MES, RAS, and FAAE (Russia), VR and KAW (Sweden), U.S. CRDF for the FSU, US-Hungary Fulbright, and US-Israel BSF.

References

[1] J.W. Cronin *et al.*, Phys. Rev. **D11**, 3105 (1975).
[2] see for example: E. Iancu, A. Leonidov, and L. McLerran, arXiv:hep-ph/0202270.
[3] H. Kowalski and D. Teaney, Phys. Rev. **D68** 114005 (2003).
[4] S. S. Adler *et al.*, Phys. Rev. Lett. **91**, 072303 (2003).
[5] S. S. Adler *et al.*, Phys. Rev. Lett. **94**, 082302 (2005).

[6] S. S. Adler *et al.*, Phys. Rev. Lett. **96**, 012304 (2006).
[7] B. Meredith *eat al.* arXiv:nucl-ex/0907.4832.

Journal of Physics: Conference Series **230** (2010) 012044 doi:10.1088/1742-6596/230/1/012044

Year-1 (Heavy-Ion) Physics with CMS at the LHC

Edwin Norbeck and Yasar Onel (for the CMS collaboration)
University of Iowa, Iowa City, IA 52242, USA
E-mail: edwin-norbeck@uiowa.edu

Abstract. The plans for the CERN LHC heavy ion program in 2010 are for collisions of lead ions during the month of November at an energy about half of the final 5.5 TeV/nucleon. The advanced preparations are mostly for extrapolations of the measurements at the maximum energy previously available, Au + Au at 0.2 TeV/nucleon. Because of the large increase in energy, surprises can be expected. One new feature made possible by the increased energy and the excellent muon energy resolution of CMS will be the study of the yields of the excited states of the upsilon meson as a function of angle and centrality. Although the main emphasis will be on the QGP formed by the overlapping parts of the Pb ions, the spectator parts and the electromagnetic field outside of the ions, γ-Pb and $\gamma\gamma$, are also important.

1. Introduction

This paper will discuss some of the plans and expectations for the heavy ion data from CMS [1]. Much of the material presented here is discussed in greater detail in [2]. Up to the time of the first Pb-Pb runs in November of 2010, the LHC [3] will feature p-p at $\sqrt{s} = 7$ TeV, which is half of the final 14 TeV. The same magnet settings for Pb will give $\sqrt{s_{\rm NN}} = 2.75$ TeV. At this energy, CMS will need to be prepared to deal with up to 572 TeV of energetic particles.

Initially the LHC Pb-Pb studies will be mostly extrapolations of the Au-Au program at RHIC. Because of the large increase in energy, a factor of 14, the experimental results may prove to be much more interesting than simple extrapolations. Examples of the extrapolations are high $p_{\rm T}$ suppression and jet quenching, azimuthal distributions (v2), and flow and temperature from energy spectra, all as functions of beam energy and impact parameter. At RHIC energies the quark-gluon plasma (QGP) is a low viscosity fluid, not a gas. The properties of the QGP at an energy 14 time larger may be quite different. New physics can also be expected from the spectator matter and from reactions of the intense photon field surrounding the Pb ions.

It is expected that the Pb-Pb collision rate during year-1 will be around 100 Hz. Although one would like a larger collision rate to be able to get better statistics for rare processes, the low rate does have advantages. The low rate will allow all minimum bias events to be written to mass storage. Unexpected event types that would be ignored by carefully prepared triggers are preserved for later discovery by enterprising physicists.

2. Ratios to p-p particle yields

Much of the RHIC data is presented as ratios of Au-Au to p-p particle yields. Accurate p-p yields at the $\sqrt{s_{\rm NN}}$ of the same energy as for the heavy-ion reactions are therefore essential to the heavy-ion program. The short, low energy p-p runs at $\sqrt{s} = 0.9$ and 2.36 TeV in 2009 are of considerable value to the heavy-ion program. They provide information about the variation with energy of transverse-momentum (p_T) and pseudorapidity (η) distributions, which are needed for the computer codes that predict the outcome of heavy-ion collisions. The 0.9 TeV data, which can be compared with other measurements, show that the CMS measurements are valid. The distributions at the energies needed for the heavy-ion calculations will be interpolated between the 2.36 TeV data and the 7.0 TeV data measured in 2010. It is fortunate for the heavy-ion program that the LHC did not start with p-p at $\sqrt{s}$ = 14.0 TeV. The Pb-Pb measurements will eventually be at $\sqrt{s_{NN}}$ = 5.5 TeV, but the measurements in 2010 are expected to be in the neighborhood of $\sqrt{s_{NN}}$ = 2.75 TeV. As was demonstrated at RHIC, it is good to have heavy ion measurements at more than one energy.

The energy for p-p will be 7 TeV for the first year or so because that energy does not require upgrading of the superconducting magnets. The $\sqrt{s_{NN}}$ = 2.75 TeV Pb-Pb experiments would use the same magnetic field settings. It is interesting to note that, in principle, the Pb-Pb luminosity could be larger at the lower energy. The Pb-Pb luminosity must be kept below the value for which the debris from the interaction point causes excessive heating of some superconducting magnet. The breakup of one Pb nucleus in the electromagnetic field of the other Pb nucleus and the pickup of an electron out of the cloud of electron-positron pairs have cross sections of hundreds of barns, and these effects increase with energy.

3. Limiting fragmentation

The spectrum of particles emitted at small angles with respect to the beam has the same shape independent of the reaction and the energy. This feature is demonstrated by plots of $\eta' = \eta$ - y_{beam} for small values of η'. The curves have the same shape for heavy-ion reactions for $\sqrt{s_{NN}}$ = 20 GeV to $\sqrt{s_{NN}}$ = 200 GeV. At the present time there is no good explanation for this phenomenon, also known as extended longitudinal scaling, so it will be most interesting to see if it holds with additional factors of 14 and 28 in energy.

4. Smaller x

The Bjorken x is essentially the momentum fraction that can be studied. Reaching small values of x requires a large energy and a small angle. As x approaches zero one might expect the number of low energy gluons to diverge. Because gluons can change into quarks, this effect would result in particle production rates to exceed unitarity at sufficiently high energy and small angles. This does not happen because, at a sufficiently high density, the gluons interact to form a classical field, described by the Color Glass Condensate (CGC) model. The values of x accessible with CMS using the full Pb beam energy will be almost two orders of magnitude smaller than at RHIC. Possible CGC effects that have been observed at RHIC will be studied in detail with CMS.

5. Hard probes

The large increase in the cross section for heavy quarks and other heavy objects can be seen in figure 1. Of these, the upsilon in its various excited state is of particular interest and will be considered in the next section. The accuracy of the CMS tracking (10 μm radially) is good enough to see the distance that a b meson travels before it decays. Even with central Pb-Pb collisions, the occupancy is expected to be at the level of 1% in the pixel detectors of the inner tracker and less than 20% in the outer silicon strip detectors [4].

Most of the hard probes have such short lifetimes that they leave the interaction region in the form of jets. A major data analysis task is the recognition of a jet and an accurate as possible measurement of its energy. Even with the high precision detectors of CMS, the energy resolution does not allow the separation of W jets from Z jets. The jets must be somehow separated from the large background, which is of the order of 5000 charged particles per unit of η. (The predicted numbers range from 1500-8000 at mid-rapidity.) There are two related

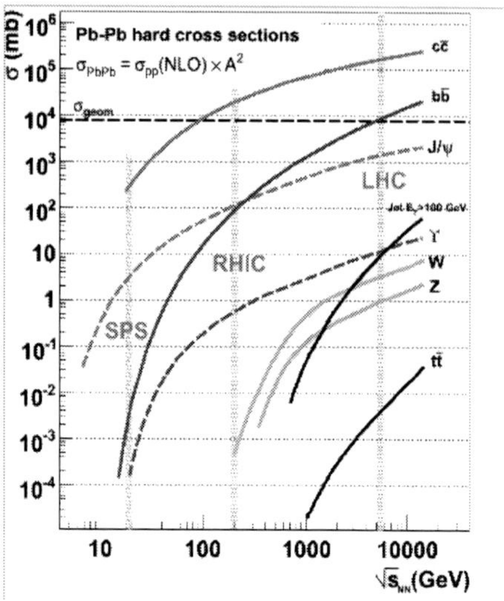

Figure 1. The cross section as a function of energy for the production of a number of interesting heavy objects using the Pb-Pb reaction.

problems. A jet can be hidden in the background, and a fluctuation in the background can be interpreted as a jet. Figure 2 shows the efficiency and purity as a function of jet energy using Monte Carlo simulations and the current jet finding programs. The efficiency is the probability that a real jet will be recognized, and the purity is the probably that a configuration that the program considers to be a jet is an actual jet and not a fake.

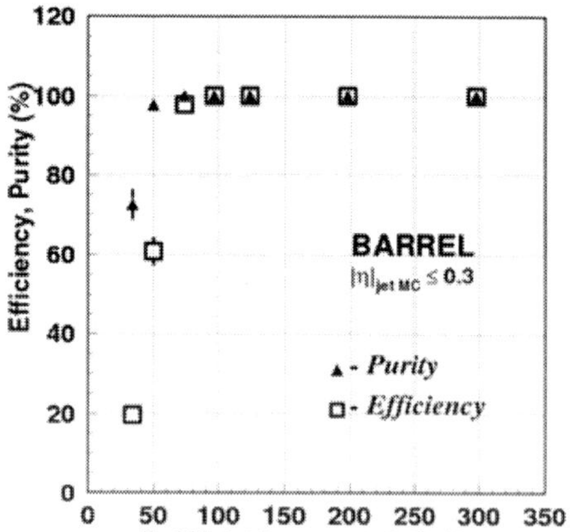

Figure 2. The efficiency and purity of the jets found by the jet finding programs assuming $dN_{ch}/d\eta = 5000$.

6. Upsilon states from muon energy spectra

The Compact Muon Solenoid (CMS) has exceptional capabilities for measuring muons. Figure 3 shows the nature of the measurements. The muons are first tracked through layers of silicon inside a homogeneous 3.8 T solenoidal magnetic field. Next they pass through the lead tungstate crystals of the electromagnetic calorimeter and then through the hadronic calorimeter, most of which is inside the magnet and consists of bronze absorbers and scintillating plastic. Beyond that is the part of CMS devoted exclusively to muons. It is made of thick layers of iron, shown in red, interleaved by a variety of detectors. The iron also provides a return path for the solenoidal magnetic field. Elaborate provisions are made to ensure that the position of every detector in

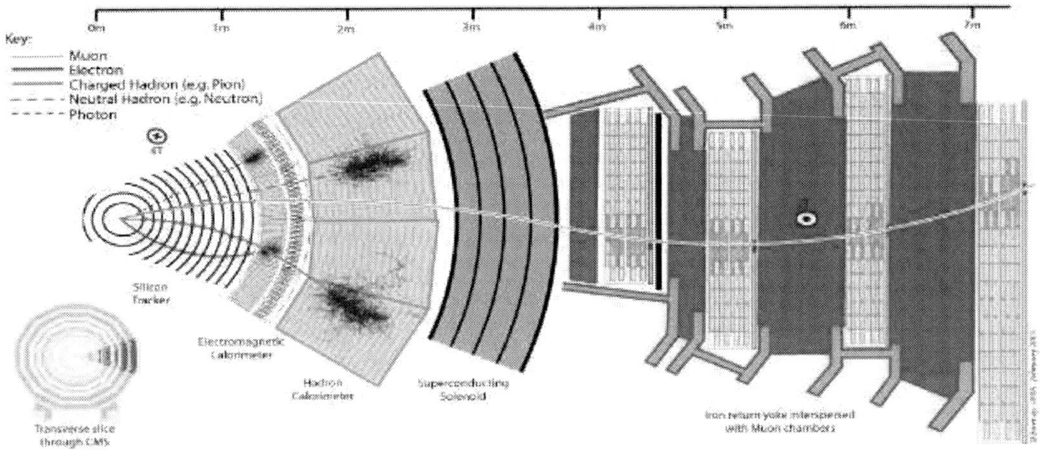

Figure 3. Schematic drawing showing how neutrons near mid rapidity are measured in CMS. The drawing does not show the part of the hadronic calorimeter just outside of the superconducting magnet.

the muon system is known to within a small fraction of a millimeter. This is made even more difficult by centimeter size elastic deformations caused by the magnetic field.

Figure 4 shows the excellent resolution of the first three upsilon states in CMS using muons from the decay $\Upsilon \to 2\mu$. The calculations made use of the known characteristics of CMS and actual measurements of cosmic ray muons. The calculations do not include the melting of the excited states in the hot quark gluon plasma. If the third state melts completely, the third peak would be only from excited upsilons formed at the surface. Much can be learned about the geometry of the hot interaction region by looking at the relative amplitude of the three peaks as a function of centrality, the angle with respect to the reaction plane, and other variables. With configurations for which there are large differences in the amplitude of the three peaks, the good resolution of the peaks is essential. Such excellent energy resolution is not available in any other experiment.

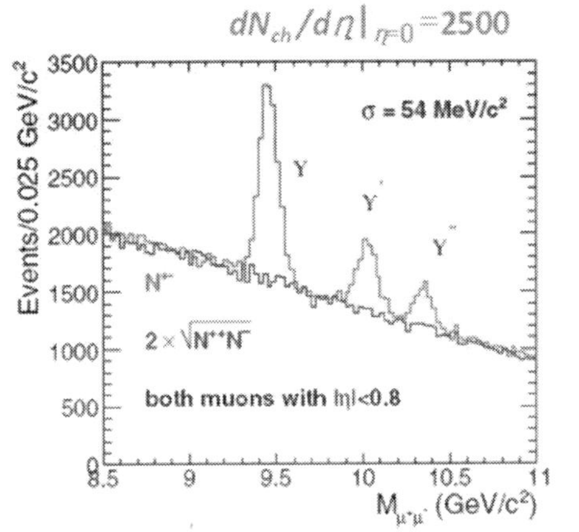

Figure 4. Invariant mass spectra of $\mu^+\mu^-$pairs in the Υ mass range. This simulation does not take into account the partial melting of the excited states in the hot quark gluon plasma.

7. Ultraperipheral reactions

There is an extremely large electromagnetic field around the Pb ion. A number that is easy to remember is 2×10^{20} gauss for the magnetic field midway between two Pb ions passing at 20 fm at the full $\sqrt{s_{NN}} = 5.5$ TeV.

The physics of ultraperipheral collisions at the LHC has been described in a 171 page document [5]. Here there is space only for a brief introduction. The Lorentz contracted electromagnetic field around a relativistic Pb ion can interact as a high energy photon. The equivalent photon spectrum flux increases with Z^2, which provides for Pb a $Z^2 \sim 7000$ enhancement factor compared to electron or proton beams. The first factor of Z is from Coulomb's law and the second from the Bio-Savart law. For photon-photon reactions the increase is proportional to Z^4. Photoproduction has been extensively studied at RHIC; but the LHC offers a much larger number of higher energy photons. Over a certain range, the radiation field of all 82 protons in the lead nucleus is coherent. The requirement of coherence places an upper limit on the maximum photon energy, which increases linearly with beam energy. At RHIC the maximum energy available from photon-photon collisions is about 6 GeV. For year-1 at the LHC the energy is 100 GeV, and with full energy Pb-Pb it is 200 GeV. The reaction $\gamma\gamma \rightarrow \Upsilon$ is well within the LHC energy range. The energy spectrum should look like figure 4, but with less background. With direct photoproduction of heavy quarks, the only contribution is from $\gamma g \rightarrow Q\bar{Q}$, thereby providing a measurement of the gluon distribution in the Pb nucleus along with nuclear shadowing effects. The total luminosity during year-1 will not be sufficient for many detailed studies, but measurements at $\sqrt{s_{NN}}$ of both 2.75 TeV and 5.5 TeV should help identify the extent of various background processes.

The large cross sections for the breakup of the Pb nucleus, mostly from γ-Pb, are sufficiently calculable that they can be used in luminosity determinations. A measurement of the rate for the production of one or more neutrons in both ZDCs (zero degree calorimeters) should provide a luminosity measurement with an accuracy of 2%[5].

8. CMS at small angles

The tracker and muon chambers cover the pseudorapidity region $|\eta| < 2.5$. Larger values of $|\eta|$ are covered by calorimeters, all of which have both electromagnetic and hadronic parts. With p-p at the design luminosity of $10^{34}\text{cm}^{-2}\text{s}^{-1}$ there can be two dozen interactions in one beam crossing. The tracker can localize a reaction to within 20 μm along the beam direction, which is more than enough to allow matching of each track to the interaction that produced it. For the calorimeters at $|\eta| > 2.5$ there is no direct way of knowing the source of a detected particle, but with heavy ions there are never two interactions in one beam crossing so that the small angle detectors are fully useful.

The construction details of these forward angle calorimeters are given in [1]. In all of them, charged particles are detected by photomultiplier tubes (PMTs) viewing Čerenkov light in pure quartz (radiation resistant synthetic fused silica). The absorber material is either iron or tungsten. As of this writing, all of these calorimeters are deciding how to deal with the occasional spurious large signal caused by charged particles hitting the front window of the PMTs.

9. Stranglets

Cosmic ray studies at mountain observatories [6] show an abundance of events that can be interpreted as light nuclei containing a number of strange quarks, so that the ratio Z/A is small compared with ordinary nuclei. These objects, sometimes called stranglets, (also called MEMOs, metastable exotic multihypernuclear objects [7])are found in the cores of cosmic ray showers and are thought to be the result of collisions of cosmic iron or nickel with nitrogen or oxygen at the top of the atmosphere. The center of mass energy per nucleon of the collisions is in the same range as with the LHC. In a heavy ion collision, there are two parts, a hot interaction region and a non-overlapping region that gives rise to spectator matter that continues undeviated from the original beam direction. A recent paper [8] has shown that there is essentially zero probability that a bound nuclear system could come out of the hot interaction region. For small Pb-Pb impact parameters the spectators are disintegrated into neutrons and protons so that the number of neutrons observed in a zero degree calorimeter (ZDC) is proportional to the amount of spectator matter. At a sufficiently large impact parameter, the number of neutrons is no longer proportional to the amount of spectator matter because some of the spectator neutrons are bound with protons into small nuclear systems. If the objects seen in cosmic rays are stranglets, they are formed from spectator nuclei that have somehow become infiltrated with strange quarks from the hot interaction region. With a large iron nucleus incident on a small nitrogen nucleus there is a large amount of spectator matter.

Stranglets would appear in the CMS ZDC as an abnormally large signal. The impact parameter can be measured by other parts of CMS so that the expected number of neutrons in the ZDC will be known. A stranglet of A nucleons would add to the ZDC signal like A additional neutrons. The ZDC sits between the beam lines, 140 m from the center of CMS, so that a stranglet would need to have Z/A well under the 92/208 of lead in order to not be swept aside by the beam magnets. The electromagnetic part of the ZDC is divided horizontally into five equal sections. The spectator neutrons will mostly be incident on the middle section. A stranglet would most likely go through the section closest to the outgoing beam and leave an easily observable amount of energy.

The observation of stranglets in the ZDC may have to wait for $\sqrt{s_{\rm NN}} = 5.5$ TeV runs. The larger energy will increase the number of strange quarks produced in the interaction region, and the relativistic γ factor will be twice as large. The lifetime of a stranglet is likely to be short—of the order of the lifetime of a Λ^0. The extra factor of two in the time dilatation effect may be necessary for an adequate number to reach the ZDC. The mountain top observatories could observe much shorter lifetimes because the γ factor for cosmic iron is huge compared with the γ factor with colliding beams at the same $\sqrt{s_{\rm NN}}$.

The ZDC of the ALICE experiment [9] may be a better device for observing stranglets. In addition to the part between the beam lines, ALICE has a part on the other side of the outgoing beam line. This outer part would be able to see stranglets with $Z/A < 1/2$ but $> 92/208$. Also, the distance that the stranglet must travel is smaller with ALICE, only 116 m instead of 140 m.

An extensive search by the STAR experiment at RHIC failed to find a single stranglet candidate [10]. These results are not surprising. The experiment looked only at the 4% most central collisions for which there would be no bound spectator nuclei. It is also possible that at the RHIC energy there are not a sufficient number of strange quarks produced to convert a spectator nucleus into a stranglet.

References

[1] CMS Collaboration 2008 *JINST* **3** S08004 doi:10.1088/1748-0221/3/08/S08004
[2] CMS Collaboration 2007 *J. Phys. G: Nucl. Part. Phys.* **34** 2307
[3] Evans L (ed.) and Bryant P (ed.) 2008 *JINST* **3** S08001 doi:10.1088/1748-0221/3/08/S08001
[4] CMS Collaboration 2006 "Detector performance and software" CERN/LHCC 2006-001 CMS TDR 8.1
[5] Baltz A *et al* 2008 *Phys. Reports* **458** 1-171
[6] Gladysz-Dziaduś E 2003 *Phys. Part. Nucl* **34** 285-347
[7] Schaffner-Bielich J, Greiner G, Diener A and Stöcker H 1997 *Phys. Rev.* C **55** 3038-46
[8] Ellis J *et al* 2008 *J. Phys. G: Nucl. Part. Phys.* **35** 115004
[9] 1999 "Zero-Degree Calorimeter Technical Design Report" CERN/LHCC 99-5
[10] STAR Collaboration 2007 *Phys. Rev.* C **76** 011901(R)

Journal of Physics: Conference Series **230** (2010) 012045 doi:10.1088/1742-6596/230/1/012045

Relativistic Shock Waves and Mach Cones in Viscous Gluon Matter

Ioannis Bouras[1], Etele Molnár[2,3], Harri Niemi[2], Zhe Xu[1,2], Andrej El[1], Oliver Fochler[1], Francesco Lauciello[1], Carsten Greiner[1] and Dirk H. Rischke [1,2]

[1] Institut für Theoretische Physik, Johann Wolfgang Goethe-Universität, Max-von-Laue-Str. 1, D-60438 Frankfurt am Main, Germany

[2] Frankfurt Institute for Advanced Studies, Ruth-Moufang-Str. 1, D-60438 Frankfurt am Main, Germany

[3] KFKI, Research Institute of Particle and Nuclear Physics, H-1525 Budapest, P.O.Box 49, Hungary

E-mail: `bouras@th.physik.uni-frankfurt.de`

Abstract.
To investigate the formation and the propagation of relativistic shock waves in viscous gluon matter we solve the relativistic Riemann problem using a microscopic parton cascade. We demonstrate the transition from ideal to viscous shock waves by varying the shear viscosity to entropy density ratio η/s. Furthermore we compare our results with those obtained by solving the relativistic causal dissipative fluid equations of Israel and Stewart (IS), in order to show the validity of the IS hydrodynamics. Employing the parton cascade we also investigate the formation of Mach shocks induced by a high-energy gluon traversing viscous gluon matter. For $\eta/s = 0.08$ a Mach cone structure is observed, whereas the signal smears out for $\eta/s \geq 0.32$.

1. Introduction

Jet quenching has been discovered in heavy-ion collisions at BNL's Relativistic Heavy Ion Collider (RHIC). In this context, very exciting jet-associated particle correlations [2] have been observed, which indicates the formation of shock waves in the form of Mach cones [3], induced by supersonic partons moving through the quark-gluon plasma (QGP). Measuring the Mach cone angle could give us the possibility to extract the equation of state of the QGP.

Shock waves can only develop in a medium which behaves like a fluid. The large elliptic flow coefficient v_2 measured at RHIC [4] implies that the QGP created could be a nearly perfect fluid with a small viscosity. Calculations of viscous hydrodynamics [5] and microscopic transport theory [6, 7] have estimated the shear viscosity to the entropy density ratio η/s to be less than 0.4. There is still an open question if this upper limit of the η/s ratio is small enough to allow the formation of Mach shocks.

In this work we address the question, whether and when relativistic shock waves and Mach cones can develop in viscous gluon matter for given η/s values. For this purpose we consider first the relativistic Riemann problem [8], which we solve within the kinetic theory and the Israel-Stewart (IS) theory of viscous hydrodynamics for comparisons. Here we employ the microscopic parton cascade BAMPS (Boltzmann Approach of MultiParton Scatterings) [9] and a solver of the

IS equations, vSHASTA (viscous SHArp and Smooth Transport Algorithm) [10]. Particularly, we demonstrate agreements between the two approaches for matter with (extreme) small η/s values and also show deviations when the η/s ratio becomes large, which implies the invalidity of the IS theory. Second, we consider a traverse of a high-energy gluon through gluon matter and investigate the formation of shocks in form of Mach cones. Preliminary results are obtained by using BAMPS.

2. BAMPS and vSHASTA

BAMPS is a microscopic transport model solving the Boltzmann equation

$$p^\mu \partial_\mu f(x,p) = C(x,p) \tag{1}$$

for on-shell particles with the collision integral $C(x,p)$. The algorithm for collisions is based on the stochastic interpretation of the transition rate [9]. In this study, we consider only binary gluon scattering processes with an isotropic cross section, which is adjusted locally at each time step to keep a constant η/s value [11, 12, 13].

Simulations of particle evolution in space and time are performed in a static box, where the whole box is divided into spatial cells with a volume $V_{\rm cell} = \Delta x\, \Delta y\, \Delta z$. Collisions of particles in same cells are simulated by Monte Carlo technique according to the individual collision probability within a time step of Δt,

$$P_{22} = v_{\rm rel} \frac{\sigma}{N_{\rm test}} \frac{\Delta t}{V_{\rm cell}}, \tag{2}$$

where σ is the total cross section, $v_{\rm rel} = (p_1 + p_2)^2/(2E_1E_2)$ denotes the relative velocity of the two incoming particles with four momenta p_1, p_2 and $N_{\rm test}$ is the testparticle number. The testparticle method is introduced to reduce statistical fluctuations and is implemented such that the mean free path is left invariant.

The relationship between the shear viscosity η and the total cross section σ is given by $\eta = 4e/(15R^{tr})$ [12], where $R^{tr} = n\langle v_{\rm rel}\sigma^{tr}\rangle = 2n\langle v_{\rm rel}\sigma\rangle/3$ is the transport collision rate in case of isotropic scattering processes. $\langle\rangle$ stands for ensemble average in local rest frame. We obtain

$$\eta = \frac{2}{3} e \lambda_{\rm mfp} \tag{3}$$

for an ultrarelativistic massless gas. Here the LRF energy density is $e = 3nT$, n is the LRF particle density, T the temperature and $\lambda_{\rm mfp} = 1/(n\langle v_{\rm rel}\sigma\rangle)$ is the particle mean free path. In kinetic equilibrium the LRF entropy density is given by $s = 4n - n \ln \lambda$, where $\lambda = n/n_0$ is the fugacity, the ratio of the LRF number density to the one in thermal equilibrium $n_0 = 16T^3/\pi^2$.

vSHASTA solves the IS equations of dissipative hydrodynamics for shear pressure and heat flow. In 1+1 dimensions the relaxation equations for heat conductivity and shear stress are

$$Dq^z = \frac{1}{\tau_q}(q^z_{NS} - q^z) - I^z_{q1} - I^z_{q2} - I^z_{q3}, \tag{4}$$

$$D\pi = \frac{1}{\tau_\pi}(\pi_{NS} - \pi) - I_{\pi1} - I_{\pi2} - I_{\pi3}, \tag{5}$$

with

$$I^z_{q1} = \frac{1}{2} q^z \left(\theta_z + D \ln \frac{\beta_1}{T}\right), \tag{6}$$

$$I^z_{q2} = -q^z v_z \gamma_z^3 \left(\partial_t v_z + v_z \partial_z v_z\right), \tag{7}$$

$$I^z_{q3} = \frac{1}{5}\left[\gamma_z^2 \left(v_z \partial_t \pi + \partial_z \pi\right) + \gamma_z \pi \left(v_z \theta_z + \gamma_z \partial_t v_z\right)\right] \tag{8}$$

and

$$I_{\pi 1} = \frac{1}{2}\pi\left(\theta_z + D\ln\frac{\beta_2}{T}\right), \quad (9)$$

$$I_{\pi 2} = \frac{10}{9}(q^z\gamma_z^2)(\partial_t v_z + v_z\partial_z v_z), \quad (10)$$

$$I_{\pi 3} = \frac{2}{9}\left(v_z\partial_t q^z + \partial_z q^z - \frac{q^z v_z}{\gamma_z}\theta_z\right), \quad (11)$$

where q^z_{NS} and π_{NS} are the Navier-Stokes values [10] and $D \equiv u^\mu\partial_\mu$. τ_q and τ_π are the relaxation times, respectively. For vanishing shear pressure and heat conductivity the IS equations reduce to the equations of ideal hydrodynamics.

3. The relativistic Riemann problem

The relativistic Riemann problem [8] is a well-known shock problem in ideal hydrodynamics. Initially matter is separated by a membrane at $z = 0$ in two regions, $z < 0$ and $z > 0$, with two different pressures P_0 and P_4. The velocities on both sides are $v_0 = v_4 = 0$. The matter is assumed to be homogeneous in transverse direction.
Removing the membrane in the case for a perfect fluid, i.e., $\eta/s = 0$, we observe (the green curve in Fig. 1) a propagation of a shock wave to the right with a larger velocity than the speed of sound and a rarefaction wave to the left exactly with the speed of sound.

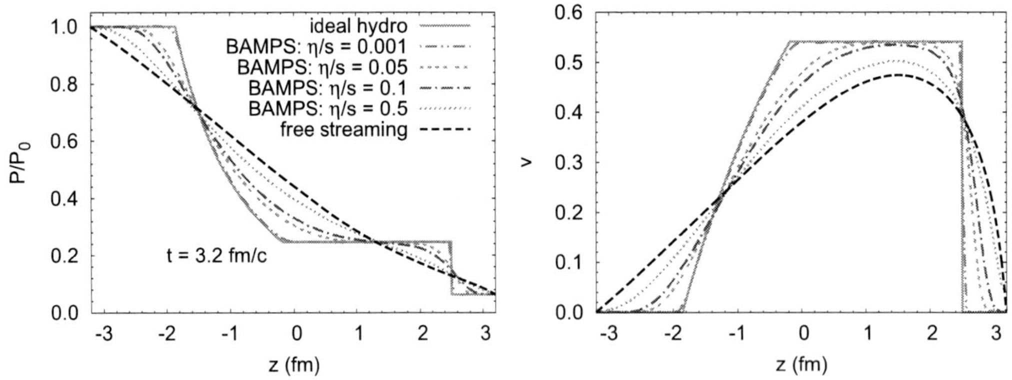

Figure 1. Spatial profile of pressure (left panel, normalized by the initial value) and collective velocity (right panel) at $t = 3.2\,\mathrm{fm/c}$.

The BAMPS solutions for various η/s are depicted in Fig. 1. In particular, the BAMPS result for $\eta/s = 0.001$ reproduces the ideal solution for a perfect fluid to a very high precision. With the increasing η/s value we see a clear transition from the formation of shock waves in ideal fluid to the smearing out in free streaming of particles [11]. The characteristic shock profiles disappear for large η/s values.

In Fig. 2 we show comparisions between BAMPS and vSHASTA for the shear pressure $\pi = \pi^{zz}/\gamma^2$ and heat flow $q^z = h\gamma^2(vN^0 - N^3)$ at $t = 3.2\,\mathrm{fm/c}$. $\pi^{\mu\nu}$ is the shear stress tensor, N^μ is the particle four-flow, $\gamma^2 = (1-v^2)^{-1}$ and $h = (e+P)/n$ is the enthalpy per particle. For $\eta/s = 0.01$ we see a very good agreement between vSHASTA and BAMPS, whereas for $\eta/s = 0.1$ deviations in the region of the shock front and rare faction fan appear. In the case of $\eta/s = 0.1$, the local system at the shock front and rare faction fan is strongly out of equilibrium and, thus, the applicability of the IS theory is questionable. The microscopic transport approach does not suffer from this drawback.

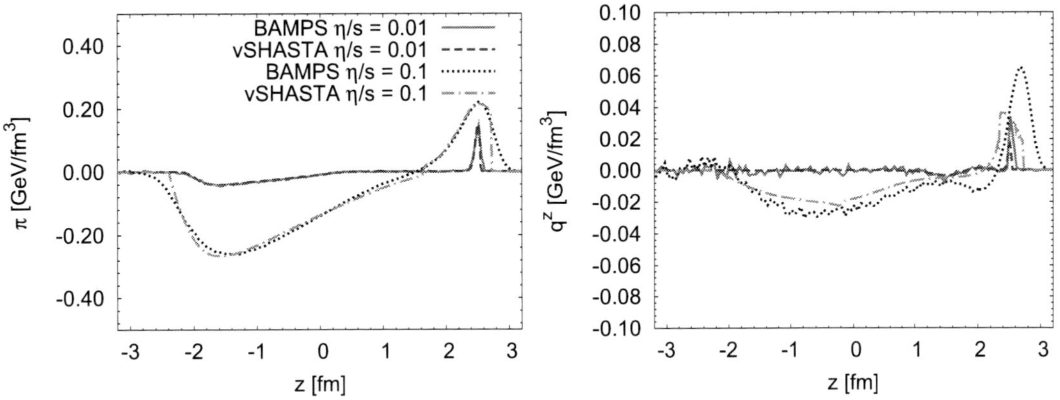

Figure 2. Comparison of BAMPS and vSHASTA for the shear pressure π and heat flow q^z.

4. Mach Cones in BAMPS

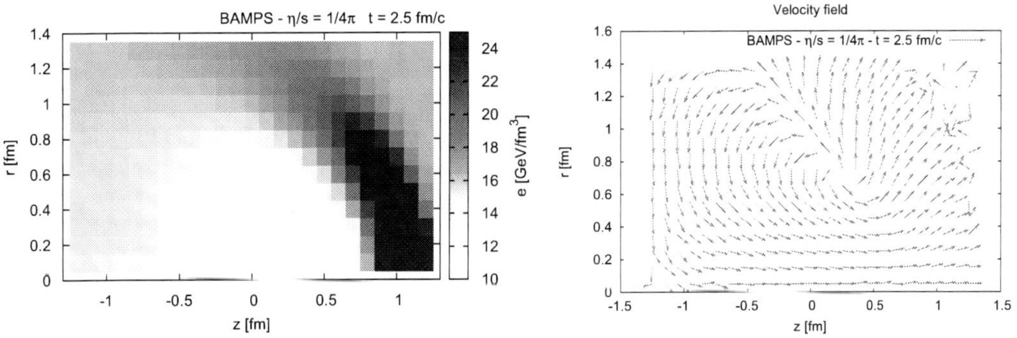

Figure 3. Spatial profile of the energy density (left panel) and the velocity field (right panel). The η/s value of the thermal medium is $1/4\pi$. The initial velocity of the high-energy gluon is in z-direction. The length of the arrays in the right panel is unit, i.e., only the direction of the velocity is shown.

Formation of 3-dimensional shock waves in form of Mach cones is investigated by shooting a gluon with energy of 20 GeV into a thermal gluon medium with a temperature of $T = 400\,\mathrm{MeV}$. The thermal medium is embedded in a static box.

Fig. 3 shows spatial profiles of the energy density and velocity field at $t = 2.5\,\mathrm{fm/c}$ for a medium with $\eta/s = 1/4\pi$. The energy, which the gluon probe lost due to interactions with the medium, creates a shock wave that propagates in form of a Mach cone. The energy density of the region behind the Mach cone is smaller than the initial energy density of the medium. This region is called a diffusion wake. Collective behavior of the medium response is also clearly seen in the profile of the velocity field. Our results agree qualitatively with those found in [14, 15] For higher values of the η/s ratio the typical Mach cone structure smears out as observed in Fig. 4. The strength of the Mach cone signal and also the lower energy density region behind the shock front become weaker because of weaker particle interactions in medium with larger η/s.

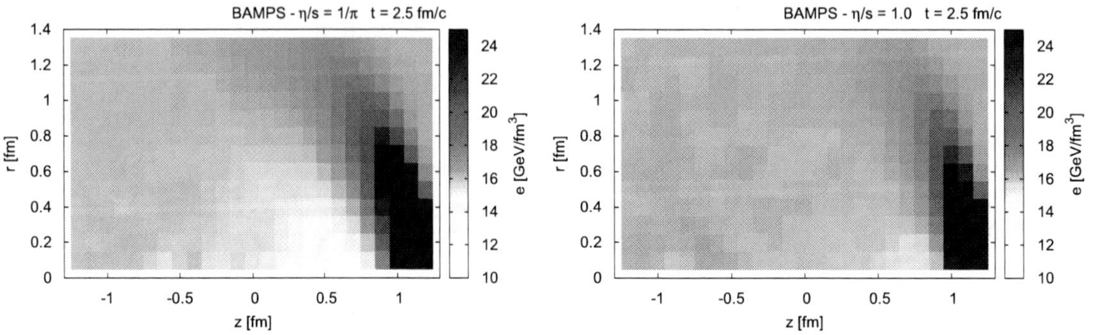

Figure 4. Same as Fig. 3, but for $\eta/s = 1/\pi$ (left panel) and 1.0 (right panel).

5. Summary

We have solved the relativistic Riemann problem using BAMPS and vSHASTA, in order to investigate the formation of shock waves in viscous matter. A transition from a characteristic shock profile in ideal fluid to a complete smearing out in particle free streaming has been observed when increasing the η/s value. For high η/s values deviations between BAMPS and vSHASTA occur, which indicates the break down of the IS hydrodynamics. We have also investigated the formation of Mach cones within BAMPS. For $\eta/s = 1/4\pi$ a Mach cone is clearly visible. For larger values of $\eta/s = 0.32$ and 1.0 the dissipative effect is so strong that the typical Mach cone structure disappears.

Acknowledgements

The authors are grateful to the Center for the Scientific Computing (CSC) at Frankfurt for the computing resources. I. Bouras is grateful to HGS Hire. E. Molnár acknowledges the Return Fellowship support by the Alexander von Humboldt foundation. The work of H. Niemi was supported by the Extreme Matter Institute (EMMI).

This work was supported by the Helmholtz International Center for FAIR within the framework of the LOEWE program launched by the State of Hesse.

References

[1] J. Adams *et al.* [STAR Collaboration], Phys. Rev. Lett. **91**, 172302 (2003); A. Adare *et al.* [PHENIX Collaboration], *ibid.* **101**, 232301 (2008).

[2] F. Wang [STAR Collaboration], J. Phys. G **30**, S1299 (2004); J. Adams *et al.* [STAR Collaboration], Phys. Rev. Lett. **95**, 152301 (2005); S. S. Adler *et al.* [PHENIX Collaboration], *ibid.* **97**, 052301 (2006); J. G. Ulery [STAR Collaboration], Nucl. Phys. A **774**, 581 (2006); N. N. Ajitanand [PHENIX Collaboration], *ibid.* **783**, 519 (2007); A. Adare *et al.* [PHENIX Collaboration], Phys. Rev. C **78**, 014901 (2008).

[3] H. Stöcker, Nucl. Phys. A **750**, 121 (2005); J. Ruppert and B. Müller, Phys. Lett. B **618**, 123 (2005); J. Casalderrey-Solana, E. V. Shuryak and D. Teaney, J. Phys. Conf. Ser. **27**, 22 (2005); V. Koch, A. Majumder and X. N. Wang, Phys. Rev. Lett. **96**, 172302 (2006).

[4] S. S. Adler *et al.* [PHENIX Collaboration], Phys. Rev. Lett. **91**, 182301 (2003); J. Adams *et al.* [STAR Collaboration], *ibid.* **92**, 052302 (2004); B. B. Back *et al.* [PHOBOS Collaboration], Phys. Rev. C **72**, 051901 (2005).

[5] M. Luzum and P. Romatschke, Phys. Rev. C **78**, 034915 (2008); H. Song and U. W. Heinz, arXiv:0812.4274.

[6] Z. Xu, C. Greiner and H. Stöcker, Phys. Rev. Lett. **101**, 082302 (2008);

[7] Z. Xu and C. Greiner, Phys. Rev. C **79**, 014904 (2009).

[8] V. Schneider *et al.*, J. Comput. Phys. **105** (1993) 92.

Journal of Physics: Conference Series **230** (2010) 012045 doi:10.1088/1742-6596/230/1/012045

[9] Z. Xu and C. Greiner, Phys. Rev. C **71** (2005) 064901 [arXiv:hep-ph/0406278].
[10] E. Molnar, H. Niemi and D. H. Rischke, Eur. Phys. J. C **65** (2010) 615 [arXiv:0907.2583 [nucl-th]].
[11] I. Bouras *et al.*, Phys. Rev. Lett. 103 (2009) 032301
[12] Z. Xu and C. Greiner, Phys. Rev. Lett. **100**, 172301 (2008);
[13] A. El *et al.*, Phys. Rev. C **79**, 044914 (2009)
[14] B. Betz *et al.*, arXiv:0812.4401;
[15] D. Molnar, AIP Conf. Proc. **1182**, 791 (2009) [arXiv:0908.0299 [nucl-th]].

Third-order dissipative hydrodynamics from the entropy principle

Andrej El[1], Zhe Xu[2,1], and Carsten Greiner[1]

[1] Institut für Theoretische Physik, Johann Wolfgang Goethe-Universität Frankfurt, Max-von-Laue-Str. 1, D-60438 Frankfurt am Main, Germany

[2] Frankfurt Institute for Advanced Studies, Ruth-Moufang-Str. 1, D-60438 Frankfurt am Main, Germany

E-mail: el@th.physik.uni-frankfurt.de

Abstract. We review the entropy based derivation of third-order hydrodynamic equations and compare their solutions in one-dimensional boost-invariant geometry with calculations by the partonic cascade BAMPS. We demonstrate that Grad's approximation, which underlies the derivation of both Israel-Stewart and third-order equations, describes the transverse spectra from BAMPS with high accuracy. At the same time solutions of third-order equations are much closer to BAMPS results than solutions of Israel-Stewart equations. Introducing a resummation scheme for all higher-oder corrections to one-dimensional hydrodynamic equation we demonstrate the importance of higher-order terms if the Knudsen number is large.

A causal theory of relativistic dissipative hydrodynamics was first formulated by Israel and Stewart [1] and has been successfully applied to study a wide range of ultra-relativistic heavy-ion collision phenomena [2, 3, 4]. However, recently presented detailed comparisons of solutions of Israel-Stewart equations and kinetic transport calculations have demonstrated that deviations between them increase with increasing strength of dissipation. Israel-Stewart equations can be derived from the divergence of the off-equilibrium entropy current, which is expanded up to second-order in dissipative fluxes. It is thus of interest to investigate whether a better agreement between hydrodynamic and kinetic transport calculations can be achieved if the entropy current is expanded one order higher than in Israel-Stewart theory.

The underlying equation of the kinetic transport theory is the Boltzmann equation

$$p^\mu \partial_\mu f(x,p) = C[f(x,p)] \tag{1}$$

which describes the space-time evolution of the phase-space particle distribution $f(x,p) = \frac{dN}{d^3pd^3x}$ due to drift and diffusion on the left hand side and the collision processes on the right hand side of the equation. A connection between the microscopic kinetic transport theory and the macroscopic theory of hydrodynamics can be established using the Grad's method, in which the off-equilibrium distribution is approximated by

$$f_{off-eq} = f_0 \left(1 + \epsilon + \epsilon_\mu p^\mu + \epsilon_{\mu\nu} p^\mu p^\nu\right) \tag{2}$$

where f_0 denotes the isotropic (equilibrium) distribution and the fields ϵ, ϵ_μ and $\epsilon_{\mu\nu}$ are related to the dissipative fluxes Π, q^μ and $\pi^{\mu\nu}$. The exact form of ϵ, ϵ_μ and $\epsilon_{\mu\nu}$ can be obtained from the definitions of dissipative currents and the matching conditions

$$u_\mu u_\nu (T^{\mu\nu} - T^{\mu\nu}_{eq}) = 0, \quad u_\nu (N^\nu - N^\nu_{eq}) = 0 \tag{3}$$

where $T^{\mu\nu}_{eq}$ and N^ν_{eq} are the energy-momentum tensor and the particle current in equilibrium.

In the following we consider a massless gas of Boltzmann particles (gluons) undergoing a boost-invariant one dimensional expansion [5]. For the considered system the equation of state is $e = 3p$. Bulk pressure and heat flux vanish identically and the local rest frame off-equilibrium distribution becomes [9, 10]

$$f(x,p) = d_g e^{-E/T} \left(1 + \frac{3}{8eT^2} \pi_{\mu\nu} p^\mu p^\nu \right) \tag{4}$$

with the energy density e and the degeneracy factor for gluons, $d_g = 16$. The matching conditions ((3)) allow to define the temperature T for an off-equilibrated system by matching its energy and particle densities e and n to a fictitious equilibrium state:

$$e = e_{eq}, \quad n = n_{eq} \tag{5}$$

For the Boltzmann gas considered here the temperature T in Eq.(4) is then

$$T = \frac{e}{3n}. \tag{6}$$

Grad's approximation, Eqs.(2) resp. (4), is essential for derivations of hydrodynamic equations from the Boltzmann equation [20, 15] resp. from the entropy principle [6, 10]. It is thus important to quantify how accurate the approximation in Eq.(4) can reproduce the off-equilibrium distribution obtained from the numerical solution of the Boltzmann Equation in the partonic cascade BAMPS [17]. BAMPS has recently been applied to investigate a wide range of phenomena such like the buildup of the elliptic flow [18], the energy loss of high energy gluons [19], the extraction of the second-order viscosity coefficient [9], and the formation and propagation of shock waves [14] in ultra-relativistic heavy-ion collisions.

For the results presented in this work we use BAMPS calculations with elastic isotropic cross section adjusted in such a way that the shear viscosity to entropy density ratio is constant throughout the evolution, as introduced in [8, 10]. The initial condition is a Boltzmann distribution with $T_0 = 500 MeV$ at the initial time $\tau_0 = 0.4 fm/c$. All results are extracted from the central rapidity bin $\eta \in [-0.1 : 0.1]$.

To quantify the deviations of the expression in Eq.(4) from the actual distribution in BAMPS we take the ratio δ^{BAMPS}_{Grad} of the transverse distributions calculated as follows:

$$\delta^{BAMPS}_{Grad} = \frac{(dN/p_T/dp_T)_{BAMPS}}{(dN/p_T/dp_T)_{Grad}} = \frac{\left\langle p_0 f_{BAMPS} \right\rangle_{y,\varphi}}{\left\langle p_0 d_g \lambda e^{\frac{-p_T \cosh y}{T}} \left(1 + \frac{3}{8T^2} \frac{\pi}{e} p_T^2 \left(\frac{1}{2} - \sinh^2 y\right)\right) \right\rangle_{y,\varphi}}. \tag{7}$$

In the latter expression π denotes the shear pressure, which is the dissipative correction to the longitudinal pressure:

$$\pi = -\pi^{33} = T^{33}_{eq} - T^{33} \tag{8}$$

For the analytic calculation using Eq.(4) the values of e and T as well as the components of $\pi_{\mu\nu} = T_{\mu\nu} - T^{eq}_{\mu\nu}$ are extracted from BAMPS. The deviation δ^{BAMPS}_{Grad} is shown in Fig.1 at different times as function of p_T. The deviations of Grad's approximation from actual distribution in BAMPS do not exceed 10% for the chosen value of η/s below $p_T \sim 3GeV$. The good agreement

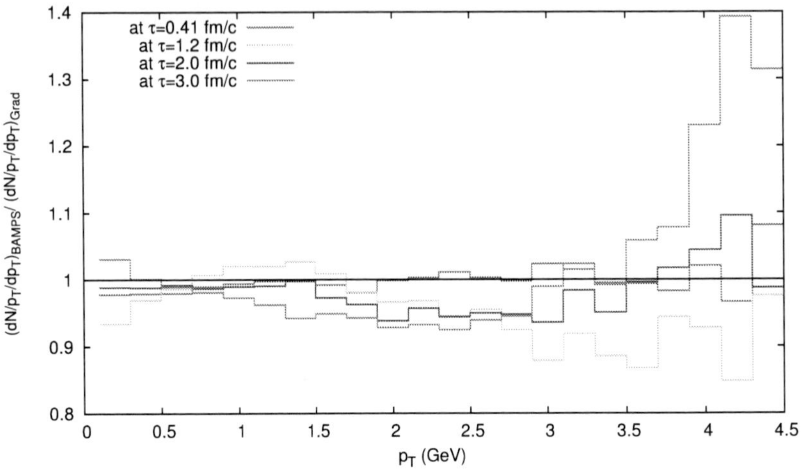

Figure 1. Ratio of transverse particle distribution from BAMPS with $\eta/s = 0.4$ to the one calculated by Eq.(4) at different times with e, T, $\pi^{\mu\nu}$ extracted from BAMPS.

of BAMPS distribution with Grad's approximation observed in Fig.1 might indicate that the analytic expression in Eq.(4) can be applied for derivations of hydrodynamic equations.

Derivation of the evolution equation for the shear stress tensor has been reported by us recently in [10]. The starting point for the derivation is the entropy current which in kinetic theory can be written as

$$s^\mu = -\int \frac{d^3p}{E} p^\mu f(\ln f - 1)\,. \tag{9}$$

Writing the off-equilibrium distribution as $f = f_0\,(1+\phi)$ (comp. Eqs. (2) and (4)) we expand the logarithm up to third order in ϕ and obtain by a direct calculation

$$s^\mu \approx s_0 u^\mu - \frac{\beta_2}{2T}\pi_{\alpha\beta}\pi^{\alpha\beta}u^\mu - \frac{8}{9}\frac{\beta_2^2}{T}\pi_{\alpha\beta}\pi^\alpha_\sigma\pi^{\beta\sigma}u^\mu \tag{10}$$

with $s_0 = 4n - n\ln\lambda$ and $\beta_2 = \frac{9}{4e}$. Note that truncating the expansion at order ϕ^2 we obtain the entropy current from the Israel-Stewart theory [1, 6].

The original Israel-Stewart equations explicitly satisfy the second law of thermodynamics since they are obtained directly from the requirement of non-negativeness of the divergence of the entropy current, $\partial_\mu s^\mu \geq 0$. Using the same argumentation with a third-order entropy current in Eq. (10), i.e. calculating its divergence and imposing a linear relation between the dissipative flux and the corresponding thermodynamic force, we obtain an extended version of Israel-Stewart's equation for $\pi^{\mu\nu}$ in which we keep only terms up to third-order in Knudsen number $Kn \sim \tau_\pi \partial_\mu u^\mu$ resp. dissipative fluxes [10]:

$$\dot{\pi}^{\alpha\beta} = -\frac{\pi^{\alpha\beta}}{\tau_\pi} + \frac{\sigma^{\alpha\beta}}{\beta_2} - \pi^{\alpha\beta}\frac{T}{\beta_2}\partial_\mu\left(\frac{\beta_2}{2T}u^\mu\right) - \frac{8}{9}\frac{T}{\beta_2}\partial_\mu\left(\frac{\beta_2^2}{T}u^\mu\right)\pi^{\langle\alpha}_\sigma\pi^{\sigma\beta\rangle} \tag{11}$$

The latter equation constitutes a novel third-order evolution equation for the shear tensor. The notation $\dot{\pi}$ denotes derivative with respect to the proper time τ. τ_π is the relaxation time and is the same as in Israel-Stewart theory:

$$\tau_\pi = 2\eta\beta_2\,. \tag{12}$$

Neglecting the last two terms in Eq. (11) the second-order Israel-Stewart equation is recovered. To second order the equation we obtain does not contain all terms found in recent works [3, 12, 13, 15]. It is an interesting task for the future to understand the differences between various formulations of dissipative hydrodynamic equations, as discussed for example in [15]. We have to stress that in our derivations, as presented here and in more detail in [10], we do not use the Boltzmann Equation directly. The terms we obtain are the full set of terms which can be obtained from the entropy principle if Grad's approximation, given by Eq. (4), is used.

For a one-dimensional system with boost-invariance Eq. (4) takes the following form

$$\dot{\pi} = -\frac{\pi}{\tau_\pi} - \frac{4}{3}\frac{\pi}{\tau} + \frac{8}{27}\frac{e}{\tau} - 3\frac{\pi^2}{e\tau}. \tag{13}$$

In the latter equation π denotes the shear pressure which reduces the longitudinal pressure p_L:

$$p_L = T_{33} = p - \pi. \tag{14}$$

Again, the Israel-Stewart equation is recovered from Eq. (13) if the last term is neglected. Eq. (13) has to be solved together with the evolution equations for the energy and particle densities. The former is obtained from the conservation of the energy-momentum tensor component, $\partial_\nu T^{\nu 0} = 0$. The latter we obtain assuming conservation of the particle number current, $\partial_\mu N^\mu = 0$, i.e. assuming a medium in which net particle number is constant. In the geometry chosen here, the evolution equation for the energy and particle densities read

$$\dot{e} = -\frac{4}{3}\frac{e}{\tau} + \frac{\pi}{\tau}\ , \quad \dot{n} = -\frac{n}{\tau}. \tag{15}$$

Before discussing the solutions of Eq. (13) resp. of the Israel-Stewart's equations we would like to discuss the effect of higher than 3rd order contributions to Eq. (13). This analysis will be presented here for a one-dimensional system. In order to include all orders of corrections into Eq. (13) we assume they all have the form $x_n \left(\frac{\pi}{e}\right)^n \frac{e}{\tau}$ with $n \geq 3$. Note that the second and third-order terms in Eq. (13) are already of this form. Thus, an *ansatz* for an equation containing all orders of corrections can be written in the following way

$$\begin{aligned}
\dot{\pi} &= -\frac{\pi}{\tau_\pi} - \frac{4}{3}\frac{\pi}{\tau} + \frac{8}{27}\frac{e}{\tau} + \sum_2^\infty x_n \left(\frac{\pi}{e}\right)^n \frac{e}{\tau} = \\
&= -\frac{\pi}{\tau_\pi} - \frac{4}{3}\frac{\pi}{\tau} + \frac{8}{27}\frac{e}{\tau} + \frac{\pi^2}{e\tau}\underbrace{\sum_2^\infty x_n \left(\frac{\pi}{e}\right)^{n-2}}_{\chi} = \\
&= -\frac{\pi}{\tau_\pi} - \frac{4}{3}\frac{\pi}{\tau} + \frac{8}{27}\frac{e}{\tau} + \chi\frac{\pi^2}{e\tau}.
\end{aligned} \tag{16}$$

The coefficient χ is supposed to be an unknown function of time. Since the equation we consider is supposed to include all orders of corrections, it should be applicable in the free-streaming, i.e. $\tau_\pi \to \infty$, limit. In the free-streaming limit the solutions for e, π and n are known. Since the gas is streaming free, longitudinal pressure cannot be built up, i.e. $p_L = 0$, which means $\pi = p = \frac{e}{3}$. Using this, one finds that the energy density evolves according to Eq. (15) as $e(\tau) = e_0\tau_0/\tau$. Using these solutions in the $\tau_\pi \to \infty$ limit in Eq. (16) one obtains

$$\chi = -\frac{5}{3}. \tag{17}$$

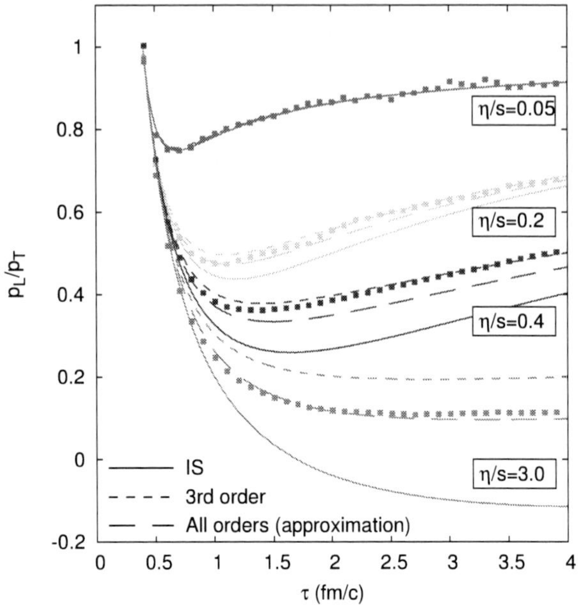

Figure 2. Time evolution of pressure isotropy. The shear pressure is calculated by Eqs. (15) and (16) using $\chi = 0$ (Israel-Stewart), $\chi = -3$ (extension of Israel-Stewart's equation to third order) and $\chi = -\frac{5}{3}$ (approximate inclusion of all order of corrections).

This value is the result of resummation of higher-order terms in the heuristic *ansatz* Eq. (16). The equation obtained this way includes all orders of corrections, but only approximately. If only third-order terms are included, one obtains $\chi = 3$, which corresponds to Eq. (13). The Israel-Stewart equation is obtained by setting $\chi = 0$.

In the following we present the solutions of hydrodynamic equation of second (Eq. (16) with $\chi = 0$) and third orders (Eq. (16) with $\chi = -3$) and of the approximation of all orders (Eq. (16) with $\chi = -\frac{5}{3}$). The observable we use to quantify the deviations from equilibrium is the pressure isotropy $\frac{p_L}{p_T} = \frac{p-\pi}{p+\frac{\pi}{2}}$. The results are compared to BAMPS calculations using thermal initial conditions with $T_0 = 0.5GeV$ and $\tau_0 = 0.4fm/c$. For the comparisons presented here only elastic processes with isotropic cross section are included in BAMPS. The cross section σ_{22} is parametrized to keep the η/s value constant, as has been already done in [8, 14, 10]:

$$\sigma_{22} = \frac{6}{5}\left(\frac{\eta}{s}\right)^{-1}\frac{T}{4n - n\ln\lambda}. \tag{18}$$

where $4n - n\ln\lambda = s$. The method of derivation of hydrodynamic equations which has been presented here does not allow to obtain an analytic expression for the shear viscosity coefficient. We thus rely on the expression for η obtained in [21] which is strictly speaking valid for the Israel-Stewart, i.e. second-order, theory.

Time evolution of the pressure isotropy is presented in Fig.2. It was demonstrated in [14] that the evolution of the system is governed by the Knudsen number $Kn \equiv \tau_\pi/\tau \approx 6\frac{1}{T\tau}\frac{\eta}{s}$. Since $T \approx T_0\left(\tau_0/\tau\right)^{\frac{1}{3}}$, the Knudsen number depends in the situation considered here on τ_0, τ, T and η/s. Since all results are obtained using the same initial conditions, the evolution depends only on the chosen value of η/s, which is thus the direct measure of the strength of dissipative effects. The deviations between the Israel-Stewart and BAMPS results become considerable already at

$\eta/s = 0.4$. At $\eta/s = 3$ the Israel-Stewart's equations ($\chi = 0$ in Eq. (16)) lead to a negative longitudinal pressure which is not physical for the considered setup (This phenomenon has been as well investigated in [7]). The absolute deviations are reduced if one extends Israel-Stewart's equations to third-order, Eq. (13) resp. $\chi = -3$ in Eq. (16). Moreover, the negative longitudinal pressure does not occur in the solutions of the third-order equations. Especially at late times, where the relaxation towards equilibrium sets in, the third-order results on pressure isotropy are in very good agreement with BAMPS solutions. At early times the third-oder and BAMPS results still deviate since there the expansion scalar $\partial_\mu u^\mu = 1/\tau$, is still large and thus all orders of corrections have to be taken into account. Indeed, at early times solution of the approximate all-order equation, $\chi = -\frac{5}{3}$ in Eq. (16), is in very good agreement with BAMPS results. This is due to the fact that the value $\chi = -\frac{5}{3}$ has been obtained from the $\tau_\pi \to \infty$ or alternatively $Kn \to \infty$ limit, to which the system is close at early times.

In this study we have presented an extension of the Israel and Stewart's entropy based approach to third order in dissipative fluxes. For this study we have considered a one-dimensional boost-invariant gas of massless Boltzmann particles (gluons). Results of hydrodynamic calculations have been compared to solutions of the Boltzmann Equation from the partonic cascade BAMPS. Although the Grad's approximation, upon which the derivation of Israel-Stewart's equation is based, has been shown to describe transverse spectra from BAMPS with remarkable accuracy, the solutions of Israel-Stewart equations demonstrate large deviations from BAMPS. Inclusion of third-order terms into the evolution equation for shear tensor reduces the deviations between hydrodynamic and BAMPS results considerably. Up to $\eta/s = 0.4$ the third-order solutions are in very good agreement with BAMPS. In order to estimate the effect of all orders of correction, we have introduced a resummation scheme which allows to represent the infinite series of higher-order terms by one single term. The solution of this equation is in very good agreement with BAMPS results even at $\eta/s = 3$, when the Knudsen number is very large, which underlines the importance of higher-order corrections at early times of evolution. It is still an important task for future studies to understand the differences between the equations obtained here from the entropy principle and the second-order equations presented in recent publications by different authors [13, 3, 20, 15] . It is as well important to investigate whether a possible extension of Grad's approximation can further improve the agreement between hydrodynamic and kinetic transport calculations.

Acknowledgements

The authors thank D. H. Rischke and P. Huovinen for fruitful discussions, comments and their interest in this work. A. E. thanks A. Muronga for discussions and acknowledges the hospitality of UCT, Cape Town, where part of this work was accomplished and acknowledges support by the Helmholtz foundation. This work was supported by the Helmholtz International Center for FAIR within the framework of the LOEWE program launched by the State of Hesse.

References

[1] W. Israel, Ann. Phys. (N.Y.) 100 (1976) 310; J.M. Stewart, Proc. R. Soc. London , Ser. A 357 (1977) 59; W. Israel, M. Stewart, Ann. Phys. (N.Y.) 118 (1979) 341.
[2] U. W. Heinz, arXiv:0901.4355 [nucl-th].
[3] P. Romatschke, arXiv:0902.3663 [hep-ph].
[4] D. A. Teaney, arXiv:0905.2433 [nucl-th].
[5] J. D. Bjorken, Phys. Rev. D 27 (1983) 140.
[6] A. Muronga, Pys. Rev. C 69 (2004) 034903.
[7] M. Martinez, M. Strickland, Phys. Rev. C 79 (2009) 044903.
[8] P. Huovinen, D. Molnar, Phys. Rev. C 79 (2009) 014906.
[9] A. El et. al., Phys. Rev. C 79 (2009) 044914.
[10] A. El, Z. Xu and C. Greiner, Phys. Rev. C 81 (2010) 041901
[11] A. Muronga, Phys. Rev. C 76 (2007) 014909.

[12] P. Romatschke, Class. Quant. Grav. 27 (2010) 025006
[13] R. Baier et. al. , JHEP 0804 (2008) 100.
[14] I. Bouras et. al., Phys. Rev. Lett. 103 (2009) 032301.
[15] B. Betz, D. Henkel, D. H. Rischke, arXiv:0812.1440.
B. Betz, D. Henkel, D. H. Rischke, J. Phys. G 36 (2009) 064029.
[16] M. Lublinsky, E. Shuryak, Phys. Rev. D 80 (2009) 065026.
[17] Z. Xu, C. Greiner, Phys. Rev. C 71 (2005) 064901; Phys. Rev. C 76 (2007) 024911.
[18] Z. Xu, C. Greiner, and H. Stocker, Phys. Rev. Lett. 101 (2008) 082302; Z. Xu and C. Greiner, Phys. Rev. C 79 (2009) 014904.
[19] O. Fochler, Z. Xu and C. Greiner, Phys. Rev. Lett. 102 (2009) 202301.
[20] G. S. Denicol, Xu-Guang Huang, T. Koide and D. H. Rischke, arXiv: 1003.0780 [hep-th].
[21] S. R. de Groot, W. A. van Leeuwen, Ch. G. van Weert, Relativistic Kinetic Theory: Principles and Applications, North Holland, Amsterdam, 1980

AUTHOR INDEX

Institute of Physics
Dirac House, Temple Back
Bristol BS1 6BE UK

CURRAN ASSOCIATES INC
proceedings.com